Communicating in Risk, Crisis, and High Stress Situations

Communicating in Risk, Crisis, and High Stress Situations

Evidence-Based Strategies and Practice

Vincent T. Covello

Center for Risk Communication
New York City and Washington, DC
IEEE PCS Professional Engineering Communication Series

IEEE PCS Professional Engineering Communication Series

IEEE PRESS

WILEY

For general information on our other products and services or for technical support, please contact our Customer Care Department within the United States at (800) 762-2974, outside the United States at (317) 572-3993 or fax (317) 572-4002.

Wiley also publishes its books in a variety of electronic formats. Some content that appears in print may not be available in electronic formats. For more information about Wiley products, visit our web site at www.wiley.com.

Library of Congress Cataloging-in-Publication Data is applied for

Paperback: 9781119027430

Cover Design: Wiley
Cover Image: © Austin Goodwin

Set in 9.5/12.5pt STIXTwoText by Straive, Pondicherry, India

10 9 8 7 6 5 4 3 2 1

Contents

A Note from the Series Editor

By Series Editor, Ryan K. Boettger Ph.D.

As our world continues to fundamentally change, I have started (and restarted) writing this editor's note several times.

Whatever I write about the COVID-19 global pandemic today (June 2021) will have changed by the time you read this. Every day, we learn about new science, new insights, and new challenges that we will no doubt continue to grapple with for decades to come. However, we appear to be on an optimistic and upward trajectory with three vaccines in circulation and the decline of new COVID-19 cases in many parts of the world.

When Dr. Vincent Covello and I first met about this project in 2019, we discussed the value of creating a book that relayed evidence-based strategies for risk, crisis, and high stress situations. Rather than lecturing to readers, Dr. Covello wanted to share his experiences – successes and failures – with other leaders besides offering materials and resources that he has developed or adapted over the years. Neither Dr. Covello nor I predicted the 2020 pandemic and its very real implications for the foundations and principles outlined in this book.

The global impacts of the pandemic motivated a complete reconceptualization of this project. In fact, we delayed the production of this manuscript because Dr. Covello's expertise in risk, high concern, and crisis communication was needed. During the pandemic, Dr. Covello worked nearly full time with the State Health Directors and Governors on their COVID-19 communications. It was important, challenging work that required nearly daily adjustments to account for changes in knowledge and policies. When he returned to working on this manuscript, Dr. Covello detailed many of those experiences for readers.

Each chapter contains at least one case diary, or a personal account of Dr. Covello's extensive tenure and experiences in the areas of risk, high concern, and crisis communication. These are stories and experiences that Dr. Covello hasn't detailed in previous works. Many of these cases are COVID-19-related and account for the various communication failures and successes our world has experienced over the last year and a half. Dr. Covello has written a timely text that you will draw insights from over decades to come. As our understanding of COVID-19 continues to grow, Dr. Covello's recommended principles and response techniques become even more relevant to our responses to the next crisis.

Vincent Covello is *the* leader in the areas of risk and crisis communication. He is the founder and Director of the Center for Risk Communication besides being a nationally and internationally recognized trainer, researcher, and consultant. As you read this book, you'll discover more about Dr. Covello's extensive professional experiences. For example, he was the founding director of the NSF program on decision-making communication about technology risks. His work here also led him to become one of the first presidents of the Society for Risk Analysis, and he helped launch the journal *Risk Analysis*, one of the leading scientific publishers of original risk communication research. In short, there is no one better to learn these strategies from than Dr. Covello. And it was a privilege to work with him on this important project.

On a personal note, this is the last title under my editorship for the *Professional Engineering Communication* series. I strived to deliver quality content for your professional development. The series, backed by Wiley-IEEE Press, remains an impressive collection of communication and professional insights from scholars. In particular, I hope the last three titles heightened your thinking about the future of STEM education.

Dr. Lydia Wilkinson (University of Toronto) will assume the editorial responsibilities of this series. I worked with Dr. Wilkinson for several years as part of IEEE-PCS, so I know first-hand the enthusiasm and competence she brings to this position. As a faculty member at the Institute for Studies in Transdisciplinary Engineering Education and Practice at U of T, Dr. Wilkinson's perspectives on STEM education will further elevate the quality of this series.

Thank you to the Wiley Press and IEEE teams for the opportunity to edit this series. Once again, the cover art contributions of Austin Goodwin were instrumental in my mission to rebrand and freshen this series. I can only perform these editorial duties because of the support from the Department of Technical Communication at the University of North Texas. Finally, to my son Liam, who brightens my world every day. . . and who is hungry again, waking from his nap as I (finally) finish writing this note.

Acknowledgments

I owe thanks to many people who have helped me bring this book to published form.

First and foremost, I am grateful to Ryan K. Boettger, Ph.D., the Series Editor, for his invaluable advice, steadfast encouragement, and inspiring ideas.

I offer deep appreciation to my long-time colleague and friend Joseph Wojtecki for his insightful suggestions. I am also grateful to Robert Coble, David Degagne, Donna Dinkin, James Gallagher, Randall Hyer, Thomas Hipper, Benjamin Morgan, Bernard Pleau, Peter Jacobs, Sean Tolnay, Steven Wolf, and Diane Yu for their thoughtful reading of draft chapter material and for their expert editorial work and suggestions. I thank Pat Levine and Esther Brumberg for their careful editorial assistance; and I thank Mary Hatcher at Wiley for her feedback, encouragement, and much appreciated vision for the project. Finally, this book could not have reached its final state without the candid and astute editorial contributions and eternal patience of my wife, Carol Mandel.

Author Biography

Dr. Vincent T. Covello, director of the Center for Risk Communication, is one of the world's leading experts and practitioners on risk, crisis, and high concern communications. He is the author of more than 150 articles in scientific journals and the author/editor of more than 20 books. Dr. Covello is a consultant, writer, speaker, researcher, and teacher. He is a frequent keynote speaker and has conducted communication skills training for thousands.

Dr. Covello has served as a communication adviser to numerous private and public organizations. His work for government includes the World Health Organization, the US Department of Health and Human Services, the US Environmental Protection Agency, the US Department of Defense, the US Centers for Disease Control and Prevention, the US Nuclear Regulatory Agency, the White House Council on Environmental Quality, and the United Nations Scientific Committee on the Effects of Atomic Radiation. His work has been applied nationally and globally to a wide range of topics, including environmental incidents, natural hazards, disease outbreaks, terrorism, industrial accidents, occupational safety, air pollution, water contamination, hazardous waste, physician–patient communications, vaccine safety, operational disruptions, and organizational change. He has worked closely with the US State Health Directors on their responses to questions from the media and the public on disease outbreaks, including Ebola, Zika, and COVID-19.

Dr. Covello has received many awards for his work and has held numerous positions, including Associate Professor of Clinical Medicine and Environmental Sciences at Columbia University, a program manager at the National Science Foundation, a study director at the National Academy of Sciences, President of the Society for Risk Analysis, and Vice-Chairperson of the Radiation Education, Risk Communication and Education Committee of the National Council on Radiation Protection and Measurements. Dr. Covello obtained his BA with honors and MA from Cambridge University (England) and his doctorate from Columbia University.

1

The Critical Role of Risk, High Concern, and Crisis Communication

---

CHAPTER OBJECTIVES

This chapter addresses the role – and necessity – of successful communication in situations involving risk, high stress concerns, or crisis. It describes the book's intent to serve both as a handbook for individuals and as a resource for training and education. At the end of this chapter, you will be able to:

- describe the professional value of learning about risk communication principles and skills,
- identify how recent changes in the social and technical environment affect communication practices, and
- relate the organization and contents of this book to your individual needs.

The single biggest problem in communication is the illusion that it has taken place.

—George Bernard Shaw

This book is about communicating with people in the most challenging circumstances: high stress situations. The ability to communicate effectively in a high stress situation is an essential communication competency. It is a competency that differs in significant ways from other generic communication skills. If done well, it can build trust and agreement, enabling beneficial solutions and constructive behaviors even in the face of fear and anxiety. In a public health or environmental hazard situation, it can save lives. Poor communication in high stress situations can have disastrous consequences, whether the loss of a business or the failure to resolve a high impact policy or operational issue. Professionals in every field can be thrust into situations demanding specialized high stress communications skills, whether they are confronting an external crisis or leading organizational change. I wrote this book so that you can be prepared.

As a manager or technical professional, you likely have a logical, research-based approach for addressing complex issues. You strive to ensure that people and communities benefit from this expertise. Yet all too often, the individuals and populations you serve do not share your trained perspective and thought processes; they do not consider your facts, judgements, and decisions persuasive, especially in situations fraught with high concern.

Enabling technical expertise to inform decisions and policy outcomes requires a body of well-researched knowledge and trained skills in risk, high concern, and crisis communication. Without this knowledge and related skills, the negative consequences can be major.

Communicating in Risk, Crisis, and High Stress Situations: Evidence-Based Strategies and Practice, First Edition.
Vincent T. Covello.
© 2022 by The Institute of Electrical and Electronics Engineers, Inc. Published 2022 by John Wiley & Sons, Inc.

1.1 Case Diary: A Collision of Facts and Perceptions

A few years ago, an established nuclear research facility hired me as a consultant. The facility housed a nuclear reactor used for high-level research. It was also near a densely populated community that sat above a protected aquifer. This aquifer was the community's sole source of local drinking water.

The site managers contacted me with concerns about a local newspaper article on this nuclear reactor. The article reported the facility's nuclear reactor had leaked radioactive water for over a decade. Site managers and engineers had reportedly known about the leak for years. The leak resulted from a hairline crack, but the amount of the leaking radioactive water was well below levels that could cause human health consequences.

Leadership did not report the leak because they feared community outrage. They believed the public could misunderstand the science and react irrationally, even though the technical facts proved there was no significant environmental impact. Revealing the leak to the surrounding community might lead to unwarranted fear and panic and give ammunition to activists who were lobbying to shut down the reactor.

Unfortunately, the article also reported an internal poll of managers and engineers at the facility, asking how they would like to spend the facility's end-of-year funding surplus. Respondents had two primary choices:

1) repair the hairline crack and stop the leak of radioactive water, or
2) support work enhancements, including refreshments for the facility's visiting speaker program.

The facility's employees – applying their scientific knowledge and logic that the hairline crack was inconsequential – chose the refreshments.

I was hired to consult *after* the publication of this newspaper article. The engineers and managers explained to me, in meticulous technical detail, the nature of the crack and why the amount of radioactive water leaking into the community's aquifer was miniscule and posed no threat to human health.

After I sat and listened to a variety of technical presentations, I conducted a training on basic principles of risk, high concern, and crisis communication. I agreed that accurate technical facts were essential for decision-making, but facts by themselves were not always sufficient. Technical facts are only one factor that influences public fears and risk perceptions. Emotional factors also drive decision-making. Trust is based on attributes, including caring and concern. People in high-stress situations need to know you care before they will listen to you. I pointed out that nuclear power and radiation is a highly emotionally charged issue and raises high levels of anxiety for the public.

I predicted the public would perceive the facility's actions as a major breach of trust, notwithstanding the actual lack of potential harm. I recommended actions the managers and engineers could still take to regain trust and counter community anger and outrage. These recommendations included a sincere apology for not communicating early and a commitment to restore trust and ensure the mistakes would not be repeated. Such actions included environmental restoration and creation of a community advisory committee with significant oversight powers.

Unfortunately, I was brought into this situation too late. An avalanche of negative stories followed the first newspaper article, triggering community outrage before leadership could implement any of the recommended actions. Outrage was further fueled by the publication of a previously unrevealed government report that cataloged a long history of environmental shortcomings at this facility. That report cited the facility for failing to respond to the discovery of a leak of radioactive water over 10 years earlier and described delays on promises to make environmental improvements.

Government agencies withdrew their support for continued operation of the reactor, citing environmental and economic concerns. Community and environmental groups pressured government representatives to deny the reactor a permit to continue to operate. The nuclear reactor was indefinitely closed, and all the scientific research it supported ended.

I began Chapter 1 with this story because it encapsulates several vital lessons:

First, effective communication is critical to the effective prevention of and response to risks, high concern issues, and crises.

Second, trust is a prerequisite for communicating successfully about controversial and emotionally charged issues.

Third, organizations and institutions interact with their environments, eco-systems, and communities. Those responsible for leading those organizations and managing them at every level must understand how stakeholders view what they do; they must seek and be prepared for stakeholder engagement, and they must build – and earn – and nurture trust from those interested or affected.

Fourth, leaders, managers, supervisors, and technical professionals require training in the principles of trust, stakeholder perceptions, and communication about risks, high concerns, and crises prior to encountering situations that require effective communication.

If the facility managers and engineers in this story had learned and applied the principles and values discussed in the chapters to follow, the research reactor would likely be in operation today.

1.2 What Will Readers Find in This Book?

This book identifies the principles underlying effective communication in situations where there is risk, crisis, or other causes of high concern. This book describes both the differences between and similarities among the situations of risk, crisis, and high concern, describing principles that underlie all such situations and practices specific to each. Previous books have written about these topics, but there are now important new fields of scientific inquiry and enormous new challenges in the communication environment. Inquiries are taking place on diverse fronts, including scientists and experts in anthropology, economics, engineering, epidemiology, law, psychology, sociology, media studies, medicine, statistics, toxicology, and neuroscience. Each discipline has generated publications related to risk, high concern, and crisis communication, and each adds to the understanding of the practice. However, with few exceptions, nearly all existing resources focus on a specific subset of the literature, on a specific area (e.g. bioterrorism, nuclear power, climate change, or genetically modified foods), or on topics of direct interest to the authors' discipline. As a result, the literature has become highly specialized and dispersed.

While this literature specialization serves experts within specific disciplines, it is less useful for the many professionals who work outside of these explicit fields and who may encounter any one of a wide range of challenges. This book offers a common framework of the major principles, strategies, and tools and shows how they relate to inform the work of communicators in high concern or emotionally charged situations.

This book provides the background and practices essential for successful communication in risk, high concern, and crisis situations. It describes what often happens when feelings and facts collide. It explains why leadership accepts some ideas for managing a risk, high concern, or crisis issue and rejects others.

Why do some projects that encounter high concern and controversy go forward and others do not? Why are some facts, information, and guidance recommended to stakeholders in high concern situations heeded more than others? Why are some presentations about high concern issues well received, while other presentations are ignored or incite anger? Why do some team interactions proceed without a hitch and others fail? Why are some interactions with upper management effective and others are ignored? Why are some people trusted with challenging communication responsibilities and others are not? Why do some meetings about high concern issues succeed while others fail? Why are some people better able to handle difficult or controversial situations than others?

This book addresses and answers these questions. The answers often come down to knowledge of risk communication principles and practices.

1.3   Why You Will Use This Book

I organized this book around two primary aims. The first is to help professionals understand the best communication practices for a high concern or emotionally charged situation. My second goal is to give readers the skills to apply these best practices in a variety of situations.

How you communicate in high concern situations will directly influence the course of the events you manage. In a crisis, stakeholders (i.e. interested and affected individuals and organizations) demand timely and accurate information. Leaders, managers, engineers, scientists, and technical professionals will be asked to take on unfamiliar roles and responsibilities. Those involved in the crisis will be surrounded by uncertainty, ambiguous information, high emotion, and upset people. Beyond the situation of an immediate crisis, competence in risk, high concern, and crisis communication is a prerequisite for navigating through the many situations where feelings and facts are at odds, whether it is concern about a health-threatening risk or high anxiety about an impending change. Performed well, high concern communication can enhance trust and confidence, calm nerves, reduce anxiety, encourage cooperative behaviors, provide information for informed decision-making, and help mitigate or reduce potential adverse outcomes. Poor, inadequate high concern communication can disrupt processes, fan emotions, undermine trust and confidence, and result in adverse outcomes.

I have organized this book to introduce theory and best practices in high concern communication, focusing specially on meeting the communication needs of those who work as engineers, technical professionals, leaders, or managers in fields that may encounter health, safety, occupational, and environmental responsibilities and challenges. Communications related to health, safety, occupational, and environmental issues are often stressful. They often raise complex technical, economic, social, political, policy, and ethical questions and then place the resulting demands on organizations.

I also wrote this book for technical professionals, leaders, and managers at all levels who desire to communicate more effectively in high concern situations within their organizations. For example, organizational change often raises concerns and emotionally charged issues from employees, including such questions as: What is the proposed organizational change and why is the change needed? Will I lose my job because of the change? How will the change affect me and my relationship to others? How technical professionals, managers, and supervisors respond to these and related questions is critical to the successful initiation and sustainability of change.

During high concern situations, ineffective communication can cause inefficiencies; disruption; low morale; and wasted time, money, and other resources. Ineffective communication results in messages being garbled by the noise, unintended adverse consequences, rejected messages, and unnecessary fear and confusion. When deployed effectively by leaders and managers at all levels, risk, high concern, and crisis communication skills serve as invaluable tools for engendering trust, protecting organizational value, and helping people make informed decisions.

Leaders, managers, supervisors, and technical professionals can benefit from effective risk, high concern, and crisis communication skills by ensuring their customers, potential customers, and the public have the information they need to evaluate the company's products and operations. Effective communication often determines why some ideas and products are accepted over others, and why some individuals are accepted as leaders over others.

By keeping internal and external stakeholders informed about potential risks, corporations can reduce adverse outcomes and protect themselves from reputational damage. Internal stakeholders are particularly important because of their high credibility, especially when their views about a risk or threat differ from those of management. Similarly, nonprofit and governmental organizations can benefit from effective risk communication by educating and informing their constituents about threats and issues. With the ever-widening array of information sources, the negative effects of misinformation are more likely and damaging. People may not be aware of critical information or may take actions based on misinformation.

Engineers, technical professionals, leaders, and managers, as well as students aspiring to these positions, can read this book, gaining a comprehensive overview of the most important aspects of the field. I wrote each chapter to be self-contained and have provided additional resources for those who want to further probe particular topics. The intent is to make it easier for readers to home in on issues, such as stakeholder engagement, communicating numbers, decision-making tools, warning systems, working with the media, theory, message development, or evaluation.

1.4　The Need for This Book – Now

Critical changes have occurred in the field and in the environment in which high concern communication occurs, changes further magnified and intensified by the coronavirus disease (COVID-19) pandemic of 2020–2021 as described in the last section of this chapter. This book draws on new, and established, valid, and reliable research and applies it to relevant, complex, and difficult communications environments, including lessons learned in the extraordinary global public health crisis of the pandemic.

1.4.1　New Literature, New Research

The scientific literature on risk, high concern, and crisis communication has expanded considerably in the past three decades. From modest beginnings, there are now more than 8,000 articles published in journals and more than 2,000 books focused on risk issues. The research spreads across fields, drawing on the work of behavioral scientists, social scientists, engineers, economists, statisticians, medical scientists, toxicologists, epidemiologists, industrial hygienists, lawyers, media studies, neuroscientists, and a host of other disciplines. The field has benefited from these new understandings and insights.

The material needs to be selected, synthesized, and interpreted for practitioners, integrating new findings into concrete recommendations for application.

I wrote this book to give you that synthesis.

1.4.2 Changes in the Communications Landscape

Technological, economic, and social changes have upended many of the traditional ways that risk-related information is communicated. Changes in communication technologies have radically transformed the way risk information is shared and transferred and how it is used.

Changes are occurring at both the societal and personal level that affect risk communication. Three of the biggest impacts of these changes are:

1) experts and authorities are less trusted;
2) whom to trust is now a central topic in virtually all risk, high concern, and crisis communications; and
3) the way the people seek information about risk, high concern, and crisis issues has shifted from traditional broadcast and print media to online sources and social networks.

Because of changes in the communications landscape, information about risks, high concern issues, and potential or ongoing crises is now readily available 24/7. The streams of information have increased exponentially. Websites of many news organizations update their information every few minutes.

On a personal level, powerful communication changes have resulted from the extensive use of social media and mobile device technologies. People exchange emails, send text and voice messages, make video calls, and share images, videos, diagrams, charts, and emoticons to express thoughts and meaning to what's going on in their world and lives. Messages posted on a vast array of social media platforms communicate instantaneously to multiple recipients or mass audiences. Mobile communications allow people to connect from almost any location. People schedule and conduct virtual meetings with anyone in the world who can connect with them through the Internet or cellular network.

The wide use of social media and virtual interactions are making communications less nuanced as there are fewer face-to-face interactions. As a result of these impersonal interactions, information communicated with nonverbal cues makes it difficult to interpret the sender's intended message. These changes are also influencing writing. For example, people are less likely to spell carefully and write complete sentences because of their increased use of text messaging and social media platforms. Their mode of communication more typically relies on short sentences or fragments, simple tenses, and a limited vocabulary, using phonetic spelling and little or no punctuation. As a result, texting and social media platforms encouraging brief messages are replacing traditional conventions in writing that enabled fuller explanation.

Changes in communications and communication technologies increase the volume of messaging about all topics. Email and texting are currently two of the most popular forms of online communication, even after discounting the large volume of spam messages sent. Beyond even normal increases based on ease of email/text use, many people are addicted to checking and sending email or texts. Billions of business and consumer emails are sent each day. Information overload increases, which also hampers communication. Dependence on continual online interaction also makes communications by individuals and organizations more vulnerable to problems such as mass power outages, disruptions, scams, identity theft, and cyberattacks.

These and related changes affect every aspect of risk communication. On a macroscale, they shape major social institutions (e.g. economics, politics, religion, family, education, science, technology, and legal systems). On a microscale, they shape values, attitudes, beliefs, and behaviors.

1.4.3 Changes in Journalism and the Perception of Facts

The profession of journalism is radically changing, in part because of changes in communication technologies. The models on which modern journalism was founded, including fact checking, freedom of speech, and freedom of the press, are no longer universal norms. Even the definition of *journalist* is evolving. These changes are forcing many news organizations to cut staffs and scale down operations. Misinformation is rising and trust in traditional broadcast and print media is declining as alternative sources of information become readily available. Basic assumptions about truth in journalism are being increasingly challenged by accusations of fake news, hyperbole, and the existence of *alternative facts*. Confirmation bias – whereby people search for "facts" that confirm what they already believe and discount information that is inconsistent with their beliefs – has become epidemic. Citizen journalism and peer-to-peer communication often replace information from professional journalists and other central or "authoritative" sources.

1.4.4 Changes in Laws, Regulations, and Societal Expectations

Right-to-know and right-to-participate laws and regulations have increased. Many public and private sector organizations have made risk and crisis communication and consultation an obligatory task of risk and crisis management. Citizens increasingly expect risk and crisis managers to recognize that (a) people and communities have a right to take part in decisions that affect their lives, their property, and the things they value; and (b) the goal of best communication practice is not to diffuse concerns or avoid action but to engage people in a dialog that produces informed individuals and organizations that are involved, thoughtful, solution-oriented, and collaborative.

1.4.5 Changes in Concerns about Health, Safety, and the Environment

Public concerns about exposures to potentially toxic substances, physical agents, and hazardous events have significantly increased in recent decades. These interests have led to increasing demands for risk information in crisis and noncrisis situations. Interest and concerns about risks have also resulted in the expansion of risk-related issues by traditional broadcast and print outlets and on social media channels.

Inequalities in health, safety, and exposures to hazards between different populations are increasingly being brought to light. The increased understanding of the harm caused by governments and organizations to marginalized, vulnerable, and minority populations has further eroded trust, increased suspicion of "authorities," and raised demand for more nuanced information and more complete data.

1.4.6 Changes in Levels of Trust

The erosion of trust in traditional experts and authorities is driving the need for more effective risk, high concern, and crisis communication. Over the past 50 years, there has been a precipitous drop in trust in institutions overall and with risk management institutions specifically.

Perceptions that undermine trust include observations that technical experts and authorities are:

- paternalistic and insensitive or dismissive of concerns and fears about risks as irrational;
- unwilling to listen, express empathy, or acknowledge the emotions people feel when facing risks;
- unwilling to be fully transparent;
- unwilling to share complete and timely information about what they know about a risk;

- unaware they are using bureaucratic or technical language and jargon that people in the public do not understand;
- more interested in protecting their positions of power than in protecting people from harm or adverse impacts;
- often inconsistent in their statements about risks;
- inconsistent in their recommendations regarding preventive and protective actions; and
- often unwilling to allow meaningful stakeholder participation and engagement in the decision-making process.

In this distrusting environment, advantages accrue to those with effective risk and crisis communication skills.

1.4.7 Changes in the Global Political Environment

In the current global political environment, debates about how to manage and control risks often become hostile. Arguments and polarization often replace compromise and joint problem-solving. Disagreements among stakeholders arise from many interconnected sources, crossing political and geographical boundaries. Small disruptions often rapidly escalate, due in part to the complexity and coupling of large, complex systems. The principles and practices of risk, high concern, and crisis communication presented here are essential to make effective policies and sound decisions.

1.4.8 The COVID-19 Pandemic and the Changed Communication Landscape

In December 2019, a new viral disease was reported in Wuhan, a city of 11 million people in Hubei Province, China. Initially, Chinese health officials reported no human-to-human cases of transmission. However, that assessment quickly changed. Human-to-human cases multiplied. Wuhan went into a near complete lockdown, but not before cases began to show up around the world. By March of 2020, the virus had spread to virtually every nation on the planet and entire nations urgently implemented stay-at-home orders. On March 11, 2020, the World Health Organization declared a global pandemic. In a little over one year, COVID-19 went on to kill more than 2.5 million people, including more than 500,000 in the United States. Hope did not appear on the horizon until the arrival of vaccines in January and February 2021.

The COVID-19 pandemic reshaped the communications landscape in profound ways. Because of the harm being caused by pandemic, the need for effective risk and crisis communication was never greater. Navigating the pandemic called for sophisticated communication skills, not just for public health officials but throughout government – in fact, through all organizations, as change and uncertainty causing high concern became the norm. Even those well-skilled in crisis communication faced unprecedented challenges. The crisis was global, and few governments were prepared for the communication challenges. Responses and messages were uncoordinated, and too often politicized. Even messages based in science were often confusing and frequently changing, as experts quickly learned more about the disease and its means of spread. And the audience for the messages – essentially everyone on earth – had difficulty hearing and understanding even clear messages, as they were experiencing high levels of stress, uncertainty, and anxiety about their health and every aspect of their lives.

COVID-19 was the first pandemic in history where social media was used on a massive scale to communicate information aimed at keeping people safe, informed, productive, and connected. Unfortunately, social media also created a communication *infodemic* – defined as an overabundance of information, both online and offline, that is overwhelming in its volume, largely unstoppable in the speed and breadth of its spread, and which includes as much, or more, unreliable, misleading, and inaccurate content as it does facts and useful advice. The COVID-19 *infodemic* undermined the global response to COVID-19 and cost lives. The communication *infodemic* jeopardized measures to control the pandemic by enabling and amplifying misinformation, i.e. incorrect information, and disinformation, i.e. information deliberately intended to deceive.

In April 2020, the UN Secretary-General launched the United Nations Communications Response initiative to combat the spread of mis- and disinformation. At the World Health Assembly in May 2020, the World Health Organization Member States passed a resolution that recognized that effective risk and crisis communication was a critical part of controlling the COVID-19 pandemic. We continue to learn from the communication failures and successes of the COVID-19 crisis, and COVID-19-related findings, examples, and case studies are contained throughout this book. The experience of the global pandemic profoundly illustrates that successful risk, high concern, and crisis communication plays a critical role in all aspects of human well-being. The principles and practices described in this book will help its readers achieve a critical positive impact through their communications.

2

Core Concepts

CHAPTER OBJECTIVES

This chapter defines the fields of risk, crisis, and high concern communication and outlines the broad and varied range of situations that require their application.

At the end of this chapter, you will be able to:

- describe how risk, high concern, and crisis communication practices can be employed for a wide and varied range of issues;
- explain the terms used in risk, high concern, and crisis communication;
- explain the defining characteristics of risk, high concern, and crisis situations; and
- determine the scope and nature of situations in which you may need to draw on risk, high concern, or crisis communication.

Good words are worth much, and cost little.

— George Herbert.

In this chapter, I define and discuss the core concepts of risk, high concern, and crisis communication. These concepts may appear familiar, but what appears to be a simple concept is often more complex in application. Core concepts provide a framework upon which you can build strategies and action plans. The concepts function as a compass, pointing the user in a particular direction.

I also include a brief history of each core concept as their meanings have morphed or changed. My hope is these brief histories not only are interesting but also suggest alternative ways of thinking about the familiar. For example, one of the original meanings of *crisis* was a "turning point in a disease; a sudden change for worse or better." This original meaning suggests a *crisis* can cause either a bad or a good outcome. The following case diary illustrates the importance of understanding high concern as a core communication concept.

2.1   Case Diary: Recognizing Change as a High Concern Issue

I was in between trips helping clients when I received an urgent call late one Friday night. On the line was the chief executive officer (CEO) of an organization for which I had often done work related to health, safety, and the environment. We knew each other well. Apologizing for the late

Communicating in Risk, Crisis, and High Stress Situations: Evidence-Based Strategies and Practice, First Edition. Vincent T. Covello.

hour of the call, the CEO said he urgently needed my help. He was about to announce a major change in his organization: he was in the final stages of negotiating the sale of his company to a much larger company.

The CEO said the offer to buy his company came suddenly. It was a generous offer and not a takeover but designed to be extremely beneficial to the mission, operations, personnel, and finances of his company. The company making the offer required a prompt response. He had quickly discussed the offer with his senior managers, the board of directors, and lawyers. All agreed they should accept and begin negotiations about details.

The CEO told me that although negotiations about the sale had to be secret, anxiety-ridden rumors were already spreading throughout his organization. The rumors were causing staff to imagine that the sale would require many employees to lose their jobs, relocate, change job functions and responsibilities, renegotiate benefits, work longer hours, travel more, and work under new bosses. He said this was not true. The draft agreement with the buyer left jobs and operations virtually the same. The CEO expressed concern about the risk of employees leaving the company based on rumors and misinformation. This situation would become a crisis if the company lost its best people. He expressed high regard for his employees' talents and believed they held his leadership in high regard. Beyond that, departures of talented staff could jeopardize the sale.

The CEO had prepared a PowerPoint presentation for the following week to announce the purchase to employees. At a minimum, the CEO expected complaints from employees about not receiving enough notice and information. At worst, some employees would leave the company.

I asked the CEO to review his presentation. The first slide in the deck announced the impending sale. It said the sale would be fantastic for the company. Everything he had hoped for the organization could now occur. The slides addressed *why* the sale was needed to remain competitive, *how* the sale would benefit stockholders, *who* would be affected by the sale, *what* changes could be expected, *where* changes would likely take place, *when* the sale would be announced, and *how* the company would implement changes.

When he finished reviewing his slides, my first thought was – this is not my field. I may need to punt. He really needs an organizational and change management consultant. But then I realized that organizational change communications were a subset of the larger field of risk, high concern, and crisis communication – my specialty.

As with communication about other risks, high concerns, or crisis situations, organizational change initiatives also introduce uncertainty about the future. As discussed in this chapter, the concept of *uncertainty* is central to the definition of the terms *risk, high concern,* and *crisis.* For employees, organizational change threatens and puts at risk many of the things they value, including their autonomy, value, and livelihood. These perceptions of threat can produce a strong emotional reaction just as in the flight-freeze-fight response that has kept humans alive for thousands of years. These reactions can severely affect people's ability to focus, solve problems, communicate, cooperate, and think logically. Communicating successfully in the context of these reactions requires knowledge and implementation of the core concepts, principles, strategies, and tools of risk, high concern, and crisis communication.

I told the CEO the communication strategies we had successfully used together to navigate through the rough waters of previous health, safety, or environmental issues also applied to these current changes. His presentation focused on *how* the company benefited from the sale, but not the concerns of the employees. It was also too complex to be absorbed by anxious employees. This was not a normal company briefing. I agreed to assemble a communications team with a cross-section of members of his organization and develop a draft strategic communications plan.

When the team met, we first agreed that providing clear and accurate factual information was a necessary but not sufficient condition for effective communication about a major organizational change.

Second, we needed to work quickly. Leaks about the sale were already occurring and rumors were spreading. We wanted to avoid the appearance we were keeping a secret until the sale papers were signed.

Third, we agreed we needed to do much more communication than a single PowerPoint presentation. We needed to communicate to employees through every available means what the impending sale would mean for them.

Fourth, we agreed the communications strategy would not just be a one-way transfer of information. We would need to establish effective means for listening to, and exchanging information with, employees. We needed to focus on the concerns of the employees and answer questions, such as what will it mean for me, my job, and my working environment?

Fifth, we agreed we could not have all the answers to questions from employees. We would be honest about what we knew.

Sixth, we agreed we would develop a wide and diverse set of communication products venues, and structures. This would include blogs, webcasts, social media platforms, walk-arounds, Q&A fact sheets, briefings cascading information down through the organization, and direct communications with employees. If someone posed a question, it should be clear who they should ask.

Finally, we decided to create a system for tracking, monitoring, and evaluating our change communications. We needed to develop effective means for receiving feedback and determining if our communications were increasing trust; providing useful information; and affecting knowledge, attitudes, beliefs, and intended behaviors.

A key component would be conversations with employees in face-to-face meetings, small group meetings, and open house sessions. The employees' immediate team supervisor would lead face-to-face and small group meetings because employees typically viewed these leaders as the most trusted part of management. Additionally, from a strategic communications perspective, the more emotionally charged or technically complex an issue, the more important it is to communicate the information on a personal level.

The CEO accepted the strategic communications plan and its implementation met with outstanding success. Almost all employees stayed with the organization after the sale. They gave high marks to the communications during this period of change.

We grounded this successful strategic plan in the core concepts and definitions provided in this chapter. These core concepts and definitions helped us to recognize the scope, nature, and challenges of the situations we were facing and to employ best practices, principles, strategies, approaches, and tools.

2.2 Defining the Concept and Term *Risk*

The first term that needs defining is the term *risk*. Unfortunately, there is no consensus among scholars about how to define the term. The scientific literature on risk and risk communication has offered numerous, competing definitions.

According to the Oxford English dictionary, *risk* is "a situation involving exposure to danger." *Risk* is inherent in virtually every action, even inaction.

A *risk* expresses the probability of an adverse outcome and uncertainty about its occurrence and/ or magnitude.[1] From the risks associated with crossing the street or eating at a restaurant to an earthquake, terrorist attack, or disease outbreak, individuals face degrees of risk each day.

A *risk* can have positive or negative consequences of varying magnitudes. However, as used in most health, safety, and environmental studies, the focus is typically on negative consequences. The term *risk* is defined as the probability of an adverse outcome. It is the probability that a potential situation will cause harm or damage to people, property, and /or the environment.

As noted by Covello and Merkhofer, *risk* is multi-dimensional.[2] At a minimum, the term *risk* includes two elements: the likelihood of something happening and the consequences if it happens. At a more complex level, *risk* is a measure of uncertainty. It involves the possibility of an adverse consequence or outcome, the probability of exposure to the occurrence, the timing of the occurrence, and the magnitude of adverse consequence or outcomes.

One source of potential confusion about the term *risk* is the difference between *risk* and *hazard*. The terms are often used interchangeably. However, from a technical point of view, they are different. In the literature on risk assessment, *hazard* is typically described as a source of risk. A hazard is a dangerous situation that could lead to loss or injury. The term *hazard* typically refers to a substance, action, or event that can cause loss, harm, or other adverse consequences.

By comparison, *risk*, from a technical perspective, refers to likelihood of loss, harm, or other adverse consequences from exposure to a hazard. This distinguishes *risk* from *hazard*.[3] *Risk* is created by a *hazard*. For example, a toxic chemical that is a hazard to human health or an endangered species does not constitute a risk unless humans or endangered species are exposed to the hazard. However, a hazard – be it radioactive, chemical, biological, mechanical, or otherwise – can pose a wide variety of risks to the environment. Since no analysis can address all potential risks of a hazard, a key element in risk analysis is to explicitly identify the specific risk of concern.

The definitions provided above assume that risks and hazards have an objective existence. As a result, a primary goal of risk communication should be to transmit objective information to nonexperts who often see risks subjectively through a veil of emotions, culture, and subjective experiences.

Many social and behavioral scientists take a broader view of the term *risk*. They view the term as a social construct, an idea that has been created and accepted by society. According to this subjectivist view, what technical and nontechnical experts mean by the word *risk* is often radically different. For technical experts, *risk* means probability multiplied by magnitude. For nontechnical experts, risk means what technical experts mean by *risk* (i.e. probability time magnitude) **plus** numerous subjective emotional and perceptual *factors*, including trust, benefits, personal control, voluntariness, dread, and familiarity. These additional factors are sometimes called "*outrage*" factors,[4] and they are seen as influential to how people respond to risks.

Social and behavioral scientists, such as Beck and Giddens, argue this broader view of risk as a social construct helps explain why risk has become the overarching obsession of the modern world and has become a focus point for modern fears and anxieties.[5] Fears and anxieties about the potential dangers of global warming, nuclear power plants, genetically modified organisms, nanotechnology, and a host of other risks and threats transcend national and international boundaries. Transboundary risks and threats are hotly debated on global stages occupied by multiple sets of players competing for attention. The players include policymakers, scientists, experts, activist groups, government agencies, corporations, political parties, the traditional broadcast and print media, social media, and the public. According to Beck and Giddens, inequalities multiply as rich and powerful players offload risks and dangers to less fortunate players.

2.3 Defining the Concept and Term *Risk Communication*

Risk communication can be defined as the transfer and exchange of information among interested parties about the nature, magnitude, significance, or control of a risk.[6] Information about risks can be communicated through a variety of channels, including, but not limited to, fact sheets, websites, webcasts, reports, texting, emails, social media postings, warning labels, billboards, bulletin boards, public meetings, and public hearings.

Modern understandings of risk and risk communication differ greatly from the past. For example, in ancient Mesopotamia, ca. 3200 BCE, there lived in the Tigris-Euphrates valley a group called the Asipu. One of their primary functions was to serve as risk, high concern, and crisis communication consultants. Members of the Asipu could be consulted about any high concern issue. Example issues included the cause of a disease outbreak, the need for a declaration of war, an alliance with another state, a change in the economic system, the selection of a leader, a proposed marriage, a suitable building site, a legal ruling, or the guilt or innocence of an alleged criminal. The Asipu would identify the important dimensions of the problem, identify alternative actions, collect information on the issue and the likely outcomes of each alternative, and consult the best data. From their perspective, the best data were signs from the gods, which the priest-like Asipu were especially qualified to interpret. The Asipu would then create a report with spaces empty for each alternative. A plus sign was added if the signs from the gods were favorable and a minus sign if unfavorable. The Asipu would communicate these results to their client, etched upon a clay tablet. The clay tablets of the Asipu appear to be among the first recorded instance of risk communication.[1]

One of the first formal definitions of *risk communication* in the health, safety, and environmental literature was offered by Covello, Slovic, and von Winterfeldt.[7] According to these authors, risk communication is the act of conveying or transmitting information among parties about levels of health, safety, or environmental risks; the significance or meaning of data about health, safety, or environmental risks; and decisions, actions, or policies aimed at managing or controlling health, safety, or environmental risks. Interested parties include government, agencies, corporations, industry groups, unions, the media, scientists, engineers, technical professionals, professional organizations, public interest groups, and individuals.

Covello, Slovic, and von Winterfeldt focused their definition of risk communication on the sharing and exchange of information about health, safety, and environmental topics. However, the authors noted their definition does not exclude the study of other risks, such as financial or legal risks. Nor does their definition exclude the study of secondary and tertiary effects triggered by the risk communication process, including psychological, social, economic, legal, and political repercussions.

In 1989, the National Academy of Sciences/National Research Council (NAS/NRC) offered one of the longest definitions of risk communication.[8] According to the NAS/NRC, risk communication:

> is an interactive process of exchange of information and opinion among individuals, groups, and institutions. It involves multiple messages about the nature of risk and other messages, not strictly about risk, that express concerns, opinions, or reaction to risk messages or to legal or institutional arrangements for risk management.

Additionally, the United Nations Food and Agriculture Organization (FAO) defined risk communication as "the exchange of information and opinions concerning risk and risk-related factors among risk assessors, risk managers, consumers and other interested parties."[9] The FAO definition, as does the NAS/NRC definition, highlights that risk communication is, ideally, an interactive, two-way, multi-dimensional exchange of information. Risk communication is therefore a process rather than a single product. It is a tool to help people make an informed decision about

---

1 Covello, V.T., Mumpower, J. (1985). "Risk analysis and risk management: An historical perspective." Risk Analysis 5 (2):103.

managing risks. The tool is effective because it creates trusting relationships, raises the level of understanding of relevant issues or actions for those interested or affected, and satisfies stakeholders that they are adequately informed within the limits of available knowledge.

On its website, the US Environmental Protection Agency (EPA) defines risk communication as "the process of informing people about potential hazards to their person, property, or community."[10] The EPA also cites the broader definition of risk communication offered by scholars in the field: "risk communication is a science-based approach for communicating effectively in situations of high stress, high concern or controversy."

In 2019, the EPA identified risk communication as one of the top priorities of the agency. The EPA administrator said:

> Risk communication goes to the heart of EPA's mission of protecting public health and the environment. We must be able to speak with one voice and clearly explain to the American people the relevant environmental and health risks that they face, that their families face and that their children face.[11]

According to the EPA, the purpose of risk communication is to help people understand the processes of risk assessment and management, to form scientifically valid perceptions of the likely hazards, and to take part in deciding how risk should be managed. The EPA points out the best risk communication occurs in contexts where the participants are informed, the process is fair, and the participants can solve whatever communication difficulties arise.

Ideally, risk communication is a two-way exchange of information and conversation in which an organization informs, and is informed by, affected community members. When the exchange goes well, risk communication provides people with timely, accurate, and credible information. It becomes the starting point for creating a public that is appropriately concerned about the risks they face and that is more likely to engage in risk-related behaviors. Effective risk communication creates a place for participation and dialog where people can engage in an interactive process that is thoughtful, solution-oriented, cooperative, and collaborative.

As shown in Figure 2.1, there are three primary goals of risk communication: (1) build trust, (2) promote knowledge, and (3) encourage supportive relationships and constructive dialog.

Building trust means building confidence and repairing trust if lost or damaged. It also includes building alliances and partnerships with those perceived to be trustworthy. *Promoting knowledge* means raising awareness and understanding of risks and dangers; promoting message consistency and transparency; and informing perceptions, attitudes, practices, beliefs, decisions, intentions, and behaviors. *Encouraging supportive relationships and constructive dialogue* means strengthening existing relationships, building new relationships, promoting participation and involvement by all interested parties, gaining consensus or agreement, promoting mutual aid, and enabling productive conversations. Lundgren and McMakin elaborate on these goals in their discussion of care and consensus risk communication.[12]

Of the three goals of risk communication, building trust is the most important. It is the first and most consequential step toward effective risk communication.

The functions of risk communication are multifold.[13] First, it must communicate the probabilities and consequences of known risks to stakeholders. Second, it must communicate to stakeholders proposals and policies for preventing, avoiding, mitigating, reducing, and managing the risk. Third, it should seek consensus among stakeholders regarding a specific course of response and mitigation.

Figure 2.1 Risk communication goals.

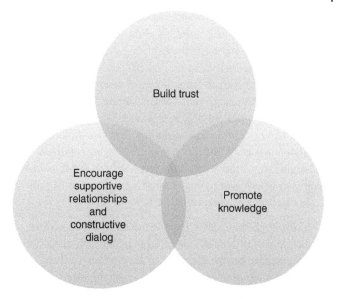

From a stakeholder perspective, Renn argues the ultimate purpose of risk communication is "to assist stakeholders and the public at large in understanding the rationale for a risk-based decision, and to arrive at a balanced judgment that reflects the factual evidence about the matter at hand in relation to their own interests and values."[14] A key challenge in risk communication is establishing communication networks and channels where stakeholders can trust each other and work together.

Risk communication is not public or health education. Public and health education requires risk communication skills, but the two tasks are distinct activities. "Education" implies a "teacher/student" relationship, in which the expert transfers and shares knowledge. Risk communication is primarily more of a peer-to-peer, two-way communication.

Risk communication also is not public relations. The typical focus of public relations is attempting to make people see issues the way the client or sponsor wants them seen. By comparison, the assumption of risk communication is that experts and nonexperts often have different perspectives on risk-related issues and that these different perspectives need to be heard, acknowledged, and respected.

At the heart of risk communication are efforts to understand and appreciate the perceptions and worldviews of others. If people perceive that their stress, concerns, worries, and fears are not being heard, acknowledged, respected, and addressed, they may lose trust in experts and risk management authorities. An effective response to these concerns is to engage in dialog, listen to concerns, and have a transparent discussion of what the scientific data about the risk show, including uncertainties. A key concept of risk communication is that the overall risk management process is seen differently from those who live with the risk than those who generate or manage the risk.

2.4 Risk Communication and Its Relationship to Risk Analysis

Risk analysis is a set of scientific methods for identifying risks, evaluating the likelihood and consequences of the risks occurring, and deciding how best to prevent, avoid, mitigate, reduce, manage, and communicate the risk.[15] Modern, formal *risk analysis* has four components: *hazard identification, risk assessment, risk management,* and *risk communication.*[16] The first component, *hazard identification,* comprises methods for identifying hazards and the conditions and events under which they potentially produce adverse consequences. The second component, *risk assessment,* comprises methods for organizing and evaluating information about the nature, strength of evidence, likelihood, and magnitude of adverse outcomes. The third component, *risk management,* comprises methods for analyzing, selecting, implementing, and evaluating actions to reduce risk. The fourth component, *risk communication*, comprises methods for communicating results from *hazard identification, risk assessment,* and *risk management.* As shown in Figure 2.2, *risk communication* interacts with all components of a *risk analysis.*

Formal quantitative risk analysis methods have been applied to a wide variety of issues. For example, health, safety, and environmental researchers have applied risk analysis principles, strategies, approaches, and methods to:

1) **Cancer risks**: Cancer risks resulting from exposures to chemicals, heavy metals, and other substances proven or suspected to be human carcinogens.
2) **Noncancer health risks**: Noncancer risks resulting from exposures to toxic substances in the environment that can cause adverse health effects on the heart, kidneys, liver, brain, and reproductive system.
3) **Ecological risks**: Ecological risks to natural ecosystems resulting from both habitat modification and environmental pollution.
4) **Natural hazard risks**: Natural hazard risks resulting from extreme events that originate in the natural environment, including (1) meteorological hazards, such as severe storms, heat waves, tornadoes, hurricanes, droughts, climate change, and wildfires; (2) hydrological hazards, such as floods, storm surges, and tsunamis; (3) geophysical hazards, such as volcanic eruptions,

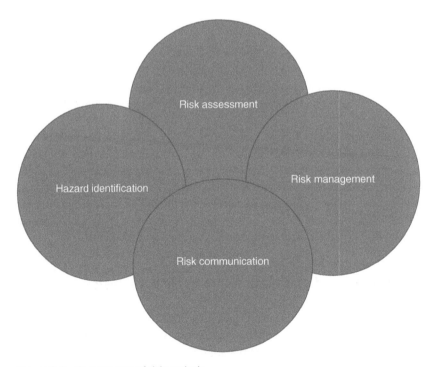

Figure 2.2 Components of risk analysis.

earthquakes, and landslides; and (4) biological hazards, such as epidemics, disease outbreaks, insect infestations, animal attacks, and food contamination incidents.

5) ***Technological risks***. Technological risks resulting from events that originate in human-controlled processes, including industrial accidents, transport accidents, dam collapses, mining accidents, and other types of technology-based accidents or incidents (e.g., accidents resulting in the release of toxic, flammable, explosive, radiological, or nuclear materials).

6) ***Human conflict risks***. Human conflict risks resulting from events such as terrorist bombing, active shooter incident, mass shooting incident, and cyberattack.

The line between these types of risks is often blurred. One example is Hurricane Katrina, which struck three US states – Louisiana, Mississippi, and Alabama. Problems caused by the natural hazard were increased by technological failures (e.g., by the failure of the levees and flood control systems); by the lack of coordination and conflict between public and private sector organizations at the federal, regional, state, and local level; and by huge operational and communication failures during the pre-crisis, crisis, and post-crisis phases of the disaster.[17]

2.5 Defining the Concepts and Terms *High Concern* and *High Concern Communication*

According to the Oxford English Dictionary, the word *concern* means "an interest or stake in something; a matter with which a person is occupied." The word comes from Latin, Anglo-Norman, Middle French, and French words meaning *to relate, regard*, or *consider*. For this book, a *high concern issue* is a problem of great interest. The problem becomes more intense when it has high consequences (stakes), occurs repeatedly (frequency), has lasted for a significant amount of time (duration), affects many people (scope or range), disrupts personal or community life (disruptive), deprives people of their perceived legal or moral rights (equity), and has negative effects perceived to be serious enough to require attention (severity).

High concern issues can vary from individual to individual, group to group, and place to place. They may also change over time because of factors, including history and sociodemographic conditions. Understanding these factors is needed to ensure that communication strategies, messages, materials, and activities are appropriately designed and implemented.

Levels of concern about an issue can be determined through a variety of means, including surveys and interviews with stakeholders. For example, the Agency for Toxic Substances and Disease Registry, an agency within the US Centers for Disease Control and Prevention, recommends the following questions to determine levels of community concern about exposure to a toxic chemical.[18]

- Is exposure to the chemical involuntary, as opposed to voluntary (e.g., an accidental chemical spill vs. a workplace exposure)?
- Is exposure to the chemical perceived to be controlled by others, as opposed to under an individual's control (e.g., in the water supply for a town vs. a place that can be easily avoided)?
- Is the exposure perceived to be unfairly distributed (e.g., affecting a certain part of town or a certain population vs. the entire town equally or randomly)?
- Is the exposure human-made and/or deliberate (e.g., the act of terrorism or vandalism)?
- Does the exposure have dramatic, long-lasting effects on the community (e.g., people can no longer live in a certain neighborhood or property was destroyed vs. something that can be cleaned up)?
- Is the source of exposure perceived to be an untrusted source (e.g., an industrial plant with a history of problems)?

- Does the exposure appear to affect children more than adults?
- Have there been deaths or serious illnesses that are perceived to be directly caused because of the chemical exposure or are deaths or serious illnesses expected?
- Does the media and/or the public perceive the event as the "first," "worst," or "biggest" of its type?
- Does the community perceive the response of public officials and others in authority to date has been inadequate or slow?
- Is a criminal investigation involved?

Concerns similar to these arise from issues other than exposures to toxic chemicals. For example, neighbors and community members often object to facilities they consider detrimental to their "backyards" or to the wider community. NIMBY (Not In My Back Yard) concerns and LULU (Locally Unwanted Land Use) concerns frequently arise following proposals to construct a new highway, casino, airport, wastewater treatment plant, garbage dump, prison, homeless shelter, wind farm, nuclear power plant, hydroelectric dam, center for the treatment of drug addiction, or half-way home for schizophrenic adults.

Concerns about inequities often create especially high levels of concern and even outrage. The perception that some people are more exposed to risks or harm more than others aggravates perceptions of risks and harm. This is especially the case if locational decisions are based on, result from, or produce social or economic inequities.

High levels of concern can produce a strong emotional reaction, such as anxiety, worry, uncertainty, apprehension, stress, fear, and outrage. When individuals experience these intense feelings, their ability to process information declines significantly.

A *high concern issue* can be external (e.g., health, social, economic, or political issue or change) or internal (e.g., work or domestic issue or change). A classic example of a high concern issue is the COVID-19 pandemic. Beginning in January and February 2020, Americans struggled to cope with the disruptions caused by COVID-19. By the beginning of 2021, more than 30 million Americans had contracted the disease and more than 500,000 had died. Globally, more than 120 million people had contracted the disease and more than 2.5 million had died.

On December 2, 2020, the director of the Centers for Disease Control and Prevention warned an anxious nation that it would face a devastating winter. He predicted that total deaths from COVID-19 could exceed half a million unless a large percentage of Americans followed precautions, including mask-wearing and social distancing. He said the next few months could be "the most difficult time in the public health history of this nation."[19]

According to the American Psychological Association, eight in ten adults identified COVID-19 as a significant source of stress in their life. Two-thirds of all adults said they had experienced increased stress over the course of the pandemic.[20]

High levels of concern about the COVID-19 pandemic were compounded by societal stressors pervasive in American society. These included mass shootings, unemployment, access to health care, racism, climate change/global warming, immigration, sexual assaults, and the opioid epidemic. More than three in five adults said the number of issues America faces currently is overwhelming to them. This marks a significant increase from 2019. And more than seven in ten Americans said 2020 was the lowest point in the nation's history they could remember.

The intensity of feelings and emotions generated by a high concern issue is determined by multiple factors. These include the perceived risk (i.e. the perceived probability and magnitude of the threat or danger) and the ability of the individual, group, or organization to cope and manage the stress associated with the issue. Feelings and emotions are also influenced by specific contextual characteristics of the perceived threat or danger, such as its intentionality.

High concern can produce a heightened state of arousal, which protects humans from threats and dangers. It is a defense and adaptive mechanism whereby parts of the brain and body typically slow down and other parts of the body and brain typically take over. It is a nervous system response that results in fight-freeze-flight behavior.[21]

Heightened arousal typically produces psychological and physiological changes. Physiological changes may include a rise in heart rate, blood pressure, a rise in body temperature, an increase in perspiration, an increase in constriction of the arteries, and secretion of neurotransmitters and hormones such as adrenaline and cortisol.

Neurological changes that most affect communications in high concern situations are (1) increased activation of the brain's amygdala and hypothalamus, which are central to the brain's system for the early detection of a threat or danger; and (2) decreased activation of the brain's frontal lobe, which is central to the brain's system for rational thought. Because of these changes, heightened arousal can decrease a person's ability to take in information, listen, and be empathic.

Heightened arousal is closely linked to what Nobel Prize winner Daniel Kahneman called *System 1 Thinking.*[22] System 1 thinking is heavily influenced by activation of the amygdala and the fight-freeze-flight response. It is fast, automatic, intuitive, and relies heavily on emotions. Kahneman contrasts System 1 thinking with *System 2 Thinking. System 2 Thinking* is heavily influenced by activation of the frontal lobes of the brain. It is slow, analytical, and relies on reasoning and logic. In high concern situations, System 1 thinking often dominates and can cause biased decision-making and dysfunctional communications.

Based on these observations, high concern communications can be defined as the transfer and exchange of information in emotionally charged situations where people are worried, upset, angry, stressed, or anxious. High concern communication encompasses virtually any situation involving risk or threat to the things people value. The goals of high concern communication are the same as the goals of risk communication: to build trust, promote knowledge, and encourage supportive relationships and constructive dialog.

At the personal level, individuals may use high concern communication practices to transfer and exchange information about rejections, discrimination, failures, major life changes (e.g., job changes, locational changes, divorce, sickness, scandals, arrests, births, deaths), and worries about health, children, and other family members. Individual high concern communications may be as basic as the exchange of information by a person communicating why they are late for a meeting or why they missed a deadline.

For groups and organizations, high concern communications may involve the exchange of information about mergers, restructuring, re-organization, budget cuts, cost overruns, missed deadlines, facility closures, performance failures, performance reviews, job interviews, reputational attacks, protests, and employee complaints. For organizations or business units (e.g., Customer Service Departments) that serve people who are worried or anxious or have an issue to resolve, high concern communications should begin at the front door. For example, at a hospital, high concern communications should begin with exchanges of information between a patient or visitor and the hospital parking lot attendant, security guard, information desk, reception desk, and nurse's desk.

For organizations, one of the major precipitators of high concern is organizational change. For example, proposed changes to the organizational structure typically lead to high levels of concerns about job security, mission, services, budget, workplace relationships, and job assignments. It may lead to any or all of the following questions: Why do we need the proposed change? Why now? Who decided these changes are needed? What is the full extent of the proposed change? What are the imperatives driving the proposed change? Why are existing procedures and strategies no longer good enough? Are you sure the proposed change will improve the situation?

The organizational change also often leads to questions of a more personal nature. For example: What will happen to my job because of the change? Will this affect my pay? Will I lose my benefits package? Will I lose my seniority? Will I need to master new skills? Will I have the same team and coworkers? Will I have to relocate? If I have to relocate, will the organization pay my relocating expenses? What will happen to me if I don't want to relocate? Who will be my supervisor? Will I have to travel more?

Equivocal answers to these questions introduce *uncertainty* about the future. The concept of *uncertainty* is central to the definition of the term *high concern*. For employees, organizational change threatens and risks many of the things they value, including the five key domains identified in the neuroscience-based SCARF model: [23]

Status – perceptions regarding their importance to others
Certainty – perceptions regarding their ability to predict the future
Autonomy – perceptions regarding their ability to control events
Relatedness – perceptions regarding how they relate to others and how safe they feel in these relationships
Fairness – perceptions regarding their being treated fairly and equitably

Neuroscience research underlying the SCARF model shows that negative perceptions regarding any of the SCARF domains can activate the same threat and reward responses in the brain as does a physical threat. Negative perceptions of the SCARF domains can also produce a strong emotional reaction and the body's release of cortisol, adrenaline, and epinephrine – chemicals associated with the flight-freeze-fight response. These chemical releases can severely affect the ability of a person to focus, solve problems, communicate, cooperate, and think rationally and logically – functions performed largely by the prefrontal cortex of the brain.

From a psychological and neurological perspective, whether and to what degree a person, group, or organization experiences high arousal, the fight-freeze-flight response, or stress from a situation is determined in part by what an individual, group, or organization defines as high concern. Based on cultural factors, people, groups, and organizations have different tendencies toward fight-freeze-flight. People, groups, and organizations use different lenses to determine what is of high concern. The level of concern felt and assigned to a situation depends on personalities, worldview, beliefs, and culture. What is defined as high concern matters to a specific individual, group, organization, culture, or society.

High concern, as in stress, is often seen as a negative. However, high concern and stress can have positive and helpful effects, such as when it motivates people to accomplish more. It is when high concern and stress become excessive and overload the capacity of a person or group to cope that it becomes mentally and physically dangerous. What makes high concern and stress excessive is often fear. It is therefore not enough to give the facts about a risk or threat. Fear is real and can keep individuals and communities from making informed decisions. However, through effective communication, fear can be channeled into productive behaviors. For example, if not excessive, fear can lead to information-seeking. People are often more accepting of fear when it is acknowledged.

2.6 Defining the Concept and Term *Crisis*

According to the Oxford English Dictionary, a *crisis* is "a time of intense difficulty, trouble, or danger." A *crisis* typically represents a decisive turning point and an unstable situation where difficult and crucial decisions must be made. The origin of the word *crisis* is revealing. The word has origins in the Latin word *crisis* (decisive moment), which comes from the Greek word *krisis* (decision, judgment) and from the Greek word *krinein* (decide, judge). It also has origins in the Old French word *crise* and the Middle English medical word *crisis*, denoting "the turning point in a disease; a sudden change for better or worse." Based on the origin of the word, a crisis is an event or turning point that brings, or has the potential to bring, great reputational, financial, psychological, and/or physical harm to an individual, group, population, organization, or institution. Examples of a crisis include a major industrial accident, a major spill of toxic materials, a major storm, or a pandemic. Crises can human-made or natural. The cause will influence how others view the situation and the response.

Coombs and Holloday offer a more complex definition of *crisis*. [24] They offer four defining attributes of a crisis: (1) unpredictability, (2) threat to stakeholder expectations, (3) impact on organizational performance, and (4) potential for negative outcomes. A similar definition comes from Ulmer, Sellnow, and Seeger, who also define *crisis* with four attributes: (1) the unexpected nature of the event, (2) the nonroutine demands on the organization, (3) the production of uncertainty, and (4) the threat to achieving organizational goals. [25]

By combining aspects from these various definitions, for purposes of this book, I offer a shorter and a longer definition of the word *crisis*. The shorter definition is: *a crisis is a significant risk manifested*. This definition is consistent with the definition offered by Heath and O'Hare: *a crisis is a threat with significant potential adverse impacts that has materialized.* [26]

A longer definition is: a crisis is a risk manifested that characteristically (1) is abrupt and unexpected, (2) exceeds the expectations of those affected, (3) disrupts normal processes, (4) places nonroutine and unique demands on the responding organizations, (5) produces high amounts of uncertainty, (6) challenges organizational performance, and (7) poses a significant chance of harm or loss to individuals and organizations.

Crises typically cause disruptions in our normal lives, high levels of stress and high concerns about adverse consequences, confusion, fear, and an active search for leadership and support. Four characteristics can cause leaders to label a situation a crisis: (1) there are imminent dangers and significant consequences, (2) resolution requires quick action, (3) they feel unprepared, and (4) there is knowledge of the event or situation by the outside world, particularly the media. Therefore, when a leader is trained and feels better able to handle a situation, it is less likely to be perceived as a crisis.

Researchers have debated the difference among a *crisis*, a *disaster*, and an *emergency*. They are often used interchangeably, although *crisis* is a broader term. For example, in his definition of the term *disaster*, Oliver-Smith noted: [27]

> Disaster is a term that is used fairly liberally in popular parlance. Many events or processes are colloquially referred to as disasters—everything from a failed social event to a regional hurricane.

One of the most widely accepted definitions of the term *disaster* in research literature is offered by the United Nations (UN). According to the UN, a disaster "is a serious disruption of the functioning of a community or a society involving widespread human, material, economic or environmental losses and impacts, which exceeds the ability of the affected community or society to cope using its own resources." [28] This definition has elements of the definitions of the term crisis offered earlier in this chapter. It also aligns with what many might call an emergency.

Despite the interchangeability of the terms in many studies, there are arguably important differences. For example, disaster and emergency can be distinguished by their familiarity and severity.

Disasters are typically characterized by large-scale direct and indirect adverse effects. These adverse effects include loss of life, loss of property, damage to infrastructure, and loss of revenue and unemployment. As pointed out by Lindell, Prater, and Perry,[29] the term *emergency* is typically used to describe:

> . . .an event involving minor consequences for a community—perhaps a few casualties and a limited amount of property damage. In this sense, emergencies are events that are frequently experienced, relatively well understood, and can be managed successfully with local resources—sometimes with the resources of a single local government agency. Emergencies are the common occurrences we see uniformed responders managing—car crashes, ruptured natural gas pipelines, house fires, traumatic injuries, and cardiac crises.

Lindell, Prater, and Perry offer another usage of the term emergency when the goal is to communicate the imminence of an event rather than the severity of its consequences. In this context, *emergency* refers to a situation where a higher than normal probability of an extreme event occurring exists. The term disaster is reserved for the actual occurrence of an event that produces casualties and damage at a level exceeding a community's ability to cope.

There is no universal definition of *crisis* or *disaster* but both share common characteristics. Crises and disasters are typically (1) sudden and abrupt; (2) cause, or have the potential to cause, significant human, material, economic or environmental harm; and (3) challenge the immediate capacity or ability of individuals, organizations, communities, or societies to respond.

Differences among researchers about core definitions, such as those described above, are not unusual. For example, Kroeber and Kluckhohn, after surveying the literature in anthropology, found 164 definitions of the term culture – a core concept in anthropology.[30] These definition differences are not without consequences. They often lead to different theories, principles, approaches, methods, and tools. For example, 20 years ago, many authors failed to clearly discriminate between the concepts of crisis prevention, crisis preparedness, crisis mitigation, and crisis management. Adding to the confusion, many authors used the same term to discuss different phases or dimensions of a crisis. For example, some authors used the term crisis management to describe only the immediate response to a triggering event. Other authors used the term crisis management to describe the immediate response to a triggering event but also to crisis prevention and preparation.

2.7   Defining the Concept and Term *Crisis Communication*

Crisis communication can be defined as the exchange of risk information about an abrupt, uncertain, nonroutine, and disruptive event that poses immediate and significant consequences. Crisis communication is ideally the planned, intentional transfer of risk information when preparing for a crisis, responding to a crisis, and recovering from a crisis. Crisis communication is primarily concerned with that part of the risk communication continuum that alerts stakeholders to an immediate threat and provides options to minimize the risk. It serves a motivational and time-sensitive purpose.

There is a large overlap between risk and crisis communication. Communications about risks and threats follow a cycle of prevention, preparedness, warning, response, and recovery. The first two steps – prevention and preparedness – have traditionally belonged to scholars and practitioners of risk communication. The latter three steps – warning, response, and recovery – have traditionally belonged to scholars and practitioners of crisis communication. Scholars and practitioners of both risk and crisis communication focus on what the human brain hears, understands, believes, and decides about a risk or threat.

Crisis communicators primarily focus on a situation, something that has just happened or is still happening. Risk communicators primarily focus on what might happen. For example, for a food contamination scenario, risk communicators might focus on questions such as: How likely is food contamination? How can people be made more aware of the potential for food contamination? How can food contamination incidents be prevented? Crisis communicators might focus on questions such as: What do people want to know about the incident? What things should people be doing in response to the incident? Where can people go for credible information? What else might go wrong?

The objectives of crisis communication are similar to the objectives of risk and high concern communication: to build trust, promote knowledge, and encourage appropriate behaviors and supportive relationships. Specific objectives change as the continuum moves from the pre-crisis preparedness stage, through the crisis event stage, and then to the recovery stage. The overarching goal of crisis communication is to reduce or eliminate harm through individual, group, organizational, or institutional action.

One of the key lessons to be learned from successful cases of crisis communication is that each phase of a crisis has a distinct set of communication objectives and each phase requires a distinct set of communication skills. For example, a key communication objective of pre-crisis preparedness communication is to provide information needed by stakeholders to avert a crisis from occurring. During the crisis itself, a key communication objective is to share the information about (1) what people can or should do to protect themselves and what they value; (2) the location of and access to crisis resources; and (3) connecting with first responders, emergency management, and family and friends. Skipping a phase, such as communications in the pre-crisis/preparation phase, seldom produces satisfactory results. Making mistakes in any phase can negate hard-won gains.

Crisis communication effectiveness can be measured by changes in knowledge, perceived trust, safety, calm, connectedness, hope, and self- and group-efficacy. Self- and group-efficacy refers to beliefs by individuals or groups that they can engage in and perform protective behaviors and actions. Self- and group-efficacy reflects confidence by individuals or groups that they can exert control over their behavior and their environment. An important point here that relates to self- or group-efficacy is in how one frames a situation. If an individual or a group believes they can handle the situation, the situation no longer is seen as a crisis.

When crisis communication is not planned and implemented effectively, a long list of negative outcomes can occur. These include confusion caused by contradictory messages, people rejecting or refusing to follow recommendations, counterproductive behaviors, loss of trust, and social disruption.

2.8 Chapter Resources

Below are additional resources to expand on the content presented in this chapter.

Andrews, R. (1999). *Managing the Environment, Managing Ourselves: A History of American Environmental Policy*. New Haven: Yale University Press.

Aufder Heide, E. (2004). "Common misconceptions about disasters: Panic, the "disaster syndrome," and looting," in *The first 72 hours: A community approach to disaster preparedness*, ed. M. O'Leary. Lincoln, NB: iUniverse Publishing.

Beck, M., Kewell, B. (2014). *Risk: A Study of Its Origins, History and Politics*. New Jersey: World Scientific Publishing Company.

Beck, U. (1992). *Risk Society: Towards a New Modernity*. Sage: London.

Beck, U. (2008) "Living in the world risk society." *Economy and Society* (35):329–345.

Becker, S. M. 2007. "Communicating risk to the public after radiological incidents." *British Medical Journal* 335(7630):1106–1107.

Bennett, P., Calman, K., eds., (1999). *Risk Communication and Public Health*. New York: Oxford University Press.

Bennett, P., Coles, D., McDonald, A. (1999). "Risk communication as a decision process," in *Risk Communication and Public Health*, eds. P. Bennett and K. Calman. New York: Oxford University Press.

Bier, V. M. (2001). "On the state of the art: Risk communication to the public." *Reliability Engineering and System Safety* 71(2):139–150.

Bostrom, A., C. Atman, Fischhoff, B., Morgan, M. (1994). "Evaluating risk communications: Completing and correcting mental models of hazardous processes, part II." *Risk Analysis* 14(5):789–797.

Centers for Disease Control and Prevention (CDC) (2012). *Emergency, Crisis, and Risk Communication*. Atlanta, GA: Centers for Disease Control and Prevention.

Chess, C., Hance, B. J., Sandman, P. M. (1986). *Planning Dialogue with Communities: A Risk Communication Workbook*. New Brunswick, NJ: Rutgers University, Cook College, Environmental Media Communication Research Program.

Coombs, T. W. (1995). "Choosing the right words: The development of guidelines for the selection of the 'appropriate' crisis-response strategies." *Management Communication Quarterly* 8(4): 447–476.

Coombs, W. (2019). *Ongoing Crisis Communications: Planning, Managing, and Responding*. Thousand Oaks, CA: Sage Publications.

Coombs, W.T. (2007). "Protecting organization reputations during a crisis: The development and application of situational crisis communication theory." *Corporate Reputation Review* 10(3):163–176.

Coombs, W.T., Holloday, S.J. (2017). *Handbook of Crisis Communication*. London: Wiley-Blackwell.

Covello V. (1993). "Risk communication and occupational medicine." *Journal of Occupational Medicine* 35:18–19.

Covello, V.T. (2003). "Best practices in public health risk and crisis communication." *Journal of Health Communication* 8(Suppl. 1):5–8; discussion, 148–151.

Covello, V.T. (2006). "Risk communication and message mapping: A new tool for communicating effectively in public health emergencies and disasters." *Journal of Emergency Management* 4(3):25–40.

Covello V.T. (2010). "Strategies for overcoming challenges for effective risk communication", in *Handbook of Risk and Crisis Communication*, eds. Heath RL, O'Hair H. New York: Routledge.

Covello V.T. (2011). "Risk communication, radiation, and radiological emergencies: Strategies, tools, and techniques." *Health Physics* 101:511–530.

Covello, V.T. (2014). "Risk communication," in *Environmental health: From global to local*, ed. H. Frumkin. San Francisco: Jossey-Bass/Wiley.

Covello, V.T., Allen, F.W. (1988). *Seven Cardinal Rules of Risk Communication*. Washington, D.C.: US Environmental Protection Agency.

Covello, V.T, Merkhofer, M. (1993). *Risk Assessment Methods: Approaches for Assessing Health and Environmental Risks*. New York: Plenum Press.

Covello, V.T., McCallum, D. B., Pavlova, M. T., eds. (1989). *Effective Risk Communication: The Role and Responsibility of Government and Nongovernment Organizations*. New York: Plenum Press.

Covello, V.T., Mumpower, J. (1985). "Risk analysis and risk management: An historical perspective." *Risk Analysis* 5 (2):103–120.

Covello, V.T., Peters, R., Wojtecki, J., Hyde, R. (2001)." Risk communication, the West Nile virus epidemic, and bio-terrorism: Responding to the communication challenges posed by the intentional or unintentional release of a pathogen in an urban setting." *Journal of Urban Health* 78(2):382–391.

Covello, V.T., Sandman, P. (2001). "Risk communication: Evolution and revolution," in *Solutions to an environment in peril*, ed. A. Wolbarst. Baltimore, MD: Johns Hopkins University Press.

Covello, V.T., Slovic, P., von Winterfeldt, D. (1986). "Risk communication: A review of the literature." *Risk Abstracts* 3(4):171–182.

Covello, V.T., Minamyer, S., Clayton, K. (2007). *Effective risk and crisis communication during water security emergencies*. EPA Policy Report; EPA 600-R07-027. Washington, D.C.: US Environmental Protection Agency.

Cvetkovich, G., Siegrist, M., Murray R., Tragesser, S. (2002). "New information and social trust asymmetry and perseverance of attributions about hazard managers." *Risk Analysis* 22(2):359–367.

Dunwoody, S. (2014). "Science journalism," in *Handbook of Public Communication of Science and Technology*, eds. M. Bucchi and B. Trench. New York: Routledge.

Earle, T.C. (2010). "Trust in risk management: A model-based review of empirical research." *Risk Analysis* 30(4):541–574.

Earle, T.C., Siegrist, M. (2008). "On the relation between trust and fairness in environmental risk management." *Risk Analysis* 28(5):1395–1414.

Fearn-Banks, K. (2007). *Crisis Communications: A Casebook Approach*, 3rd ed. Mahwah, NJ: Lawrence Erlbaum Associates.

Finn Frandsen, F., Johansen, W. (2017). *Organizational Crisis Communication*. Thousand Oaks, California: Sage Publications.

Fitzpatrick C., Mileti DS. (1994) "Public risk communication," in *Disasters, Collective Behavior, and Social Organization*, eds. R.D. Dynes, K.J. Tierney. Newark, DE: University of Delaware Press.

Fischhoff, B. (1995). "Risk perception and communication unplugged: Twenty years of process." *Risk Analysis* 15(2):137–145.

Fischhoff, B. (2013). "The science of science communication." *Proceedings of the National Academy of Sciences of the United States of America* 110: 14033–14039.

Fischhoff, B., Davis, A. L. (2014). "Communicating scientific uncertainty." *Proceedings of the National Academy of Sciences of the United States of America* 111(Suppl. 4):13664–13671.

Fischhoff, B., Kadvany, J. (2011). *Risk: A Very Short Introduction*. New York: Oxford University Press.

Flynn, J., Slovic, P., Mertz, C. K. (1994). "Gender, race, and perception of environmental health risks." *Risk Analysis* 14(6):1101–1108.

Glik D.C. (2007). "Risk communication for public health emergencies." *Annual Review of Public Health* 28(1):33-–54.

Haight, J.M., ed. (2008). *The Safety Professionals Handbook: Technical Applications*. Des Plaines, IL: The American Society of Safety Engineers.

Hance, B. J., Chess, C., Sandman, P. M. (1990). *Industry Risk Communication Manual*. Boca Raton, FL: CRC Press/Lewis Publishers.

Heath, R., O'Hair, D., eds. (2009). *Handbook of Risk and Crisis Communication*. New York: Routledge.

Heath, J., O'Hair, D. (2009). "The significance of crisis and risk communication," in *Handbook of Risk and Crisis Communication*, eds. R.L. Heath and H.D. O'Hair. New York: Taylor and Francis/Routledge.

Hyer, R.N., Covello, V.T. (2007). *Effective Media Communication During Public Health Emergencies: A World Health Organization Handbook*. Geneva: World Health Organization Publications.

Kahneman, D. (2011). *Thinking, Fast and Slow*. New York: Macmillan Publishers.

Kahneman, D., Slovic, P., Tversky, A., eds. (1982). *Judgment Under Uncertainty: Heuristics and Biases*. New York: Cambridge University Press.

Kahneman, D., Tversky, A. (1979). "Prospect theory: An analysis of decision under risk." *Econometrica* 47(2):263–291.

Kasperson, R.E. (1986). "Six Propositions on public participation and their relevance for risk communication." *Risk Analysis* 6(3):275–281.

Kasperson, R.E., Golding, D., Tuler, D. (1992). "Social distrust as a factor in sitting hazardous facilities and communicating risks." *Journal of Social Issues* 48(4):161–187.

Kasperson, R. E., Renn, O., Slovic, P., Brown, H. S., Emel, J., Goble, R., Kasperson, J. X., Ratick, S. (1988). "Social amplification of risk: A conceptual framework." *Risk Analysis* 8(2):177–187.

Kasperson R.E., Palmlund I. (1989) "Evaluating risk communication," in *Effective Risk Communication. Contemporary Issues in Risk Analysis*, eds. Covello V.T., McCallum D.B., Pavlova M.T., vol 4. Boston: Springer.

Kovoor-Misra, S. (2009). *Crisis Management: Resilience and Change*. Thousand Oaks, CA: Sage Publications.

Kroeber, A., Kluckhohn, C. (1952). *Culture: A Critical Review of Concepts and Definition*. New York: Vintage Books.

Lindell M., Perry R. (1992). *Behavioral Foundations of Community Emergency Planning*. Washington, DC: Hemisphere Publishing Co.

Lindell, M.K, Prater, C.S., Perry, R.W. (2006). *Fundamentals of Emergency Management*. Washington, D.C.: Federal Emergency Management Agency. https://training.fema.gov/hiedu/aemrc/booksdownload/fem/

Lindenfeld, L., Smith, H., Norton, T., Grecu, N. (2014). "Risk communication and sustainability science: Lessons from the field." *Sustainability Science* 9(2):119–27.

Lundgren, R., McMakin, A. (2018). *Risk Communication: A Handbook for Communicating Environmental, Safety, and Health Risks*. Hoboken, NJ: IEEE Press.

McComas, K. A. (2006). "Defining moments in risk communication research: 1996–2005." *Journal of Health Communication* 11(1):75–91.

Mileti, D.S., Sorenson J. (1990). *Communication of Emergency Public Warnings*. ORNL-6609. Oak Ridge, TN: Oak Ridge National Laboratory.

Mileti, D.S. (2017). Public Response to Disaster Warnings. http://swfound.org/media/82620/PUBLIC%20RESPONSE%20TO%20DISASTER%20WARNI NGS%20-%20Dennis%20S.%20Mileti.pdf. Accessed May 25, 2017.

Mileti, D., Nathe, S., Gori, P., Greene, M., Lemersal, E. (2004). *Public Hazards Communication and Education: The State of the Art*. Boulder, CO: Natural Hazards Center.

Morgan, M.G., Fischhoff, B., Bostrom, A., Atman, C. J. (2002). *Risk Communication: A Mental Model Approach*. Cambridge, UK: Cambridge University Press.

National Research Council (1989). *Improving Risk Communication*. Washington, D.C.: National Academy Press.

National Research Council (1996). *Understanding Risk: Informing Decisions in a Democratic Society*. Washington, DC: National Academy Press.

National Research Council. (2013). *Public Response to Alerts and Warnings Using Social Media: Report of a Workshop on Current Knowledge and Research Gaps*. Washington, DC: National Academies Press.

National Oceanic and Atmospheric Administration. US Department of Commerce. (2016). *Risk Communication and Behavior. Best Practices and Research*. Washington, DC: National Oceanic and Atmospheric Administration.

National Academy of Sciences/National Research Council. (2008). Public Participation in Environmental Assessment and Decision Making. Panel on Public Participation in Environmental Assessment and Decision Making, eds. T. Dietz and P.C. Stern. *Committee on the Human Dimensions of Global Change, Division of Behavioral and Social Sciences and Education*. Washington, DC: National Academies Press.

National Academy of Sciences (2014). *The Science of Science Communication II: Summary of a Colloquium*. Washington, DC: The National Academies Press.

National Academy of Sciences (2017) *Communicating Science Effectively*. Washington, D.C.: The National Academies Press

Oliver-Smith, A., Hoffman, S., eds. (1999). *"What is a Disaster?" in The Angry Earth: Disaster in Anthropological Perspective*. New York: Routledge.

Peters, E. (2012). "Beyond comprehension: The role of numeracy in judgments and decisions." *Current Directions in Psychological Science* 21(1):31–35.

Peters, R., McCallum, D., Covello, V. T. (1997). "The determinants of trust and credibility in environmental risk communication: An empirical study." *Risk Analysis* 17(1):43–54.

Pidgeon, N., Kasperson, R., Slovic, P. (2003). *The Social Amplification of Risk*. Cambridge University Press.

Renn, O. (2008). *Risk Governance: Coping with Uncertainty in a Complex World*. London, UK: Earthscan.

Renn, O. (2009). "Risk communication," in *Handbook of Risk and Crisis Communication*, eds. R.L. Heath, H.D. O'Hair. New York: Taylor and Francis/Routledge.

Renn, O., Levin, D. (1991). "Credibility and trust in risk communication," in *Communicating risks to the public*, eds. R. Kasperson, P. Stallen. Dordrecht, The Netherlands: Kluwer Academic Publishers.

Reynolds, B. (2014). *Crisis and Emergency Risk Communication*. Atlanta, GA: US Centers for Disease Control and Prevention.

Reynolds, B., Seeger, M. W. (2005). "Crisis and emergency risk communication as an integrative model." *Journal of Health Communication* 10:43–55.

Rock, D. (2008). "SCARF: A brain-based model for collaborating with and influencing others." *Neuro Leadership Journal* 1:1–7.

Rodrıguez, H., Dıaz, W., Santos, J., Aguirre, B. (2007). "Communicating risk and uncertainty: Science, technology, and disasters at the crossroads," in *Handbook of Disaster Research*, eds. H. Rodríguez, E.L. Quarantelli, R.R. Dynes. New York: Springer.

Rowan, K. E. (1991). "Goals, obstacles, and strategies in risk communication: A problem-solving approach to improving communication about risks." *Journal of Applied Communication Research*, 19:300–329.

Sandman, P.M. (1989). "Hazard versus outrage in the public perception of risk," in *Effective Risk Communication: The Role and Responsibility of Government and Non-Government Organizations*, eds. V.T. Covello, D. B. McCallum, M.T. Pavlova. New York: Plenum.

Scarlett, H. (2019). *Neuroscience for Organizational Change: An Evidence-based Practical Guide to Managing Change* (2nd Edition). London and New York: Kogan Page Limited Publishing Co.

Sedej, T., Justinek, G. (2017). Effective tools for improving employee feedback during organizational change, In *Organizational Productivity and Performance Measurements Using Predictive Modeling and Analytics*, eds. M. Tavana, K. Szabat, K. Puranam. Hershey, PA: IGO Global.

Seeger, M.W. (2006). "Best practices in crisis communication: An expert panel process." *Journal of Applied Communication Research* 34(3):232–44.

Seeger, M. W., Sellnow, T. L., Ulmer, R. R. (2003). *Communication and Organizational Crisis*. Westport, CT: Prager

Sheppard, B., Janoske, M., Liu, B. (2012). "Understanding Risk Communication Theory: A Guide for Emergency Managers and Communicators." *Report to Human Factors/Behavioral Sciences Division, Science and Technology Directorate*. College Park, MD: US Department of Homeland Security.

Slovic, P. (1987). "Perception of Risk." *Science* 236 (4799):280–285.

Slovic, P. (1999). 'Trust, emotion, and sex." *Risk Analysis* 19(4): 689–701.

Slovic, P. (2000). *The Perception of Risk*. London, UK: Earthscan.

Slovic, P. (2016). "Understanding perceived risk: 1978-2015." *Environment: Science and Policy for Sustainable Development* 58(1):25–29.

Slovic, P., M. Finucane, L., Peters, E., MacGregor, D.G. (2004). "Risk as analysis and risk as feelings: Some thoughts about affect, reason, risk, and rationality." *Risk Analysis* 24(2): 311–322.

Sellnow, T.L., Ulmer, R.R., Seeger, M.W. (2009). Effective Risk Communication: A Message-Centered Approach. New York, NY: Springer.

Sorenson, J.H. (2000). "Hazard warning systems: Review of 20 years of progress." *Natural Hazards Review* 1:119–125.

Stallen, P. J. M., Tomas, A. (1988). "Public concerns about industrial hazards." *Risk Analysis* 8(2):235–245.

Steelman, T. A., and McCaffrey, S. (2013). "Best practices in risk and crisis communication: Implications for natural hazards management." *Natural Hazards* 65(1):683–705.

Tversky, A., Kahneman, D. (1974). "Judgment under uncertainty: Heuristics and biases." *Science* 185(4157):1124–1131.

Ulmer, R., Sellnow, T., Seeger, M. (2011). Effective Crisis Communication: Moving from Crisis to Opportunity. Thousand Oaks, CA: Sage Publications.

United Nations International Strategy for Disaster Reduction. (2004). Living with Risk: A Global Review of Disaster Reduction Initiatives. New York: United Nations International Strategy for Disaster Reduction.

US Centers for Disease Control and Prevention (2014). Crisis and Emergency Risk Communication. Atlanta, GA: US Centers for Disease Control and Prevention.

US Environmental Protection Agency. (2005). Superfund Community Involvement Handbook. EPA 540-K-05-003. Washington, D.C.: US Environmental Protection Agency.

US Department of Health and Human Services. (2006). Communicating in a Crisis: Risk Communication Guidelines for Public Officials. Washington, D.C.: US Department of Health and Human Services.

US Occupational Safety and Health Administration. (2020). Hazard Communication. Washington, D.C: US Occupational Safety and Health Administration. Occupational Safety and Health. Accessed at: https://www.osha.gov/hazcom

Walaski, (Ferrante), P. (2011). Risk and Crisis Communications: Methods and Messages. Hoboken, NJ: Wiley.

Weinstein, N. D. (1987). Taking Care: Understanding and Encouraging Self-Protective Behavior. New York: Cambridge University Press.

Wojtecki, J.G., Peters, R.G. (2000). "Organizational Change: Information Technology Meets the Carbon Based Employee Unit." The 2000 Annual: Volume 2, Consulting. San Francisco, CA: Jossey-Bass/Pfeiffer

World Health Organization/Food and Agriculture Organization (2006). Food safety risk analysis: a guide for national food safety authorities. Rome, Italy: World Health Organization/Food and Agriculture Organization.

Wood, M.M., Mileti, D.S., Kano, M., Kelley, M.M., Regan, R., Bourque, L.B. (2011). "Communicating actionable risk for terrorism and other hazards." *Risk Analysis* 34(4):601–615.

Zimmerman, R. (1987). "A process framework for risk communication." *Science, Technology, and Human Values* 12 (Summer/Fall): 131–137.

Endnotes

1 Covello, V.T., Merkhofer, M. (1993). *Risk Assessment Methods. Approaches for Assessing Health and Environmental Risks.* New York: Plenum Press.

2 Covello, V.T, Merkhofer, M. (1993). *Risk Assessment Methods: Approaches for Assessing Health and Environmental Risks.* New York: Plenum Press.

3 See, e.g., Sandman, P.M. (1989). "Hazard versus outrage in the public perception of risk," in *Effective Risk Communication: The Role and Responsibility of Government and Non-Government Organizations*, eds. V.T. Covello, D. B. McCallum, M.T. Pavlova. New York: Plenum. See also Covello, V. T., Sandman, P. (2001). "Risk communication: Evolution and revolution," in *Solutions to an environment in peril*, ed. A. Wolbarst. Baltimore, MD: Johns Hopkins University Press.

4 See, e.g., Sandman, P.M. (1998). "Hazard versus outrage in the public perception of risk" in *Effective Risk Communication: The Role and Responsibility of Government and Nongovernment Organization*, eds. V.T. Covello, D.B. McCallum, M.T. Pavlova. New York: Plenum Press. See also: Covello, V.T., Sandman, P.M. (2001). "Risk communication: Evolution and revolution" in *Solutions to an Environment in Peril*, ed. A. Wolbarst. Baltimore: Johns Hopkins University Press.

5 See, e.g., Giddens, A. (1991). *Modernity and Self-Identity, Self and Society in the Late Modern Age.* Cambridge, UK: Polity. Beck, U. (1992) Risk society. *Towards a new modernity.* Sage, London; Beck U (2008) "Living in the world risk society." *Economy and Society* (35):329–345.

6 See, Covello, V. (1992). *"Risk communication: an emerging area of health communication research" in Communication Yearbook 15*, ed. S. Deetz, 359–373. Newbury Park, CA: Sage.

7 Covello, V. T., Slovic, P., von Winterfeldt, D. (1986). "Risk communication: A review of the literature." *Risk Abstracts* 3(4):171–182.

8 National Academy of Sciences/National Research Council (1989). *Improving Risk Communication.* Washington, DC: National Academies Press. p. 21

9 United Nations Food and Agriculture Organization (1995). Joint FAO/WHO Expert Consultation on the Application of Risk Communication to Food Standards and Safety Matters. Section 3: Elements and Guiding Principles of Risk Communication. Accessed at: http://www.fao.org/3/x1271e/X1271E03.htm

10 Accessed at: https://www.epa.gov/risk/risk-communication

11 Accessed at: https://www.eenews.net/stories/1060089069

12 Lundgren, R., McMakin, A. (2018). *Risk Communication: A Handbook for Communicating Environmental, Safety, and Health Risks.* Hoboken, NJ: Wiley/IEEE Press.

13 Rowan, K. E. (1991). "Goals, obstacles, and strategies in risk communication: A problem-solving approach to improving communication about risks." *Journal of Applied Communication Research* 19:300–329

14 Renn, O. (2009) "Risk communication," in *Handbook of Risk and Crisis Communication*, eds. R.L. Heath and H.D. O'Hair. New York: Taylor and Francis/Routledge p. 80.

15 Covello, V.T., Mumpower, J. (1985). "Risk analysis and risk management: an historical perspective." *Risk Analysis* 5 (2):103–120.

16 See, e.g., Covello, V.T, Merkhofer, M. (1993). *Risk Assessment Methods.* New York: Plenum Press.

17 See Cole, T.W, Fellows, K.L. (2008) "Risk communication failure: a case study of New Orleans and Hurricane Katrina." *Southern Communication Journal*, 73(3):211–228. See also the case study of Hurricane Katrina in Fearn-Banks, K. (2007). *Crisis communications: A casebook approach.* Mahwah, NJ: Lawrence Erlbaum Associates, Publishers.

18 Agency for Toxic Substances and Disease Registry (2020). Community Concern Assessment Tool. Accessed at: https://www.atsdr.cdc.gov/communications-toolkit/documents/08_community-concern-assessment-tool_508.pdf

19 New York Times, December 2, 2020. "Covid-19 Live Updates: C.D.C. Director Warns Winter May Be 'Most Difficult Time' in US Public Health History." Accessed at: https://www.nytimes.com/live/2020/12/02/world/covid-19-coronavirus?name=styln-coronavirus&region=TOP_BANNER&block=storyline_menu_recirc&action=click&pgtype=LegacyCollection&impression_id=9875aa71-34e7-11eb-937d-ad0013f652b2&variant=1_Show

20 American Psychological Association (2020). Stress in America 2020: A National Mental Health Crisis. Accessed at: https://www.apa.org/news/press/releases/stress/2020/report-october; See also American Psychological Association (2021). Stress in America: 2021. One year later, a new wave of pandemic health concerns. Accessed at: https://www.apa.org/news/press/releases/stress/2021/one-year-pandemic-stress

21 See, e.g., Gottman, J. (2011). *The Science of Trust*. New York: W.W. Norton & Co. pp. 19–20

22 Kahneman, D. (2011). *Thinking, Fast and Slow*. New York: Macmillan Publishers.

23 Rock, D. (2008). "SCARF: a brain-based model for collaborating with and influencing others." *NeuroLeadership Journal*, (1), 1–7. See also Scarlett, H. (2019). *Hilary Scarlett (2019, Neuroscience for Organizational Change: An Evidence-based Practical Guide to Managing Change* (2nd Edition). London and New York: Kogan Page Limited Publishing Co.

24 Coombs, W.T., Holloday, S.J. (2017). *Handbook of Crisis Communication*. London. Wiley-Blackwell

25 Ulmer, R., Sellnow, T., Seeger, M. (2011). *Effective Crisis Communication: Moving from Crisis to Opportunity*. Thousand Oaks, CA: Sage Publications

26 Heath, J., O'Hair (2009). "The Significance of Crisis and Risk Communication," in *Handbook of Risk and Crisis Communication*, eds. R.L. Heath and H.D. O'Hair. New York: Taylor and Francis/Routledge.

27 Oliver-Smith, A., (1999). *"What is a disaster?" in The Angry Earth: Disaster in Anthropological Perspective*, eds Oliver-Smith, A., Hoffman, S. New York: Routledge, p. 19.

28 United Nations International Strategy for Disaster Reduction. (2009), UNISDR Terminology on Disaster Risk Reduction, p. 9. Accessed at: https://www.unisdr.org/files/7817_UNISDRTerminologyEnglish.pdf

29 Lindell, M.K, Lindell, M.K, Prater, C.S., Perry, R.W. (2006). *Fundamentals of Emergency Management*. Washington, D.C.: Federal Emergency Management Agency. Accessed at: https://training.fema.gov/hiedu/aemrc/booksdownload/fem/ p.5.

30 Kroeber, A., Kluckhohn, C. (1952). *Culture: A Critical Review of Concepts and Definition*. New York: Vintage Books.

3

An Overview of Risk Communication

CHAPTER OBJECTIVES

This chapter offers a general understanding of the field of risk, high concern, and crisis communication and how you might communicate risk information to audiences. At the end of this chapter, you will be able to:

- identify the situations that require effective risk communication;
- remember what makes risk perception a social construct;
- employ the factors that help – and inhibit – an individual's ability to understand and evaluate risk information;
- remember each risk communication principle and guideline is based on trust, transparency, cooperation, dialog, and respect; and
- demonstrate a basic understanding of how risk communication theory translates into good practice.

This chapter begins with a puzzle. The puzzle illustrates several key principles of effective risk communication in practice.

3.1 Case Diary: Complex Issues Destroy Homes

A government agency hired me to work with them on a mysterious problem. Homeowners in scattered locations were complaining that green, slimy, and foul-smelling substances were oozing through cracks into the basements of their homes. Their basements served as family rooms, and children played in the contaminated basements.

The community was poor. Many homeowners were unemployed and could not afford to move or pay for toxicity testing. Homeowners asked the government to test the substances for toxicity. When the results revealed that the substances were highly toxic, homeowners asked the government agency for assistance in relocation and in addressing possible health problems.

The government's environmental detectives were perplexed. They shared this perplexity with me and asked if I could help. What was the source of the toxic green substances? The environmental detectives had applied principles of epidemiology, first identified by John Snow, who is often considered the father of field epidemiology, in his now classic investigations of the sources of

Communicating in Risk, Crisis, and High Stress Situations: Evidence-Based Strategies and Practice, First Edition. Vincent T. Covello.
© 2022 by The Institute of Electrical and Electronics Engineers, Inc. Published 2022 by John Wiley & Sons, Inc.

cholera outbreaks in London in the nineteenth century.[1] But, the houses in this case study community with contaminated basements were few in number and many miles apart from each other. No industrial facilities were near the contaminated houses. None of the houses were built on top of an old industrial or hazardous waste site. Investigators could not find an explanatory pattern.

Investigators did determine that the toxic substances found in the basements matched toxic substances found at an abandoned industrial site more than 50 miles away. The site was surrounded by a tall, barbed wire fence, and posted every 25 yards were signs with the following messages: "Hazardous Waste Site. Danger from Possible Exposure to Toxic Substances. Absolutely No Trespassing. Trespassers Will Be Prosecuted." The site was also posted with the traditional warning of danger image, particularly regarding poisonous substances – the skull-and-crossbones symbol (☠). No additional information was provided on the signs.

Despite the long distance from the contaminated homes to the abandoned industrial site, the government environmental detectives tested the groundwater under the industrial site for leaks. Numerous monitoring wells were drilled, but no leaks were found. The government tested the groundwater and well water in areas nearby the industrial facility but found no evidence of toxic substances. Interviews with homeowners provided no suspected sources. The abandoned industrial site sat in a lowland area. A leak of toxic waste from the area would need to travel uphill for more than 50 miles – a highly unlikely physical phenomenon. Discussions between the government agency and the homeowners had reached an impasse.

I met with the residents. I listened to their concerns about their health, their children's health, and their diminished property values. In our chats, the homeowners often talked about their gardens and the happiness their gardens gave them. I had noticed the gardens when first entering the homes. They were typically much nicer than their neighbors. I engaged them by asking them their secrets of gardening, noting that I had a garden and had once written a book on one of my hobbies: a specialized type of Japanese gardening. Oddly, their responses to my questions about gardening were convoluted and confusing.

I revisited the abandoned industrial site where the scientists had found the same toxic substances oozing into the homeowner's basements. I noticed something unusual. Many of the plants growing at the abandoned industrial site were identical to those in the gardens of the homeowners with contaminated basements. I did research on the plants. To my surprise, I found that some plants grow well in toxic soil, such as soil containing arsenic. A light bulb went on in my head. Since it was exceedingly unlikely that the toxic substances at the abandoned industrial site had moved uphill and underground for more than 50 miles to several isolated homes, I adopted a principle I remembered from the casebooks of Sherlock Holmes: once you eliminate the impossible, whatever remains, no matter how improbable, is likely to be true.

I communicated my hypothesis to the homeowners: perhaps the soil at their homes was the same type of soil found at the abandoned industrial site. If my hypothesis was correct, how did the soil get into their backyards? The homeowners grudgingly said I was correct and told me they had taken the soil and plants for their gardens from the abandoned industrial site. They did not believe the signs. How could healthy plants be growing in toxic soil?

In my conversations with the homeowners, I empathized with their reasoning. They agreed to work with me and talk with the government agency. The government agency took responsibility for not engaging the community in discussing what was at the abandoned industrial site and not explaining the dangers it presented. The homeowners acknowledged responsibility for breaking into the site and taking the soil and plants. After considerable negotiation, the homeowners and government agency agreed to a clean-up plan that was acceptable to all.

This case diary illustrates the complexity of stakeholder engagement and the need to engage in active listening, demonstrate understanding, and pursue constructive dialog. It also illustrates the importance of using risk communication principles and strategies in developing warnings and involving the community in developing education programs about potential dangers.

3.2 Challenges and Difficulties Faced in Communicating Risk Information

Over the last five decades, the literature on risk communication has skyrocketed. Hundreds of articles and books have been published. Many of the first studies focused on risks associated with exposures to health, safety, environmental, and occupational risk agents. However, it quickly became apparent that the focus on health, safety, and environmental issues was too narrow. Researchers expanded the boundaries of the field to include risks and threats created by virtually any high stress or an emotionally charged issue or situation.

A substantial part of the risk communication literature has focused on challenges and difficulties faced by leaders; risk managers; and technical, engineering, and scientific professionals in effectively communicating technical information to nonexperts. I have organized these challenges and limitations into five categories: (1) characteristics and limitations of scientific and technical data about risks; (2) characteristics and limitations of spokespersons in communicating scientific and technical information about risks; (3) characteristics and limitations of risk management regulations and standards; (4) characteristics traditional media channels in communicating information about risks; (5) characteristics of social media in communicating risk information; and (6) characteristics and limitations of people in evaluating and interpreting risk information.

3.2.1 Characteristics and Limitations of Scientific and Technical Data about Risks

One source of difficulty in communicating information about risks is the uncertainty and complexity of the data generated by risk assessments. Risk assessments are the theories and methods used to determine the risks posed by a particular hazard or event that may have a negative impact on individuals, groups, or the environment. Many organizations conduct risk assessments to characterize the nature and magnitude of health, safety, and environmental risks. Organizations also conduct risk assessments to characterize the nature and magnitude of legal and financial risks.

Despite their strengths, risk assessments seldom provide exact answers. Because of limitations in scientific understanding, data, models, and methods, the results of most assessments are approximations. The resources needed to resolve these uncertainties are seldom adequate to the task. Table 3.1 lists sources of uncertainty in risk assessments.

These uncertainties invariably affect communication. For example, uncertainties in risk assessments often lead to radically different estimates of risk and the outcomes of efforts taken to reduce the severity of the harm or adverse effects. Debates about risks often derive from these uncertainties.

3.2.2 Characteristics and Limitations of Spokespersons in Communicating Information about Risks

> *Men wanted for hazardous journey. Small wages. Bitter cold. Long months of complete darkness. Safe return doubtful. Honor and recognition in case of success.*
>
> *— Newspaper ad attributed to Sir Ernest Shackleton to solicit participants for his fateful 1914 Imperial Trans-Antarctic Expedition.*

Table 3.1 Sources of uncertainty in risk assessments.

- Have the study results or claims been successfully tested using more than one method?
- Have the results or claims been re-evaluated using different measurement or statistical techniques?
- Do the study results or claims test high for statistical significance?
- Is the probability so small that the same effect could have occurred by chance alone?
- What is the statistical strength of the study result or claim?
- How substantial is the strength of association?
- Are the claims of a strong or clear effect supported by a strong strength of association?
- Are the study results or claims specific as to health effects of the risk agent or are they general in nature?
- Can the study results or claims be explained by confounding factors or other relationships?
- What is the amount of detail in describing data and possible weaknesses in the study?
- What types of data are missing and how important are the gaps?
- What variables are missing?
- How significant are the missing data or variables?
- What are the greatest sources of uncertainty in the results?
- What does the investigator feel is known well and what is not known?
- Are the conclusions clearly stated?
- Are the conclusions of the study justified by the findings and substantiated by the evidence presented? Which ones are? Which ones are not?
- Are the conclusions linked to the original objectives of the study?
- Are the generalizations confined to the populations from which the sample was drawn? If not, why not?
- What are the implications of the study?
- What action does the study suggest?
- What additional studies are needed?
- Could this study be replicated? If not, why not?
- Has the study been peer-reviewed by qualified professionals?
- Has the study been published or accepted for publication in a scientific journal?
- If the study has not been submitted for publication in a scientific journal, will it be submitted?

A central question addressed by risk communication researchers is why some individuals and organizations are trusted as sources of risk information while others are not. The Antarctic expedition leader Sir Ernest Shackleton (quoted above) achieved much of his success and admiration through the high levels of trust he engendered in his expedition crews. He communicated, through both words and actions, essential trust-building characteristics: listening, caring, empathy, conviction, honesty, and optimism. Shackleton's risk management and communication skills were vividly demonstrated during the 1914 Imperial Trans-Antarctic expedition, saving the lives of all 28 crew members after the ship was crushed by ice. Although Shackleton was personally concerned about the outcome, he concealed his anxiety to ensure that it did not spread. He communicated his conviction that the crew would survive and get home. It was a message he repeated frequently.

Unfortunately, many technical, engineering, and scientific professionals, together with government and industry authorities – among the most visible, prominent, and important sources of risk information – lack effective risk communication skills. Leaders, risk managers, and technical experts are frequently insensitive to, or unaware of, the information needs of interested and affected parties.

Numerous examples can be cited. For example, Secretary of Defense Donald Rumsfeld, at a news briefing on weapons of mass destruction in Iraq, famously said:

> As we know, there are known knowns; there are things we know we know.
> We also know there are known unknowns; that is to say we know there are some things we do not know. But there are also unknown unknowns—the ones we don't know we don't know (Secretary of Defense Donald Rumsfeld, 12 February 2002).

In another example, outrage swept through Canada in 2003 when it was revealed the beloved Tim Hortons fast-food restaurant chain had been freezing its doughnuts rather than serving them fresh, undermining the company's "Always fresh" marketing motto. A media spokeswoman for the company unwisely said: "Until I confirm or deny anything, it simply doesn't exist." Unfortunate statements such as this communicate a lack of caring and undermine trust, key building blocks of effective risk communication.

Despite the vulnerability of a nation's food supply to terrorism as indicated by the thousands of foodborne illnesses that occur accidentally, Tommy Thompson, in his farewell address as US Secretary of Health and Human Services, shockingly said:

> I, for the life of me, cannot understand why the terrorists have not attacked our food supply because it is so easy to do. (Secretary of Health and Human Services Tommy Thompson, 3 Dec. 2004).

Unfortunately, risk communication spokespeople often make similar errors. Many of these errors occur because many risk communications are unplanned. One major result of the lack of communication skills among spokespeople is the loss of trust and credibility in experts and risk management authorities.

Many risk managers and spokespersons lack the skills needed to effectively communicate information about risk. A partial listing of these skills is shown in Tables 3.2–3.4. As one example, many use complex and difficult technical language and jargon in communicating information about risks to the media and the public. Technical language or jargon is not only difficult to comprehend but creates perceptions that the person is unresponsive, uncaring, or evasive.

Experts often operate on the assumption that they share a common framework with their audience for evaluating and interpreting risk information. However, this is seldom the case. People consider complex emotional, psychological, cultural, qualitative, and quantitative factors when defining, evaluating, and acting on risk information. Different assumptions, considerations, and definitions – such as what dimensions (values) to consider in an analysis or how to measure a particular consequence – produce different evaluations. People inherently trust their own evaluations and distrust evaluations by others. One of the costs of mistrust is the reluctance to believe risk information provided by leaders, risk managers, and technical experts, especially those from government and industry. Efforts to overcome such mistrust require, at a minimum, a commitment to enhanced risk assessment, management, and communication.

3.2.2.1   Case Study: "Go Hard, Go Early": Risk Communication Lessons from New Zealand's Response to COVID-19

New Zealand Prime Minister Jacinda Ardern took on the role as leader of New Zealand's response to COVID-19 and became the national spokesperson for the crisis. On 26 March 2020, she announced

Table 3.2 Risk communication skills, traps, and pitfalls.

Category	Dos	Don'ts
Jargon, technical terms, and acronyms	Avoid using technical jargon. Define technical terms and acronyms; limit their use and explain those you do or must use.	Do not use undefined jargon, technical terms, or acronyms.
Absolutes	Avoid absolutes – never say never in high-stress situation.	Do not offer guarantees; do not use the terms such as "every" or "all."
Truthfulness	Tell the truth; be open and transparent.	Do not lie or present half-truths.
Negatives	Use positive or neutral terms.	Do not use terms with strong negative associations; do not use highly charged analogies; do not repeat the words used in an allegation.
Defensiveness	Respond to issues, not personalities; stay calm and collected.	Do not let your temper interfere with your ability to communicate.
Clarity	Confirm understandings.	Do not assume understandings.
Visuals	Use graphics, examples, metaphors, and analogies to aid understanding.	Do not talk in abstractions.
Attack	Focus attacks on issues.	Do not make peripheral attacks against persons or organizations.
Promises	Promise only what you are certain will occur or what you can deliver.	Do not make promises you cannot back up or keep.
Speculation	Stick to the facts: state what you know, what you don't know, and what is being done to answer the question.	Do not discuss extreme worst-case scenarios.
Humor	The benefits of humor are seldom worth the risks in high-stress situations.	Do not use, as the audience may be offended, may think you are not taking the issue seriously, or may think you don't care, don't have empathy, or don't have compassion.
Blame	Accept your fair share of responsibility; focus on how problems can be solved and how challenges can be overcome.	Do not point your finger at others as a means of dodging responsibility.
Risk comparisons	Use comparisons to gain perspective; cite trustworthy sources of data; compare the same risk at two different times or circumstances; compare with a standard or regulation that is understood by the listener.	Do not use comparisons for gaining acceptance of a risk.

New Zealand would enter its second lockdown to eliminate the spread of COVID-19. Her 26th March announcement and the communications that followed demonstrated best practices of risk, high concern, and crisis communication: be first, be right, be credible, be clear, listen, express empathy, promote action, show respect, and involve stakeholders as partners.

New Zealand thought it had eliminated the COVID-19 after its first lockdown. However, an outbreak in Auckland, New Zealand's largest city, indicated that COVID-19 was not defeated. Ardern believed a second, stricter lockdown was needed if the country wanted to eliminate the virus.

Table 3.3 Risk communication skills of spokespersons.

Category	Skills/profile
Interpersonal communication	<ul><li>Is able to convey empathy</li><li>Is an effective active listener</li><li>Is respectful of the emotions, concerns, values, and beliefs of others</li><li>Uses personal pronouns and words such as "I," "we," and "our"</li><li>Is able to talk about shared responsibilities</li><li>Can be eloquent, creative, innovative, and imaginative</li></ul>
Knowledge	<ul><li>Is able to answer basic questions about the issue in question</li><li>Is able to convey clear, accurate, and factual messages with confidence</li><li>Is able to reinforce messages with visuals and experience</li><li>Communicates a workable strategy</li></ul>
Trust/credibility	<ul><li>Is associated with a respected organization or institution</li><li>Is able to use nonverbal communication to enhance trust and credibility</li><li>Is able to make a personal connection with the target audience</li><li>Is able to communicate hope and optimism</li><li>Is able to communicate self-sacrifice, determination, and restraint</li></ul>

Table 3.4 Eight ways to avoid mistakes with the media.

Follow These Guiding Principles
Your words have consequences – think about them in advance and make sure they are the right ones.
Know what you want to say, say it, and then say it again; don't over-reassure or offer guarantees.
If you don't know what you are talking about, stop talking; never say anything you don't want to see as a media headline.
Focus on informing people, not impressing them.
Use everyday language.
Be the first to share bad news.
Don't speculate, guess, or assume. When you don't know something, say so. Don't say "No comment" as you will look as if you are hiding something.
Don't get angry; when you argue with the media, you are likely to lose – and you will lose publicly.

In shaping her strategy, Ardern recognized that New Zealand enjoyed advantages not shared by other countries in its response to COVID-19. These included geographical isolation, a civil political environment, recent experience with crises and disasters, and a small population. She described the lockdown as part of a larger COVID-19 elimination strategy that would prevent additional illnesses and deaths. The lockdown would help the country shift focus from managing community transmission to prevention, control, vaccines, and therapeutics. The elimination strategy would also help eliminate further income and ethnic inequities in the number of COVID-19 hospitalizations and deaths.

To gain attention and raise public awareness of the government's change in strategy, she needed a motto. In a 14th March speech, she said: "We must go hard, and go early, and do everything we can to protect New Zealanders' health." Multiple traditional and social media outlets repeated her first few words. Combined with strict border controls and high compliance with lockdown measures, Ardern's motto became, "Go hard, Go early." Her goal was to stamp out the virus that causes COVID-19 wherever and whenever it comes back. Her strict controls worked.

Ardern's COVID-19 strategy was grounded in risk and crisis communication fundamentals. First, she announced the government would engage in an aggressive public communication program focused on creating clear, concise, and consistent messages about the need for frequent hand washing, cough etiquette, mask wearing, and social distancing. Similarly, instructions for lockdowns were shared through an emergency alert prior to the lockdowns. She created a COVID-19-dedicated website so people could easily find information. She had COVID-19 information translated into 28 languages in an attempt to reach diverse communities. She ensured the communication materials had excellent visuals and graphics.

Second, Ardern committed to move fast. She said her strategy would be "evidence-and science-based;" "Go hard, Go early" was not just a sound bite. Facts and science would drive politicians like herself. She reminded people that when COVID-19 first surfaced in New Zealand, she acted quickly to contain the virus, despite consequences for agriculture and tourism and blowback from politicians and the public. She committed to this course until no active cases remained. This was despite considerable political pressure from businesses and members of her own governing coalition. She demonstrated an understanding of the importance of speaking quickly.

Third, as the nation's leader, Ardern committed herself and her government to transparency and honesty. She admitted her policies would have large negative consequences, and there were large uncertainties about these consequences. Ardern also admitted her policies were likely to have disproportionate consequences for disadvantaged populations and large uncertainties remained about these consequences. She demonstrated the importance of showing caring and understanding.

Fourth, Ardern acknowledged her COVID-19 strategy might not work as planned. Ardern listened carefully to her Health Department and the 11 members of the Department's expert advisory committee. She explored the costs, risks, and benefits of options and alternatives to strict control with these experts. The consensus was that none would work. The floodgates needed to be closed, recognizing the ultimate elimination of COVID-19 would depend on developing effective vaccines.

Fifth, Ardern communicated her COVID-19 strategy required a wide, diverse, and often painful array of control measures. She would listen carefully to stakeholders so these control measures could be tailored to local needs.

Sixth, despite myths about panic in disasters, she expressed confidence in the ability of her nation to come together. She created a campaign built on the theme, "Unite Against COVID-19." The campaign emphasized the need for people to band together to defeat the virus. The battle would require personal sacrifices, particularly some personal freedoms. Public opinion polls consistently showed Ardern a supermajority of the population supported her actions, including the lockdown.

Seventh, Ardern admitted to errors and mistakes, apologized for them, and committed to stop them from reoccurring. For example, testing for COVID-19 in New Zealand had gotten off to a slow start, the initial Māori COVID-19 response plan was flawed, communications for arriving passengers at the borders were unclear, and testing for security forces at the borders were inadequate.

Eighth, Ardern made extensive use of social media platforms to communicate information about COVID-19. For example, she used multiple social media channels to communicate her message of social responsibility. During the lockdown, she encouraged everyone, including tourists, to stay home. The message was a twist from the usual tourism message encouraging visitors to New Zealand. Her postings about social responsibility were fueled by user-generated content. Ardern used an advertising method, the slogan "Go Hard, Go Early" to create buzz on social media platforms. She created a simple, clear message that led with her values. This gave her messages an emotional impact.

Ardern was particularly effective in using her Facebook and Instagram postings from her home to show solidarity with her constituents. In one posting, she began: "Kia ora, everyone. I'm standing against a blank wall in my house – because it's the only view in my house that is not messy." She spoke directly to viewers using her phone at the end of each day, inviting viewers into her home. She wore a sweatshirt, her hair was messy, and she looked tired. New Zealander's could easily identify with her.

When talking about COVID-19 on social media, there was an obvious change in her tone, projecting herself as a listening, caring, and empathic mother. She was often joined by public health experts. Just as President Franklin Roosevelt used his radio Fireside Chats in the 1930s and 1940s to explain issues and policy to Americans, Ardern used social media to explain COVID-19 issues and her COVID-19 policies.

3.2.3 Characteristics and Limitations of Risk Management Regulations and Standards

Compounding nearly every problem that affects risk communication are beliefs by the public that regulations and standards are often too weak, that risk management authorities have, in the past, overpromised and under-delivered, and that government and industry have often done a poor job in managing risk and communicating risk information. Perceptions of poor risk management and communication often lead to people overreacting or under-reacting, taking inappropriate actions, and losing trust in risk management authorities.

Several factors compound these perceptions and problems.

3.2.3.1 Debates and Disagreements

Many technical, engineering and scientific professionals have often engaged in highly visible debates and disagreements about the reliability, validity, and meaning of the results of risk assessments. Many times, equally prominent experts have taken diametrically opposed positions on the risks associated with a diverse set of hazards and events. For example, experts disagree about the risks associated with nuclear power plants, hazardous waste sites, asbestos, electric and magnetic fields, lead, radon, PCBs (polychlorinated biphenyls), arsenic, dioxin, genetically modified organisms, industrial emissions, childhood vaccination, climate change, terrorism, and contagious diseases like influenza and COVID-19. While such debates can be constructive for the development of scientific knowledge, they often undermine public trust and confidence.

3.2.3.2 Limited Resources for Risk Assessment and Management

Resources for risk assessment and management are seldom adequate to meet demands for definitive findings and rapid regulatory action. Stakeholders are typically not satisfied by authorities' explanations that generating valid and reliable data is expensive and time-consuming, or that regulatory actions are constrained by resource, technical, statutory, legal, or other limitations.

3.2.3.3 Underestimating the Difficulty of and Need for Risk Communication

Resources for risk communication training, planning, implementation, and evaluation are seldom adequate. Resources include funding, staff, space, and equipment. Several reasons explain the lack of resources. First, many risk managers see little value added by risk communication activities, especially when compared to the value added by collecting additional technical data. Second, many risk managers do not see risk communication as a complex skill, let alone a science- and evidence-based discipline. Third, risk communication is often seen as a relatively simple task, requiring little more than the ability to put an understandable sentence together. Fourth, many risk managers believe incorrectly that technical facts speak for themselves. They see the main job of managers is to get the technical facts right, tell people the facts, and explain what the facts mean. Fifth, many risk managers believe that credentials and experience alone (e.g., "I've done this all my life") are sufficient preparation for producing an effective risk communication.

3.2.3.4 Lack of Coordination and Collaboration

Coordination and collaboration among risk management authorities are seldom adequate. In many debates about risks, for example, lack of coordination and collaboration has severely undermined public trust and confidence. Compounding such problems is the lack of consistency in approaches to risk assessment and management by authorities at the local, regional, national, and international levels. For example, only limited requirements exist for regulatory agencies to develop coherent, coordinated, consistent, and interrelated plans, programs, and guidelines for managing risks. As a result, risk management systems are often highly fragmented. This fragmentation often leads to jurisdictional conflicts about which organization or agency, or level of government has the ultimate responsibility for assessing and managing a particular risk. Lack of coordination, different mandates, and confusion about responsibility and authority also lead to the production of multiple and competing estimates of risk. A commonly observed result of such confusion is the erosion of trust, confidence, and acceptance.

3.2.4 Characteristics and Limitations of Traditional Media Channels in Communicating Information about Risks

Because traditional broadcast and publishing media – newspapers, television, radio, and magazines – are in the business of selling news, journalists favor stories that attract readers, viewers, and listeners. Stories about conflicts, disagreements, and inconsistencies attract media attention. Journalists also favor stories that contain dramatic material, especially dramatic stories with clear villains, victims, and heroes. Much less attention is typically given to daily occurrences.

In reporting about risks, journalists often focus on the same characteristics of a risk that raise public concerns, including a lack of trust in risk management organizations, the potential for adverse outcomes, a lack of familiarity with the risk, scientific uncertainty, risks to future generations, unclear benefits, inequitable distributions of risks and benefits, and potentially irreversible effects. Traditional media coverage of risks is frequently deficient in that many stories contain oversimplifications, distortions, and inaccuracies. For example, traditional media reports on cancer risks often provide few statistics on general cancer rates for comparison; often provide little information on common forms of cancer; rarely address public misperceptions about cancer; and provide little information about detection, treatments, and other protective measures.

These problems often stem from characteristics of the traditional media and the constraints under which reporters work. Many reporters work under tight deadlines that limit the time for research that yields valid and reliable information. Reporters also rarely have adequate time or space to present the complexities and uncertainties surrounding many risk issues.

Journalists achieve objectivity in a story by balancing opposing views. Truth in journalism is often different from truth in science. In journalism, there are different or conflicting views and claims to be covered ideally as evenly as possible. At the same time, general assignment reporters do not always have the scientific background needed to evaluate the data and disagreements related to a particular risk.

Journalists are source dependent. Under the pressure of deadlines and other constraints, reporters often rely heavily on sources that are easily accessible and willing to speak out. Sources of information that are difficult to contact, hard to draw out, or reluctant to provide interesting and nonqualified statements are often left out, even if these sources have fulsome information.

Effective risk communication depends in part on understanding the constraints and needs of traditional media and adapting one's behavior and information to meet these needs. Given the continuing importance of traditional media in the transfer and exchange of information about risks and threats, Chapter 12 of this book discusses the constraints and needs of traditional media in detail. The chapter points out, for example, that few journalists have science or technical backgrounds and that many traditional sources of media have had to cut staff due to changes in where people get information.

3.2.5 Characteristics and Limitations of Social Media Channels in Communicating Information about Risks

As discussed in Chapter 13 of this book, the Internet and social media channels create new opportunities for people to share and exchange risk information. Tens of millions of people now share and exchange risk information directly with one another, unfiltered, and free from dependence on official and traditional sources.

Despite the growing impact of the Internet and social media on risk communication, much of the scientific and technical information about risks that people receive still originates from traditional journalism. The Internet and social media nonetheless play an extremely important and increasingly influential role in personal risk decision-making. Through sites and platforms such as Wikipedia, blogs, YouTube, Facebook, Twitter, Snapchat, Reddit, WhatsApp, and Instagram, social media platforms significantly affect people's knowledge, perceptions, and understanding of risks in a dynamic and highly competitive communication environment.

Social media are relatively new areas of study for risk communicators and create both opportunities and challenges. For example, as shown in Table 13.1 in Chapter 13, social media channels have several benefits for risk communication purposes. Despite these benefits and potential opportunities, social media channels create challenges, such as identity theft, the potential to be hacked, and disinformation campaigns.

Research is emerging to fill gaps in science-based knowledge about the following key issues:

1) How do individuals and decision makers access, evaluate, and use information about risks acquired through social media channels?
2) How do people evaluate and use risk information that is not vetted through traditional sources of risk information, such as technical, engineering, and scientific professionals?
3) How effective are social media channels in reaching those that do not have the knowledge, resources, skills, or motivation to access and use traditional sources of risk information?
4) To what extent do the social networks that are created by social media affect people's risk-related perceptions, understandings, attitudes, beliefs, and behaviors?

5) How can research help organizations struggling with updating their social media policies – such as their employee's use of social media?
6) How can research help organizations struggling with changes caused by the expanding reality of teleworking?
7) How can research help organizations struggling with the changes caused by the expanding reality of distance learning?

Researchers have also focused on how to overcome new challenges associated with the rapid spread of misinformation, disinformation, and "fake news" about risks by social media channels. For example, in a speech about COVID-19 in February 2020, Tedros Adhanom Ghebreyesus, Director-General of the World Health Organization (WHO), said WHO is "not just fighting an epidemic; we're fighting an *infodemic*. Fake news spreads faster and more easily than this virus, and is just as dangerous."[2]

WHO defined *infodemic* as "an overabundance of information, both online and offline. It includes deliberate attempts to disseminate wrong information to undermine the public health response and advance alternative agendas of groups or individuals. Mis- and disinformation can be harmful to people's physical and mental health; increase stigmatization; threaten precious health gains; and lead to poor observance of public health measures, thus reducing their effectiveness and endangering countries' ability to stop the pandemic. Misinformation costs lives."[3]

In response to the COVID-19 *infodemic*, a team of WHO "mythbusters," working with social media companies, mounted a campaign to counter the spread of rumors, misinformation, and disinformation about COVID-19. On WHO's website, the mythbusters team posted a rapidly expanding list of COVID-19 rumors, misinformation, and disinformation being spread by social media. These included rumors that taking vitamin and mineral supplements could cure COVID-19; that water or swimming could transmit the COVID-19 virus; that coronavirus disease was caused by bacteria and treated with antibiotics that in fact do not work against viruses; that drinking alcohol protects you against COVID-19; that thermal scanners can detect COVID-19; that adding hot pepper to your soup or other meals prevented or cured COVID-19; that COVID-19 was transmitted through houseflies; that 5G mobile networks can spread COVID-19; that exposing yourself to the sun or temperatures higher than 25°C protected you from COVID-19; that being able to hold your breath for 10 seconds or more without coughing or feeling discomfort means you were free from COVID-19; that taking a hot bath could prevent COVID-19; that COVID-19 virus could be spread through mosquito bites; that hand dryers were effective in killing the COVID-19 virus; that vaccines against pneumonia and seasonal influenza protected you against the COVID-19 virus; that eating garlic could prevent COVID-19; and that spraying and introducing bleach or another disinfectant into your body would protect you against COVID-19.[4]

Researchers are testing various strategies for overcoming fake news, such as those identified by the International Federation of Library Associations and Institutions in Figure 3.1.

3.2.6 Characteristics and Limitations of People in their Ability to Evaluate and Interpret Risk Information

As shown in Table 3.5 and described below, at least 13 factors interfere with the ability of people to evaluate and interpret risk information. Individually and collectively, these factors often lead to uniformed decisions. One of the challenges faced by risk managers and communicators is to identify and address the specific interference factor or factors affecting informed decision-making.

Figure 3.1 Strategies for overcoming false information *Source*: International Federation of Library Associations and Institutions (2021). "How to Spot Fake News." Accessed at: https://www.ifla.org/publications/node/11174.

Table 3.5 Factors that affect the ability of people to make informed decisions about risks.

- Inaccurate perceptions of risk
- Difficulties in understanding statistical or complex scientific information related to unfamiliar activities or technologies
- Strong emotional responses to risk information
- Desires and demands for scientific certainty
- Strong beliefs and opinions that resist change and distort understanding
- Weak beliefs and opinions that can be manipulated by the way information is presented and framed
- Ignoring or dismissing risk information because of its perceived lack of personal relevance
- Using risks as proxies or surrogates for other personal, societal, economic, cultural, or political agendas and concerns

---

1) **Inaccurate perceptions of risk:** People often overestimate some risks and underestimate others. For example, people often overestimate the risks of dramatic or sensational causes of death, such as accidents at manufacturing plants, but underestimate the risks of less dramatic causes of death, such as asthma, emphysema, and diabetes. Adverse consequences of risk overestimation include dysfunctional behaviors, stress, anxiety, dread, confusion, hopelessness,

helplessness, and misallocation of risk-reduction resources. Adverse consequences of risk underestimation include apathy and denial, which can lead to the failure to take appropriate actions. Overestimation or underestimation of risks is caused in part by the tendency for risk judgments to be influenced by the memorability of past events and by the ability to imagine future events. A recent disaster, intense media coverage, or a vivid film can heighten the perception of risk. Conversely, risks that are not memorable, obvious, palpable, tangible, or immediate are underestimated.

2) **Difficulties understanding statistical or complex scientific information related to unfamiliar activities or technologies**: A variety of cognitive biases and related factors hamper people's understanding of probabilities. This difficulty hampers discussions about risks between experts and nonexperts. For example, risk experts are often confused by the public's rejection of the argument that a risk from a new activity or technology is acceptable if the risk is smaller than the ones people face in their daily lives.

3) **Personalization:** People often personalize the risk. Sample question: What if I am the person who is harmed or adversely impacted?

4) **Trustworthiness:** People often raise questions of trust. Sample question: Why should I believe you on this issue given you previously made mistakes or changed your mind about risks and threats?

5) **Cumulative Risks**: People often raise concerns about cumulative risks. Sample question: I already have enough risks in my life. Why should I take on even one more?

6) **Benefits**: People often question whether the risks are worth the benefits. Sample question: Will the benefits of the new activity or technology significantly outweigh the risks?

7) **Ethics**: People often raise ethical questions. Sample question: Who gave you the right to make decisions that violate the moral principles and rights of others? Complicating the perception of fairness is the difficulty people have understanding, appreciating, and interpreting small probabilities, such as the difference between 1 chance in 100,000 and 1 chance in 1,000,000. These same problems hamper discussions between technical experts and nonexperts about what is remotely possible and probable. Given these difficulties, to be effective, risk communication strategies must address the experiences, attitudes, beliefs, values, and culture of those receiving information about a risk or threat. Effective risk communication skills are built on a foundation of understanding how people perceive risks.

8) **Strong emotional responses to risk information:** Strong feelings of fear, worry, anger, outrage, and helplessness are often evoked by exposure to unwanted or dreaded risks. These emotions often make it difficult for leaders, risk managers, and technical experts to engage in constructive discussions about risks in public settings. Emotions are most intense when people perceive the risk to be involuntary, unfair, not under their personal control, managed by untrustworthy individuals or organizations, and offering few benefits. More extreme emotional reactions often occur when the risk affects children, when the adverse consequences are particularly dreaded, and when worst-case scenarios are imagined. Strong emotional responses to risk information are not necessarily wrong or contrary to knowledge. They can be based on practical or experiential knowledge and emphasize what people value, such as fairness and equity. Strong emotional responses are also not necessarily opposed to reason. That rationality and emotions are opposed to each other derives from the belief that the human brain perceives reality in two distinct ways: one is emotional, instinctive, intuitive, spontaneous, while the other is rational, analytical, statistical, and occurred later in human evolution. However, strong emotional responses to risk information can be rational and prevent people from engaging in dangerous activities.

9) **Desires and demands for scientific certainty:** People often display a marked aversion to uncertainty. They use a variety of coping mechanisms to reduce the anxiety generated by uncertainty. This aversion frequently translates into a clear preference for statements of fact over statements of probability, which is the language of risk assessment. People often demand that technical experts tell them exactly what will happen, not what might happen. For example, the changing recommendations during the COVID-19 pandemic on things such as whether face masks were effective or whether a person without symptoms could spread the disease caused many people to become frustrated and caused them to distrust science.

10) **Strong beliefs that resist change**: People tend to seek out information that confirms and supports their beliefs and often ignore evidence that contradicts their beliefs. Beliefs often operate on a polarized scale of True or False, with little gray in-between. Opinions often operate on a different scale – Favorable or Unfavorable. According to the Four Hit Theory of Belief Formation, once formed, a belief is difficult or impossible to change. On average, four unanswered risk communication messages (hits) from trustworthy sources can crystalize into a belief. Less than four risk communication messages (hits) are typically still an opinion. A hit from one side can be negated by a hit from the other side.

Strong beliefs about risks or threats, once formed, change slowly and are extraordinarily persistent even in the face of contrary evidence. Initial beliefs about risks structure the way subsequent evidence is interpreted. Fresh evidence – e.g., data provided by a technical expert – appears reliable and informative only if it is consistent with the initial belief; contrary evidence is dismissed as unreliable, erroneous, irrelevant, or unrepresentative.

11) **Opinions can be manipulated by how information is presented:** When people lack strong prior beliefs, subtle changes in the way risk information is presented and framed can have a major impact on opinions. For example, two groups of physicians were asked to choose between two therapies – surgery or radiotherapy.[5] Each group received the same information; however, the probabilities were expressed either in terms of dying or surviving. Both numbers expressed the same probability, but the different presentations resulted in dramatic variation in the choice of therapy. Here, physicians received the survival data better. However, the effects of information framing are modified by factors, including risk aversion, experience, beliefs, level of risk, type of risk, or costs of risk mitigation.

12) **Ignoring or dismissing risk information because of its perceived lack of personal relevance:** Risk data often relates to society. These data are often of minimal interest to individuals, who are more concerned about risks to themselves rather than risks to society.

13) **Using risks as proxies or surrogates for other personal, societal, economic, cultural, or political agendas and concerns:** The specific risks that people focus on reflect their beliefs about values, social institutions, and moral behavior. Risks and crises may be exaggerated or minimized under their personal, societal, economic, cultural, or political agendas, priorities, and concerns. Debates about risks often serve as proxies or surrogates for debates about high concern issues. The debate about nuclear power, for example, is sometimes less about the specific risks of nuclear power than about other issues such as the proliferation of nuclear weapons, the adverse effects of nuclear waste disposal, the value of large-scale technological progress and growth, and the centralization of political and economic power in the hands of a technological elite.

Cultural factors, such as values, norms, social networks, group memberships, and loyalties, have a profound effect on agenda setting and what risks are important. Cultural factors are the web of meaning shared by members of a society. This web is a shared system of beliefs, values, customs, behaviors, and artifacts that group members used to cope with their world and with one another. Integrating diversity and cultural differences into the risk communication process can be challenging. To meet the challenge, risk communicators must develop cultural competence. Cultural competence requires communicators to be aware of their own cultural biases, develop in-depth knowledge of other cultures, and have a willingness to accept and respect diversity.

3.3   Changes in How the Brain Processes Information Under Conditions of High Stress

The brain processes information differently in low-stress and high-stress situation. Neuroscience and behavioral science research studies show that when people are fearful, stressed or upset, they typically:

1) **Want to "know that you care before they care what you know."**
   - Perceptions of caring, empathy, and listening account for as much as 50% of how people determine whether they trust an individual or organization.
   - Trust is often determined in the first 9-30 seconds.
   - Once lost, trust is difficult to regain.
2) **Have difficulty hearing, understanding, and remembering information.**
   - Fear, stress, and anxiety can reduce the ability to process information by up to 80%-100%.
   - Ninety-five percent of the questions and concerns that cause fear, stress, and anxiety can be anticipated and prepared for in advance.
3) **Receive information best when presented in small digestible chunks and bytes.**
   - Key messages ideally contain no more than 140 characters, 27 words, and 3-5 messages, with each message supported and expanded by 3-5 facts or additional information.
4) **Are more likely to recall information they hear first and last.**
   - Provide the most important information first.
   - Provide the second most important last.
   - Prepare for people to ignore or forget messages not announced first or last.
   - Repeat the first and last messages several times.
5) **Process information at a grade level substantially below their formal educational attainment.**
   - Keep initial messages short and simple, often four grade levels below formal educational attainment.
   - Use a variety of tools, such as visuals, to simplify risk information.
6) **Will focus more on negative information than positive.**
   - Negative information typically needs to be balanced by three to five pieces of positive or constructive information.
   - Avoid negative absolute statements (e.g., statements that contain the words *never, nothing,* or *none.*)
   - Avoid words or phrases with high negative imagery (these typically go to the visual part of the brain for processing and "stick").
7) **Judge risks to a large extent based on perceptions of trust, benefits, personal control, dread, fairness, and voluntariness.**
   - As much as 95% of fear, anxiety, and stress caused by risks can be traced back to factors such as perceived trust, benefits, personal control, dread, fairness, and voluntariness.

8) **Actively look for visual information to support verbal messages about risks.**
   - People often give greater weight to nonverbal cues and visual information than verbal information.
   - People in high concern and high-stress situations often assign a negative interpretation to nonverbal cues, such as body language.
   - A significant amount of risk and high concern information is processed in primitive parts of the brain (the lizard or reptilian brain) that focuses on nonverbal information and determines the response of fight, freeze, or flight.

3.4   Risk Communication Theory

An understanding of risk communication theory is helpful in understanding the various risk communication strategies and messaging. As shown in Figure 3.2, there are four basic risk and high concern communication theories: trust determination theory, negative dominance theory, mental noise theory, and risk perception theory. These four theories are the foundation stones on which risk communication rests.[6]

3.4.1   Trust Determination Theory

Trust determination theory states that trust is the most powerful factor influencing how people make risk-related decisions. The more people trust the information source, the more they accept the messages, messengers, and channels for acquiring information. As shown in Figure 3.3, the basic components of trust are caring and empathy; competence and expertise; openness and honesty; and other factors such as dedication and commitment. Based on context, the relative weight given to these factors can shift.

Several variables predict a higher level of trust. These include:

- acknowledging uncertainty;
- transparency, including openness and candidness about negative information;
- speed in disseminating risk information;
- disseminating technical information that is easy to understand;
- seeking input from stakeholders and encouraging constructive dialogue;
- ensuring coordination of communication activities within and among risk management authorities;
- avoiding the dissemination of conflicting information;
- disseminating risk information through multiple traditional and social platforms.

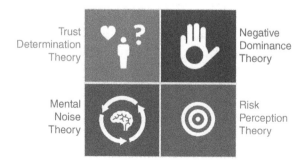

Figure 3.2   Risk and high concern communication theories.

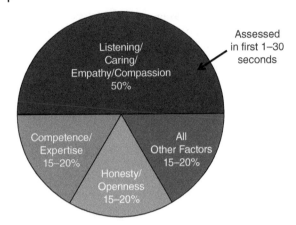

Figure 3.3 The basic components of trust.

3.4.2 Negative Dominance Theory

Negative dominance theory states that people under stress put much more weight on negative information than on positive information. Three to five positive messages are typically needed to offset negative information. Information provided to stakeholders should not contain unnecessary negatives and it should emphasize what is being done over what is not.

3.4.3 Mental Noise Theory

Mental noise theory states the ability of stressed and upset individuals to process risk-related information is reduced by as much as 80%-100%. Mental noise distracts individuals from the task at hand and diminishes their ability to effectively hear, understand, and remember messages. Constructing and delivering effective messages in high stress, and high concern situations are radically different from constructing and delivering effective messages in low stress, low concern situations. At a minimum, messages in high stress, high concern situations must initially be kept short and simple (the KISS principle, Keep It Short and Simple).

3.4.4 Risk Perception Theory

Risk perceptions are the subjective judgments people make about the characteristics and severity of a risk. Risk perceptions are also subject to the beliefs a person holds regarding a risk, including the definition, probability, and outcome of the risk. Risk perception theory recognizes that risk is not an objective phenomenon perceived in the same way by all interested parties. Instead, it is a social and cultural construct with its roots deeply embedded in personal experiences and a specific social, economic, political, and cultural context. A variety of scientific, psychological, social, economic, political, and cultural factors determine which risks will ultimately be selected for individual, group, and societal attention and concern.

Evidence about the magnitude of possible adverse consequences is only one of many factors influencing public decisions about the acceptability of a risk. The level of risk, as assessed by technical, engineering, and scientific professionals, is only one among several variables that determines acceptability. Deciding which risks are acceptable is typically based more on risk perception factors than on scientific facts. Much of the risk communication literature focuses on these risk perceptions factors

Risk perception factors profoundly affect the ability of people to make informed decisions about risk-related issues. Table 3.6 contains a list of the 20 most important risk perception factors. Each factor is described in more detail below.

1) **Trust in responsible authorities and institutions**: People are often more concerned about activities or actions where the responsible assessor or manager is perceived to be untrustworthy (e.g., individuals, organizations, or institutions that have a clear conflict of interest) than they are about activities or actions where the responsible assessor or manager is perceived to be trustworthy (e.g., trust in first responders, such as fire department personnel).

2) **Voluntariness**: People are often more concerned about activities and actions that are perceived to be involuntary, coerced, or imposed on them (e.g., exposure to chemicals or radiation from an accident at an industrial facility) than about activities or actions that are perceived to be voluntary or chosen (e.g., smoking, sunbathing, talking on a cellphone while driving, or mountain climbing).

3) **Scope/catastrophic potential**: People are typically more concerned about activities and actions perceived to be cataclysmic and where harm, fatalities, and injuries are grouped in time and space (e.g., harm, fatalities, and injuries resulting from a major release of toxic chemicals or radiation) than about activities or actions where harm, fatalities, and injuries are scattered, occur over a long period, or are random in time and space (e.g., automobile accidents).

4) **Familiarity/exotic**: People are typically more concerned about activities or actions perceived to cause harm and perceived to be unfamiliar (e.g., leaks of chemicals or radiation from waste disposal facilities; outbreaks of unfamiliar infectious diseases such as Zika, West Nile Virus, and Ebola) than about activities or actions that are familiar and routine (e.g., household accidents).

5) **Understanding/visibility**: People are often more concerned about activities or actions perceived to be characterized by invisible or poorly understood exposure mechanisms or processes (e.g., long-term exposure to low doses of toxic chemicals or radiation) than about activities or actions perceived to be characterized by visible and apparently well-understood exposure mechanisms or processes (e.g., pedestrian accidents or slipping on ice).

6) **Uncertainty**: People are often more concerned about activities or actions that are perceived to cause harm and are perceived to have unknown causes or uncertain risks (e.g., mysterious outbreaks of illnesses; risks from a radioactive waste facility designed to last 20,000 years) than about activities or actions that are perceived to cause harm and that are perceived to have known causes and for which there are relatively certain risk-related data (e.g., actuarial data on automobile accidents).

7) **Controllability (personal)**: People are often more concerned about activities or actions that are perceived as outside their control (e.g., flying in an airplane; exposure to releases of toxic chemicals or radiation from an accident at an industrial facility) than about activities or actions that are perceived to be under their personal control (e.g., driving an automobile or riding a bicycle).

8) **Effects on children:** People are often more concerned about activities or actions that are perceived to adversely affect children or specifically put children in the way of harm or risk (e.g., asbestos in school buildings; milk contaminated with radiation or toxic chemicals; children's food contaminated with pesticide residues; pregnant women exposed to radiation or toxic chemicals) than about activities or actions engaged in by adults and that are not

Table 3.6 Risk perceptions (fear factors).

Factor	Conditions associated with higher perceived risks, increased concerns, greater fears	Conditions associated with lower perceived risks, decreased concerns, and greater fears
Trust	Lack of trust in responsible persons	Trust in responsible persons
Voluntariness	Involuntary/coerced/imposed	Voluntary/chosen
Scope/catastrophic potential	High catastrophic potential	Low catastrophic potential
Familiarity	Unfamiliar/exotic	Familiar/routine
Understanding/visibility	Invisible/mechanisms or process not understood	Visible/mechanisms or process understood
Uncertainty	Effects and outcomes unknown or uncertain	Effects and outcomes known
Controllability (personal)	Effects and outcomes uncontrollable by the person	Effects and outcomes controllable by the person
Effects on children	Children specifically at risk	Children not specifically at risk
Effects manifestation	Delayed effects	Immediate effects
Effects on future generations	Significant threat to future generations	Little or no threat to future generations
Victim identity/specificity	Identifiable and/or specific person or victims	Nameless, faceless, or statistical victims
Pleasurable/dreaded	Outcomes and effects not pleasurable/dreaded	Outcomes and effects pleasurable/not dreaded
Awareness/media attention	Much awareness/media attention	Little awareness/media attention
Fairness/equity	Inequitable distribution of risks and benefits	Equitable distribution of risks and benefits
Benefits	Unclear benefits	Clear benefits
Reversibility	Effects and outcome irreversible	Effects and outcomes reversible
Personal stake	Direct and significant perceived personal risk or threat	Little or no perceived significant personal risk or threat
Nature of evidence	Evidence from human studies	Evidence from laboratory studies
Morality	Immoral/callous/unethical	Moral/ethical
Origin	Caused by human actions or failures	Caused by acts of nature or God

perceived to specifically put children in the way of harm or risk (indoor air pollution in office buildings).

9) **Effects manifestation:** People are often more concerned about risks that have delayed effects (e.g., the development of cancer after exposure to low doses of chemicals or radiation) than about risks that have immediate effects (e.g., poisonings).

10) **Effects on future generations:** People are often more concerned about activities or actions perceived to pose significant risks to future generations (e.g., genetic effects related to exposure to toxic chemicals or radiation) than activities or actions perceived to pose no special significant risks to future generations (e.g., skiing accidents).

11) **Victim identity**: People are often more concerned about activities or actions that are perceived to cause harm and that impact identifiable victims or a named person (e.g., a worker exposed to high levels of toxic chemicals or radiation; a child who has fallen in a well) than about risks that are statistical and impact persons that are nameless or faceless (e.g., statistical deaths related to automobile accidents).

12) **Pleasurable/Dreaded**: People are often more concerned about activities or actions that are perceived as unpleasant, dreaded, or evoke a response of fear, terror, or anxiety (e.g., exposure to radiation or chemicals that can cause cancer or birth defects) than to activities or action risks that are perceived to be pleasurable, not especially dreaded, or do not evoke a special response of fear, terror or anxiety (e.g., using recreational drugs, common colds, or household accidents).

13) **Awareness/Media attention**: People are often more concerned about activities or actions that are perceived to cause potential harm and for which there is high public awareness and media attention (e.g., cancer, airplane crashes, hazardous waste sites, accidents at industrial or nuclear power facilities) than about activities or actions that are perceived to cause harm but for which there are little awareness and media attention (e.g., on-the-job accidents).

14) **Fairness**: People are often more concerned about activities or actions that are perceived to be characterized by an inequitable or unfair distribution of risks, costs, and benefits (e.g., inequities related to the siting of waste disposal or industrial facilities) than about activities or actions perceived to be characterized by an equitable distribution of risks, costs, or benefits (e.g., flu vaccination).

15) **Benefits**: People are often more concerned about activities or actions that are perceived to have unclear, questionable, or diffused benefits (e.g., waste disposal facilities) than about activities or actions that are perceived to have obvious benefits (e.g., elective surgery).

16) **Reversibility:** People are often more concerned about activities or actions that are perceived to have potentially permanent adverse outcomes or effects (e.g., nuclear war) than about activities or actions perceived to have potentially reversible adverse outcomes or effects (e.g., injuries from most sports or household accidents).

17) **Personal stake**: People are often more concerned about activities or actions that they perceive place them, or their families or friends, personally and directly in the way of harm or risk (e.g., living near an industrial facility with potentially hazardous air emissions) than about activities or actions that do not place them or their families and friends personally and directly in the way of harm or risk (e.g., disposal of hazardous waste in remote places).

18) **Nature of evidence**: People are often more concerned about activities or actions that are based on risk assessments from human studies (e.g., risk assessments based on adequate exposure data of humans) than about activities or actions based on risk assessments from nonhuman studies (e.g., laboratory studies of the effects of potentially hazardous chemicals using mice or rats).

19) **Morality:** People are often more concerned about activities or actions that are perceived to violate culturally based principles of morality and ethics (e.g., raising the price of a life-saving prescription drug to a very high level) than about activities or actions that are perceived to be consistent with culturally based principles of morality and ethics (e.g., lying about the nationality or ethnicity of a child to protect the child from those who want to do harm because of child's nationality or ethnicity).

20) **Human vs. natural origin:** People are often more concerned about activities and actions that are perceived to cause harm and are perceived to have their origin in human actions and failures (e.g., accidents, leaks, and spills at waste disposal or industrial sites caused by negligence, inadequate safeguards, inadequate supervision, or operator error) than about activities and actions that are perceived to cause harm and that are perceived to be caused by acts of nature or God (e.g., exposure to sunshine or cosmic rays).

Risk perception theory states that risks are more worrisome, more fearful, and less acceptable if they are perceived as having the characteristics listed in Table 3.6 and described above. Risk perception theory counters the conventional notion that "facts speak for themselves." People commonly accept high risks but also become outraged over much less likely risks.

Risk perception factors can change concerns, perceptions of risk, fear, and perceived dangers exponentially. They explain the aversion of parts of the public toward activities and technologies such as nuclear power, required childhood vaccinations, and genetically modified food.

Perception factors also help to explain phenomena, such as the "not in my back yard" (NIMBY) and the "locally unwanted land use" (LULU) responses to many chemicals, nuclear, and other industrial facilities. For example, residents in communities where industrial facilities exist or are planned often become outraged if they believe government and industry officials:

1) have excluded them from meaningful participation in the decision-making process;
2) have denied them the resources needed to evaluate or monitor health, safety, or environmental risks;
3) have denied them the opportunity to give their "informed consent" to management decisions that affect their lives or property;
4) have imposed or want to impose upon them facilities that provide few local economic benefits;
5) have imposed or want to impose upon them facilities that entail high costs to the community (e.g., adverse health, safety, wildlife, recreational, tourism, property value, traffic, noise, odor, scenic view, and quality of life effects);
6) have imposed or want to impose on them facilities that provide most of the benefits to those other than the community hosting the facility; and
7) have dismissed their opinions, fears, and concerns as irrational and irrelevant.

Critical to resolving NIMBY, LULU, and related risk-related controversies is the recognition that a fairly distributed risk is more acceptable than an unfairly distributed one. A risk entailing significant benefits to the affected parties is more acceptable than a risk with no benefits. A risk where no alternatives exist is more acceptable than a risk an alternative technology could eliminate. A risk the affected have control over is more acceptable than a risk beyond their control. A risk that the parties at risk assess and decide voluntarily to accept is more acceptable than a risk that is imposed upon them.

Risk is multidimensional; size is only one of the relevant dimensions. If the validity of this point is accepted, then many risk communication strategies present themselves. Factors such as fairness, benefits, and voluntariness are often equal or more important to the public in judging the acceptability of a risk than the risk data provided by technical, engineering, and scientific professionals. Therefore, efforts to engage stakeholders and make a risk more fair, beneficial, or voluntary are as appropriate in making a risk more acceptable as efforts to make a risk technically smaller. Similarly, because control is important in determining the acceptability of a risk, efforts to engage people and share power, such as by establishing citizen or worker advisory committees or by supporting third-party research, audits, inspections, and monitoring, are as appropriate in making a risk more acceptable as efforts to make a risk technically smaller.

In summary, deciding what level of risk ought to be acceptable is not a technical question but a social, cultural, and value question. People vary in how they assess risk acceptability. They weigh the various factors according to their own values, sense of risk, and perceived stake in the outcome. Because acceptability is a matter of values and opinions, and because values and opinions differ, debates about risk are often debates about values, accountability, and control.

3.5 Risk Communication Principles and Guidelines

Given increasing demands by stakeholders for timely, credible, accurate, and relevant information about risks and threats, the development of improved risk communications strategies and practical tools will continue to be the focus of increasing attention in years to come. The observations reported here and in later chapters are samples of the growing area of risk communication research and practice.

From the literature on risk communication, several crosscutting principles and guidelines can be extrapolated. They are described below. Although many of these principles and guidelines may seem obvious, they are so often violated in practice that it is useful to ask why they are so frequently not followed.

Figure 3.4 is a toolbox for putting these risk communication principles into practice. The toolbox contains templates and tools that will be discussed in later chapters of this book.

3.5.1 Principle 1. Accept and Involve All Interested and Affected Persons as Legitimate Partners

Two basic tenets of risk communication in a democracy are generally understood and accepted. Communities have a right to participate in decisions that affect their lives, property, and the things they value. Next, the primary goal of risk communication should not be to diffuse concerns or avoid action. It should be to produce informed individuals, groups, and populations that are solution-oriented and collaborative. Success is measured largely by whether information has been exchanged and whether stakeholders have been informed and understand each other's points of view.

Guidelines:

1) Show respect and sincerity by involving those interested or affected early, before important decisions are made.
2) Clarify you understand the appropriateness of basing decisions about risks on factors beyond only the magnitude of the risk.
3) Involve all interested parties and stakeholders.

3.5.2 Principle 2. Plan Carefully and Evaluate Performance

Different goals, audiences, and media require different risk communication strategies and practical tools. Risk communication will be successful only if carefully planned and designed for the specific situation and audience.

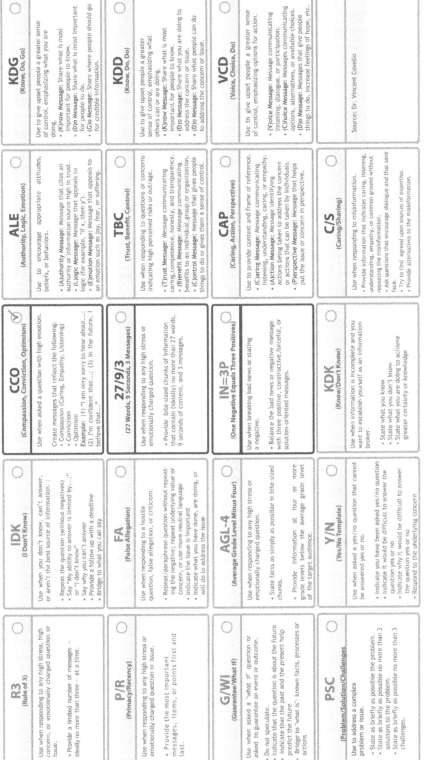

R3 (Rule of 3)

Use when responding to any high stress, high concern, or emotionally charged question or issue.
- Provide a limited number of messages – ideally no more than three – at a time.

IDK (I Don't Know)

Use when you don't know, can't answer, or aren't the best source of information.
- Repeat the question (without negatives).
- Say "My ability to answer is limited by...," or "I don't know"
- Say why you can't answer
- Provide a follow up with a deadline
- Bridge to what you can say

CCO (Compassion, Conviction, Optimism)

Use when asked a question with high emotion.
Create messages that reflect the following:
- Compassion (Caring, Empathy, Listening)
- Conviction
- Optimism
Example: (1) "I am very sorry to hear about...; (2) I'm confident that...; (3) In the future, I believe that...."

ALE (Authority, Logic, Emotion)

Use to encourage appropriate attitudes, beliefs, or behaviors.
- (Authority) Message: Message that cites an authority or information source high in trust.
- (Logic) Message: Message that appeals to logic (for example, "If x, then y").
- (Emotion) Message: Message that appeals to an emotion such as joy, fear, or suffering.

KDG (Know, Do, Go)

Use to give upset people a greater sense of control, emphasizing what you are doing.
- (K)now Message: Share what is most important for people to know.
- (D)o Message: Share what is most important for people to do.
- (G)o Message: Share where people should go for credible information.

P/R (Primacy/Recency)

Use when responding to any high stress or emotionally charged question or issue.
- Provide the most important messages, items, or points first and last.

FA (False Allegation)

Use when responding to a hostile question, false allegation, or criticism.
- Repeat/paraphrase question without repeating the negative; repeat underlying value or concern, or use more neutral language
- Indicate what you have done, are doing, or will do to address the issue

27/9/3 (27 Words, 9 Seconds, 3 Messages)

Use when responding to any high stress or emotionally charged question.
- Provide bite-sized chunks of information that contain (ideally) no more than 27 words, 9 seconds of content, and 3 messages.

TBC (Trust, Benefit, Control)

Use when responding to questions or concerns indicating high perceived risks or outrage.
- (T)rust Message: Message communicating caring, competence, honesty, and transparency.
- (B)enefit Message: Message communicating benefits to an individual, group, or organization.
- (C)ontrol Message: Message that gives people things to do or gives them a sense of control.

KDD (Know, Do, Do)

Use to give upset people a greater sense of control, emphasizing what others can or are doing.
- (K)now Message: Share what is most important for people to know.
- (D)o Message: Share what you are doing to address the concern or issue.
- (D)o Message: Share what people can do to address the concern or issue.

G/WI (Guarantee/What If)

Use when asked a "what if" question or asked to guarantee an event or outcome.
- Do not speculate.
- Indicate that the question is about the future
- Indicate that the past and the present help predict the future
- Bridge to "what is": known facts, processes or actions

AGL-4 (Average Grade Level Minus Four)

Use when responding to any high stress or emotionally charged question.
- State facts as simply as possible in bite-sized chunks.
- Provide information at four or more grade levels below the average grade level of the target audience.

IN=3P (One Negative Equals Three Positives)

Use when breaking bad news or stating a negative.
- Balance the bad news or negative message with three positive, constructive, forceful, or solution-oriented messages.

CAP (Caring, Action, Perspective)

Use to provide context and frame of reference.
- (C)aring Message: Message communicating listening, understanding, caring, or empathy.
- (A)ction Message: Message identifying actions being taken to address the concern or actions that can be taken by individuals.
- (P)erspective Message: Message that helps put the issue or concern in perspective.

VCD (Voice, Choice, Do)

Use to give upset people a greater sense of control, emphasizing options for action.
- (V)oice Message: Message communicating listening, dialogue, or participation.
- (C)hoice Message: Messages communicating options, alternatives, or available choices.
- (D)o Message: Messages that give people things to do, increase feelings of hope, etc.

PSC (Problem/Solution/Challenges)

Use to address a complex problem or issue.
- State as briefly as possible the problem.
- State as briefly as possible no more than 3 solutions to the problem.
- State as briefly as possible no more than 3 challenges.

Y/N (Yes/No Template)

Use when asked a yes/no question that cannot be answered yes or no.
- Indicate you have been asked yes/no question
- Indicate it would be difficult to answer the question yes or no
- Indicate why it would be difficult to answer the question yes or no
- Respond to the underlying concern

KDK (Know/Don't Know)

Use when information is incomplete and you want to establish yourself as an information broker.
- State what you know
- State what you don't know
- State what you are doing to achieve greater certainty or knowledge

C/S (Caring/Sharing)

Use when responding to misinformation.
- Provide information that indicates caring, listening, understanding, empathy, or common ground without repeating the misinformation.
- Ask questions that encourage dialogue and that save face.
- Try to find agreed upon sources of expertise.
- Provide alternatives to the misinformation.

Source: Dr. Vincent Covello

Figure 3.4 Communication templates and tools for risk, crisis, and high stress situations.

Guidelines:

1) Begin your planning with explicit objectives, such as: informing decision-making by individuals and groups; motivating individuals and groups to engage in constructive action; or contributing to conflict or dispute resolution.
2) Evaluate the information you have about risks and know its strengths and weaknesses.
3) Identify different subgroups among your target audience and customize information to address their information needs.
4) Determine the specific subgroups in your audience and design your communication for each.
5) Recruit spokespersons with strong presentation skills, emotional intelligence, and personal interaction skills.
6) Provide risk communication training and skills for your leaders, managers, and technical staff.
7) Reward outstanding performance.
8) When possible, pretest your messages.
9) Carefully evaluate your efforts and learn from your mistakes.

3.5.3 Principle 3. Listen to Your Audience

People are often more concerned about issues such as trust, credibility, control, competence, voluntariness, fairness, caring, and compassion than about mortality statistics and the details of quantitative risk assessment. If you do not listen to people, you cannot expect them to listen to you. Communication is most effective when there is an exchange of information and active listening.

Guidelines:

1) Don't assume what people know, think, or want to be done.
2) Determine what people are thinking: use techniques such as interviews, focus groups, face-to-face meetings, open houses, and surveys.
3) Acknowledge all interested parties and stakeholders.
4) Acknowledge the legitimacy of people's emotional response to risks issues.
5) Provide feedback to people on what you heard and ask for confirmation.
6) Recognize the "hidden agendas," symbolic meanings, and broader cultural, economic, or political considerations that often underlie and complicate risk and crisis communication.

3.5.4 Principle 4. Be Honest, Frank, and Open

Honesty and transparency play a large role in trust determination. Trust is your most precious asset when communicating information about risks or threats. Trust is difficult to earn, and once lost, difficult to regain.

Guidelines:

1) State your credentials but do not ask or expect to be trusted.
2) Disclose risk and crisis information as soon as possible, emphasizing appropriate reservations about reliability.
3) Do not minimize or exaggerate the level of risk.
4) Speculate only with great caution.
5) If in doubt, lean toward sharing more information, not less.

6) Discuss data uncertainties, strengths, and weaknesses, including those identified by credible sources.
7) Identify worst-case estimates, citing ranges of risk estimates when appropriate.

3.5.5 Principle 5. Coordinate and Collaborate with Other Credible Sources

Allies and partners can be critically effective in helping you communicate risk information in crisis and noncrisis situations. Few things make risk and crisis communication more difficult than conflicts or public disagreements with other credible sources.

Guidelines:

1) Coordinate all interorganizational and intraorganizational risk and crisis communications efforts.
2) Devote effort and resources to the slow, hard work of building bridges, and relationships with important allies and partners.
3) Use credible and authoritative intermediaries to communicate risk and crisis information.
4) Consult with allies and partners about who is best able to answer questions.
5) Try to issue joint communications with trustworthy sources.

3.5.6 Principle 6. Meet the Needs of Traditional and Social Media

Traditional media outlets – such as radio, television, newspapers, and magazines – and social media outlets are prime sources of risk information. They play a critical role in setting agendas and in determining outcomes.

Guidelines:

1) Be accessible.
2) Respect deadlines.
3) Provide information tailored to the needs of each type of media.
4) Prepare in advance and provide background material on complex issues.
5) Follow up media stories with praise or criticism, or respectful corrections, as warranted.
6) Try to establish relationships of trust with specific editors, reporters, writers, commentators, and bloggers.
7) Be proactive, first, accurate, and credible.

3.5.7 Principle 7. Speak Clearly and with Compassion

Technical language and jargon are useful as professional shorthand, but they are barriers to the successful risk and crisis communication.

Guidelines:

1) Use language appropriate to the target audience.
2) Use vivid, concrete images that communicate information about risks on a personal level.
3) Use stories, examples, and anecdotes that make technical risk-related data and information come alive.
4) Avoid distant, abstract, unfeeling language about deaths, injuries, illnesses, or harm; acknowledge – and say – that any illness, injury, or death is a tragedy.

5) Acknowledge, and respond to, both in words and with actions, the emotions people are feeling and expressing, including anxiety, fear, anger, and outrage.

6) Acknowledge and respond to the factors that people view as important in evaluating and accepting risks.

7) Use comparisons, especially comparisons to regulatory or professional standards, to help put risk information in perspective, improve understanding of a risk, and improve the adoption of protective behaviors, but be careful of comparisons that ignore factors that people consider important in evaluating and accepting risks.

8) Include a discussion of risk-reduction and control actions, including what people can do to increase feelings of self-efficacy and control or reduce their exposures to risks.

9) Promise only what you can deliver with confidence and do what you promise; guarantee processes rather than outcomes.

Research and analysis of case studies have shown that these principles and guidelines form the basic building blocks for effective risk communication. Each principle and guideline recognizes (differently) that effective risk communication is a process based on mutual trust, stakeholder engagement, and respect. Each principle and guideline also recognize that effective risk communication is central to informed decision-making and is a complex art and skill that requires substantial knowledge, training, and practice.

3.6 Key Takeaway Concepts and Conclusions from this Overview Chapter

1) Risk communication is a science-based discipline.
2) High concern situations change the rules of communication.
3) The key to risk communication success is anticipation, preparation, and practice (APP).
4) Opinions about a risk or threat not addressed effectively can morph into unchangeable beliefs. People under stress:
5) have difficulty hearing, understanding, and remembering information
6) want to know that you care before they care what you know
7) focus much more on negative information
8) focus most of what they hear first and last
9) process information well below their educational level
10) actively seek out additional sources of information to reduce stress and risks,

3.7 Chapter Resources

Below are additional resources to expand on the content presented in this chapter.

Árvai, J., and Rivers, L. III., eds. (2014). *Effective Risk Communication*. London: Earthscan.

Arvai, J., and Campbell-Arvai, V. (2014). "Risk Communication: Insights from the Decision Sciences." in *Effective Risk Communication*, eds. J. Arvai and L. Rivers III. London: Taylor and Francis.

Andrews, R. (1999). *Managing the Environment, Managing Ourselves: A History of American Environmental Policy*. New Haven: Yale University Press.

Aufder Heide, E. (2004). "Common misconceptions about disasters: Panic, the "disaster syndrome," and looting," in *The First 72 Hours: A Community Approach to Disaster Preparedness*, ed. M. O'Leary. Lincoln, NB: iUniverse Publishing.

Beck, U. (1992). *Risk Society: Towards a New Modernity*. Los Angeles, CA: Sage Publications.

Beck, M., and Kewell, B. (2014). *Risk: A Study of Its Origins, History and Politics*. New Jersey: World Scientific Publishing Company.

Becker, S. (2004). Emergency communication and information issues in terrorist events involving radioactive materials. *Biosecurity and Bioterrorism: Biodefense Strategy, Practice, and Science* 2(3):195–207.

Bennett, P., and Calman, K., eds. (1999). *Risk Communication and Public Health*. New York: Oxford University Press.

Bennett, P., Coles, D., and McDonald, A. (1999). "Risk communication as a decision process," in *Risk Communication and Public Health*, eds. P. Bennett and K. Calman. New York: Oxford University Press.

Bier, V.M. (2001). "On the state of the art: risk communication to the public." *Reliability Engineering and System Safety* 71(2):139–150.

Bohnenblust, H., and Slovic, P. (1998). "Integrating technical analysis and public values in risk based decision making." *Reliability Engineering & System Safety* 59:151–159.

Boholm, Å. (2019). "Risk communication as government agency organizational practice." *Risk Analysis* (39)8:1695–1707.

Boin, A., Rhinard, M., and Ekengren, M. (2014). "Managing transboundary crises: the emergence of European Union Capacity." *Journal of Contingencies and Crisis Management* 22(3):131–142.

Boin, A., Hart, P., Stern, E. and Sundelius, B. (2005). *The Politics of Crisis Management: Public Leadership Under Pressure*. Beverly Hills, CA: Sage.

Bostrom, A. (2003). "Future risk communication." *Futures* 35:553–573.

Bostrom, A., Atman, C., Fischhoff, B., and Morgan, M.G. (1994). "Evaluating risk communications: completing and correcting mental models of hazardous processes, Part II." *Risk Analysis* 14 (5):789–797.

Breakwell, G.M. (2007). *The Psychology of Risk*. Cambridge, UK: Cambridge University Press.

Centers for Disease Control and Prevention (2019). Crisis and *Emergency Risk communication*. Atlanta, GA: Centers for Disease Control and Prevention.

Chess, C., Hance, B.J., and Sandman, P. M. (1986). *Planning Dialogue with Communities: A Risk Communication Workbook*. New Brunswick, NJ: Rutgers University, Cook College, Environmental Media Communication Research Program.

Chess, C., Hance, B.J., and Sandman, P.M. (1988). *Improving Dialogue with Communities: A Short Guide to Government Risk Communication*. Trenton, NJ: New Jersey Department of Environmental Protection.

Chess, C., Hance, B.J., and Sandman, P.M. (1989). *Planning Dialogue with Communities. A Risk Communication Workbook*. New Brunswick, NJ: Rutgers University, Cook College, Environmental Communication Research Program.

Chess, C., Salomone, K.L., and Hance, B.J. (1995). "Improving risk communication in government: research priorities." *Risk Analysis* 15 (2):127–135.

Chess, C., Salomone, K.L., Hance, B.J., and Saville, A. (1995). "Results of a national symposium on risk communication: next steps for government agencies." *Risk Analysis* 15 (2):115–120.

Cvetkovich, G., Vlek, C.A., and Earle, T.C. (1989). "Designing technological hazard information programs: towards a model of risk-adaptive decision making," in *Social Decision Methodology for Technical Projects*, eds. C.A.J. Vlek, G. Cvetkovich. Dordrecht: Kluwer Academic.

Coombs, W.T. (1998). "An analytic framework for crisis situations: Better responses from a better understanding of the situation." *Journal of Public Relations Research* 10(3):177–192.

Coombs, W. (1999). *Ongoing Crisis Communications: Planning, Managing, and Responding*. Thousand Oaks, CA: Sage Publications, Inc.

Coombs, W.T. (2007). "Protecting organization reputations during a crisis: The development and application of situational crisis communication theory." *Corporate Reputation Review* 10(3):163–176.

Covello, V. (1992). "Risk communication, trust, and credibility." *Health and Environmental Digest* 6(1):1–4.

Covello, V. (1993). "Risk communication, trust, and credibility." *Journal of Occupational Medicine* 35:18–19.

Covello, V.T (1993). "Risk communication and occupational medicine." *Journal of Occupational Medicine* 35:18–19.

Covello, V.T. (2003). "Best practices in public health risk and crisis communication." *Journal of Health Communication*, 8(Supplement):5–8.

Covello, V.T. (2006). "Risk communication and message mapping: a new tool for communicating effectively in public health emergencies and disasters." *Journal of Emergency Management* 4(3):25–40.

Covello, V.T. (2005). "Risk communication". In *Environmental Health: From Global to Local*, ed. Frumkin, H. San Francisco: Jossey-Bass/Wiley:988–1008.

Covello, V.T. (2014). "Risk communication," in *Environmental Health: From Global to Local*. 5th edition), ed. H. Frumkin. San Francisco: Jossey-Bass/Wiley.

Covello, V., and Allen, F. (1988). *Seven Cardinal Rules of Risk Communication*. Washington, DC: U.S. Environmental Protection Agency, Office of Policy Analysis.

Covello, V., and Merkhofer, M. (1993). *Risk Assessment Methods: Approaches for Assessing Health and Environmental Risks*. New York: Plenum Press.

Covello, V., McCallum, D., and Pavlova, M., eds. (1989) *Effective Risk Communication: The Role and Responsibility of Government and Nongovernment Organizations*. New York: Plenum Press.

Covello, V., Peters, R., Wojtecki, J., and Hyde, R. (2001). "Risk communication, the West Nile virus epidemic, and bio-terrorism: Responding to the communication challenges posed by the intentional or unintentional release of a pathogen in an urban setting." *Journal of Urban Health* 78(2):382–391.

Covello, V., Sandman, P., and Slovic, P. (1988). *Risk Communication, Risk Statistics, and Risk Numbers*. Washington, DC: CMA.

Covello, V., and Sandman, P. (2001). "Risk communication: Evolution and revolution," in *Solutions to an Environment in Peril*, ed. A. Wolbarst. Baltimore, MD: Johns Hopkins University Press.

Covello, V., Slovic, P., and von Winterfeldt, D. (1986). "Risk communication: A review of the literature." *Risk Abstracts* 3(4):171–182.

Covello, V., Minamyer, S., and Clayton, K. (2007). *Effective Risk and Crisis Communication during Water Security Emergencies. EPA Policy Report*; EPA 600-R07-027. Washington, D.C.: U.S. Environmental Protection Agency.

Covello, V., and Sandman, P.M., (2001). Risk communication: Evolution and revolution, in Solutions to an environment in peril, ed. A. Wolbarst (Ed.; in press). Baltimore, MD: John Hopkins University Press: 166–178.

Covello, V., Slovic, P., and von Winterfeld, D. (1987). *Risk Communication: A Review of the Literature*. Washington, DC: National Science Foundation.

Cox, Jr., A.L. (2012). "Confronting deep uncertainties in risk analysis." *Risk Analysis* 32 (10):1607–1629.

Cvetkovich, G., Siegrist, M., Murray R., and Tragesser, S. (2002). "New information and social trust asymmetry and perseverance of attributions about hazard managers." *Risk Analysis* 22(2):359–367.

Dietz, T. (2013). "Bringing values and deliberation to science communication." *Proceedings of the National Academy of Sciences* 110:14081–14087.

Dunwoody, S. (2014). "Science journalism," in *Handbook of Public Communication of Science and Technology*, eds. M. Bucchi and B. Trench. New York: Routledge.

Earle, T.C. (2010). "Trust in risk management: A model-based review of empirical research." *Risk Analysis* 30(4):541–574.

Earle, T.C., and Siegrist, M. (2008). "On the relation between trust and fairness in environmental risk management." *Risk Analysis* 28(5):1395–1414.

Federal Emergency Management Agency (2021). *FEMA Flood Risk Communication Toolkit for Community Officials*. Washington, DC: Federal Emergency Management Agency.

Fearn-Banks, K. (2007). *Crisis Communications: A Casebook Approach*, 3rd ed. Mahwah, NJ: Lawrence Erlbaum Associates.

Fischhoff, B. (1989). "Helping the public make health risk decisions. In Effective Risk Communication: The Role and Responsibility of Government and Nongovernment Organizations", eds. Covello, V., McCallum, D.B., Pavlova, M.T. New York, NY: Plenum Press: 111–116.

Fischhoff, B., Slovic, P., Lichtenstein, L., Read, S., and Combs, B. (1978). "How safe is safe enough? A psychometric study of attitudes towards technological risks and benefits." *Policy Sciences* 9:127–152.

Fischhoff, B. (1995a). "Risk perception and communication unplugged: twenty years of process." *Risk Analysis* 15(2):137–145.

Fischhoff, B. (1995b). "Strategies for risk communication. Appendix C." in *National Research Council: Improving Risk Communication*. Washington, DC: National Academies Press.

Fischhoff, B. (2012). *Judgment and Decision Making*. New York: Earthscan.

Fischhoff, B. (2013). "The Sciences of Science Communication," in *Proceedings of the National Academy of Sciences* 110 (Supplement 3):14033–14039.

Fischhoff, B., Brewer, N.T., and Downs, J.S., eds. (2011). *Communicating Risks and Benefits: An Evidence-based User's Guide*. Washington, DC: Food and Drug Administration.

Fischhoff, B., and Davis, A.L. (2014). "Communicating scientific uncertainty," in *Proceedings of the National Academy of Sciences of the United States of America* 111 (Suppl. 4): 13664–13671.

Fischhoff, B., and Kadvany, J. (2011). *Risk: A Very Short Introduction*. New York: Oxford University Press.

Fischhoff, B., Lichtenstein, S., Slovic, P., and Keeney, D. (1981). *Acceptable Risk*. Cambridge, MA: Cambridge University Press.

Flynn, J., Slovic, P., and Mertz, C.K. (1994). "Gender, race, and perception of environmental health risks." *Risk Analysis* 14(6):1101–1108.

Giddens, A. (1991). *The Consequences of Modernity*. Cambridge, UK: Polity Press.

Glik D.C. (2007). "Risk communication for public health emergencies." *Annual Review of Public Health* 28(1):33–54.

Hance, B. J., Chess, C., and Sandman, P. M. (1990). *Industry Risk Communication Manual*. Boca Raton, FL: CRC Press/Lewis Publishers.

Halvorsen, P. A. (2010). "What information do patients need to make a medical decision?" *Medical Decision Making* 30 (5 Suppl):11S–13S.

Haight, J.M., ed. (2008). *The Safety Professionals Handbook: Technical Applications*. Des Plaines, IL: The American Society of Safety Engineers.

Heath, R., and O'Hair, D., eds. (2009). *Handbook of Risk and Crisis Communication*. New York: Routledge.

Hess, R., Visschers, V.H.M., Siegrist, M., and Keller, C. (2011). "How do people perceive graphical risk communication? The role of subjective numeracy." *Journal of Risk Research* 14(1):47–61.

Hyer, R.N., and Covello, V.T. (2007). *Effective Media Communication During Public Health Emergencies: A World Health Organization Handbook*. Geneva: World Health Organization Publications.

Jardine, C.G., Driedger S.M. (2014). "Risk communication for empowerment: an ultimate or elusive goal?" in *Effective Risk Communication*, eds J. Arvai and L. Rivers III. London: Earthscan.

Hyer, R., and Covello, V. (2007). *Effective Media Communication during Public Health Emergencies: A World Health Organization Handbook*. Geneva, Switzerland: World Health Organization.

Johnson, B.B., and Covello, V. (1987). *The Social and Cultural Construction of Risk: Essays on Risk Selection and Perception*. Dordrecht, Holland: D.Reidel Publishing.

Joslyn, S., and LeClerc, J. (2012). "Uncertainty forecasts improve weather related decisions and attenuate the effects of forecast error." *Journal of Experimental Psychology* 18:126–140.

Joslyn, S., Nadav-Greenberg, L., Taing, M.U., and Nichols, R.M. (2009). "The effects of wording on the understanding and use of uncertainty information in a threshold forecasting decision." *Applied Cognitive Psychology* 23:55–72.

Kahneman, D. (2011). *Thinking, Fast and Slow*. New York: Macmillan Publishers.

Kahneman, D., Slovic, P., and Tversky, A., eds. (1982). *Judgment Under Uncertainty: Heuristics and Biases*. New York: Cambridge University Press.

Kahneman, D. Tversky, A. (1979). "Prospect theory: An analysis of decision under risk." *Econometrica*, 47(2):263–291.

Kasperson, R.E. (1986). "Six Propositions on public participation and their relevance for risk communication." *Risk Analysis* 6(3):275–281.

Kasperson, R.E. (2014). "Four questions for risk communication." *Journal of Risk Research* 17 (10):1233–1239.

Kasperson, R.E., Golding, D., and Tuler, S. (1992). "Social distrust as a factor in sitting hazardous facilities and communicating risks." *Journal of Social Issues* 48(4):161–187.

Kasperson, R. E., Renn, O., Slovic, P., Brown, H.S., Emel, J., Goble, R., Kasperson, J. X., and Ratick, S. (1987). "Social amplification of risk: A conceptual framework." *Risk Analysis* 8 (2):177–187.

Kasperson, R., Kasperson, J.X., and Golding, D. (1999). "Risk, Trust, and Democratic Theory," in *Social Trust and the Management of Risk*, eds. G. Cvetkovich, R. Löfstedt. London: Earthscan.

Kasperson, R. E., and Stallen, P.J.M. eds. (1991). *Communicating Risks to the Public: International Perspectives*. Dordrecht: Kluwer.

Krimsky, S., and Plough, A. (1988). *Environmental Hazards: Communicating Risks as a Social Process*. Dover, MA: Auburn House.

Löftstedt, R. (2003). "Risk communication. pitfalls and promises." *European Review* 11 (3):417–435.

Liess, W. (1996). "Three phases in the evolution of risk communication practice." *Annals of the American Academy of Political Social Science* 545 (1):85–94.

Löfstedt, R. (2005). *Risk Management in Post-Trust Societies*. New York: Palgrave Macmillan.

Löfstedt, R.E., and Bouder, F. (2014). "New transparency policies: Risk communication's doom?" in *Effective Risk Communication*, eds J. Àrvai and L. Rivers III. London: Earthscan.

Lindenfeld, L., Smith, H., Norton, T., and Grecu, N. (2014). "Risk communication and sustainability science: lessons from the field." *Sustainability Science* 9 (2):119–127.

Lundgren, R.E., and McMakin A.H. (2018). *Risk Communication: A Handbook for Communicating Environmental, Safety, and Health Risks*. Hoboken, NJ: Wiley-IEEE Press.

McComas, K.A. (2006). "Defining moments in risk communication research: 1996–2005." *Journal of Health Communication* 11(1):75–91.

McComas, K.A., Arvai, J., and Besley, J.C. (2009). "Linking public participation and decision making through risk communication," in *Handbook of Risk and Crisis Communication*, eds. R. Heath and D. O'Hair. New York: Routledge.

Mileti, D., Nathe, S., Gori, P., Greene, M., and Lemersal, E. (2004). *Public Hazards Communication and Education: The State of the Art*. Boulder, Colo.: Natural Hazards Center.

Morgan, B., B. Fischoff, A. Bostrom, L. Lave, and C.J. Atman (1992). "Communicating risk to the public". *Environmental Science and Technology* 26:2048–2056.

Morgan, M.G., Fischhoff, B., Bostrom, A., and Atman, C.J. (2002). *Risk Communication: A Mental Models Approach*. Cambridge, UK: Cambridge University Press.

National Research Council (1989). *Improving Risk Communication*. Washington, D.C.: National Academies Press.

National Research Council (1996). *Understanding Risk: Informing Decisions in a Democratic Society*. Washington, DC: National Academies Press.

National Research Council) (2008). *Public Participation in Environmental Assessment and Decision Making*. Washington, DC: The National Academies Press.

National Oceanic and Atmospheric Administration, US Department of Commerce. (2016). *Risk Communication and Behavior. Best Practices and Research*. Washington, DC: National Oceanic and Atmospheric Administration.

National Academy of Sciences/National Research Council (2008). "Public participation in environmental assessment and decision making," in *Panel on Public Participation in Environmental Assessment and Decision Making*, eds. T. Dietz and P.C. Stern. Committee on the Human Dimensions of Global Change, Division of Behavioral and Social Sciences and Education. Washington, DC: The National Academies Press.

National Academy of Sciences (2014). *The Science of Science Communication II: Summary of a Colloquium*. Washington, DC: The National Academies Press.

National Academy of Sciences (2017). *Communicating Science Effectively*. Washington, D.C.: The National Academies Press.

Olsson, E. (2014). "Crisis communication in public organizations: dimensions of crisis communication revisited." *Journal of Contingencies and Crisis Management* 22(2):113–125.

Olsson, E. (2015). "Transboundary crisis networks: the challenge of coordination in the face of global threats." *Risk Management* 17(2):91–108.

Peters, E. (2012). "Beyond comprehension: The role of numeracy in judgments and decisions." *Current Directions in Psychological Science* 21(1): 31–35.

Peters, R., McCallum, D., and Covello, V. T. (1997). "The determinants of trust and credibility in environmental risk communication: An empirical study." *Risk Analysis* 17(1):43–54.

Pidgeon, N., Kasperson, R., and Slovic, P. (2003). *The Social Amplification of Risk*. Cambridge, UK: Cambridge University Press.

Renn, O. (1992). "Risk communication: towards a rational discourse with the public." *Journal of Hazardous Materials* 29: 465–579.

Renn, O. (2008). *Risk Governance: Coping with Uncertainty in a Complex World*. London, UK: Earthscan.

Renn, O., and Levin, D. (1991). "Credibility and Trust in Risk Communication," in *Communicating Risks to the Public*, eds. R. Kasperson and P. Stallen. Dordrecht, The Netherlands: Kluwer Academic Publishers.

Reynolds, B. (2014). *Crisis and Emergency Risk Communication*. Atlanta, GA: US Centers for Disease Control and Prevention.

Reynolds B., and Seeger, M.W. (2005). "Crisis and emergency risk communication as an integrative model." *Journal of Health Communication* 10:43–55.

Ripley, A. (2009). *The Unthinkable: Who Survives When Disaster Strikes - And Why*. New York: Harmony Books.

Rodrıguez, H., Dıaz, W., Santos, J., and Aguirre, B. (2007). "Communicating risk and uncertainty: science, technology, and disasters at the crossroads," in *Handbook of Disaster Research*. New York: Springer.

Rogers, E.M. (1996). "The field of health communication today: An up-to-date report." *Journal of Health Communication* 1 (1):15–23.

Sandman, P.M. (1989). "Hazard versus outrage in the public perception of risk," in *Effective Risk Communication: The Role and Responsibility of Government and Non-Government Organizations*, eds. V.T. Covello, D.B. McCallum, M.T. Pavlova. New York: Plenum Press.

Sandman, P.M (1993). *Responding to Community Outrage: Strategies for Effective Risk Communication.* Fairfax, Va.: AIHA Press,

Siegrist, M., Cvetkovich, G., and Roth, C. (2000). "Salient value similarity, social trust, and risk/benefit perception." *Risk Analysis* 20(3):353–361.

Seeger, M.W. (2006). "Best practices in crisis communication: an expert panel process." *Journal of Applied Communication Research* 34(3):232–44.

Sheppard, B., Janoske, M., and Liu, B. (2012). "Understanding Risk Communication Theory: A Guide for Emergency Managers and Communicators." Report to Human Factors/Behavioral Sciences Division, Science and Technology Directorate, US Department of Homeland Security. College Park, MD: US Department of Homeland Security.

Slovic, P. (1987). "Perception of risk." *Science* 236 (4799):280–285.

Slovic, P. (1993). "Perceived risk, trust, and democracy." *Risk Analysis* 13(6):675–682.

Slovic, P. (1999). "Trust, emotion, and sex." *Risk Analysis* 19(4):689–701.

Slovic, P. ed. (2000). *The Perception of Risk*. London, UK: Earthscan.

Slovic, P., ed. (2010). *The Feeling of Risk: New Perspectives on Risk Perception*. London: EarthScan.

Slovic, P. (2016). "Understanding perceived risk: 1978-2015." *Environment: Science and Policy for Sustainable Development* 58(1):25–29.

Slovic, P., M. Finucane, L., Peters, E., and MacGregor, D.G. (2004). "Risk as analysis and risk as feelings: Some thoughts about affect, reason, risk, and rationality." *Risk Analysis* 24(2):311–322.

Slovic, P., Fischhoff, B., and Lichtenstein, S. (2001). Facts and fears: Understanding perceived risk. In *The Perception of Risk*, ed. Slovic, P. (pp. 137–153). London: Earthscan Publications Ltd.

Slovic, P., Krauss, N., and Covello, V. (1990). What should we know about making risk comparisons. *Risk Analysis* 10:389–392.

Stallen, P.J.M., and Tomas, A. (1988). "Public concerns about industrial hazards." *Risk Analysis* 8(2):235–245.

Steelman, T. A., McCaffrey, S. (2013). "Best practices in risk and crisis communication: Implications for natural hazards management." *Natural Hazards* 65(1):683–705.

Substance Abuse and Mental Health Services Administration (2019). *Communicating in a Crisis: Risk Communication Guidelines for Public Officials*. SAMHSA Publication No. PEP19-01-01-005. Rockville, MD, Substance Abuse and Mental Health Services Administration.

Thompson, K.M., and Bloom, D.L. (2000). "Communication of risk assessment information to risk managers." *Journal of Risk Research* 3:333–352.

Tuler, S.P., and Kasperson, R.E. (2014). "Social distrust and its implications for risk communication: an example of high-level radioactive waste management," in *Effective Risk Communication*, eds. J. Arvai and L. Rivers III. London: Earthscan.

Tversky, A., and Kahneman, D. (1974). "Judgment under uncertainty: Heuristics and biases." *Science* 185(4157):1124–1131.

Tversky A, and Kaheman, D. (1986). "Rational choice and the framing of decisions." *Journal of Business* 59:5251–5278.

US Environmental Protection Agency. (2005). *Superfund Community Involvement Handbook*. EPA 540-K-05-003. Washington, D.C.: US Environmental Protection Agency.

US Environmental Protection Agency (2007). *Risk Communication in Action*. Washington, D.C. US Environmental Protection Agency.

U.S. Department of Health and Human Services. (2006). *Communicating in a Crisis: Risk Communication Guidelines for Public Officials*. Washington, DC: U.S. Department of Health and Human Services.

U.S. Occupational Safety and Health Administration. (2020). *Hazard Communication*. Accessed at: https://www.osha.gov/hazcom

U.S. Navy Environmental Health Center (NEHC) (2002): *Risk Communication Primer*, 2nd edition. Norfolk, Va.: Navy Environmental Health Center.

Walaski, (Ferrante), P. (2011). *Risk and Crisis Communications: Methods and Messages*. Hoboken, NJ: Wiley.

Weinstein, N. D. (1987). *Taking Care: Understanding and Encouraging Self-Protective Behavior*. New York: Cambridge University Press.

Weinstein, N.D. (1980). "Unrealistic optimism about future life events". *Journal of Personality and Social Psychology* 39:106–120.

Weinstein, N.D. (1982). "Unrealistic optimism about susceptibility to health problems". *Journal of Behavioral Medicine* 5:441–460.

Wildavsky, A., and Dake, K. (1990). "Theories of Risk Perception: Who Fears What and Why". *Daedalus* 112:41–60.

Wildavsky, A., and Douglas, M. (1983). *Risk and Culture: An Essay on the Selection of Technological and Environmental Dangers*. Berkeley, CA: University of California Press.

Wood, M.M., Mileti, D.S., Kano, M., Kelley, M.M., Regan, R., and Bourque, L.B. (2011). "Communicating actionable risk for terrorism and other hazards." *Risk Analysis* 34(4):601–615.

World Health Organization (2004). *Outbreak Communication: Best Practices for Communicating with the Public during an Outbreak*. Geneva: World Health Organization.

World Health Organization (2017). *Communicating Risk in Public Health Emergencies: A Who Guideline for Emergency Risk Communication (Erc) Policy and Practice*. Geneva: World Health Organization.

Yang, H. Pang, X., Zheng, B., Wang, L., Wang, Y. Du, S., and Lu, X, (2020). "A strategy study on risk communication of pandemic influenza." *Risk Management and Health Policy* 13:1447–1458.

Endnotes

1 For a full description of epidemiological methods, see the textbook *"Risk Assessment Methods: Approaches for Assessing Health and Environmental Risks"* written by V.T. Covello and M.W. Merkhofer.

2 World Health Organization (2020). Munich Security Conference. 15 February 2020. Accessed at: https://www.who.int/dg/speeches/detail/munich-security-conference.

3 World Health Organization (2020). "Managing the COVID-19 Infodemic: Promoting healthy behaviors and mitigating the harm from misinformation and disinformation." Accessed at: https://www.who.int/news/item/23-09-2020-managing-the-covid-19-infodemic-promoting-healthy-behaviours-and-mitigating-the-harm-from-misinformation-and-disinformation.

4 World Health Organization (2020). "Coronavirus disease (COVID-19) advice for the public: Mythbusters." Accessed at: https://www.who.int/emergencies/diseases/novel-coronavirus-2019/advice-for-public/myth-busters.

5 McNeil, B.J., Pauker, S.G., Sox, H.C., and Tversky, A. (1982). "On the elicitation of preferences for alternative therapies." *New England Journal of Medicine* 306:1259–1262. See also: McGettigan, P., Sly, K., O'Connell, D., Hill, S., Henry, D. (1999). "The effects of information framing on the practices of physicians." *Journal of General Internal Medicine.*14 (10):633–642.

6 See, e.g., Covello, V., Peters, R., Wojtecki, J., and Hyde, R. (2001). "Risk communication, the West Nile virus epidemic, and bio-terrorism: Responding to the communication challenges posed by the intentional or unintentional release of a pathogen in an urban setting." *Journal of Urban Health* 78(2):382–391. See also Covello, V. (2003). "Best practices in public health risk and crisis communication." *Journal of Health Communication* 8 (Suppl. 1): 5–8; discussion, 148–151; Covello, V. (2014). "Risk communication," in Environmental Health: From Global to Local, ed. H. Frumkin. San Francisco: Jossey-Bass/Wiley.

4

Development of Risk Communication Theory and Practice

CHAPTER OBJECTIVES

This chapter presents a concise overview of key stages, discoveries, and developments in the field of risk communication. As each development has built directly upon prior work, the chapter emphasizes essential principles through the story of the field.

   At the end of the chapter, you will be able to:

- describe the basic principles of the field, and
- give examples of how these principles were derived from actual high-impact experience coupled with research and evaluation.

4.1 Case Diary: Origin Story

Every year, I present on topics related to risk, high concern, and crisis communication. One of the most frequent questions I am asked is – how did I get started in the field?

The year 1981 was an important year for my career and for the field of risk, high concern, and crisis communication. At the time, I was a program director for the Technology Forecasting and Social Change program at the National Science Foundation (NSF). One morning I received a call to report immediately to the Foundation director. I was provided few specifics to prepare; only being told the meeting was urgent and related to a congressional inquiry.

My brain's fight or flight system activated, and I felt adrenaline and cortisol – the brain's stress hormone – flooding through my bloodstream. Had I done something seriously wrong? Had one of my projects been awarded a Golden Fleece Award?[1] Had someone filed a claim against me for bias in awarding grants? Was this meeting a turning point in my career? I was experiencing a high concern situation.

To prepare for the meeting with the Foundation Director, I wrote several talking points related to my imagined worst-case scenarios. I arrived at his office. He was cordial and friendly. In the room were other Foundation program directors.

The Director briefed us that Congress had appropriated several million dollars per year to the NSF to start a new program of interdisciplinary research on decision-making communications about technological risks. The Congressional appropriation arose in part because of concerns among some members of Congress that the public was overreacting to industrial accidents, such as the nuclear power plant accident at Three Mile Island in Pennsylvania. These overreactions

Communicating in Risk, Crisis, and High Stress Situations: Evidence-Based Strategies and Practice, First Edition. Vincent T. Covello.
© 2022 by The Institute of Electrical and Electronics Engineers, Inc. Published 2022 by John Wiley & Sons, Inc.

could negatively affect the development of important new technologies. The Director said one of us needed to drop whatever they were doing, personal and professional, and get the new program up and running. He identified three tasks that needed attention.

The first task of the new program director would be to write a white paper identifying gaps in knowledge about communications related to technological risks, to target where research was needed, and to describe how the Foundation would be spending Congressional monies to improve the knowledge base. This white paper would need to be delivered to Congress the following week.

The second task would be to develop a report for Congress on the quality of work by Federal government agencies and offices related to communications about technological risks. This would require the new program director to relocate for weeks, and perhaps months, to various agencies, including the White House Council on Environmental Quality, the Office of the Science Advisor to the President, the Food and Drug Administration, the Nuclear Regulatory Commission, the Department of Health and Human Services, and the US Environmental Protection Agency (EPA). Reports about the work of these agencies would need to be delivered to Congress by the end of the year. The Director speculated that it was unlikely these Federal agencies and offices would welcome this oversight.

The third task would be to determine how to spend the millions of dollars about to be appropriated. Support would be needed for creating a new professional society and a new peer-reviewed scientific journal. Grants would need to be awarded quickly for quality work. All research grants would be reviewed personally by the NSF Director and likely would receive scrutiny by Congress and the White House.

The Director turned his gaze on the group and asked for a volunteer. There was silence in the room. There was lots of fidgeting. He asked again for a volunteer. Again, silence.

His eyes fixed on me. He said I was already the director of the closest existing Foundation program; that I had an interdisciplinary academic background, which was needed for the job; and that he had reviewed the evaluations of my work at the Foundation and they proved I was a person who could do the job. He pointed out that I had a head start: I had already done a case study on communications in the aftermath of industrial accidents, including the nuclear power plant accident at Three Mile Island.

I volunteered. Little did I know that I was present at the founding of what would become the field of risk communication. I would become one of the first presidents of the Society for Risk Analysis and I would help launch the journal *Risk Analysis*, one of the leading scientific publishers of original risk communication research.[2] This chapter relates the history of this ever-developing field as new interdisciplinary research builds upon theory and expands our understanding of how individuals and society understand and communicate risk.

4.2 Introduction

The story of the development of the theory and application of risk communication reveals the basis for current best practice. In one of the most comprehensive recent reviews to date of this history, Lundgren and McMakin (2018) summarized the history of risk communication as one that moved from a rather simplistic, linear, and mechanistic, one-way, and send-to-receiver approach to a much more sophisticated approach that emphasized the importance of two-way communication; exchanges of information; stakeholder engagement, involvement, participation; relationships and partnership building; consensus building; and constructive dialog.[3] This more sophisticated approach drew from the work of communication researchers and practitioners from a diverse set of disciplines, including the behavioral sciences, social sciences, engineering, medicine, public health, industrial hygiene, linguistics, and neuroscience.

In what is now a classic article of the history of risk communication, Fischhoff (1995) identified (somewhat tongue in cheek) eight historical phases or stages in the development of risk management and risk communication practice.[4] They were:

Stage 1: "All we have to do is get the numbers right."
Stage 2: "All we have to do is tell them the numbers."
Stage 3: "All we have to do is explain what we mean by the numbers."
Stage 4: "All we have to do is show them that they've accepted similar risks in the past."
Stage 5: "All we have to do is to show them it's a good deal for them."
Stage 6: "All we have to do is treat them nice."
Stage 7: "All we have to do is make them partners."
Stage 8: "All of the above."

Fischhoff noted that all the principles, strategies, and tools developed in Stages 1–7 have some degree of merit. For example, it is important to know how best to present risk data and facts. Data and facts do not speak for themselves, especially when opposed by strongly held beliefs.

In a review of the history of risk communication practice, Leiss (1996) identified three historical phases in risk communication practice.[5]

4.2.1 Historical Phase 1: Presenting Risk Numbers

In the United States, Phase 1 covered the period from approximately the mid-1970s to the mid-1980s. Researchers and decision-makers focused on how best to present numerical estimates of risk to people with limited background in numerical concepts. Researchers and decision-makers also focused on how best to use risk comparisons for establishing risk management priorities.

Leiss's Phase 1 corresponds to Fischhoff's Stages 1 and 2. Historical Phase 1 included attempts to communicate risk-related numbers in 1979 at Love Canal – a neighborhood in Niagara Falls, New York, infamously known as the location of a 70-acre landfill and massive environmental pollution – and at Three Mile Island (1979) – a location near Middletown, Pennsylvania, where a nuclear power plant partially melted down, creating the most serious accident in US commercial nuclear power plant operating history.

4.2.2 Historical Phase 2: Listening and Planning

Phase 2 covered the period from approximately the mid-1980s to the mid-1990s. Researchers and decision-makers focused on the characteristics of effective risk and crisis communication, such as anticipating, planning, practice, trust, message clarity, risk perceptions, the effective use of delivery channels, and listening to the needs of those interested and affected. Leiss's Phase 2 corresponds to Fischhoff's Stages 3-6.

Phase 2 also marked the introduction of the term *risk communication* into the research literature. The first use of the term *risk communication* describing a formal discipline of scientific inquiry appeared with publications by Covello, the Conservation Foundation, and the National Research Council/National Academy of Sciences (NRC/NAS).[6] Th NRC/NAS report was one of the first documents identifying risk communication as a science.

Many techniques for effective risk communication came from research conducted in Phase 2. However, researchers also uncovered challenges and defects. One of the most important defects was failing to recognize the importance of stakeholder engagement. Another was the adoption, often without reflection, by risk communicators and managers of persuasive communication techniques developed by consumer marketers. Using such techniques raises moral questions, especially when these techniques are used to manipulate emotions or hide facts.

4.2.3 Historical Phase 3: Stakeholder Engagement

Phase 3 covers the period from the mid-1990s onward. Researchers and decision-makers shifted focus to consensus building; meaningful interaction with stakeholders; social and cultural factors influencing risk perceptions; and how trust can be built, maintained, or repaired through constructive dialog. Frequent beginning observations about trust in Phase 3 research include: Trust is situation specific and difficult to achieve. Trust accumulates and deposits over time in trust "accounts" that are based on words, interactions, relationships, and responsible actions. Trust is hard to gain and easy to lose. The importance of trust varies by hazard and respondent group.

Leiss's Phase 3 corresponds to Fischhoff's Stages 7 and 8. Leiss points out that each phase emerged in response to the earlier one. The earlier ones do not become irrelevant; instead, they are incorporated into the later phases. Each has value.

4.2.4 Covello and Sandman's Four Stages of Risk Communication

In their review of the history of risk communication, Covello and Sandman (2001) identified phases in the history of risk communication relating to health, safety, and environmental issues.[7] Covello and Sandman noted that for more than 60 years nations and organizations witnessed a tremendous take-back of power and control by people over risk management policies. In the mid-twentieth century, people were largely content to leave the management and control of risks in the hands of established authorities. However, by the end of the twentieth century and beginning of the twenty-first century, people around the world had legislatively asserted their claim to be involved in setting and implementing policies regarding risk management and decision-making. People became visibly upset and even outraged when they felt excluded from the risk management decision-making process.

Covello and Sandman argued it was in this crucible that the basic tenets of risk and crisis communication were born. The tenets were created, in part, to guide the new partnerships and dialogs with those affected by risks and crises. They provided science-based guidance for confronting the dilemma that perception equals reality, or becomes reality, and that which is perceived as real is real in its consequences.

Recognizing this dilemma, researchers examined whether there were science-based and ethical ways to calm people and provide reassurance, particularly when data showed a risk was not large but people perceive it to be large and are upset, angry, or outraged. Researchers also examined whether there were science-based and ethical ways to overcome apathy when data shows a risk is large and yet people perceive it to be insignificant and not deserving their attention.

This is the general context for the interest in risk communication that began in the 1960s and continues to this day. In part, risk communication principles and tools were identified and developed to overcome communication obstacles. These obstacles included inconsistent, overly complex, confusing, or incomplete messages about risks; the lack of trust in information sources; selective reporting by the media; and psychological and social biases that affect how risk-related information is processed.

Covello and Sandman described four evolutionary stages in the development of risk communication in terms of the changes in guiding philosophy and approach. The philosophical changes are core to understanding current practice and research.

4.2.4.1 Stage 1: Ignore the Public

The first stage was simply to ignore the public. This stage was prevalent in the United States until the late 1960s. The approach was built on the assumption that many people were apathetic about most health, safety, and environmental risk issues. As a result, they were willing to delegate decision-making to technical, engineering, and scientific professionals. Under these conditions, the best communication strategy was to ignore the public, or, when necessary, mislead them. Protect the public's health, safety, and environment, but do not let them take part in the decision-making.

For a long time, the public was content to be ignored regarding risk issues. But this approach stopped working, particularly in the late 1960s and early 1970s. Movements focused on taking back power and participating more directly in risk-related decision-making. Increasingly, when risk management authorities ignored the public, controversies became larger. This was the experience for many industries, including the nuclear power, pharmaceutical, and chemical industries. For example, in the early 1960s, thalidomide, a widely used drug for the treatment of nausea in pregnant women, was taken off the market due to massive pressure from the press and public. Experts estimated that thalidomide led to deaths and serious birth defects in thousands of children.

4.2.4.2 Stage 2: Explaining Risk Data Better

Since ignoring the public did not work, risk managers and communicators advanced to the second stage: learning how to explain risk data better. This is where many organizations are still today. Researchers developed new techniques for explaining risk assessment concepts, such as parts per billion, how assessments are conducted, and how management decisions are made. Researchers developed improved methods for interactions with the media, for reducing or eliminating the use of jargon, and for developing and using visual aids.

Most important, risk managers and communicators discovered listening and motivation were keys to learning. While risk communication materials can be developed for an average person to comprehend, people will process the information only if they are motivated and feel their concerns have been heard. When people are sufficiently motivated and feel heard, they can understand even complex technical material. For example, average people can process the probabilities associated with gambling and the complexities of mortgage rates.

For some risk problems, when the risk is large and the controversy is minimal, doing a better job with explaining risk information is the most important piece of the puzzle. For example, when people believe they have control over a particular risk, when the risk is perceived to be voluntary, and when risk management institutions are trusted, explaining data better often leads to improved decision-making. For other risk problems, such as when experts claim the risk is not significant or large, but when people are extremely concerned or outraged, explaining risk information is seldom effective in calming people or reducing outrage. Researchers and risk managers recognized solutions had to be found elsewhere.

Several events influenced developments during Stage 2. One was the passage of multiple laws requiring public participation, consultation, and the right-to-know. A second were events that resulted in public outrage. One such event was the nuclear power plant accident at Three Mile Island in Pennsylvania. The accident was the most significant accident in US commercial nuclear power plant history. It raised awareness of the dangers associated with industrial facilities and concerns about caring, competence, honesty, and transparency by industry. Two other significant events in Stage 2 were the ban by the US Environmental Protection Agency (EPA) on the use of the pesticide dichloro-diphenyl-trichloroethane (DDT)[8] and the hazardous waste crisis at Love Canal, New York.[9] President Jimmy Carter's declaration of a State of Emergency at Love Canal was made following an increase in skin rashes, miscarriages, and birth defects among residents.

Both the DDT and Love Canal events raised public awareness of, and concerns about, the risks associated with agricultural chemicals and the unregulated dumping of hazardous waste. These and related events also heightened awareness of the difficulties and challenges of risk communication and presenting risk data to emotionally charged audiences.

During Stage 2, scientists developed many techniques for explaining and putting risk data in perspective. One of the most popular of these techniques was to present risk comparisons.[10] Empirical evidence and theory suggested that risk comparisons could improve public understanding of risks and encourage the adoption of protective behaviors. Researchers hypothesized that risk comparisons could be especially useful in helping people make informed decisions about low-probability, high-consequence events, such as major flood or earthquake.[11]

Researchers also hypothesized that risk comparisons might become counterproductive if the public suspects that they are used to minimize or magnify a problem.[12] I provide several examples of risk comparison studies below.

Food risk comparisons. To gain perceptive and improved understanding of the risks posed by food, numerous studies have compared the risks posed by different foods, food products, and food additives.[13] One of the earliest and best-known comparative analyses of the risks of this type were the studies on food risks, diet, and cancer by Professor Bruce Ames and his colleagues at the University of California, Berkeley. These studies compared the cancer risks of foods that contain synthetic chemicals (e.g., food additives and pesticide residues) with the risks of natural foods. An important conclusion of the research was that synthetic chemicals represent only a tiny fraction of the total carcinogens in foods. The researchers pointed out that natural foods are not necessarily benign. Large numbers of potent carcinogens (e.g., aflatoxin in peanuts) and other toxins are present in foods that contain no synthetic chemicals. Natural carcinogens are part of a plant's natural defense system. Human dietary intake of these natural carcinogens can be as much as 10,000 times greater than the dietary intake of potentially carcinogenic synthetic chemicals in food. However, the many natural anticarcinogens also in food provide partial protection against natural carcinogens in food.

Critics of the work of Ames and his colleagues have argued that his risk estimates are inflated. Critics have also argued against the implicit, and sometimes explicit, argument and risk communication that natural carcinogens in foods deserve greater societal and regulatory attention and concern than synthetic chemicals.[14]

Energy risk comparisons. One of the earliest and the best-known studies of energy technologies was a study conducted by Inhaber (1978) for the Atomic Energy Control Board of Canada.[15] The study compared the total occupational and public health risks of different energy sources for the complete energy production cycle – from the extraction of raw materials to energy end use. The study examined the risks of eleven methods of generating electricity: coal, oil, nuclear, natural gas, hydroelectricity, wind, methanol, solar space heating, solar thermal, solar photovoltaic, and ocean thermal. Two types of risk data were analyzed: data on public health risks from industrial sources or pollutants and data on occupational risks derived from statistics on injuries, deaths, and illnesses among workers. Alternative sources of energy were compared on the basis of the calculated number of person-days that would be lost per megawatt year of electricity produced. Total risk for the energy source was calculated by summing the risks for the seven components of the complete energy production cycle: materials acquisition and construction, emissions from materials acquisition and energy production, operation and maintenance, energy backup system, energy storage system, transportation, and waste management.

Inhaber's report came to the following conclusions:

Most of the risk from coal and oil energy sources is due to toxic air emissions arising from energy production, operation, and maintenance.

Most of the risk from natural gas and ocean thermal energy sources is due to materials acquisition.

Most of the risk from nuclear energy sources is due to materials acquisitions and waste disposal.

Most of the risks from wind, solar thermal, and solar energy sources arise from the large volume of construction materials required for these technologies and the risks associated with energy backup systems and energy storage systems.

The most controversial aspect of Inhaber's report was the widely communicated conclusion that nuclear power carries only slightly greater risk than natural gas and less risk than all other energy technologies. Inhaber reported, for example, that coal-based energy has a 50-fold larger worker death rate than nuclear power. The report also communicated that, contrary to popular opinion, (1) nonconventional energy sources, such as solar power and wind, pose substantial risks; and (2) the risks of nuclear power are significantly lower than those of nonconventional energy sources.

Inhaber's report can be criticized from several perspectives. For example, the study mixed risks of different types, used risk estimators of dubious validity, made questionable assumptions to cover data gaps, failed to consider future technological developments, made arithmetic errors, and double-counted labor and backup energy requirements. Perhaps the most important criticism of Inhaber's study was methodological inconsistencies. For example, while the study considered materials acquisition, component fabrication, and plant construction in the analysis of unconventional energy sources and of hydropower, the study did not follow the same approach for coal, nuclear power, oil, and gas. Furthermore, the labor figures for coal, oil, gas, and nuclear power included only on-site construction, while those for the renewable energy sources included on-site construction, materials acquisition, and component manufacture.

Despite these criticisms, Inhaber's research represented a landmark effort in the literature on risk communication and risk comparisons. It made a significant conceptual contribution by attempting to compare, and communicate, the risks of alternative technologies intended to serve the same purpose. Also important was Inhaber's observation that risks occur at each stage in processes and product development, from raw material extraction, manufacturing, and use, to disposal. Inhaber's central argument was that risks from each stage in an industrial process or in product development need to be calculated and communicated to achieve an accurate estimate and understanding of the total risk.

Cancer risk comparisons. Doll and Peto (1981) conducted one of the earliest and best-known studies to put cancer risks in perspective.[16] The research team analyzed data for a variety of causes of cancer, including industrial products, pollution, food additives, tobacco, alcohol, and diet. Results of the study provided a comparative perspective on cancer risks. The study found, for example, that the combined effect of food additives, occupational exposures to toxic agents, air and water pollution, and industrial products account for only about 7% of US cancer deaths.

The results suggested that removing all pollutants and additives in the air, water, food, and the workplace would result in only a small decrease in cancer mortality. However, Doll and Peto pointed out that even this small percentage represents a substantial number of lives. Doll and Peto also pointed out that associations and correlations, no matter how powerful or large, do not mean causation. Only when all other available information is brought into the picture can a true causal relationship be shown.

EPA's "Unfinished Business" Risk Comparison Study. In 1987, the EPA published a landmark study of the risks associated with the thirty-one risk problems regulated by the agency titled *Unfinished Business: A Comparative Assessment of Environmental Problems*.[17] The purpose of the study was to determine if the agency could be more effective in its risk decision-making and management activities.

EPA staff were assigned to four working groups: the Cancer Risk Working Group, the Non-Cancer Health Effects Working Group, the Ecological Effects Working Group, and the Welfare Effects Working Group. Each working group looked at the same set of 31 risk problems and attempted to estimate the risks for each in their assigned areas. The results were then integrated to provide a basis for comparing the seriousness of the different risk problems. Risk problems that received relatively high rankings in three of the four working group categories, or at least medium rankings in all four, included outdoor air pollutants (for example, carbon monoxide, nitrogen oxides, and sulfur dioxide), stratospheric ozone depletion, and pesticide risks, including residues on food.

Risk problems that ranked relatively high on health but low on ecological or welfare effects (or that by definition were not considered an ecological problem) included radon, toxic air pollutants, indoor air pollution (other than radon), drinking-water contamination, pesticide application, consumer products, and worker exposure to chemicals. Risk problems that ranked high on ecological and welfare effects but low or medium on health effects included global warming, sources of surface-water pollution, physical alteration of aquatic habitats (including estuaries and wetlands), and mining wastes. Areas related to groundwater consistently ranked medium or low. Two problems for which information was particularly scarce – biotechnology and new chemicals-were considered very difficult to rank.

The EPA report found that its current risk management priorities did not correspond well with these risk rankings by its risk assessment experts. For example, the agency identified the following problems as "relatively high risk/low agency effort": indoor radon; indoor air pollution; nonpoint sources of surface-water pollution; discharges into estuaries, coastal waters, and oceans; other pesticide risks; accidental releases of toxics; consumer products; and worker exposures. Conversely, areas of high EPA effort but relatively low risk included Superfund hazardous waste sites, underground storage tanks, and municipal landfills.

The EPA report noted these divergences were not necessarily inappropriate. Some of the risk problems ranked as low in risk were low precisely because of efforts by the agency to reduce them. Other risks, such as those involving consumer products and worker exposures, are primarily the statutory responsibilities of other agencies, such as the Consumer Product Safety Commission and the Occupational Safety and Health Administration. Perhaps most importantly, the EPA report noted that risk estimates are only one of several other factors that determine EPA risk management priorities. These factors included the economic or technical controllability of the risks; the social, cultural, political, and psychological aspects of the risks (such as the degree to which the risks are perceived to be voluntary, controllable, familiar, or equitable); and the benefits of the activities that generate the risk.

The report noted that the EPA's risk management priorities corresponded well with public opinion. Survey data indicated that the public identified hazardous waste disposal, industrial accidents, and air pollution as high risks. The public ranked oil spills, worker exposures, pesticides, and drinking-water contamination as medium risks. The public ranked indoor air pollution, consumer product risks, genetically modified organisms, radiation (other than nuclear power plants), and global warming as relatively low risks.

In a follow-up article authored by an EPA official (Allen, 1987), the disparities between expert rankings of risk and public rankings of risk were explored in greater depth. For example, the article noted that beginning in the 1970s and 1980s, climate change and global warming was increasingly being recognized by scientists as a serious environmental problem. As a result, the EPA working group ranked global warming a relatively high risk. However, the public ranked global warming as a relatively low risk. Allen noted:

> The EPA task force ranked it high because of the massive potential implications for the entire world. The most probable explanations of the low public ranking are the following: 1) the consequences are very much in the future and hard for many to imagine because they extend beyond ordinary experience, 2) the problem is diffuse and there are many causes (i.e. there is no one person or thing to blame), and 3) there is simply a general lack of public familiarity with the issue.[18]

4.2.4.3 Stage 3: Stakeholder Engagement

The third stage of risk and crisis communication is built around stakeholder engagement and dialog. The publication of the *Seven Cardinal Rules of Risk Communication* by the EPA in 1988 as an official policy guidance document was an important third-stage event.[19] Two central premises of the EPA document were that people have a right to participate in decisions that affect their lives, and what people often mean by risk is much more complicated than what technical experts often mean by risk. The document argued that risk communication messages must be based on an understanding of how people obtain information, what values guide their interpretations of information, what role emotion plays in forming perceptions, how people make trade-offs, and how people arrive at their final attitude and behavior toward a risk.

In the third stage of risk communication, a profound paradigm shift in thinking took place. Risk was seen as consisting of two almost independent, basic elements: technical risk and emotion. Understanding risk required understanding the interaction of these two elements as they intertwined. The advantage of the "technical risk + emotion" concept was that it served to reframe the problem. It allowed risk managers and decision-makers to consider the many factors included in the public's definition of risk, such as trust, benefits, voluntariness, control, and fairness. This new, expanded concept of risk also pointed to the need for authentic and meaningful dialog among all interested parties. It led to the then revolutionary idea that the essence of risk communication is not just explaining risk numbers but listening, engaging in constructive dialog, and negotiating in good faith. Third-stage success requires that it is not enough for risk managers and communicators to acknowledge people's emotions and concerns by listening to them, they must also actively communicate their understanding that people are entitled to be emotional and concerned, and why.

An excellent example of third-stage risk and crisis communication occurred in the late 1980s. Medical waste was floating up on the shorelines of the northeastern United States. The public's response was powerful. In several states, people were told the medical waste did not pose a significant health threat. However, the public kept on insisting that it was still disgusting and frightening. Battles erupted.

In response to public outrage, several states, including Rhode Island, took a different approach. Public health authorities went public and said (in essence): "This is an outrage; this is unacceptable. The people in our communities will not, and should not, tolerate any medical waste or hypodermic syringes washing up on our shores. We are going to do absolutely everything in our power to stop it even though there is a negligible risk to health. We are going to turn our budget priorities upside down if needed." Many members of the public replied (in essence), "Thank you for your response, for listening, and acknowledging our concerns. Maybe we should wait. If it's really a negligible risk, how much of our money are you really planning to spend?"

Psychologists call this approach "getting on the other side of the resistance." When you share, and even exaggerate, people's concerns and emotions, and acknowledge the legitimacy of their emotions and concerns, it frees them up to feel other things and think the problem through more carefully. Many people came to see that the risk was extremely small and began questioning whether it was worth spending a lot of money to solve the problem.

Revised thinking about risk communication in Stage 3 led many organizations to identify risk communication as a core competency for those involved in risk, high concern, and crisis communications. In a large number of fields, ranging from public health, medicine, and nursing to epidemiology and engineering, a broad consensus developed on the need for risk communication training programs.[20] While many professional associations and organizations had long recognized the importance of risk communication skills for high concern situations, they now recognized the need to integrate evidence-based risk communication principles, strategies, and methods into training and standardized practices.

4.2.4.4 Stage 4: Empowerment

Stage 4 comes from a fundamental shift in the communicating organization's value system and culture. Stage 4 involves treating stakeholders as full partners. Only limited progress has been made toward achieving this goal. Stage 4 risk, high concern, and crisis communications are difficult for several reasons.

First, it is hard for individuals and organizations to change. Habit and inertia propel people in the direction of old behaviors. A second reason for limited Stage 4 progress is many of those who choose to work in technical, engineering, and scientific professions are, by disposition, people who typically prefer clear boundaries, logical approaches, and unemotional situations. They rarely felt comfortable listening to, engaging in dialog with, and negotiating with nonexperts, especially people who hold beliefs based more on emotion than on reason.

A third reason involves strongly held convictions of those in an organization. Many of those who choose to work in technical professions chose to do so because they want to make life better and protect people from risks and threats for their well-being. They are often convinced they have the knowledge to do this and resist what they perceive as competing, nonscience-based views, especially views they see as irrational. They want to protect people in a scientific, factual way, and do not perceive that dealing with people and their emotions helps them in their mission to protect.

A fourth reason for limited Stage 4 progress is organizational culture. Engaging in meaningful, respectful, and frank dialog with stakeholders involves changes in basic values and organizational culture. Effective stakeholder engagement also takes time away from other activities. Long-time employees in organizations, especially bureaucracies, are also often adept at distinguishing actual policies from those that are merely rhetoric. Is the organization's commitment to stakeholder engagement and dialog sincere? If the dialog process fails, will it harm my career? Will it affect my performance appraisal? Will I be able to get the time, the staff, the training, and the budget to do the job well?

A fifth and perhaps the most important reason for limited progress is power and comfort level. At the core of Stage 4 is empowerment. Many people and organizations resist attempts to usurp power. Sharing power and control with nonexperts – especially with people who appear to be angry, hostile, and unappreciative of all the work that goes into risk assessment and management – can also feel uncomfortable. Like so many others, risk managers frequently put a premium on protecting their own comfort level.

Except for the first stage (ignoring the public), the various stages of risk communication build on one another; they do not replace one another. New directions for risk communication research include studies on using social media; correcting false information; the impact of personal factors such as gender, ethnicity, race, age, economic status, religion, worldviews, and political orientation; and methods for increasing trust and enhancing transparency. Researchers and practitioners are also exploring the enhanced use of digital public participation and stakeholder engagement tools – such as Zoom, GoToMeeting, Microsoft Teams, Facebook groups, Neighborhood, and Nextdoor – prompted in part by the COVID-19 pandemic and the need for social distancing. With good design and support, these new digital tools have the potential to allow policymakers and practitioners to meaningfully engage and communicate with the public-at-large but also with people who have long been left out of stakeholder engagement activities. These groups could include the elderly, citizens reentering public life after incarceration, people currrently with limited access to the Internet or with limited computer literacy, immigrants, homeless people, people with physical and mental disabilities, people with low incomes, people working several jobs or working during nontraditional hours, people who are English-language learners, and people who have often been left out of stakeholder engagement activities because of racial inequality and income inequality.[21]

4.3 Summary

Risk communication is a rapidly developing field of interdisciplinary science and research. Risk communication research has gone through several evolutionary stages, with each stage placing greater importance on social and cultural context; stakeholder engagement; the impact of emotions, experiences, and cognitive biases on risk perceptions; two-way versus one-way communication; environmental justice, inequality, and risks to vulnerable and minority populations; shared decision-making and dialog; the role of economic factors in risk decision-making; the impact of social media; and combatting false information.

Risk communication research has expanded our understanding of how decisions about risks reflect different processes for valuing and weighing losses and gains and why and how disconnects often occur in the way technical experts and the public view and understand particular risks. Risk communication research has also expanded our understanding that responses by the public to risks and threats are driven more by emotions and experiences than by detailed deliberative evaluation.

4.4 Chapter Resources

Below are additional resources to expand on the content presented in this chapter.

Abraham T. (2009). "Risk and outbreak communication: Lessons from alternative paradigms." *Bulletin of the World Health Organization* 87(8):604–607.

Alaszewski, A. (2005). "Risk communication: Identifying the importance of social context." *Health, Risk & Society* 7:101–105.

Alexander, C., and Sheedy, E. (2004). *The Professional Risk Managers' Handbook: A Comprehensive Guide to Current Theory and Best Practices*. Wilmington, DE: Professional Risk Managers' International Association.

Allen, F. (1987). "Towards a holistic appreciation of risk: The challenge for communicators and policymakers," *Science, Technology & Human Values* 12(3):138–143.

Andrews, R. (1999). *Managing the Environment, Managing Ourselves: A History of American Environmental Policy*. New Haven. Yale University Press.

Arabie, P., and Maschmeyer C. (1988). "Some current models for the perception and judgment of risk." *Organizational Behavior and Human Decision Processes* 41:300–329.

Arvai, J., and Rivers L. (2014). *Effective Risk Communication*. New York: Routledge.

Aven, T. (2019). "The call for a shift from risk to resilience: What does it mean?" *Risk Analysis* 39:1196–1203.

Aven, T. (2020). *The Science of Risk Analysis*. New York: Routledge.

Balog-Way, D., McComas, K., and Besley J. (2020). "The evolving field of risk communication." *Risk Analysis* 40(S1):2240–2262.

Bostrom, A., Atman, C.J., Fischhoff B., and Morgan M.G. (1994). "Evaluating risk communications: Completing and correcting mental models of hazardous processes." *Risk Analysis* 4(5):789–798.

Brown, J., and Campbell, E. (1991). "Risk communication: Some underlying principles." *International Journal of Environmental Studies* 38(4):297–303.

Burns, W.J., and Slovic, P. (2012). "Risk perception and behaviors: Anticipating and responding to crises." *Risk Analysis* 32(4):579–582.

Centers for Disease Control and Prevention (CDC) (2018). *Crisis Emergency and Risk Communication*. Atlanta, Georgia: Centers for Disease Control and Prevention.

Chess, C., and Salomone, K.L. (1992). "Rhetoric and reality: Risk communication in government agencies." *The Journal of Environmental Education* 23(3): 28–33.

Chess, C., Salomone, K.L., Hance, B.J., and Saville A. (1995). "Results of a national symposium on risk communication." *Risk Analysis* 15(2):115–125.

Cho, H., Reimer, T., and McComas, K. eds. (2015). *The SAGE Handbook of Risk Communication*. Thousand Oaks, CA: Sage.

Cohrssen, J., and Covello, V.T. (1989). *Risk Analysis: A Guide to Principles and Methods for Analyzing Health and Environmental Risks*. Washington, D.C.: White House Council on Environmental Quality.

Coombs, W.T. (2019). *Ongoing Crisis Communication (Fifth Edition)*. Thousand Oaks, CA: Sage

Covello, V. (1992). "Risk communication, trust, and credibility." *Health and Environmental Digest* 6(1):1–4.

Covello, V.T (1993). "Risk communication and occupational medicine." *Journal of Occupational Medicine* 35:18–19.

Covello, V.T. (2003). "Best practices in public health risk and crisis communication." *Journal of Health Communication* 8(1):5–8.

Covello, V.T. (2006). "Risk communication and message mapping: A new tool for communicating effectively in public health emergencies and disasters." *Journal of Emergency Management* 4(3):25–40.

Covello, V.T. (2014). *Risk communication*. In Frumkin, H. (ed.) *Environmental Health: From Global to Local*. San Francisco: Jossey-Bass/Wiley.

Covello, V.T. (2011). "Risk communication, radiation, and radiological emergencies: Strategies, tools, and techniques." *Health Physics* 101(5): 511–530.

Covello, V.T., and Allen, F. (1988). *Seven Cardinal Rules of Risk Communication*. Washington, DC: US Environmental Protection Agency, Office of Policy Analysis.

Covello, V.T., Peters, R.G., Wojtecki, J.G., and Hyde, R.C. (2001). "Risk communication, the West Nile virus epidemic, and bioterrorism: Responding to the communication challenges posed by the intentional or unintentional release of a pathogen in an urban setting." *Journal of Urban Health* 78(2):382–391.

Covello, V.T. and P. M. Sandman (2001). "Risk communication: Evolution and revolution," in *Solutions to an Environment in Peril*, ed. A. Wolbarst. Baltimore: John Hopkins University Press, pp. 164–178.

Covello, V. T., D. B. McCallum and M. Pavlova (1989). Principles and guidelines for improving risk communication. Effective risk communication. Pp. 3–16 in V. T. Covello, D. McCallum, and

M. Pavlova (eds.), *Effective Risk Communication: The role and responsibility of government and nongovernment organizations*. New York: Plenum Press.

Covello, V.T., D. McCallum, and M. Pavlova (eds.), *Effective Risk Communication: The Role and Responsibility of Government and Nongovernment Organizations*. New York: Plenum Press.

Covello, V. T., P. Slovic, and D. v. Winterfeldt (1986). *Risk Communication: A Review of the Literature*. Washington, D.C.: National Emergency Training Center.

Dickmann, P., T. Abraham, S. Sarkar, P. Wysocki, S. Cecconi, F. Apfel, and Ü. Nurm, (2016). "Risk communication as a core public health competence in infectious disease management: Development of the ECDC training curriculum and programme." *EuroSurveillance* 21(14):30188.

Doll, R. and Peto, R. (1981). "The causes of cancer. Quantitative estimates of avoidable risks of cancer in the United States today." *Journal of the National Cancer Institute* 66:1195–1308.

Eiser, J. R., S. Miles, and L.J. Frewer (2002). "Trust, perceived risk, and attitudes toward food technologies." *Journal of Applied Social Psychology* 32(11):2423–2433.

European Centre for Disease Prevention and Control (2011). *Literature Review on Trust and Reputation Management in Communicable Disease in Public Health*. Stockholm: European Centre for Disease Prevention and Control.

European Centre for Disease Prevention and Control (2013). *A Literature Review on Effective Risk Communication for the Prevention and Control of Communicable Diseases in Europe*. Stockholm: European Centre for Disease Prevention and Control.

European Centre for Disease Prevention and Control (2014). *Social Marketing Guide for Public Health Managers and Practitioners*. Stockholm: European Centre for Disease Prevention and Control.

European Food Safety Authority (2018). *When Food is Cooking Up a Storm Proven Recipes for Risk Communications*. Brussels: European Food Safety Authority.

Fedorowicz, M. (2020). *Community Engagement during the COVID-19 Pandemic and Beyond A Guide for Community-Based Organizations*. Washington, D.C.: Urban Institute.

Glik, D. C. (2007). "Risk communication for public health emergencies." *Annual Review of Public Health* 28:33–54.

Greenberg, M., Haas, C., Cox, L.A., Lowrie, K., McComas, K., and North, W. (2012). "Ten most important accomplishments in risk analysis, 1980–2010." *Risk Analysis* 32:771–781.

Greenberg, M., Cox, A., Bier, V., Lambert, J., Lowrie, K., North, K., Siegrist, M., and Wu, F. (2020). "Risk analysis: Celebrating the accomplishments and embracing ongoing challenges." *Risk Analysis* 40(1):2113–2127.

Gurabardhi, Z., Gutteling, J.M., and Kuttschreuter, M. (2004). "The development of risk communication: an empirical analysis of the literature in the field." *Science Communication* 25(4):323–349.

Hampel, J. (2006). "Different concepts of risk–A challenge for risk communication." *International Journal of Medical Microbiology* 296:5–10.

Holly, K. (2016). *Principles of Equitable and Include Engagement*. Columbus, Ohio: Kirwin Institute Kirwan Institute for the Study of Race and Ethnicity, Ohio State University

Hyer, R.N., and Covello, V.T. (2017). "Breaking bad news in the high-concern, low trust setting. *Health Physics* 112(2):111–115.

Infanti, J., Sixsmith, J., Barry, M., Núñez-Córdoba, J., Oroviogoicoechea-Ortega, C., and Guillén-Grima, F. (2013). *A Literature Review on Effective Risk Communication for the Prevention and Control of Communicable Diseases in Europe*. Stockholm: European Centre for Disease Prevention and Control (ECSC).

Jaeger, C.C., Webler, T., Rosa, E.A. and Renn, O. (2013). *Risk, Uncertainty and Rational Action*. New York: Routledge.

Johnson, B.B., and Swedlow, B. (2021). "Cultural theory's contributions to risk analysis: A thematic review with directions and resources for further research." *Risk Analysis* 41(3):429–455.

Kasperson, J., Kasperson, R., Pidgeon, N., and Slovic, P. (2002). "The social amplification of risk: Assessing fifteen years of research and theory." Pp. 13–46 in: N. Pidgeon, R. Kasperson, and P. Slovic, eds. *The Social Amplification of Risk*. Cambridge: Cambridge University Press.

Kellens, W., Terpstra, T., Schelfaut, L., De Maeyer, P. (2013). "Perception and communication of flood risks: A literature review." *Risk Analysis* 33 (1):24–49.

Kunreuther, H., Novemsky, N., and Kahneman, D. (2001). "Making low probabilities useful." *Journal of Risk and Uncertainty* 23(2):103–120.

Leiss, W. (2004). "Effective risk communication practice." *Toxicology Letters* 149(1):399–404.

Linkov, I. and Trump, B. (2019). *The Science and Practice of Resilience*. New York: Springer.

Jagiello, R.D. and Hills, T.T. (2018). "Bad news has wings: Dread risk mediates social amplification in risk communication." *Risk Analysis*, 38(10):2193–2207.

Littlejohn, S.W., and Foss, K.A. (2010). *Theories of Human Communication*. Long Grove IL: Waveland Press.

Löfstedt, R., and Renn, O. (1997). "The Brent spar controversy: An example of risk communication gone wrong." *Risk Analysis* 17(2):131–136.

Lundgren, R.E., and McMakin, A.H. (2018). *Risk Communication: A Handbook for Communicating Environmental, Safety, and Health Risks*. Hoboken, NJ: John Wiley & Sons.

Manuele, F.A., and Main, B. (2002). "On acceptable risk." *Occupational Hazards* 64(1):57–60.

McComas, K.A. (2006). "Defining moments in risk communication research: 1996–2005." *Journal of Health Communication* 11(1):75–91.

Morgan, M.G., Fischhoff, B., Bostron, A., and Atman, C. (2002). *Risk Communication: A Mental Models Approach*. London: Cambridge University Press.

National Research Council/National Academy of Sciences (1989). *Improving Risk Communication*. Washington, DC: National Academies Press.

Parvanta, C., and Bauerle Bass, S. (2020). *Health Communication: Strategies and Skills for a New Era*. Burlington, Maine: Jones and Barlett Learning.

Perez-Floriano, L., Flores-Mora J., and MacLean J. (2007). "Trust in risk communication in organisations in five countries of North and South America." *International Journal of Risk Assessment and Management* 7(2):205–223.

Renn, O. (1992). "Risk communication: Towards a rational discourse with the public." *Journal of Hazardous Materials* 29(3):465–519.

Rich, R.C., Conn W.D., and Owens W.L. (1992). "Strategies for effective risk communication under SARA Title III: Perspectives from research and practice." *The Environmental Professional* 14(3):200.

Richard, L. (2021). "Pragmatic and (or) constitutive? On the foundations of contemporary risk communication research." *Risk Analysis* 41(3):466–479.

Rohrmann, B. (1992). "The evaluation of risk communication effectiveness." *Acta Psychologica* 81(2):169–192.

Rohrmann, B., Wiedemann, P.M., Helmut U., and Stegelmann, H.U., eds. (1990). *Risk Communication: An Interdisciplinary Bibliography*. Jülich, Germany: Research Center Jülich.

Sadar, A., and Shull, M. (1999). *Environmental Risk Communication: Principles and Practices for Industry*. Boca Raton, FL: CRC Press.

Sandman, P.M. (1987). "Risk communication: Facing public outrage." *EPA Journal* 13:21–22.

Siegrist, M. (2021). "Trust and risk perception: A critical review of the literature." *Risk Analysis* 41 (3):480–490.

Slovic, P. (1987). "Perception of risk." *Science*, 236(4799):280–285.

Stern, P. (1991). "Leaning through conflict: A realistic strategy for risk communication. *Policy Sciences* 24:99–114.

Suzuki, A. (2014). "Managing the Fukushima challenge." *Risk Analysis* 34(7):1240–1256.

US Public Health Service (1995). "Risk communication: Working with individuals and communities to weigh the odds." *Prevention Report* February-March:1–2.

World Health Organization (WHO) (2005). *WHO Outbreak Communication Guidelines.* Geneva: World Health Organization.

Zipkin, D.A, Umscheid, C.A., Keating, N.L., Allen, E., Aung, K., Beyth, R., Kaatz, S., Mann, D.M., Sussman, J.B., Korenstein, D., Schardt, C., Nagi, A., Sloane, R., Feldstein, D.A. (2014). "Evidence-Based Risk Communication: A Systematic Review." *Annals of Internal Medicine* 161(4):270–280.

Endnotes

1 "Golden Fleece awards" were given by Senator William Proxmire to publicize projects he believed were wasteful research.

2 As a result of my position as a program director at the National Science Foundation, I had the opportunity to coauthor a number of publications on risk communication with colleagues from other government agencies. For example, with Frederick Allen at the US Environmental Protection Agency (EPA), I co-authored the EPA policy document *The Seven Cardinal Rules of Risk Communication.* With a team of colleagues at EPA, I co-authored a document titled *Communicating Radiation Risks.* With John Cohrssen at the White House Council on Environmental Policy, I co-authored the document *Risk Analysis: A Guide to Principles and Methods for Analyzing Health and Environmental Risks.*

3 Lungren, R. and McKakin, A. (2018). *Risk Communication: A Handbook for Communicating Environmental, Safety, and Health Risks 6th Edition.* Hoboken, New Jersey. IEEE/John Wiley & Sons.

4 Fischhoff, B. (1995). "Risk perception and communication unplugged: Twenty years of process." *Risk Analysis* 15(2):137–145.

5 Leiss, W. (1996). "Three phases in the evolution of risk communication practice." *Annals of the American Academy of Political and Social Science* 545(May):85–94.

6 See, e.g., Covello, V.T. (1983). "The perception of technological risks: A literature review". *Technological Forecasting and Social Change,* 23:285–297; Conservation Foundation (1985). *Risk Assessment and Risk Control.* Washington, D.C.: Conservation Foundation; Covello, V.T. 1985. "Uses of social and behavioral research on risk." *Environment International* 198410:541–545; Covello, V.T., Slovic P., and Winterfeldt D.v. (1986). *Risk Communication: A Review of the Literature.* Washington, D.C., National Emergency Training Center; Covello, V.T., von Winterfeldt, D., and Slovic, P. (1986). "Communicating scientific information about health and environmental risks: Problems and opportunities from a social and behavioral perspective," *Uncertainties in Risk Assessment and Risk Management,*eds. V. Covello, A. Moghissi, and V.R.R. Uppuluri. New York: Plenum Press. 1986; Covello, V.T., Sandman P., and Slovic P. (1988). *Risk Communication, Risk Statistics, and Risk Comparisons.* Washington, D.C.: CMA; Covello, V., McCallum, D., and Pavlova M., eds. 1988b. *Effective Risk Communication: The Role and Responsibility of Government and Nongovernmental Organizations.* New York: Plenum; National Research Council/National Academy of Sciences (1989). *Improving Risk Communication.* Washington, DC: National Academies Press

7 Covello, V.T., and Sandman, P. (2001). "Risk communication: Evolution and revolution," in *Solutions to an Environment in Peril*, ed A. Wolbarst. Baltimore, MD: Johns Hopkins University Press.

8 Since 1945, DDT was available as an agricultural pesticide and as a household insecticide. Over the next 25 years, numerous researchers expressed concerns about the risks associated with DDT, citing dangers such as killing beneficial insects by bringing about the death of fish, birds (e.g., eagles), and other forms of wildlife either by their feeding on insects killed by DDT or directly by ingesting DDT. Outrage was engendered by the publication of a best-selling book *Silent Spring* by naturalist Rachel Carson. In 1972, the US Environmental Protection Agency cancelled most uses of DDT, exempting public health uses under specific conditions.

9 Love Canal is a neighborhood in Niagara Falls, New York. It is the location of a 70-acre landfill that became the cause of a massive environmental pollution crisis culminating in an extensive hazardous waste cleanup operation.

10 See, e.g., Covello, V.T., Sandman, P.M., and Slovic, P. (1988). *Risk Communication, Risk Statistics, and Risk Comparisons.* Washington, DC: CMA; Fischhoff, B. (2009). "Risk perception and communication," *Oxford Textbook of Public Health, Volume 2.* Oxford: Oxford University Press, pp. 940–953; Johnson, B. B. (2004). "Risk comparisons, conflict, and risk acceptability claims." *Risk Analysis* 24(1):131–145; Keller, C. (2011). "Using a familiar risk comparison within a risk ladder to improve risk understanding by low numerates: A study of visual attention." *Risk Analysis* 31(7):1043–1054; Keller, C., Siegrist, M., and Visschers, V. (2009). Effect of risk ladder format on risk perception in high- and low-numerate individuals. *Risk Analysis* 29(9):1255–1264; Kunreuther, H., Novemsky N., and Kahneman, D. (2001). "Making low probabilities useful." *Journal of Risk and Uncertainty* 23(2):103–120; Lipkus, I.M. (2007). "Numeric, verbal, and visual formats of conveying health risks: Suggested best practices and future recommendations." *Medical Decision Making* 27(5):696–713; Peters, E. (2020). *Innumeracy in the Wild.* Oxford: Oxford University Press; Peters, E., M.K. Tompkins, M.A.Z. Knoll, S.P., Ardoin, B. Shoots-Reinhard, and A.S. Meara, (2019). "Despite high objective numeracy, lower numeric confidence relates to worse financial and medical outcomes." *Proceedings of the National Academy of Sciences of the United States of America*,116(39):19386–19391 Peters, E., Västfjäll, D., Slovic P., Mertz C.K., Mazzocco, K., and Dickert, S. (2006). "Numeracy and decision making." *Psychological Science* 17(5):407–413; Pighin, S., Savadori L., Barilli, E., Rumiati, R., Bonalumi, S., Ferrari, M., and Cremonesi, L. (2013). "Using comparison scenarios to improve prenatal risk communication." *Medical Decision Making* 33(1):48–58; Roth, E., Morgan, M.G., Fischhoff, B., Lave, L., and Bostrom, A. (1990). "What do we know about making risk comparisons?" *Risk Analysis* 10(3):375–387.

11 See, e.g., Kunreuther, H., Novemsky, N., and Kahneman, D. (2001). "Making low probabilities useful." *Journal of Risk and Uncertainty* 23(2):103–120; See also Kellens, W., Terpstra, T., Schelfaut, L., and De Maeyer, P. (2013). "Perception and communication of flood risks: A literature review." *Risk Analysis* 33(1):24–49.

12 See, e.g., Johnson, B. B. (2004). "Risk comparisons, conflict, and risk acceptability claims." *Risk Analysis* 24(1):131–145.

13 See, e.g., Hohl, K., and Gaskell, G. (2008). "European public perceptions of food risk: Cross-national and methodological comparisons." *Risk Analysis* 28(2):311–324.

14 See, e.g., Ames B.N., and Gold, L.S. (1998). "The causes and prevention of cancer: the role of environment." *Biotherapy* 11(2-3):205–220.

15 Inhaber, H. (1978). *Energy Risk Assessment.* Ottawa, Quebec, Canada: Atomic Energy Control Board.

16 Doll, R., and Peto, R. (1981). "The causes of cancer. Quantitative estimates of avoidable risks of cancer in the United States today." *Journal of the National Cancer Institute* 66:1195–1308.

17 US Environmental Protection Agency (1987). *Unfinished Business: A Comparative Assessment of Environmental Problems and Major Risks.* Washington. DC: Environmental Protection Agency.

18 Page 140 in Allen, F. (1987). "Towards a holistic appreciation of risk: The challenge for communicators and policymakers.," *Science, Technology & Human Values* 12(3):138–143.

19 Covello, V., and Allen, F. (1988). *Seven Cardinal Rules of Risk Communication. EPA Policy Document OPA-7-020*. Washington, DC: US Environmental Protection Agency.

20 See, e.g., Dickmann, P., Abraham, T., Sarkar, S., Wysocki, P., Cecconi, S., Apfel, F., and Nurm, Ü. (2016). "Risk communication as a core public health competence in infectious disease management: Development of the ECDC training curriculum and programme." *EuroSurveillance* 21(14):30188; Bondy, S.J., Johnson, I., Cole, D.C., Bercovitz, K. (2008). "Identifying core competencies for public health epidemiologists." *Canadian Journal of Public Health* 99(4):246–251; Joint Institution Group on Safety Risk (2012). *Risk Communication and Professional Engineers*. Accessed at: https://www.theiet.org/media/1431/risk-comms.pdf; Centers for Disease Control and Prevention/Environmental Health Services (2001). "Environmental Health Competency Project: Recommendations for Core Competencies for Local Environmental Health Practitioners." Accessed at: https://www.cdc.gov/nceh/ehs/corecomp/corecompetencies.htm; Franziska Baessler, F., Zafar, A., Ciprianidis, A., Wagner, F.L., Bettina Klein, S., Schweizer, S., Bartolovic, M., Roesch-Ely, D., Ditzen, B., Nikendei, C., and Schultz, J. (2020). "Analysis of risk communication teaching in psychosocial and other medical departments." *Medical Education Online*, 25:1. Accessed at: https://www.tandfonline.com/doi/full/10.1080/10872981.2020.1746014

21 See, e.g., Holly, K. (2016). *Principles of Equitable and Include Engagement*. Columbus, Ohio: Kirwin Institute Kirwan Institute for The Study Of Race And Ethnicity, Ohio State University; See also Fedorowicz, M. (2020). *Community Engagement during the COVID-19 Pandemic and Beyond: A Guide for Community-Based Organizations*. Washington, D.C.: Urban Institute.

5

Stakeholder Engagement and Empowerment

CHAPTER OBJECTIVES

This chapter explains the critical need for meaningful stakeholder engagement and presents the methods, tools, and approaches to achieve it.
 At the end of this chapter, you will be able to:

- Demonstrate clear understanding of the nature and value of partnering with stakeholders
- Discuss tools and methods used for stakeholder engagement
- Determine where to find information for learning more about specific tools and methods
- Illustrate how the goal of stakeholder engagement is not to win, but rather to learn, to listen carefully, and openly explore ideas
- Predict the factors that enable successful, constructive stakeholder meetings

---

5.1 Case Diary: A Town Hall Public Meeting Goes Very Wrong

I was sitting in a large auditorium for a town hall public meeting. People were quickly filing into the auditorium. Reporters from local media outlets were gathering in the front of the room. The meeting was organized by the owner of a large local industrial facility. I was asked to attend and then advise on the facility's work with the community.

The purpose of the public meeting was to discuss community concerns about a possible link between illnesses among residents and contaminants found in a recent study of the community's water. Community residents claimed discharges from the company's industrial facility had polluted their drinking water and were causing people to get sick. The company had assigned a senior technical manager who was well versed in risk assessment and who had studied the discharge issues to run the meeting and serve as spokesperson.

Following 1.5 hours of PowerPoint presentations by company executives and technical experts, long lines of people queued up to speak at two open standing microphones located at each side of the auditorium. Company executives were sitting stone faced behind a long table on a stage at the front of the auditorium.

Emotions were high. The first person to speak at the open microphone was a woman who appeared to be very angry and frustrated. She was holding a young child in her arms and yelled at the people sitting on the stage. She told them she was sick all the time and her children were sick

Communicating in Risk, Crisis, and High Stress Situations: Evidence-Based Strategies and Practice, First Edition.
Vincent T. Covello.
© 2022 by The Institute of Electrical and Electronics Engineers, Inc. Published 2022 by John Wiley & Sons, Inc.

all the time. She blamed the contaminants in the water. She appealed to the people on the stage and in the audience to help her pay her medical bills.

The designated spokesperson responded to the women and others who had spoken at the open microphone with four messages. First, the community needs to understand that people often get sick, that a lot of the time the reason people get sick is something they did to themselves, and people are always looking for someone else to blame for their problems.

Second, the people who had spoken were letting their emotions get in the way of logical thinking. He said there is unquestionably no danger to the community from drinking the water. For technical experts such as himself, the "dose makes the poison." Since there has been no exposure, there is no risk.

Third, the company has a permit to discharge the pollutants into the water. The permits are issued by government regulatory authorities. The total amount of pollutants discharged into the water by the facility is strictly regulated by the government. The company meets all government regulations for discharges. If people in the community do not like what government regulations allow, they should write to their congressperson.

Fourth, the company had only limited resources and would not be able investigate each and every complaint. If people have a concern, even after hearing the technical presentations, they should consider hiring a lawyer.

The spokesperson's presentation was greeted with boos, foot-stomping, and catcalls. A person stood on a chair and started yelling through a megaphone. The town meeting was clearly getting out of hand. The moderator adjourned the meeting and asked security to clear the auditorium.

The company was unhappy with the outcome of the public meeting. It was especially unhappy about the negative coverage given to the meeting by the press. They asked me what I would recommend.

I met for coffee with the manager who had spoken at the Town Hall meeting. Before I had a chance to say anything, he said "I blew it, right? Why are these people so irrational? I don't know how to deal with people who don't base their opinions on facts. What should I do?" I told him this was a familiar problem, not only because different people think differently, but also because people typically hear information, including factual information, differently when they are fearful and under stress. Fortunately, there is a large body of research and tested practice on exactly how to deal with this – and holding one contentious town hall meeting was not the way to solve the problem.

I recommended to the company that they needed a stakeholder engagement strategy and outlined the steps they would need to take. Ideally, the company would have developed a community/stakeholder relations program long before community anger reached this boiling point, but it was never too late to begin to listen and interact productively. One company executive was skeptical – in his view, the technical, research-based facts of the matter were plain and just needed to be better publicized. I responded that the research and evidence-based facts of stakeholder engagement were also well known, and the company was experiencing first-hand the negative results when those facts were ignored.

The company followed my advice. A stakeholder engagement program was built into the company's operations. While some individuals in the community will likely always be convinced that their illness was caused by the company, evaluation data indicated that the company's stakeholder engagement activities produced a major and positive turnaround in community relations. The program also produced fruitful exchanges of information, constructive dialogue, and negotiated agreements.

5.2   Introduction

Stakeholders are individuals or groups who are interested in or affected by an issue or by an organization and its activities. Stakeholders may also include individuals or groups who affect and influence the views of others, even if that individual or group has no direct relationship to the issue at hand.

Stakeholder engagement is the systematic process by which the needs and values of stakeholders are listened to, acknowledged, and integrated into the decision-making process. It is ideally an interactive and two-way process that results in better and more informed decisions.

Basic principles of stakeholder engagement in private and public sector organizations can be traced back in part to management and stakeholder theories from the 1930s to debates about the responsibilities of an organization to its internal and external stakeholders, especially for complex and volatile issues.[1] Stakeholder engagement theory argued that corporations should create value for all stakeholders, not just shareholders. Stakeholder engagement is now standard practice among many private and public sector organizations.

Stakeholder engagement in the United States, as it relates to the public, can also be traced back in part to principles laid out in the following quote from President Thomas Jefferson.

> *I know no safe depositary of the ultimate powers of the society but the people themselves; and if we think them not enlightened enough to exercise their control with a wholesome discretion, the remedy is not to take it from them, but to inform their discretion by education.*
>
> —Thomas Jefferson, 1820

The modern age of stakeholder engagement, as it relates to health, safety, and the environment, in the US, can be traced in part to actions and communications by William Ruckelshaus as Administrator of the US Environmental Protection Agency (EPA). Ruckelshaus became the EPA's first administrator when the agency was formed in 1970. One of his first acts was to participate with other agencies in filing a lawsuit against companies discharging substantial quantities of toxic substances into the Cuyahoga River near Cleveland, Ohio. The massive fire that burned these substances in the Cuyahoga River created a national outcry.

Ruckelshaus laid the foundation for the EPA by hiring its leaders, defining its mission, deciding priorities, and creating its organizational structure. He also oversaw the implementation of the Clean Air Act of 1979 and the ban on the pesticide dichloro-diphenyl-trichloroethane (DDT). At the beginning of his second term as EPA Administrator in 1983, Ruckelshaus invoked the two core principles of stakeholder engagement. Stakeholders (1) have a right-to-know; and (2) have a right to be involved in decisions that affect their lives.[2]

Since the time of Ruckelshaus's term at the EPA, the goals of stakeholder engagement have become embedded in far-reaching right-to-know, right-to-participate, and right-to-be consulted health, safety, and environmental legislation. In the United States, among the most notable pieces of legislation requiring stakeholder engagement are *RCRA* (the Resource Conservation and Recovery Act), *CERCLA* (the Comprehensive Environmental Response, Compensation, and Liability Act, or Superfund*), TSCA* (the Toxic Substances Control Act), *NEPA* (the National Environmental Policy Act), *EPCRA* (the Emergency Planning and Community Right-To-Know Act), and *SARA* (the Superfund Amendments and Reauthorization Act). In a period of only decades, stakeholder engagement evolved from a vague ideological and philosophical concept to codified legislation. The intended purpose of stakeholder engagement, e.g., with communities, state and local governments, tribes, academia, private industry, federal agencies, and non-profit organizations, was to help stakeholders meaningfully participate in the decision-making process.

The expanding concept of stakeholder engagement and recognition of its importance reached well beyond environmental issues. For example, in healthcare systems and medicine, stakeholder engagement was a force in shaping the development and promotion of person-centered healthcare, and patient and family engagement in healthcare choices.[3] In 2001, for example, the US Institute of Medicine included "patient-centeredness" as one of its six aims of healthcare quality.[4] In Great Britain, the concept of person and family engagement began appearing with increasing regularity in UK health policy. For example, in 2008, the highly influential Darzi report, *High Quality Care for All,* described changing public expectations of services, including the importance of people being involved in decisions about their care.[5] The goal of person-centered healthcare was to focus on things that matter the most to people in healthcare settings. Although healthcare providers had often sought and incorporated input from individuals, many healthcare providers played a traditional role as the primary decision-maker. Person-centered care meant that an individual's values and preferences are identified, and once expressed, help guide healthcare decisions. Elements of person-centered care and engagement included demonstrating empathy, compassion, and respect for the patient's dignity; clear communications; a focus on quality of life; involving people in decisions about their healthcare; attention to, respect for, and understanding of personal preferences and needs; emotional support; involvement of and support for family and other care-givers; offering coordinated care, support, and treatment; enabling self-care; and performance measurement and quality improvement using feedback from the person and caregivers. Research showed that when patients are actively involved and supported in their care: (1) they are more satisfied with their care and feel more empowered; (2) they take greater responsibility for their own care; (3) they are more likely to stick to their treatment plans; (4) they choose treatments consistent with their values and preferences; (5) they are more likely to choose less costly treatments; (6) they have more knowledge, skills, and confidence to manage their health and use time efficiently; (7) they are more likely to engage in positive health behaviors; and (8) they have better health outcomes. The overarching goals of person/patient engagement is not what the healthcare provider can do for the patient but what they can do together. Studies also found that as patient engagement increased, staff performance and morale often increased.[6] Challenges to patient-centered care included increased workload for healthcare providers; identifying appropriate indicators of success; ethical, safety, and legal concerns; and developing a mindset that values and supports patients to be partners in their care.

Paralleling the rise of stakeholder engagement as a key element in risk, high concern, and crisis communication has been the erosion and decline in public trust and confidence in traditional risk, high-concern, and crisis management institutions, especially government, industry, and experts.[7] Reinforcing this point, a 2018 review of confidence in institutions by the Gallup Polling organization found:

> At least for the time being, Americans' average confidence in the nation's major governmental, economic and societal institutions has leveled off at a historical low point . . . Congress, the media (both television and print), and the criminal justice system – all entities facing significant scrutiny in the news or across social media in recent years – receive much higher negative than positive confidence ratings, serving as the poster institutions for what Americans think is wrong in the country.[8]

Similarly, public confidence in government in the United States remains near historical lows. When pollsters began asking about trust in government in 1958, about three-quarters of Americans trusted the federal government to do the right thing almost always or most of the time. Trust in government began eroding during the 1960s amid the escalation of the Vietnam War. The decline continued in the 1970s with the Watergate scandal and high inflation. Since 2007, the percentage of people saying they can trust the government has rarely surpassed 30%.[9] In 2020, only 20% of US adults said they trust the government in Washington to "do the right thing" just about always or most of the time.[10] In 2020, only 13% of Americans said they had a great deal or a lot of confidence in the US Congress.[11] Globally, governments briefly emerged as the most trusted institution in May 2020, when people entrusted them with leading the fight against COVID-19 and restoring economic health. But governments failed the test and squandered gains in trust in the next six months of 2020.[12]

According to a 2019 poll of more than 140,000 people in more than 140 countries, fewer than one in five people worldwide have a high level of trust in scientists. While 18% of people have high trust in scientists, 54% have a medium level of trust, 14% have low trust, and 13% said they "don't know." A third of people in Northern Europe, Central Asia, and Australia and New Zealand have high trust, while it is about one in 10 in Central and South America.[13] Unfortunately, the loss of trust in science had a major adverse effect on the enactment and implementation of public health policies during the 2020 COVID-19 pandemic.[14]

The key factors identified as related to an individual's level of trust in scientists are the individual's level of education and their trust in state institutions. Those educated to university level are much more likely to have trust in science. Trust in science also correlates strongly with trust in government, the judiciary, and the military.

More than half the world's population (57%) think that they do not know much – if anything – about science, and almost one in five (19%) believe that it does not benefit them personally. In high-income countries, those who find it "difficult to get by" financially are about three times as likely as people who say that they are "living comfortably" to be skeptical about whether science benefits society as a whole, or them personally.

Corresponding to the decline in institutional trust has been the rise of local and national citizen activist groups demanding formal stakeholder engagement. The dramatic growth of these groups since the 1970s has accompanied the decline of public trust in traditional risk, high-concern, and crisis management institutions. A major institutional shift in society regarding trust has occurred.

5.3 Levels of Stakeholder Engagement

Stakeholder engagement can take place through discussion or dialogue. In his book *The Fifth Discipline*, Peter Senge makes a distinction between dialogue and discussion.[15] In a typical discussion of a controversial issue, people present their views and defend them against those who disagree. The goal is to win rather than listen and learn. In dialogue, people listen carefully to what others are saying and openly explore ideas without prejudgment. The goal is not to win but to listen, seek, and acknowledge value added, encourage participants to offer new and different ideas, identify areas of agreement or "common ground," ask clarifying questions, and encourage the free expression of ideas.

For stakeholder discussion to become stakeholder dialogue, Senge argues that three basic conditions must be met. First, stakeholders must "suspend" their assumptions. Second, stakeholders must regard one another as colleagues and not as opponents or competitors. Third, stakeholder dialogue should ideally be guided by facilitators who can guide discussion and help bring the discussion back on track when it goes astray.

The earlier an organization or policy maker introduces stakeholder engagement, the more productive the engagement will be. As Schmeer points out in one of the most comprehensive guideline documents on stakeholder engagement, "policymakers and managers can use a stakeholder analysis to identity the key actors and to assess their knowledge, interests, positions, alliances, and importance related to the policy. This allows policymakers and managers to interact more effectively with key stakeholders. The goal of stakeholder engagement is not to convince people you are right. It is to learn, identify, and implement solutions together. Also, if policymakers and managers obtain input and engage stakeholders early in the decision-making process, it can lead to better solutions and increased support for a given policy or program. When stakeholder analysis is conducted before a policy or program is implemented, policymakers and managers can detect and act to prevent potential misunderstandings about and/or opposition to the policy or program. When a stakeholder analysis and other key tools are used to guide the implementation, the policy or program is more likely to succeed."[16]

Stakeholder engagement can be formal or informal. Formal stakeholder engagement is that which is mandated by legislation, rules, and regulations. For example, many health, safety, and environmental laws require public meetings, public consultation, and public comment periods. Prime examples of formal stakeholder engagement are the public hearings, public meetings, and public comment periods required for projects involving environmental impacts. These requirements for stakeholder engagement are set out legally in the US National Environment Protection Act (NEPA) and other legislation. Informal stakeholder engagement involves participation and involvement not required by legislation. It includes community meetings, open houses, petitions, protest marches, rallies, and workshops.

There are significant differences between jurisdictions and legislation requiring formal stakeholder engagement. Each jurisdiction – local, state, provincial, regional, national, and international – has its own history and cultural traditions regarding required types of stakeholder engagement. There are also significant differences among jurisdictions in how they interpret legal requirements for stakeholder engagement. For example, Conrad et al. (2011) and Eiter et al. (2014) used a five-step grading measure of stakeholder engagement.[17] They found large differences in interpretations of what formal stakeholder engagement means and what is required. The researchers noted one reason for this discrepancy is the large number of ambiguities and uncertainties in laws requiring stakeholder engagement.

These ambiguities and uncertainties allow room for radically different interpretations of requirements for stakeholder engagement. For example, it is often unclear at what stage stakeholder engagement is required in decision-making. Some scholars argue that stakeholders should be engaged before any important decisions are made. Other scholars argue that stakeholders should be engaged at selected strategic points in the decision-making process. Risk management authorities should decide when these strategic points occur.

Significant problems and conflict often happen when stakeholder engagement takes place late in the decision-making process. Stakeholders likely will not participate if they perceive (1) most of the important decisions have already been made, or (2) efforts to change those decisions are unlikely to be successful. Significant problems also arise if stakeholder engagement is perceived to be inauthentic and insincere. Stakeholder engagement is sometimes seen as duplicitous, as a means for an organization to fulfill legal requirements and to check off the bureaucratic box requiring stakeholder engagement. Stakeholder engagement efforts may be perceived as inauthentic attempts to put a democratic patina on what are essentially decisions already made.

If stakeholders do not believe their opinions are being heard or are given serious consideration in the decision-making process, people will often decline invitations to participate. They will not attend public hearings, public meetings, and open houses. They will refuse to join advisory committees. They will decline requests for comment or suggestions. They will set up their own participation processes or engage in symbolic and disruptive acts, such as protest marches, sit-ins, boycotts, graffiti, heckling, and mass resignations from advisory committees. They will shift stakeholder engagement to online and virtual engagements, which are rapidly becoming the "new norm." Most importantly, they will not follow or support decisions made or actions recommended.

Creighton identified six types of stakeholders: (1) apathetics, i.e. people who will not participate and have little interest; (2) observers, i.e. people who will keep abreast of an issue but who will generally not participate; (3) commenters, i.e. people who will be very interested, will attend meetings, or send letters; (4) technical reviewers, i.e. people who will spend the time reviewing documents; (5) active participants, i.e. people who are willing to commit a substantial amount of time and energy in order to influence decisions; and (6) co-decision makers, i.e. people who will participate as partners in decision-making, whose participation is essential to resolution of the issue, and who will approve or veto a decision. Distributions of each type will vary from issue to issue.[18]

Stakeholders can also be identified by specific dimensions or characteristics, such as by:

- **responsibility**: stakeholders who have legal, financial, operational, moral, or ethical responsibilities related to an issue or to the group affected by the issue.
- **influence**: stakeholders who are able to influence, through formal or informal means, the ability of an individual, group, organization, or institution to meet its goals.
- **proximity**: stakeholders that interact most with the individual, group, organization, or institution.
- **dependency**: stakeholders that are most dependent on the individual, group, organization, or institution for their well-being.
- **representation**: stakeholders that are entrusted to represent an individual, group, organization, or institution.

5.3.1 Types of Stakeholder Engagement

There are many typologies of stakeholder engagement. In one of the most widely cited models of stakeholder engagement and participation, Arnstein (1969) proposed a typology of stakeholder engagement called the "ladder of citizen participation."

As shown in Figure 5.1, the ladder, from the top down, includes rungs for citizen control, delegation, partnership, placation, consultation, informing, therapy, and manipulation. The top three rungs represent what Arnstein calls "Citizen Control." The next three rungs represent "Tokenism." The bottom three rungs represent "Nonparticipation." The ladder was designed to highlight the "critical difference between going through the empty ritual of participation and having the real power to affect the outcome of the process."[20] The higher up the ladder, the greater the degree of citizen empowerment. For example, at rung 4 of the ladder – consultation – citizens are invited to share their opinions but are offered no assurance their views will be heeded. The type of stakeholder engagement Arnstein holds up as an ideal is full citizen participation and stakeholder control. However, the ideal is rarely achieved.

Interest by Arnstein and others in stakeholder engagement led to the creation in 1990 of the International Association of Public Participation Practitioners (IAP2) and other interest groups. Researchers and practitioners identified seven core values and best practices of stakeholder engagement: (1) stakeholder engagement is based on the belief that those who are affected by a decision have a right to be involved in the decision-making process; (2) stakeholder engagement includes the promise that the contribution of stakeholders will influence the decision;

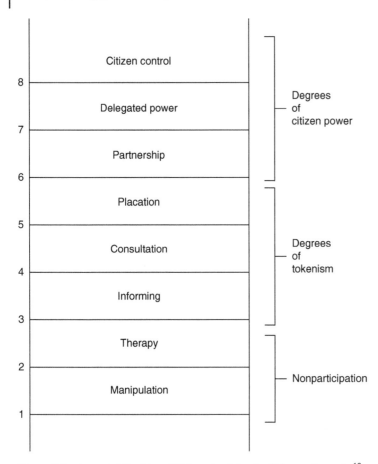

Figure 5.1 Arnstein's "Ladder of Citizen Participation." *Source:* Arnstein.[19]

(3) stakeholder engagement promotes sustainable decisions by recognizing and communicating the needs and interests of all participants, including decision-makers; (4) stakeholder engagement seeks out and facilitates the involvement of those potentially affected by or interested in a decision; (5) stakeholder engagement seeks input from participants in designing how they participate; (6) stakeholder engagement provides participants with the information they need to participate in a meaningful way; and (7) stakeholder engagement communicates to participants how their input affected the decision.

Researchers and practitioners in stakeholder engagement continually refine the five layers of stakeholder participation and engagement: *inform, consult, involve, collaborate*, and *empower*.[21] Each layer gives more influence to stakeholders over decisions. *Inform* means to provide stakeholders with balanced and objective information to help them understand the problem, options, alternatives, opportunities, possible solutions, and challenges. Examples of stakeholder engagement techniques include fact sheets, websites, 1-800 phone lines, email, information repositories, videos, podcasts, factsheets, infographics, social media postings, advertising, newsletters, posters, brochures, reports, letters, presentations, live-streaming, displays, exhibits, site visits/tours, public meetings, and open houses. *Consult* means to obtain feedback from stakeholders on the problem, options, alternatives, opportunities, possible solutions, and challenges. Examples of stakeholder consultation techniques include interviews, focus groups, on-line commenting, polls, surveys, voting, public meetings, on-line forums, social media discussions, workshops, door-to-door visits, kitchen table talks, public hearings, comment boxes, and open houses.

Involve means to work directly with stakeholders through each step of the assessment and decision-making process to ensure stakeholder concerns and expectations are well understood and considered. Several of the most frequently used techniques for meeting this objective are workshops, shared storytelling, charrettes, and citizen panels.

Collaborate means to partner with stakeholders at each step in the decision-making process. Several of the most frequently used techniques for meeting this objective are advisory committees, document co-creation, and working groups.

Empower means to share decision-making power with stakeholders. Empowering does not mean giving up power, although this might be the perception. Empowering does not undermine the authority of a proponent for a proposed action to make a final decision. Empowering, when carefully planned and executed, can lead to sustainable decisions and actions that have a greater likelihood of garnering support than opposition. Empowering means refining proposed decisions, plans, and actions in cooperation with those most affected by the proposed decisions, plans, or action. Several of the most frequently used techniques for meeting this objective are citizen committees, citizen juries, and engagements that put stakeholders at the center of the decision-making process and encourage them to use what they already possess.[22]

5.4   Benefits of Stakeholder Engagement

Risk, high-concern, and crisis communication researchers and practitioners have increasingly focused their attention on stakeholder engagement in part because of the benefits. This change in focus represents a major paradigm shift in thinking within the risk, high-concern, and crisis communication community. It represents a shift away from an earlier focus on how to improve one-way communication to how to improve two-way communication, information exchanges, listening, and interactive communication. Within this new paradigm, stakeholders are treated as collaborators and partners in the decision-making process.

Benefits of stakeholder engagement fall into three basic types: *substantive benefits, normative benefits*, and *instrumental benefits*. *Substantive benefits* include improvements in the quality of decision-making by capturing knowledge. Stakeholder engagement facilitates the exchange and two-way communication of general and local knowledge between leaders, managers, and members of stakeholder groups. Stakeholder engagement provides a means for leaders, managers, and members of stakeholder groups to receive and respond to knowledge updates.

Normative benefits include fulfilling a basic principle of democracy that people have a right to participate in decisions that affect their lives. Stakeholder engagement allows decision-makers and experts to listen to and incorporate local values and culture into the decision-making process.

Instrumental benefits include reducing the potential for conflict and controversy. Instrumental benefits include identifying potential conflicts early in the process, validating assumptions, testing ideas, and legitimizing policies and decisions.

Well-conducted stakeholder engagement activities produce several specific benefits. First, stakeholder engagement can result in the establishment of long-term, trusting relationships. Trust is typically built slowly over time by actions and communications. It is enhanced by constructive dialogue and frequent interactions, especially interactions where stakeholders can be treated as individuals with unique personalities. As trust increases, credits are deposited in a "trust bank" and used as needed. For example, when a crisis strikes, or when mistakes are made and apologies needed, or when things go wrong, "withdrawals" can be made from a trust bank, and apologies and corrections accepted.

Second, as a result of stakeholder engagement, leaders, managers, technical professionals, and communicators can acquire a better portrait and understanding of stakeholder perspectives, opinions, perceptions, beliefs, values, and behaviors. Knowledge about these characteristics plays a critical role in designing effective communication strategies and messages.

Third, stakeholder engagement can provide a fresh perspective and insights on issues. It can lead to more informed decision-making.

Fourth, stakeholder engagement helps reduce the chance for surprises. It helps identify potential risks, costs, and benefits that might otherwise be missed.

Fifth, as a result of stakeholder engagement, stakeholders can acquire a better understanding of the reasoning upon which decisions are based. Stakeholders can acquire detailed information about alternative approaches and about the complex factors that influence decisions. Knowledge about critical thinking plays a central role in stakeholder choices to support or oppose policies and decisions.[23] Critical thinking is the unbiased analysis of facts, data, or evidence to form a judgment about an issue or situation. Its purpose is to evaluate factual information without the influence from personal feelings, opinions, or biases.

Sixth, stakeholder engagement generates ownership of decisions. Ownership, in turn, can enhance the sustainability of results and outcomes.

Seventh, stakeholder engagement in its many forms has been linked to a variety of positive outcomes on the individual level (e.g., job satisfaction, organizational commitment, and job performance) and the organizational level (e.g., profitability, turnover, and sales).[24]

5.5 Limitations and Challenges of Stakeholder Engagement

There are several important limitations and challenges to stakeholder engagement. Several of the most important are identified below.

First, stakeholder engagement can be expensive, resource intensive, and time consuming. Stakeholder engagement activities – be they newsletters, social media postings, open houses, public meetings, expert availability workshops, information exchanges, neighborhood meetings, and advisory committees – all have costs, including money, resources, and time. These costs can be substantial when activities are planned out and done well. In some cases, these costs can be reduced by the use of digital technologies and social media platforms.

Second, stakeholder engagement activities pose challenges in that they require skilled staff to plan, implement, supervise, coordinate, monitor, and evaluate activities.

Third, stakeholder engagement activities often result in only limited participation by stakeholders. Those who engage and participate may only be those most focused on the issue, most upset, worried, concerned, or angry. It is often difficult to engage new people and a diversity of voices. Once again, digital technologies (e.g., virtual town hall meetings and webinars), as well as social media platforms, offer opportunities to expand the number and diversity of those involved in stakeholder engagement.

Fourth, people participating in stakeholder engagement activities may have unrealistic expectations about what can or will be achieved. They can get frustrated if they expect their viewpoints will directly translate into policy and decisions. Unrealistic expectations can be addressed in part by emphasizing the primary goal of stakeholder engagement: to improve the decision-making process and create opportunities for feedback, the exchange of information, and meaningful and constructive dialogue.

Fifth, stakeholders may perceive stakeholder engagement as a delaying tactic. They may also perceive it as an unacceptable substitute for action.

Sixth, many technical, engineering, and scientific experts resist engaging with non-expert stakeholders. By disposition, many technical, engineering, and scientific experts prefer to work in environments where there are clear boundaries, and where logic and technical facts prevail. They often feel uncomfortable communicating with non-experts in emotionally charged environments. They often dislike having to listen to views they feel are not science-based.

Seventh, people and organizations may resist what they believe are attempts to take away or share power. As discussed in the case diary at the beginning of this chapter, building stakeholder engagement into an organization's value system and operations can be challenging.

5.6 Techniques and Approaches for Effective Stakeholder Engagement

In concert with the rise of community expectations that risk management authorities will engage and consult with them has been the development of techniques and methods for effective stakeholder engagement. Researchers and practitioners have documented more than 100 methods, tools, and approaches for stakeholder engagement.[25] These techniques and approaches range from focus groups and surveys to open houses and town hall meetings. Numerous guidebooks describe the logistics, benefits, costs, and challenges, and complementarity of each technique or approach. Answers to the questions in Table 5.1 help inform decisions about which technique or approach to use.

Risk-related stakeholder engagement techniques and approaches can be categorized into two basic types based on the primary flow of information: (1) information flowing from risk management organizations to stakeholders; and (2) information flowing from stakeholders to risk management organizations. Listed in Tables 5.2 and 5.3 are examples from each type in alphabetical order.

Feedback and two-way flows of information can be encouraged in virtually any type of stakeholder engagement. For example, advisory committees and workshops allow people to solve issues together and encourage a third type of flow of information and engagement: joint problem solving and collaboration.

Table 5.1 Examples of questions to address before engaging stakeholders.

---

1) EXAMLES OF WHO QUESTIONS (Questions that relate to people)
- Who needs to be on the stakeholder engagement planning team?
- Who are the target stakeholders?
- Who are the points of contact in key stakeholder groups?
- Who are the formal and informal leaders of key stakeholder groups?
- Who feels that they have a right, need, or obligation to participate and be engaged?
- Who among the stakeholders has been most actively involved?
- Who is the most influential stakeholder?
- Who should contact the target stakeholders?

(Continued)

Table 5.1 (Continued)

2) EXAMPLES OF WHAT QUESTIONS (Questions that relate to things)

- What issues and concerns are most important for stakeholder engagement?
- What issues and concerns would benefit most from stakeholder engagement?
- What is the level of controversy and emotion related to the issue for stakeholders?
- What do we want to accomplish at each stage of engagement?
- What are our stakeholder engagement objectives?
- What do stakeholders need to know to participate effectively?
- What do we hope to achieve and learn from stakeholder engagement?
- What special circumstances affect the choice of stakeholder engagement techniques?
- What stakeholder engagement techniques are likely to work best?
- What stakeholder groups are most important for each scenario, situation, or issue?
- What issues are most important to key stakeholder groups?
- What resources can stakeholder groups bring to addressing the issue?
- What are the strengths and weaknesses of different stakeholder groups?
- What credibility does the participation of stakeholder groups bring to the situation?
- What stakeholder groups are likely to commit to collaborating and cooperating?
- What stakeholder groups will or may commit to joint communications?
- What is the historical and current relationship with different stakeholder groups? (For example, is the relationship apathetic, neutral, supportive, non-supportive, critical, adversarial, or ambivalent.)
- What specific issues are likely to be points of agreement or disagreement with different stakeholder groups?
- What are stakeholder expectations regarding engagement and how can it be improved?
- What is the perceived mission, values, and goals of different stakeholder groups?
- What methods of communication are preferred by different stakeholder groups?
- What locations are available for engaging with key stakeholders?
- What issues are non-negotiable for different stakeholder groups?
- What creative methods exist for continuing engagement and face-to-face meetings in virtual spaces and environments?
- What digital meeting platforms and tools exist for stakeholder engagement?
- What are the experiences of those who have used digital meeting platforms and tools for stakeholder engagement?
- What engagement techniques and approaches exist for including those with limited technical skills?
- What platforms exist to support effective small group conversations and "break out" rooms?
- What platforms exist that include visual note-taking or collaboration?
- What platforms exist with facilitation tips and tools for effective online conversations?

3) EXAMPLES OF WHERE QUESTIONS (Questions that relate to place)

- Where do members of target stakeholder groups meet?
- Where is it best to meet with target stakeholder groups?

4) EXAMLES OF WHEN QUESTIONS (Questions that relate to time)

- When should stakeholder engagement begin?
- When should stakeholder engagement end?
- When were stakeholders last contacted?
- When are the best times to meet with stakeholders?
- When are the worst times to meet with stakeholders?
- When do members of target stakeholder groups meet with each other?

Table 5.1   (Continued)

5) EXAMPLES OF WHY QUESTIONS (Questions that relate to the reason for something)

- Why are members of target stakeholder groups interested in this issue?
- Why are we engaging stakeholders?
- Why are stakeholders silent, uncooperative, angry, frustrated, or unhappy?
- Why should stakeholder groups meet with us?
- Why should stakeholder groups not meet with us?
- Why should stakeholder groups collaborate or partner with us?
- Why should stakeholder groups not collaborate or partner with us?
- Why have stakeholders not been engaged in the past?

6) HOW QUESTIONS (Questions that relate to choices and methods)

- How do members of target stakeholder groups typically acquire information about the issue at hand?
- How can the stakeholder groups contribute to solving problems?
- How much influence do stakeholder groups have with other stakeholders?
- How will joint communications with stakeholder groups be issued?
- How should the effectiveness of stakeholder engagement be monitored and measured?
- How should target stakeholders be contacted?
- How can we improve relationships with target stakeholders?

Table 5.2   List of stakeholder engagement activities where information flows primarily or traditionally from organizations to stakeholders.

(in alphabetical order)

1) Accessible information repositories
2) Blogs
3) Billboards
4) Briefings
5) Exhibits and displays
6) Expert availability workshops
7) Feature stories placed through media outlets
8) Mailing out key technical reports or documents
9) Mass mailings (electronic or paper)
10) Media interviews and appearances on talk shows
11) Media kits
12) News conferences and media briefings
13) Newsletters (electronic)
14) Newspaper inserts
15) News releases
16) Open houses
17) Paid advertisements
18) Panels
19) Presentations to community groups
20) Public service announcements
21) Symposia
22) Social media postings
23) Town hall meetings
24) Websites

Table 5.3 List of stakeholder engagement activities where information flows from stakeholders to organizations or which create two-way communication opportunities.

(in alphabetical order)

1) Advisory committees/groups
2) Blogs
3) Citizen juries
4) Coffee klatch sessions
5) Computer-aided negotiation
6) Consensus conference
7) Expert availability sessions
8) Field trips
9) Focus groups
10) Hotlines
11) Information workshops and exchanges
12) Interviews
13) Negotiation
14) Mediation
15) Open houses
16) Protests
17) Public meetings
18) Public hearings
19) Retreats
20) Sit-ins
21) Task forces
22) Town hall meeting
23) Social media postings
24) Websites with feedback capabilities (e.g., through survey questions, feedback forms, feedback buttons, feedback incentives, and live chat)

Key decisions need to be made for each technique or approach used for stakeholder engagement. These include decisions about (1) who will be invited, (2) will the engagement activity be open or closed, (3) how will comments from people be solicited and recorded, (4) what physical and technological resources will be required for the activity, (5) what internal staff will be involved in organizing and implementing the activity, (6) what partners will be invited to the activity, (6) how frequently will activities occur, (7) how much information will be made available to participants, (8) how will consensus and agreements be achieved and measured, and (9) how will outcomes be measured, evaluated, and communicated.

5.7 Meetings with Stakeholders

Among all stakeholder engagement techniques and approaches, meetings are one of the most frequently used. Knowing how to plan and run an effective meeting is central to successful stakeholder engagement. A primary key to successful meetings is setting clear expectations for purposes and outcomes of the meeting.

Many sources provide detailed guidance on how to plan, run, and evaluate meetings with stakeholders.[26] These documents describe different types of meetings with stakeholders, including open houses, public hearings, public comment meetings, all-hands meetings, town hall meetings, briefings, question and answer sessions, panels, workshops, conferences, symposiums, and expert availability workshops. Each type of meeting format has an associated large set of benefits, costs, and risks.

Meetings with stakeholders have four primary goals: (1) to build a trusting relationship; (2) to deliver and exchange information with stakeholders to inform decision-making; (3) to answer questions; and (4) to identify, listen to, and understand stakeholder concerns and expectations early in the decision-making process.

Most guidance documents on meetings with stakeholders strongly recommend interactive meeting formats for stakeholder engagement. These formats include open houses, face-to-face meetings, and information workshops. Controlling for other variables, the more emotionally charged or technically complex the information that needs to be communicated to stakeholders, the more important it is to share and exchange the information in face-to-face or small group settings.

Successful interactive meetings with stakeholders can include activities that create a trusting environment for dialogue, enabling interactions where participants share their own perspectives and are open to and respect the existence of other, different perspectives.

5.7.1 Town Hall Meetings

For centuries, organizations, especially government agencies, have used the "town hall meeting" format for stakeholder engagement related to risk, high-concern issues, and crisis. In the traditional town hall meeting set-up, organizational representatives and experts sit at a front table, facing an audience sitting classroom style next to open microphones. Most town hall meetings are structured and formal in nature, open to all stakeholders, include presentations by authorities and experts, and include opportunities to ask questions and receive feedback. However, some town hall meetings are structured only to receive questions and comments.

When concerns are high, when trust is low, and when the situation is emotionally charged, the formal town hall meeting format can be especially challenging. For success to be achieved, town hall meetings require careful planning and implementation. They are seldom the best method for stakeholder engagement if other options are available. The negatives associated with the dynamics of a town hall meeting can be considerable. Several of the most important disadvantages of the town hall meeting format are listed in Table 5.4.

Table 5.4 Examples of disadvantages of town hall meetings for stakeholder engagement

---

Town hall meetings:

1) can be perceived by participants as "us against them," "we against they."
2) limit the number of questions and concerns stakeholders can express within the time allocated for the meeting.
3) create discomfort among those attending who are not comfortable asking questions or expressing opinions in public or in front of others (Note: This discomfort is often present when the viewpoint being expressed by a person attending the meeting is in opposition to the viewpoints expressed by authorities or other audience members. Dale Carnegie observed over 100 years ago that public speaking is one of humankind's greatest fears.).
4) do not allow stakeholders to discuss topics in depth or at a level consistent with their knowledge and expertise.
5) can be difficult to control, especially when emotions are high.
6) have the potential to encourage feedback from stakeholders that is not constructive, even aggressive, adversarial, emotionally charged, uncivil, or impolite.
7) can result in the oversimplification of complex topics in the attempt to make information understandable to all.
8) can be hijacked and exploited for grandstanding or the pursuit of hidden or ulterior motives and agendas.

Despite these challenges and disadvantages, the formal town hall meeting still has a place in the portfolio of stakeholder engagement techniques. As shown in Table 5.5, the town hall meeting format has several benefits.

Table 5.5 Examples of benefits of town hall meetings for stakeholder engagement.

Town hall meetings provide those stakeholders attending opportunities to:
1) receive the same information at the same time, including information about goals, values, and actions.
2) acquire firsthand, up-to-date information, thereby avoiding having to receive the information secondhand.
3) ask questions and provide feedback.
4) voice their concerns in front of others.
5) participate through internet-based meeting technologies.
6) have access to and interact directly with leaders, risk managers, and experts.
7) hear and react to the views of other stakeholders.
8) feel they are participating in the decision-making process and democracy in action.

5.7.2 Open House Meetings/Information Workshops

The open house format with poster stations and information workshops is often far more effective than town hall meetings as it enables stakeholder engagement and constructive dialogue while avoiding the many disadvantages of town hall meetings. Figure 5.2 – a ground level and bird's eye view of an open house – illustrates configurations of an open house meeting with poster stations.

If needed, the open house format and the town hall meeting format can be combined in one meeting. For example, the open house can be conducted before and after a town hall meeting.

5.7.3 Tips for Meetings with Stakeholders

Here are proven tips for increasing the success of meetings with stakeholders.

Do:
1) Hold meetings at times convenient for stakeholders.
2) Hold meetings in locations convenient and easily accessible to stakeholders, including the differently abled.
3) Provide announcements in advance of the meeting describing the timing and format of the meeting.
4) Avoid holding meetings at times of the year that are inconvenient, such as holidays, vacation periods, or dates that compete with other major events.
5) Hold meetings in locations that have adequate, convenient, well-lighted parking, can accommodate more attendees than expected, and have adequate lighting and ventilation.
6) Plan, prepare, and practice delivering messages, making presentations, and showing visuals.
7) Train presenters in verbal and nonverbal risk, high-concern, and crisis communication skills, including cultural competence.
8) Train presenters on how to handle emotionally charged, difficult situations (such as disruptions) and difficult questions.

Don't:
1) Use meetings as the *only* way to engage stakeholders.
2) Attempt to conduct a meeting without practice and a dry run.
3) Forget to anticipate and prepare for questions and emotional responses.

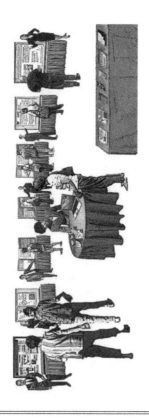

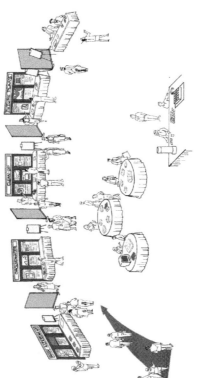

Figure 5.2 A ground level and bird's eye view of an open house meeting with poster stations.

There are additional tips for successful town hall meetings:

1) Schedule the town hall meeting to last no longer than 1.5 hours to 2 hours.
2) Have a strong moderator or facilitator who sets out ground rules and codes of conduct designed to achieve orderly meetings with maximum fairness to all participants.
3) Have unobtrusive security present.
4) Start the meeting on time.

There are additional tips for successful open house meetings:

1) Have a minimum of three poster stations and a maximum of seven poster stations (Note: A poster station consists of one to three posters. It ideally also includes a presenter to explain what is on the posters and handouts.)
2) Have three key messages for each poster.
3) Have visuals on each poster to maximize attention, understanding, and recall (e.g., maps, photographs, charts, and other graphics).
4) Consider having a pre-meeting with representatives of the media and politicians.
5) Invite partners to have their own poster stations.
6) Have simple refreshments available.
7) Greet people as they arrive.
8) Have rovers available to intervene if needed, such as when there is crowding.
9) Conduct rehearsals and dry runs, including all information providers and those who manage logistics.

5.8 Chapter Resources

Andrews, R. (1999). *Managing the Environment, Managing Ourselves: A History of American Environmental Policy*. New Haven, CT: Yale University Press.

Arnstein S. (1969). "Eight rungs on the ladder of citizen participation." *American Institute of Planning Journal* 35:216–224.

Balog-Way, D., McComas K., Besley J. (2020). "The evolving field of risk communication." *Risk Analysis* 40 (S1):2240–2262.

Bansal, P., Bingemann, T.A., Greenhawt, M., Mosnaim, G., Nanda, A., Oppenheimer, J., Sharma, H., Stukus, D., Shaker, M. (2020). "Clinician wellness during the COVID-19 pandemic: extraordinary times and unusual challenges for the allergist/immunologist." *Journal of Allergy and Clinical Immunology* 8(6):1781–1790.

Bidwell, D., Schweizer P.J. (2020). "Public values and goals for public participation." *Environmental Policy and Governance* 31 (4):257–269.

Callahan, K. (2007). "Citizen participation: Models and methods." *International Journal of Public Administration* 30 (11):1179–1196.

Chess C. (1999). "A model of organizational responsiveness to stakeholders." *Risk: Issues Health, Safety and Environment.* 10:257–267.

Chess C., Purcell K. (1999). "Public participation and the environment: do we know what works?" *Environmental Science and Technology* 33:2685–2692.

Cho, H. T. Reimer, McComas K., eds. (2015). *The SAGE Handbook of Risk Communication*. Thousand Oaks, CA: Sage.

Conrad, E., Cassar, L.F., Jones, M., Eiter, S., Izaovičova, Z., Barankova, Z.,Chriestie, M., Fazey, I. (2011). "Rhetoric and reporting of public participation in landscape policy." *Journal of Environmental Policy and Planning* 13(1):23–47.

Creighton, J.L. (2005). *The Public Participation Handbook: Making Better Decisions Through Citizen Involvement*. San Francisco, CA. John Wiley and Sons.

Devine-Wright, P. ed. (2011). *Renewable Energy and the Public: From NIMBY to Participation*. London: Earthscan.

Dryzek, J. (2005). *The Politics of the Earth: Environmental Discourses*. 2nd ed. Oxford: Oxford University Press.

Eiter, S., Lange Vik, M. (2014). "Public participation in landscape planning: Effective methods for implementing the European landscape convention in Norway." *Land Use Policy* 44:44–53.

Faga, B. (2006). *Designing Public Consensus: The Civic Theater of Community Participation for Architects, Landscape Architects, Planners and Urban Designers*. Hoboken, NJ: John Wiley and Sons.

Fagotto, E., Fung, A. (2006). "Empowered participation in urban governance." *International Journal of Urban and Regional Research*. 30(3):638–655.

Fast, S., Mabee, W. (2015). "Trust-building and place-making: the influence of policy on host community responses to wind farms." *Energy Policy* 81:27–37.

Fidler, C., Hitch, M. (2007). "Impact and benefit agreements: a contentious issue for environmental and aboriginal justice." *Environments Journal* 35(2):49–69.

Fischer, F. (2000) "Democratic prospects in an age of expertise.*" Citizens, experts and Environment*. Durham, NC: Duke University Press.

Flynn, J., Slovic, P., Mertz, C. (1994). "Decidedly different: expert and public views of risks from a radioactive waste repository." *Risk Analysis* 6:643–648.

Freudenburg, W. R. and Pastor, S. K. (2010). *Strategic Management: A Stakeholder Approach*. Cambridge: Cambridge University Press.

Freudenburg, W., Pastor, S. (1992.) "NIMBYs and LULUs: stalking the syndromes." *Journal of Social Issues* 48(4):39–61.

Funtowicz, S., Ravetz, J. (1993). "Science for the post-normal age." *Futures*. September: 739–755.

Greenberg M. (1998). "Understanding the civic activities of the residents of inner-city neighborhoods: two case studies." *Urban Geography* 19:68–76.

Greenberg M, Burger J, Gochfeld M, Kosson, D, Lowrie K, Mayer H, Powers CW, Volz CD, Vyas V. (2005). "End-state land uses, sustainably protective systems, and risk management: a challenge for remediation and multigenerational stewardship." *Remediation* 17:91–105.

Greenberg M, Burger J, Powers C, Leschine T, Lowrie K, Friedlander B, Faustman E, Griffith W, Kosson D. (2002). "Choosing Remediation and Waste Management Options at Hazardous and Radioactive Waste Sites." *Remediation* 13:39–58.

Greenberg M, Lowrie K, Burger J, Powers C, Gochfeld M, Mayer H. (2007). "Preferences for alternative risk management policies at the united states major nuclear weapons legacy sites." *Journal of Environmental Planning and Management* 50:187–209.

Greenberg, M., Lowrie, K., Burger, J., Powers, C., Gochfeld, M., Mayer, H. (2007). "Preferences for alternative risk management policies at the United States major nuclear weapons legacy sites." *Journal of Environmental Planning and Management* 50:187–209.

Glucker A, Driessen P, Kolhoff A, Runhaar H. (2014). "Public participation in environmental impact assessment: why, who and how?" *Environmental Impact Assessments Review* 43:104–111.

Kunreuther H, Easterling D, Desvousges W, Slovic P. (1990). "Public attitudes toward siting a high level nuclear waste repository in Nevada." *Risk Analysis* 10:469–484.

Innes, J.E. (2004). "Consensus building: clarifications for the critics." *Planning Theory and Practice* 3(1):5–20.

Innes, J., Booher, D. (2004). "Reframing public participation: strategies for the 21st century." *Planning Theory and Practice* 5(4):419–436.

Institute for Local Government (2005). *Getting the Most out of Public Hearings: Ideas to Improve Public Involvement.* Accessed at: http://www.ca-ilg.org/sites/main/files/file-attachments/2005_-Getting_the_Most_out_of_Public_Hearings-w.pdf.

Karl, H.A., Susskind, L.E., Wallace, K.H. (2007.) "A dialogue, not a diatribe: effective integration of science and policy through joint fact finding." *Environment: Science and Policy for Sustainable Development* 49(1):20–34.

Kasperson, R.E. (1986). "Six propositions on public participation and their relevance for risk communication." *Risk Analysis* 6(3):275–281.

Klinke, A., Renn O. (2019). "The coming of age of risk governance." *Risk Analysis* 41(3):544–557.

Laird, F.N. (1989). "The decline of deference: The political context of risk communication." *Risk Analysis* 9(2):543–550.

Maier, K. (2001). "Citizen participation in planning: Climbing the ladder." *European Planning Studies* 9(6):707–719.

Mitchell J. (1992). "Perception of risk and credibility at toxic sites." *Risk Analysis* 1:19–26.

Mitchell R.G., Peterson D., Rousch D., Brooks R.W., Paulus L.R., Martin D.B., Lantz B.S. (1997). *Idaho National Engineering and Environmental Laboratory Site Environmental Report.* Idaho Falls, ID, Environmental Science and Research Foundation.

National Research Council. (2008). *Public Participation in Environmental Assessment and Decision-Making.* Washington (DC): National Academies Press.

Navy and Marine Corps Public Health Center (NMCPH). (2017). *Guide to Public Meetings.* Accessed at: https://www.med.navy.mil/sites/nmcphc/Documents/environmental-programs/risk-communication/Appendix-G-Guide-to-Public-Meetings.pdf

Nichols, T. (2017). *The Death of Expertise: The Campaign against Established Knowledge and Why it Matters.* Boston, MA, Oxford University Press.

O'Faircheallaigh, C. (2010). "Public participation and environmental impact assessment: purposes, implications, and lessons for public policy making." *Environmental Impact Assessments Review* 30:19–27.

Parkins, J., Mitchell, R. (2005). "Public participation as public debate: a deliberative turn in natural resource management." *Society and Natural Resources* 18(6):529–540.

Petts, J., Leach, B. (2000). *Evaluation Methods for Public Participation: A Literature Review.* Bristol, UK. Environmental Agency.

Pew Research Institute. (2017). *Public Trust in Government 1958-2017.* Accessed at: http://www.people-press.org/2017/05/03/public-trust-in-government-1958-2017

PlannersWeb. (2014). *Holding Effective Meetings.* Accessed at: http://plannersweb.com/2014/10/holding-effective-public-meetings/

Presidential/Congressional Commission on Risk Assessment and Management (PCCRAM) (1997). *Report of Presidential/Congressional Commission on Risk Assessment and Management.* Washington, DC: US Government Printing Office.

Reynolds, B., Seeger, M.W. (2005). "Crisis and emergency risk communication as an integrative model." *Journal of Health Communication* 10:43–55.

Rowe, G., Frewer, L. (2005). "Public participation methods: a framework for evaluation." *Science, Technology, and Human Values* 25(3):317–348.

Ruckelshaus, W.D. (1983). "Science, risk, and public policy." *Science* 221(4615):1026–1028.

Sadar, A.J., Shull, M.D. (2000). *Environmental Risk Communication: Principles and Practice for Industry.* Boca Raton, FL: CRC Press.

Schmeer, K (1999). *Guidelines for Conducting a Stakeholder Analysis.* Bethesda, MD: Partnerships for Health Reform, Abt Associates Inc. Accessed at: https://www.who.int/management/partnerships/

overall/GuidelinesConductingStakeholderAnalysis.pdf and at https://www.who.int/
workforcealliance/knowledge/toolkit/33.pdf

Senge, P.M. (1990). *The Fifth Discipline: The Art and Practice of the Learning Organization*. New York: Doubleday

Siegrist, M., Cvetkovich, G. (2000). "Perception of hazards: the role of social trust and knowledge." *Risk Analysis* 20:713–719.

Slovic, P. (1987). "Perception of risk." *Science, New Series* 236(4799):280–285.

Slovic, P. (1993). "Perceived risk, trust, and democracy." *Risk Analysis* 13:675–682.

Slovic, P., Layman, M., Flynn, J. (1991). "Risk perception, trust, and nuclear waste: Lesson from Yucca Mountain." *Environment: Science and Policy for Sustainable Development* 33(3):6–30.

Trettin, L., Musham, C. (2000). "Is trust a realistic goal of environmental risk communication?" *Environmental Behavior* 32:410–426.

US Environmental Protection Agency (2001). *Stakeholder Involvement & Public Participation at the US: EPA Lessons Learned, Barriers, & Innovative Approaches*. Policy Document EPA-100-R-00-040. Washington, DC. US Environmental Protection Agency. Accessed at: https://www.epa.gov/sites/production/files/2015-09/documents/stakeholder-involvement-public-participation-at-epa.pdf

US Navy and Marine Corps Public Health Center, *Guide to Public Meetings*. Accessed at: https://www.med.navy.mil/sites/nmcphc/Documents/environmental-programs/risk-communication/Appendix-G-Guide-to-Public-Meetings.pdf

Webler, W. and S. Tuler (2021). "Unpacking the idea of democratic community consent-based siting for energy infrastructure." *Journal of Risk Research* 24(1):94–109.

Webler, W. and S. Tuler (2021). "Four decades of public participation in risk decision making." *Risk Analysis* 41(3):503–518.

Young, S. F., L.A. Steelman, M.D. Pita, and J. Gallo, J. (2020). "Role-based engagement: scale development and validation." *Journal of Management & Organization*. Published online by Cambridge University Press: 16 November 2020: 1–21.

Endnotes

1 See, e.g., Freeman, R.R. (2010). *Strategic Management: A Stakeholder Approach*. Cambridge: Cambridge University Press.

2 Ruckelshaus, W.D. (1983). "Science, risk, and public policy." *Science* 221(4615):1026–1028.

3 See, e.g.,Taylor, C., Lynn, P., and Bartlett, J. (2019). *Fundamentals of Nursing: The Art and Science of Person-Centered Care*. 9th edition. New York: Wolters Kluwer; Centers for Medicare & Medicaid Services (2018). *Person and Family Engagement Strategy: Sharing with our Partners*. Accessed at: https://www.cms.gov/Medicare/Quality-Initiatives-Patient-Assessment-Instruments/QualityInitiativesGenInfo/Downloads/Person-and-Family-Engagement-Strategy-Summary.pdf;Brownie, S., Nancarrow, S. (2013). "Effects of person-centered care on residents and staff in aged-care facilities: A systematic review." *Clinical Interventions in Aging* 8:1–10; The Health Foundation (2016). *Person-Centred Care Made Simple: What Everyone Should Know about Person-Centred Care*. Accessed at: https://www.health.org.uk/publications/person-centred-care-made-simple; Sillner, A.Y., Madrigal, C., and Behrens, L. (2021). "Person-centered gerontological nursing: An overview across care settings." *Journal of Gerontological Nursing* 47(2):7–12.

4 Institute of Medicine (2001). *Crossing the Quality Chasm: A New Health System for the 21st Century*. Washington, DC: National Academies Press

5 Darzi A. (2008). *High Quality Care for All: National Health Service (NHS) Next Stage Review Final Report*. Norwich: TSO

6 See, e.g., The King's Fund (2012). *Leadership and Engagement for Improvement in the NHS: Together We Can*. London: The King's Fund. See also Brownie, S., Nancarrow, S. (2013). "Effects of person-centered care on residents and staff in aged-care facilities: A systematic review." *Clinical Interventions in Aging* 8:1–10.

7 Nichols, T. (2017). *The Death of Expertise: The Campaign against Established Knowledge and Why it Matters*. Boston: Oxford University Press; See also Laird, F.N. (1989). "The decline of deference: The political context of risk communication." *Risk Analysis* 9(2):543–550.

8 Source: https://news.gallup.com/poll/236243/military-small-business-police-stir-confidence.aspx

9 Pew Research Institute. (2017). *Public Trust in Government 1958-2017*. Accessed at: http://www.people-press.org/2017/05/03/public-trust-in-government-1958-2017

10 See, e.g., https://www.pewresearch.org/politics/2020/09/14/americans-views-of-government-low-trust-but-some-positive-performance-ratings/

11 See, e.g., https://news.gallup.com/poll/1597/confidence-institutions.aspx

12 See, e.g., https://www.edelman.com/trust/2021-trust-barometer

13 https://www.timeshighereducation.com/news/global-poll-shows-only-18-cent-have-high-trust-scientists

14 See, e.g., Devine, D., J. Gaskell, W. Jennings, and G. Stoker (2020). "Trust and the Coronavirus Pandemic: What are the Consequences of and for Trust? An Early Review of the Literature." *Political Studies Review* 1-12. Accessed at: https://journals.sagepub.com/doi/pdf/10.1177/1478929920948684

15 Senge, P.M. (1990). *The Fifth Discipline: The Art and Practice of the Learning Organization*. New York: Doubleday

16 Schmeer, K (1999). *Guidelines for Conducting a Stakeholder Analysis*. Bethesda, MD: Partnerships for Health Reform, Abt Associates Inc. Page 3. Accessed at: https://www.who.int/workforcealliance/knowledge/toolkit/33.pdf

17 Conrad, E., Cassar, L. F., Jones, M., Eiter, S., Izaovičova, Z., Barankova, Z., Chriestie, M., and Fazey, I. (2011). "Rhetoric and reporting of public participation in landscape policy." *Journal of Environmental Policy and Planning*. 13(1):23–47; Eiter, S., Lange Vik, M. (2014). "Public participation in landscape planning: Effective methods for implementing the European landscape convention in Norway." *Land Use Policy* 44:44–53.

18 Creighton, J. L. (2005). *The Public Participation Handbook: Making Better Decisions Through Citizen Involvement*. San Francisco. John Wiley and Sons.

19 Arnstein S. (1969). "Eight rungs on the ladder of citizen participation." *American Institute of Planning Journal* 35:216.

20 Arnstein S. (1969). "Eight rungs on the ladder of citizen participation." *American Institute of Planning Journal* 35:216.

21 IAP2 Spectrum of Public Participation. IAP2's Spectrum of Public Participation is designed to assist with the selection of the level of participation that defines the public's role in any public participation process. The IAP2 Spectrum is found in public participation plans around the world and has evolved over time, putting great emphasis on the importance of evidence-based techniques and methods for listening and engaging stakeholders through meaningful dialog and deliberation. The Spectrum can be accessed at: https://cdn.ymaws.com/www.iap2.org/resource/resmgr/pillars/Spectrum_8.5x11_Print.pdf

22 See, e.g., Tamarack Institute (2020). *Index Of Community Engagement Techniques*. Waterloo, Ontario: Tamarack Institute, University of Waterloo. Accessed at: https://cdn2.hubspot.net/hubfs/316071/Resources/Tools/Index%20of%20Engagement%20Techniques.pdf?__hstc=16332726

7.49b1d0b32c3c682873a2744e8e298d14.1618516171484.1618516171484.1618516171484.1&__hss
c=163327267.2.1618516171484&__hsfp=3461249301&hsCtaTracking=c
ee0990e-2877-474b-93f7-c21defcae9b5%7C0769d43e-10f2-41a2-ab08-4c5c9fc8c4ba

23 See, e.g., Pherson, K.H., and Randolf, H. (2021). *Critical Thinking for Strategic Intelligence.* 3rd
    edition. Thousand Oaks, CA: CQ Press/Sage; see also Fisher, A. (2011). *Critical Thinking.*
    Cambridge: Cambridge University Press

24 See., e.g., Macey, W., and B. Schneider, B. (2008). "The Meaning of Employee Engagement."
    Industrial and Organizational Psychology, 1(1): 3–30.

25 See, e.g., Creighton, J. L. (2005). *The Public Participation Handbook: Making Better Decisions Through
    Citizen Involvement.* San Francisco. John Wiley and Sons; Rowe, G., Frewer, L. (2005). "Public
    participation methods: a framework for evaluation." *Science, Technology, and Human Values*
    25(3):317–348; the publications of the International Association for Public Participation Practitioners
    (https://iap2usa.org); and the many public participation and community involvement tools published
    by the US EPA, such as those accessed at: https://www.epa.gov/superfund/superfund-community-
    involvement-tools-and-resources; See also Young, S. F., L.A. Steelman, M.D. Pita, and J. Gallo,
    J. (2020). "Role-based engagement: scale development and validation." *Journal of Management &
    Organization.* Published online by Cambridge University Press: 16 November 2020: 1-21

26 See, e.g., Creighton, J. L. (2005). *The Public Participation Handbook: Making Better Decisions
    Through Citizen Involvement.* San Francisco. John Wiley and Sons; Institute for Local Government
    (2005). Getting the Most out of Public Hearings: Ideas to Improve Public Involvement. Accessed
    at: http://www.ca-ilg.org/sites/main/files/file-attachments/2005_-_Getting_the_Most_out_of_
    Public_Hearings-w.pdf; Navy and Marine Corps Public Health Center (NMCPH). (2017). Guide to
    Public Meetings. Accessed at: http://www.med.navy.mil/sites/nmcphc/Documents/
    environmental-programs/risk-communication/Appendix_G_NMCPHCGuidetoPublicMeetings.pdf

6

Communicating in a Crisis

CHAPTER OBJECTIVES

This chapter covers the principles and execution of communications needed in the three phases of crisis communication: precrisis/preparedness, crisis/response, and postcrisis/recovery. At the end of this chapter, you will be able to:

- remember how to develop a crisis communication plan, understanding all the necessary components of that plan;
- identify the best spokesperson who should be prepared for a crisis;
- define the principles of emergency message design;
- explain when issues of blame, responsibility, and apology are relevant to a crisis situation and how to address them; and
- remember what elements are needed for communication in a postcrisis situation.

> *The best way to manage a crisis is to prevent one. The second-best way to manage a crisis is to prepare for one.*
>
> —Author, unknown

> *Almost every crisis contains within itself the seeds of success as well as the roots of failure. Finding, cultivating, and harvesting that potential success is the essence of crisis management. And the essence of crisis mismanagement is the propensity to take a bad situation and make it worse.*
>
> —Norman R. Augustine

This chapter summarizes the crisis communication literature as it applies to the three phases of the life cycle of a crisis: *the precrisis preparedness phase, the crisis response phase, and the postcrisis recovery phase.* As the Centers for Disease Control and Prevention points out, understanding the pattern of a crisis can help communicators anticipate problems and respond effectively. For communicators, it is vital to know that every emergency, disaster, or crisis evolves in phases. Communication, too, must evolve through these changes.[1]

Communicating in Risk, Crisis, and High Stress Situations: Evidence-Based Strategies and Practice, First Edition. Vincent T. Covello.
© 2022 by The Institute of Electrical and Electronics Engineers, Inc. Published 2022 by John Wiley & Sons, Inc.

6.1 Case Diary: The Challenge of Partnership in a Crisis

My team members and I were being briefed by emergency responders in Asia. The atmosphere in the room was tense. The briefers told us the situation was dire. Because of an explosion at an industrial plant, toxic gases were escaping, and the situation was getting worse by the day as explosions were still occurring. Additional gas releases could be stopped only by turning off equipment located deep inside the basement of the plant.

The briefers told us they had warned the plant manager and workers that entry into the plant was exceedingly dangerous. Workers would be at risk of exposure to high levels of toxic substances that would likely cause serious health problems and even death.

In response to the warning, and to the surprise of the briefers, the plant manager and several dozen workers volunteered to return to the site to do the needed shutdown tasks. They had jointly decided they had no choice but to return to the site and regain control over the stricken plant. They would work together in shifts. As the plant manager and workers left the crisis management headquarters, company executives, employees, firefighters, police officers, and government officials lined up to salute them.

Upon returning to the site, the plant manager and workers found the situation to be even worse than anticipated. The scale of the disaster left the workers short of personal protective gear and vital emergency equipment. What little equipment was found at the site had been destroyed or badly damaged in the explosion.

Our team communicated to the company that we possessed the needed equipment, and we could get it to the workers quickly. To our surprise, the company summarily rejected our offer, with no explanation.

Our team huddled together to evaluate options and try to better understand the company's rejection of our offer of help. We determined that a different mode of communication could, we hoped, make a difference. We identified cultural and political considerations as key. Several of us had spent significant amounts of time in the Asian country and were well familiar with the culture and political context.

We hypothesized that company officials were likely deeply offended by widely publicized negative comments by critics from other nations about their safety policies and emergency response efforts. The officials were also likely to be upset by the lack of acknowledgment of their technical accomplishments, industrial safety record, and the dedication of their workforce. Because of these perceived insults, officials were likely feeling that accepting technical assistance from outsiders would be humiliating. It would cause a loss of face and honor, critically important values in their country's culture.

We drafted a memorandum containing three messages and sent it to the company officials.

The first message congratulated the plant manager and workers who had volunteered to return to the site of the explosion for their extraordinary acts of valor, courage, and sacrifice. We noted that few organizations in the world could gain such dedication from workers. We also cited parallel acts of valor, courage, and sacrifice by citizens in their nation as recounted in many films and theatrical performances.

The second message congratulated the company for its technical achievements, safety record, and dedicated workforce.

The third message congratulated emergency responders for their teamwork and heroic efforts.

The memorandum finished with a request by our group to become a part of the emergency response team. The company accepted the offer. Their acceptance memorandum was followed shortly by another memorandum listing the emergency equipment our team members might bring with them. The equipment enabled the heroic workers to avoid the most serious hazards.

This case diary illustrates the importance of communication and coordination among partners in a crisis. It also illustrates one of the bedrock principles of crisis communication: for stakeholders, including those who are partners in the response effort, trust is at the heart of hearing, believing, and acting. If partners are not convinced you have their best interests in mind, they are unlikely to hear your message, and they are even less likely to cooperate and engage in constructive dialogue. Building, strengthening, and, if necessary, repairing trust is an immutable prerequisite for successful crisis communication. Building trusted relations with partners and stakeholders well before a potential event would have enabled a much faster response to an urgent situation.

6.2 The Three Phases of a Crisis

There are three phases of the life cycle of a crisis: the *precrisis preparedness phase*, the *crisis response phase*, and the *postcrisis recovery phase*. Although these phases are often presented as a linear progression, the reality is that they intersect continuously with one another through a complex network of feedback loops.

In the *precrisis preparedness phase*, skillful communication promotes constructive dialogue, engagement, and awareness. In the *crisis response phase*, skillful communication assists people in making informed and responsible decisions. In the *postcrisis recovery phase*, skillful communication creates the shared knowledge needed to recover, be resilient, evaluate, and learn from mistakes.

The topics your team may discuss in each phase will vary to meet these communication goals. As an example, Tables 6.1, 6.2, and 6.3 list various communication topics for hurricanes. I developed this list of message topics and the resulting messages as part of a joint crisis communication effort sponsored by the government of Puerto Rico and the Centers for Disease Control and Prevention following the devastation caused by Hurricane Maria in 2017.

What happens during each phase of a crisis is difficult to predict and flexibility is essential. However, proactive communication strategies can be planned for each phase. Planning increases the likelihood that messaging, and the larger communication effort, will contribute positively to building trust, informing decisions, and gaining agreement. Well conceptualized and practiced communications also reduce the spread of rumors and misinformation. For example, during Hurricane Maria in Puerto Rico rumors and misinformation spread rapidly about every topic in Tables 6.1, 6.2, and 6.3.

The Chinese have a word for the crisis that some scholars describe as made up of two words: *danger* and *opportunity*. Independent of debates among Chinese linguistic scholars about the exact meaning and origins of the two words, a *crisis*, and preparing for a *crisis,* can present opportunities to learn more about prevention, preparedness, protection, response, recovery, and resilience.

Table 6.1 Phase-based public health messaging for a hurricane: precrisis/preparedness phase.

Preceding landfall: messages focused on readiness and prevention

- Evacuation guidance
- Flood safety
- Power outage risks
- Household preparation for those sheltering in place (e.g., restocking their kits, checking generators)
- Securing homes
- Developing and communicating a hurricane response plan with family members, friends, and neighbors

---

Table 6.2 Phase-based public health messaging for a hurricane: crisis/response phase.

Storm and poststorm: messages focused on protection

- Power outage and electrical risks
- Carbon monoxide risks
- Food risks
- Water risks
- Injury risks
- Cleaning and sanitation risks
- Medication and medication storage risks

Table 6.3 Phase-based public health messaging for a hurricane: postcrisis/recovery phase.

Post storm: messages for recovery and resilience

- Mental health, including coping with trauma
- Mold remediation
- Rebuilding and reconnecting
- Consultation services
- Protection from exposure to electricity
- Protection from exposure to toxic substances
- Protection from disease carrying mosquitos
- Cleanup safety precautions
- Volunteering
- Recovery assistance
- Transportation
- Sheltering/evacuation
- Power outages
- Social service needs
- Finding gasoline
- Location of charging stations
- Donations
- Rumors
- Staying in touch with family
- Locating loved ones
- Medical supplies
- Food safety
- Safe Water
- Drugs exposed to water
- Looting
- Prescription safety
- Flood water safety
- Driving through water

Crisis communicators focus primarily on what just happened or is still happening, while risk communicators primarily focus on what might happen. For example, for an accident scenario, risk communicators might focus on questions such as how likely is an accident? How can we alert and make people aware of the potential for an accident to occur? What might be the consequences of an accident? How might accidents be prevented? For the same scenario, crisis communicators will typically focus on the incident or event itself, collecting and verifying as many facts as possible. What happened and where? When did it happen? Who was involved? Who is responding? How did it happen? How did it unfold? What is currently being done? What went wrong? What else might go wrong? What is being done to ensure it does not happen again? Who is to blame? What lessons have been learned? As the crisis unfolds, crisis communicators will continually be asking: Do we have all available facts? What other information do we need to put the incident or event into perspective? Have facts about the situation been confirmed? How credible is available information? Is information coming from different sources consistent? What do we want people to know? What do people already know? What are people sharing with others? What do people want to know? What things should people be doing? Where can people go for trusted information? As the crisis subsides, crisis managers and communicators will be asking: What have we learned that can prevent future incidents and build resilience? How can we best address after-effects such as damage and mental health problems? How can we improve our response and communication preparedness?

In crisis situations, the human brain goes through a five-step sequential cognitive process that shapes knowledge, attitudes, perceptions, intentions, actions, and behaviors: (1) hearing, (2) understanding, (3) believing, (4) evaluating, and (5) deciding. Crisis communication needs to address what people will hear, understand, believe, and do.

6.3 Communication in the *Precrisis Preparedness Phase*

Next week there can't be any crisis. My schedule is already full.

—Henry Kissinger

You don't climb mountains without being fit.
You don't climb mountains without being prepared.
You don't climb a mountain by accident.
It has to be intentional.

—Mark Udall, US Senator (Colorado)

Communication in the *precrisis preparedness* phase is focused on preventing, avoiding, and preparing for a crisis. A primary goal of *precrisis* communication is to build trusting relationships, provide the knowledge needed to increase awareness, create appropriate levels of concern, encourage appropriate protective behaviors, and increase resilience.

Communication in the precrisis phase is most effective when grounded in strategic preparation. As shown in Table 6.4, strategic crisis communication preparation has three key elements: anticipation, preparation, and practice.

Anticipation means identifying what can go wrong, who will be involved or affected, and what questions and concerns will be raised.

Table 6.4 Strategic crisis communication in the pre-crisis/preparedness phase: the APP (anticipate, prepare, practice) strategy.

Anticipate

– Potential crises and scenarios
– Stakeholders
– Questions and concerns

Prepare

– A crisis communication plan
– Messages
– Messengers (spokespersons)
– Means (communication channels)

Practice

– Test messages
– Test messengers
– Test means/channels for communication
– Rehearse and test the crisis communication plan through role-plays, drills, and exercises

---

Preparation means developing a crisis communication plan, developing prioritized lists of anticipated questions, developing answers to anticipated questions, training spokespersons and team members, and choosing the most effective ways to deliver and exchange information.

Practice means testing messages, messengers, and methods for communication; delivering messages; and rehearsing the crisis communication plan through role-plays, dry runs, simulations, drills, and exercises, such as table-top, games, and operational simulations.

A key decision needed in the *precrisis preparedness phase* is how much to anticipate, prepare, and practice. For example: Which crisis scenarios are most important? Should only high-probability, high-consequence scenarios be considered? Who are the most important stakeholders? How comprehensive should the list of anticipated questions be? Which questions should be prioritized and receive the most attention? How frequently should the crisis communication plan be updated and practiced?

Answers to these and related questions depend on existing and available financial and other resources. Answers to these questions can also be determined through formal diagnostic tools, both: (1) *assessment* tools and (2) *risk perception* tools.

Assessment tools help determine, from a technical perspective, which crisis scenarios are most important. For example, quantitative risk assessment tools measure (1) the *probability* of a specific crisis scenario – how likely the crisis is to occur, and (2) the *magnitude* of a specific crisis scenario – how bad the outcomes or consequences of the crisis are. Formal risk assessment tools to quantitatively measure probability and magnitude are described in detail in the textbook by Covello and Merkhofer, *Risk Assessment Methods*.[2] Researchers have also determined other ways to identify potential crisis scenarios. For example, Mitroff (2004) suggested preparing for six different types of crises including economic, informational, physical, human resource, psychopathic acts, and natural disasters.[3]

Risk perception tools are a second way to determine which crisis scenarios are most important. Risk perception tools measure the psychological and emotional impacts of a crisis. They help crisis managers and communicators assess how much concern, fear, stress, worry, upset, and outrage are likely to be caused by a crisis. Risk perceptions are determined in part by the probability and magnitude of the risks – the technical hazard itself – as well as by psychological, emotional, and sociocultural characteristics of the risk, such as whether the risk is perceived to be unfamiliar or familiar, involuntary or voluntary, and unfair or fair.

For the general public, psychological, emotional, and sociocultural factors, which can cause extreme distress, are often more important in determining the importance of a crisis than facts about the probability and magnitude of the risk.[4] Nonexperts rarely estimate and judge risks the same way as scientific and technical experts. For example, objective estimates of risk may be judged acceptable by scientific and technical experts (e.g., a risk of one in one million).[5] However, nonexperts may judge these levels as too risky and unacceptable.

Using risk perception diagnostic tools, potential crisis scenarios can be rated on scales that measure stakeholder perceptions on factors such as:

- natural vs. man-made/industrial
- familiar vs. not familiar
- not dreaded vs. dreaded
- trust in leaders and authorities vs. distrust of leaders and authorities
- personally controlled vs. controlled by others
- voluntary exposure vs. involuntary exposure
- fair vs. unfair
- certain adverse outcomes vs. uncertain adverse outcomes

Communication activities in the *precrisis/preparedness phase* revolve around the following nine activities:

1) identify potential crises
2) identify crisis communication goals and objectives
3) develop a crisis communication plan
4) identify and train the lead and backup crisis spokespersons
5) engage stakeholders
6) identify and prioritize anticipated questions and concerns
7) draft answers to prioritized anticipated questions and concerns
8) test messages, messengers, and methods for communication
9) test the crisis communication plan.

Each activity is described in the following sections.

6.3.1   Precrisis Communication Activity: Identifying Potential Crises

The primary purpose of this task is to identify what can go wrong. This means identifying vulnerabilities, credible events, and failure scenarios that may cause harm to people, organizations, property, or the environment. It means identifying adverse events, estimating the probability of events, listing possible impacts of the adverse events (e.g., financial, reputational, or health and safety), and estimating the costs and benefits of prevention.

As noted in Chapter 2, a *crisis* is *risk* manifested. A longer definition of the word *crisis* is a risk manifested that characteristically (1) is abrupt and unexpected, (2) exceeds the expectations of those affected, (3) disrupts normal processes, (4) places nonroutine and unique demands on the responding organizations, such as needing assistance external to the organization (5) produces high amounts of uncertainty, (6) challenges organizational performance, and (7) poses a significant chance of harm or loss to individuals and organizations.

Major types of crises include crises caused by natural hazards or industrial accidents; health, safety, and environmental crises; economic crises; reputational crises; political crises; crises caused by the acts of terrorists; and informational crises caused by the spread by harmful rumors or false information.

From a communication perspective, a useful tool for identifying potential crises is *signal detection*. *Signal detection* functions much like a smoke detector, delivering early warnings. Signals for a potential crisis can be identified through a variety of means, including monitoring the content and spread of information produced by traditional and social media channels; establishing a network of listening outposts; establishing open lines of communication with important stakeholders; and monitoring trends in the larger legal, political, and socioeconomic environment. Although signal detection can be helpful in identifying and communicating potential crisis scenarios, it is often not undertaken. Reasons include overconfidence, limited resources, and beliefs that crises are unavoidable, and failing to recognize it can take only one person to plunge an entire organization into a crisis.

Signals can be strong or weak. Strong signals include indicators, such as complaints and reports of near misses. Weak signals include actions and events that may go unnoticed but are at odds with established procedures, protocols, guidelines, norms, or values.[6] The explosion of the space shuttle *Challenger* is a prime example of a crisis where early warning signals and communications were ignored. The Report of the President's Commission on the Space Shuttle Accident identified numerous memos and strong signals before the event pointing out that design flaws in the rocket could cause a catastrophic failure.[7]

Mitroff (2004) identified several principles of signal detection, including: crises are preceded by early warning signals; signal detection is a direct reflection of priorities; different crises require different detectors: not all signals are alike; every signal detector needs a system for monitoring; signals have to be communicated to the right people at the right time in the right place.[8]

6.3.2   Case Study: The 2010 BP Deepwater Horizon Oil Spill

A prime example of ignoring signals and warning signs was the 2010 BP Deepwater Horizon Oil Spill in the Gulf of Mexico. BP had hundreds of safety violations over the three years before the 2010 spill.

Many things went wrong at the Deepwater Horizon oil rig in the weeks before the disaster. After nearly every disaster, investigators find warning signs that were ignored. So why did BP apparently ignore the strong signals and warning signs? That BP officials would ignore strong signals and warning signs is especially surprising given their advertised safety culture and worker risk communication policies. As noted by Sandman (2010),

> Everyone aboard Deepwater Horizon – Transocean people and Halliburton people as well as BP people – had stop work authority (SWA), a right and an obligation long enshrined in offshore drilling operations. That is, anyone who believed a practice to be dangerous was entitled to force a halt without risking retribution (at least formal retribution), even if the brass wanted the work to proceed. . ..[9]

In a 234-page report, BP described 8 main causes for the explosion, which killed 11 men and created an environmental disaster.[10] The report cited, among other causes, unusual pressure test readings as warning signs of a breach of the cement seal and casing at the bottom of the well. Tony Hayward, BP's chief executive officer at the time of the explosion, said in a statement: "To put it simply, there was a bad cement job."[11] Mark Bly, BP's head of safety and the leader of BP's internal investigation, told reporters at a press conference in Washington, DC: "Given everything that came before, there probably should have been more risk assessment. They probably should have been more careful."[12] The report showed that Transocean, the rig owner, and Halliburton, which carried out cement work related to the drilling operation, shared much of the responsibility.

Following the release of the report, Transocean and Halliburton accused BP of attempting to shift attention from its own mistakes of bad oil well design and disregard of safety procedures. Members of Congress and environmentalists also criticized the report, claiming the report was more about protecting BP's financial interests than getting to the cause of the disaster.

Even as the crisis began, BP continued to ignore strong signals and warning signs. Instead, BP offered what appeared to be overly optimistic statements and reassurances. For example, BP initially estimated that between 1,000 and 5,000 barrels of oil were spilling into the Gulf of Mexico each day. These estimates were at the extremely low range of estimates. After a month and a half of spilling oil, Deepwater Horizon had become the biggest oil spill in US history. The spill dwarfed the *Exxon Valdez* disaster of the coast of Alaska in 1989.[13]

In addition to what appeared to be obtuseness to signals and warning signs, the 2010 BP Deepwater Horizon Oil Spill is a classic case study of how not to communicate in a crisis. Crisis communication errors and mistakes were rampant not only before but also during the crisis. Most importantly, these included what appeared to be an abject lack of caring and empathy. For example, BP's chief executive officer, Tony Hayward, made the following statements:[14]

- On 3 May 2010, Mr. Haywood said: "The drilling rig was a Transocean drilling rig. It was their rig and their equipment that failed, run by their people and their processes."
- On 14 May 2010, Mr. Hayward said: "The Gulf of Mexico is a very big ocean. The amount of volume of oil and dispersant we are putting into it is tiny in relation to the total water volume."
- On 17 May, Mr. Hayward said: "I think the environmental impact of this disaster is likely to be very, very modest. It is impossible to say and we will mount, as part of the aftermath, a very detailed environmental assessment as we go forward. We're going to do that with some of the science institutions in the US. . . But everything we can see at the moment suggests that the overall environmental impact of this will be very, very modest."
- On 30 May, Mr. Hayward said: "I'm sorry. We're sorry for the massive disruption it's caused their lives. There's no one who wants this over more than I do. I would like my life back." (This statement in particular was widely criticized given that 11 workers died in the drilling platform explosion that caused the spill. Mr. Hayward later apologized for this remark.)
- On 31 May 2010, Mr. Haywood said, in response to claims about a large underwater oil plume threatening the Gulf of Mexico ecosystem: "There aren't any plumes."
- On 1 June 2010, Mr. Hayward offered a reason as an alternative to oil fumes for why cleanup workers were getting sick: "Food poisoning is clearly a big issue."
- On 17 June 2010, Mr. Hayward said, at a US Congressional hearing on the BP spill: "I can't possibly know why the decisions were made [on the rig]. . . I can't answer because I wasn't there. . .. That was a decision I was not party to."[15]
- On 19 June 2010, Mr. Hayward took part in the JP Morgan Asset Management Round the Island yacht race on the Isle of Wight, sailing on his own yacht with his crew.
- On 27 July 2010, BP announced Mr. Hayward would soon be redeployed to its operations in Siberia.[16]

It is interesting to note that in an interview before he was named BP's chief executive officer in 2007, Mr. Hayward talked about the effect the death of a worker under his command in Venezuela had on him:

> I went to the funeral to pay my respects. At the end of the service, his mother came up and beat me on the chest. "Why did you let it happen?" she asked. It changed the way I think about safety. Leaders must make the safety of all who work for them their top priority.[17]

6.3.3   Precrisis Communication Activity: Identify Goals and Objectives

Goals describe a desired end state. Objectives ideally describe measurable steps to achieve goals. Objectives are important because they focus on organizations. An effective way to set objectives is through the popular acronym SMART: specific, measurable, achievable/assignable, realistic/reasonable, and time dependent. An example of a time-dependent objective would be the percentage of the target audience that by a set time has heard a crisis communication message or has taken a recommended protective action. Crisis communication plans that have SMART objectives are more likely to be successful because they begin with clear objectives about what they are trying to achieve and how they will do it. SMART objectives can be set for all precrisis, crisis, and postcrisis communication activities. Examples of core objectives include: providing accurate, timely information to all targeted internal and external stakeholders; providing guidance on preventive and protective measures; engaging stakeholders in constructive dialogue; keeping stakeholders updated and informed; making sure employees understand key roles; and demonstrating a commitment to the well-being of all stakeholders.

One concern too often mistakenly identified as a core objective in a crisis communication plan is how to prevent panic. In the crisis phase, concerns about panic are often cited as the reason for secrecy and lack of transparency. Panic describes an intense contagious fear causing individuals to think only of themselves. Panic differs from the high levels of psychological stress, organizational stress, and fear often observed in a crisis. Despite the importance attached to preventing panic by many leaders, crisis managers, and crisis communicators, panic is largely a myth. Panic behavior is only infrequently observed in crisis situations. Except in rare situations, most people respond cooperatively and adaptively in a crisis.

The dominant human behaviors in a crisis are to help others and to seek information. For example, in the evacuation of the World Trade Center following the terrorist attack in 2001, panic was conspicuous by its absence. As Wessely (2005) points out: "A building was on fire, about to collapse and the emergency services not yet present. Yet there was no panic, but an orderly evacuation perhaps aided by pre-existing social networks."[18]

Not that panic never occurs, as it clearly does. For example, it occurs when there is a surge of people trying to exit through a narrow, single exit point from a crowded subway, theater, or arena. Panic clearly occurred in 2005 when hundreds of Hindu pilgrims attending an annual pilgrimage to a temple in India's Maharashtra district died in a stampede. There was no guidance from authorities through narrow, crowded streets as masses of people tried to escape a fire and the explosion of gas canisters in nearby shops.

As shown in Table 6.5, for panic to occur several risk factors need to be present.

Table 6.5 Factors contributing to panic.

- being in an enclosed area
- confronting bottlenecks, congestion, and dead ends
- loss of trust in leaders or crisis managers
- believing there is:
  - an immediate and clear source of death or serious injury
  - insufficient time to escape
  - only a small or no chance of escaping without injury
  - no easily identifiable or accessible escape routes
  - scarce or limited resources for assistance
  - a "first come, first served" system for assistance
  - inadequate or ineffective crisis leadership and management
  - a benefit to conforming to the panic behavior of those around you

---

In a comprehensive review of the research literature on public response to warnings of a nuclear power plant accident, Mileti and Peek point out:

> It cannot be overemphasized that the public simply does not panic in response to warning of impending disasters, including nuclear power plant accidents. This myth is largely the result of movie producers who depict masses of screaming, fleeing, and completely panicked individuals in dangerous scenarios.[19]

In summary, a central goal for crisis communicators is to give people the information they are seeking and need because people want to make informed decisions and act constructively.

6.3.4 Precrisis Communication Activity: Develop a Crisis Communication Plan

Effective crisis communication requires a written crisis communication plan that is endorsed by senior management. The plan needs to be a substantive part of the overall crisis management plan. An effective crisis communication plan allows for a proactive, quick, and effective response during a crisis. It helps expedite action because it will have ensured that many crisis communication decisions will already have been made. If carefully designed, a crisis communication plan can save precious time. It enables leaders and spokespersons to focus on the quality, accuracy, and speed of their responses. A plan also enables the organization to focus on the key planning elements, expressed by the emergency management acronym PPOST: priorities, problems, objectives, strategies, and tactics.

The crisis communication plan should be distributed to management, employees, and partner organizations as both an electronic and a hard copy. It should be updated and regularly practiced through table-top and other exercises. Among the most important parts of the crisis communication plan are the written statements that describe policies and procedures, the people who handle specific communication tasks, the questions likely to be asked for different scenarios, the messages that will be offered in response to expected questions, the methods for disseminating messages, and the people who need to be notified and contacted. A crisis communication plan should be a living document that can be easily accessed, referenced, and updated. It should not be allowed to gather dust and become outdated.

A crisis communication plan can be as short as a few pages or as long as several hundred pages.[20] From a practical application standpoint, a crisis communications plan does the following:

- Defines and assigns the crisis communications team.
- Outlines the roles and responsibilities of each member of the crisis communications team.
- Details communication steps to take during a crisis event.
- Indicates whom to contact, resources that are available, and procedures to follow.
- Provides a platform for training, testing, and improvement.

The crisis communication plan must be informed by and supportive of the crisis management plan. The plan can be developed for multiple crisis scenarios or a specific crisis scenario.[21] For example, Covello developed a highly detailed crisis communication plan for radiological emergencies for the US Nuclear Regulatory Commission.[22] The stated purpose of the document was to support the delivery of understandable, timely, accurate, consistent, and credible information to the public, the media, and other stakeholders during a radiological crisis. The document contains dozens of sample statements for use or adaptation by crisis response organizations in developing their own crisis communication plans. The document is linked to another NRC document by Covello (2011) listing frequently asked questions in a radiological crisis along with sample message maps.[23]

A comprehensive crisis communication plan should have the elements listed in Table 6.6.

Table 6.6 Key elements in a crisis communication plan.

- Signed endorsement from senior leadership
- Designated responsibilities for members of the crisis communication team
- Internal information verification and expedited clearance procedures
- Agreements on information-release authorities (who releases what, when, and how)
- Regional and local media contact lists, including after-hours news desks
- Procedures to coordinate with crisis communication teams in other organizations, including procedures for developing a Joint Information Center if needed
- Designated spokespersons for different crisis scenarios
- Internal crisis communication and crisis response team members' after-hours contact numbers
- Contact lists for external crisis response partners, such as fire department, police department, emergency management agencies, and the Red Cross.
- Agreements and procedures to join a Joint Information Center, if activated
- Procedures to secure needed resources such as space, equipment, and personnel, to operate information and media operation during a crisis 24 hours-a-day, 7-days-a-week, if needed
- Lists of likely or key stakeholders, ways to reach them, with demographic and background descriptions
- Information dissemination methods that can be used to communicate to stakeholders, and partners during a crisis, including the following:
  - Websites
  - Social media channels, such as Twitter, Facebook, YouTube, Instagram, and Snapchat
  - E-mail lists
  - Listservs
  - Broadcast media
  - Press releases

The following sections should be part of the crisis communication plan:

Management and Organizational Structure Section: This section describes the chain of command and how decisions related to messaging, coordination, and implementation will be made.

Human Resources Section: This section describes the crisis communication team, including tasks, roles, responsibilities, training requirements, and contact information. Major crisis communication tasks are listed in Table 6.7.

The human resources section of the plan also describes policies and procedures designed to prevent staff burnout during a crisis. In many crises, communications staff members work long hours without relief. An effective and supportive crisis communication plan will lay out work schedules and contact information for relief staff.

Financial Resources Section: This section describes the estimated budget for developing, implementing, and maintaining the crisis communication plan and where the funding will be obtained.

Physical Resources Section: This section describes the resources needed for the communication team to carry out their tasks, including space requirements and technical equipment. This section of the plan should also identify what steps should be taken if there is a communication breakdown. For example, when all the electronic methods of communication are down and the team decides to print and disseminate large numbers of printed communications on issues such as how to stay safe and how to locate loved ones, the plan must identify how this will be accomplished.

Spokespersons Section: This section describes spokesperson skills, tasks, responsibilities, training requirements, and contact information. This section of the plan also identifies who will be the lead spokesperson and backups for different crisis scenarios.

Communication Channels Section: This section of the plan describes preferred techniques and tools for the dissemination of information through communication channels and who has access to these channels. This section needs to consider both journalistic outlets and social media.

Monitoring and Evaluation: This section describes how internal and external communications will be monitored, reviewed, and evaluated.

Table 6.7 Major crisis communication tasks.

---

9) Leadership tasks

10) Traditional media tasks (e.g., newspapers, magazines, radio, and television)

11) Social media tasks

12) Message development and materials development tasks

13) Partner and stakeholder outreach tasks

14) Web site development and maintenance tasks

15) Administrative and technical support tasks

16) Studio and broadcast support tasks

17) Research tasks

18) Media monitoring task

19) Hotline tasks

20) Workforce communications tasks

21) Subject matter expert support tasks

22) Legal support tasks

23) Information management and technology tasks

Messaging Section: This section describes how messages will be prepared, reviewed, and tested. It also contains lists of anticipated questions for different crisis scenarios and messages drafted to answer these questions. Lists of anticipated questions and drafted messages save precious time in a crisis. They allow leaders and communicators to deliver preapproved messages and thereby avoid disagreements, errors, delays, and communication regrets.

The messaging section of the plan should also contain templates and drafted communication materials, including fact sheets, brochures, website content, blog content, biographical profiles, news/press releases, prepared statements, social media content (e.g., Twitter, Facebook, and Instagram postings), telephone hotline scripts, graphics (e.g., charts, diagrams, tables, maps, infographics, photos), videos, video scripts, and podcasts.

Contact and Notification List Section: This section contains updated lists of stakeholders who should be contacted or notified for different crises. The lists should contain as much information as possible, including organization name, contact name, organizational telephone number, mobile numbers (day and night), social media addresses, and email addresses. This information needs to be updated regularly – routinely at least once every month. Contact lists for all internal and external stakeholders should be immediately accessible to members of the crisis communications team. Contact information should also be secured to protect confidential information and be available only to authorized users. Electronic lists can be hosted on a secure server for remote access with a web browser; hard copies of lists should also be available at an accessible location.

6.3.5 Precrisis Communication Activity: Identify, Train, and Test Crisis Communication Spokespersons

Spokespersons are the public face of the organization, and especially during a crisis. Crisis spokespersons must be able to reduce the gap between perceptions and reality. A primary goal for the crisis spokesperson is to establish trust with key stakeholders. Key stakeholders include internal audiences – such as employees and leadership – and external audiences – such as the media, first responders, partners, regulatory agencies, politicians, victims, victims' families, customers, contractors, shareholders, and the general public.

Because crisis spokespersons vary greatly in their verbal and nonverbal communication skills, and because different crisis situations require different communication skills, the lead spokesperson is not always the most senior leader of an organization. However, the spokesperson must have sufficient authority or expertise to be accepted as speaking on behalf of the organization during a crisis.

Identifying and preparing crisis spokespersons are tasks that should not be underestimated, as these individuals must succeed in difficult circumstances. These individuals should be trained and given support to anticipate questions, prepare messages, rehearse for presentations and interviews, and receive feedback on both their verbal and nonverbal communication. Specific skills that will be demanded of the spokesperson in a crisis are listed in Table 6.8.

6.3.6 Precrisis Communication Activity: Engaging Stakeholders

An essential activity in the precrisis/preparedness phase is developing a knowledge base and building positive relationships with key stakeholders. As described in the sections of this book on stakeholder engagement, stakeholders are those who are interested, who are affected, or who influence

Table 6.8 Specific skills of a crisis communication spokesperson.

An effective crisis communication spokesperson should be able to:

- promote or defend effectively a point of view
- respond to sensitive or hostile questions in a professional, respectful, and sensitive manner
- stay on message, bridge repeatedly to key messages, yet remain flexible
- offer key messages supported by evidence, examples, anecdotes, and stories
- remain calm and composed under pressure
- defer, delegate, and redirect questions to subject matter experts and trusted sources when needed

the attitudes and beliefs of others. Stakeholders also include those who might serve as allies, partners, and subject-matter experts.

One key to successful precrisis communication is understanding who the most important stakeholders are, what they care about, what they know, what they want to know, and what they need to know. Table 6.9 lists the characteristics of various stakeholders that communicators need to understand in order to be effective in stakeholder engagement. Differences in perceptions, opinions, beliefs, and behaviors are often grounded in differences in these characteristics.

One of the most challenging aspects of stakeholder engagement is how wide a net is to cast. In general, the wider the net of stakeholders, the more effective the crisis communication. Stakeholders who feel that they have been left out of the conversation may feel resentful and later act on this resentment.

As noted in the chapter of this book on stakeholder engagement, several benefits flow from positive relationships with stakeholders formed in the precrisis stage. One benefit is the knowledge that comes from listening to stakeholders. A second benefit is a trust that can develop from regular and authentic information exchanges, dialogue, and contact. That trust forms an important base for later communications.

Interactions with key stakeholders should include exchanges of information about potential crises, levels of concern about potential crises, scenarios, sources of trusted information, and trust in information sources. Interactions with stakeholders should also include a review of existing messages, background materials, and proposed dissemination methods. Since people can get and receive crisis information from many places, one objective of crisis communication is to become the "go-to" place or trusted source for information.

Table 6.9 Stakeholder characteristics.

- Demographic, cultural, economic, social, historical characteristics
- Political landscape
- Preferred communication approaches, methods, and channels (e.g., in-person, visuals, social media)
- Timing preferences (e.g., providing information prior to a meeting)
- Language and venue preferences for communications
- Trusted information sources and locations
- Pre-existing knowledge related to the topic
- Potential perceptions, reactions, and responses to information provided
- Ability to integrate and implement recommendations and potential impediments

If precrisis stakeholder engagement is episodic, or if relationships are initiated only after a crisis occurs, attempts to forge positive relationships during a crisis are likely to fail or fall short. A crisis is the poorest possible time to make introductions and establish new relationships. Nor is a crisis the best time to create a trust or establish trusting relationships. A common dictum in crisis communication is "a crisis is not the time to hand out business cards." People and organizations are more likely to respond to requests for help in a crisis if a trusting relationship already exists. Potential allies and partners include all internal and external parties who have the potential to cooperate, share information, and share resources.

Precrisis stakeholder engagement and relationship building also include proactively building positive relationships with key persons in traditional and social media. Positive relationships with editors, reporters, bloggers, and other media players are a cornerstone of successful crisis communication in the modern world. The quality, tone, and content of traditional and social media coverage in a crisis are strongly influenced by preexisting relationships. Communication activities and interactions with traditional and social media in the precrisis phase include meeting with editors, reporters, and bloggers; being accessible and available for providing background information; and creating and maintaining updated contact lists.

6.3.7   Precrisis Communication Activity: Identifying Stakeholders' Questions and Concerns

One of the benefits arising from stakeholder engagement is a list of questions and concerns raised by key stakeholders. These can be revealed through direct stakeholder engagement, including focus groups, face-to-face meetings, interviews with subject-matter experts, and interviews with leaders and members of key stakeholder groups. Questions and concerns can also be identified through searches of public documents, including public meeting records, media reports, entries in encyclopedias (e.g., Wikipedia and the *Encyclopedia Britannica*), blogs, podcasts, YouTube, and postings on relevant social media platforms.

Experience from prior crises of a similar nature provides considerable information about what questions can be anticipated. Research on prior cases can pay off in developing lists of anticipated questions and crafting messages for response. Two examples of questions to anticipate and expect from stakeholders in a crisis are contained in the appendices to chapter 9. One example contains the most frequently asked questions following an active shooter incident. The other example lists the most frequently asked questions by journalists in a disaster.

6.3.8   Drafting Messages for Anticipated Stakeholder Questions and Concerns

To the extent possible, messages for anticipated questions and concerns should be drafted before a crisis. One of the most powerful tools for drafting clear and concise crisis messages is the *message map*. As described later in this book, a message map is a visual aid that consists of layered and hierarchically organized information that can be used to respond to anticipated questions or concerns. Message maps are beneficial in a number of ways. They can tell communicators what questions to anticipate and expect. They can provide crisis communicators with vetted responses to these questions. They can provide message consistency between multiple communicators who are working across diverse platforms. And they can also provide a way for communicators to tick off the key points they want to make and minimize their chances of experiencing the communicator's regret – i.e., regretting saying something or regretting not saying something.

Message maps should ideally be developed in draft form for every anticipated stakeholder question or concern in response to identified risks. Alternatively, message maps can be developed for a prioritized subset of questions and concerns.

Message maps can be presented in a box or bullet format. Both formats enable crisis communicators to respond quickly with timely, accurate, clear, concise, consistent, credible, and relevant information. The top level of the message map template identifies the stakeholder and the question that the map is intended to address. The second level of the message map template contains the three to five overarching key messages that address the question. The last section contains supporting information that amplifies the key messages. Ideally, each message in the map should be able to stand alone for purposes of quotation. (Message maps are described more fully in Chapter 9 of this book.)

The content of a message map, especially the key messages, can be repurposed and used effectively to develop content for social media platforms in a crisis. For example, the Social Media Library developed by researchers at the School of Public Health at Drexel University contains more than 4,000 crisis-related messages.[24] The Social Media Library is a web-based resource containing vetted social media messages for different crisis scenarios. The library contains message content for the preparedness, response, and recovery phases of the crisis scenarios. The Social Media Library website is illustrated in Figures 6.1 and 6.2.

The Social Media Library features message templates for Twitter, Facebook, and Instagram. The Social Media Library includes content relevant to more than 20 different public health crisis scenarios, including natural hazards, infectious disease outbreaks, industrial accidents, and terrorism. The Social Media Library also contains vetted messages for individuals with special needs in a crisis, such as those with physical disabilities and communication difficulties. The messages in the library were reviewed by subject-matter experts in the content areas. They were designed to

Figure 6.1 Social Media Library.

Social media library

Natural disasters
Extreme heat
Flood
Hurricane
Tornado
Winter storm/extreme cold

Accidental disasters
Hazardous materials release
Nuclear power plant
Utility interruption

Infectious disease outbreak
Pandemic flu
Coronavirus (MERS-CoV)

Intentional disasters

Biological
Anthrax
Botulism
Plague
Smallpox
Tularemia
Viral hemorrhagic fevers (ebola)

Chemical
Cyanide
Sulfur mustard (H)
Nitrogen mustard (HN)
Lewisite (L)
Ricin
Sarin

Radiological
Rdd/dirty bomb

Figure 6.2 Contents of Social Media Library.

help public information officers and other spokespersons create and deliver timely, accurate, and consistent information likely to be needed during a public health crisis.

Message maps are particularly useful in a crisis where timing is critical and where social media platforms may play a dominant role in the communication field. When used for crisis communication, message maps are like the prescripted templates and short list of talking points developed and used by many organizations for interactions with the media in a crisis. However, message maps differ from these prescripted templates and talking points in that: (1) they are constructed to adhere strictly to the principles of risk, crisis, and high-stress communication; (2) they can be repurposed for a wide variety of communication channels; and (3) they can be developed to address, as comprehensively as resources allow, questions from diverse stakeholders for a selected crisis scenario.

6.3.9 Precrisis Communication Activity: Conducting Exercises to Test the Crisis Communication Plan

Training and exercises are particularly important in the precrisis phase. Because of a variety of factors, many crisis communication spokespersons and team members tend to learn on the job. Their first experience communicating in a crisis is when a crisis occurs, and they likely would have greatly benefited from prior training and practice.

Tabletop exercises create an opportunity for all members of the crisis communication team to walk each other through how they would respond during a simulated crisis. It is often through thoughtful dialogue that opportunities to adjust and strengthen the crisis communication plan emerge.

Functional exercises differ from a tabletop exercise in that rather than talking about how members of the crisis communication team would respond, they must actually respond to a simulated scenario. This type of simulation can help strengthen the organization's level of preparedness by identifying gaps in process or roles.

6.3.10 Precrisis Communication Activity: Incident Command System (ICS) and the Joint Information Center (JIC)

Some agencies and organizations need to be prepared for administering or participating in major interagency communication initiatives in a crisis. Many organizations use the Incident Command System (ICS) and Joint Information Center (JIC) for responding to inquiries from mainstream broadcast and print media outlets in a major crisis. ICS and JIC manuals recommend all responses to inquiries from mainstream broadcast and print journalists come through the ICS structure and the JIC.

ICS is a military-style management structure created by government agencies to respond effectively to a major crisis. It is described in detail in various government publications, including documents produced by the United States, Canada, and other governments.[25] ICS was initially developed to address problems of inter-agency responses to wildfires in California and Arizona. In the United States, it has evolved into a central component of the National Incident Management System. It is currently used for all major crises, ranging from mass casualty active shooting incidents to oil spills.

A JIC is either a physical or "virtual" location where communications staff representing organizations responding to a major crisis come together to coordinate and disseminate information. The JIC structure was initially designed to meet the needs of mainstream broadcast and print media. However, the structure evolved to address the communication needs of social media outlets and all interested or affected populations. A JIC is designed to disseminate information through news/press conferences, press releases, media interviews, media advisories, social media channels, websites, and other communication means as needed.

A JIC functions as a centralized crisis communications room or "war room." Public and news media inquiries are handled through the JIC. As such, the JIC serves as the nerve center for crisis communications. It is the home for the crisis communication team, either in person or virtually. The main JIC can be supplemented or supported by local JICs, a dedicated space at the emergency operations center, or a trailer at the actual site of a crisis, or through links to remote sites.

Within the JIC, risk and crisis communication functions are fully under the control of a single public information officer (PIO). The PIO is a member of the command staff. The PIO reports to the incident commander, the leader of the ICS structure, who also often acts as the lead spokesperson during the crisis. The PIO coordinates all communication activities related to the crisis. Such activities are ideally conducted in concert with information officers and staff from all organizations responding to the crisis.

Several factors determine whether the JIC will function effectively in a crisis. First, the JIC should be located as close as possible to the scene of the major crisis, such as a disaster or emergency, but out of the way of harm. Second, mainstream broadcast and print media outlets will accept an invitation to come to a JIC only if they believe they will receive important information and have their questions answered. Third, every organization participating in response to a major crisis should be encouraged to have a representative at the JIC. Fourth, to the extent possible and reasonable, organizations participating in the response to a major crisis should refer broadcast and print media journalists to the JIC for answers to questions. Fifth, at least one person should be available round the clock at the JIC to respond to inquiries from the media and other inquiries about the crisis.

Detailed instructions for establishing and running a JIC are found in Covello (2011).[26] The nearly 200-page document contains detailed information related to JIC functions, structure, policies, tasks, logistics, staffing, information dissemination, communication products (e.g., media advisories, news/press releases, news conferences, and media interviews), spokesperson skills, emergency broadcasts, call centers, telephone hotlines, social media interactions, and communication products for special or vulnerable populations, such as disabled, transient, low literacy, and institutionalized populations.

6.4 Communications in the Crisis Response Phase

Tell the truth, tell it well, tell it early, tell it all, and tell it fast.

—Anonymous

Communications in the *crisis response phase* are focused primarily on protecting people and the things they value. Messaging revolves around answering questions about the five Ws (and one H) – who, what, where, when, why, and how.

Social media and traditional media (broadcast and publishing media) play a key role in the crisis/response phase. For example, social media postings from multiple sources often occur within seconds and minutes after a crisis begins. If the crisis calls for a news conference, the first news conference is especially critical and will set the tone for subsequent interactions. It will have a large and potentially irreversible effect on the success or failure of the overall crisis management and communication effort.

In the crisis response phase, multiple questions and concerns need to be quickly addressed, including information about what is known, what is not known, what actions have or will be taken, and where people can access timely and credible information. As the crisis progresses, questions often shift to who is to blame, whether the crisis could have been prevented, and how the crisis could be prevented from happening again. If questions are not quickly addressed, it can create a vacuum causing people to seek answers from other, perhaps less credible, sources or to attempt to fill the void by developing their own speculative answers or stories. Expectations regarding rapid answers often increase exponentially. Those interested and affected may demand information immediately and continually.

One of the most important messaging failures in the *crisis response phase* is failing to be transparent, truthful, and caring. A failure in these areas can have a critical influence on perceptions, attitudes, beliefs, and behaviors. For example, messages and actions that fail to communicate transparency and truthfulness can result in a significant erosion of trust. Individuals and organizations typically experience less reputational harm from a crisis when sincere and authentic statements of caring, concern, and empathy are offered. Such statements could also help reduce the number and size of claims made against an organization following a crisis.

Messages in the crisis/response phase are also more effective when they are

- clear,
- repeated,
- specific,
- authoritative,
- timely,

- originate from trusted sources,
- acknowledge the needs of special populations,
- acknowledge emotions,
- acknowledge uncertainties, and
- give people things to do.

The training and abilities of crisis spokespersons are critical to effective crisis communication; the delivery of a message can be as important as the content of the message itself. Crisis spokespersons should have highly developed verbal and nonverbal communication skills as described in the following sections.

Listening/Caring/Empathy Skills: Crisis spokespersons should be able to communicate verbally and nonverbally with sincere and authentic listening, caring, empathy, and compassion. When confronting a threat, risk, or high-stress situation, people typically want to know you care before they care what you know. Sincerity and authenticity mean, among other things, listening, saying back what you hear, and trying to understand the perceptions and realities of the other person. Insincerity and inauthenticity mean, among other things, fake looks or statements of caring and compassion, words not backed up by actions, and not really answering questions. For example, the statement made by BP CEO, Tony Haywood about the oil rig crisis, "I just want my life back" is an example of a message that was perceived as not showing caring.

Listening skills include the active listening skills listed in Table 6.10. Most people do not use these skills and fully listen. They are typically preoccupied with what they want to say next. As a result, much is lost. Active listening makes a person feel they are truly understood and that the speaker is attentive to what they are saying. Active listening is just as important, albeit often more challenging, in a town hall meeting or media briefing as in one-on-one conversations.

Active listening is more than simply hearing what is being said by the other person. It also includes paying attention to nonverbal signals – such as voice cues (how the person says something), eye cues (what the eyes are doing), posture cues (what is being done with the body), and hand cues (what is being done with the hands). It involves an empathic response to what is heard, such as nonjudgmental, expansive questions.

Unfortunately, active listening skills are difficult to master and are rarely taught in schools. Schools offer training in giving presentations but seldom teach students how to carefully listen.

Table 6.10 Active listening skills.

When interacting with others, crisis spokespersons should demonstrate:

1) concentration skills: fully concentrating on what is being said and how it is being said (nonverbal cues)
2) paraphrasing skills: re-stating and giving non-judgmental feedback to check understanding
3) clarifying, probing, and encouraging skills: asking open-ended questions requesting clarification (including empathetic questions) or further explanation.
4) summarizing skills: drawing together or validating main points.
5) attentiveness skills: maintaining eye contact, slightly leaning forward, nodding of the head, offering a small smile, allowing for short periods of silence, and vocal expressions of agreement (such as "yes," "I see," "Mmmm," and "hmm")
6) mirroring skills: whereby one person imitates the nonverbal signals – e.g., gestures, speech patterns, or posture, eye contact -- which can lead to building rapport with others. (From a neuroscience perspective, mirror neurons are activated in the person experiencing mirroring, allowing the individual who is being mirrored to feel a stronger connection with the other individual.)

Plain Language/Presentation Skills: Crisis spokespersons should be able to translate complex scientific, technical, and other information into easily understandable concepts. The spokesperson must be able to convey this complex information in simply worded, clear, and accurate messages about risks and protective actions. People in a crisis typically have difficulty processing information – hearing, understanding, and remembering information. Information, therefore, needs to be offered in clear, layered messages with key messages offered in less than 27 words or 9 seconds, with the most important messages placed first and last, and with language kept jargon free and easy to comprehend. (Chapter 9 provides guidance on message development.)

Simply worded, clear, and accurate messages are necessary in a crisis but not sufficient. If the information provided is too limited, it can create an information vacuum and may not sufficiently inform protective actions. People facing a crisis will typically want to know as much as possible. They will also want to confirm what they know with others. Also, as shown in Figure 6.3, when communicating with the public, the crisis spokesperson should present information using the inverted pyramid to communicate complex information: that is, start with the "bottom line" (often referred to as BLUF, or bottom line up front) and tell people what is most important to know. As also shown in the figure, communicating with nonexperts differs from experts communicating with each other.

Positive Messaging Skills: Crisis spokespersons should be skilled in problem solving, quickly analyzing a situation, and sharing messages focused on what is being done or needs to happen – and not on what is not being done or did not happen. People in a crisis will typically focus more on negatives than positives, thus missing important and constructive information. The crisis spokesperson must therefore be able to balance negative information and bad news with three to four positive or constructive messages.

Emotional Intelligence Skills: Crisis spokespersons should be able to share messages that acknowledge and focus on perceptual and emotional factors such as fear, anger, helplessness, trust, fairness, and personal control. People in a crisis will often focus more on perceptual and emotional factors than on technical facts and data. Emotional intelligence skills also include remaining calm and composed, even when under extreme pressure.

Negotiation Skills: Crisis spokespersons should be able to draw on their own negotiation skills, or on the skills of trained and experienced negotiators, to navigate the many difficult and delicate conversations that accompany many discussions about risks. Most commonly, negotiation skills are needed by risk communicators to resolve conflicts and negotiations about issues such as the

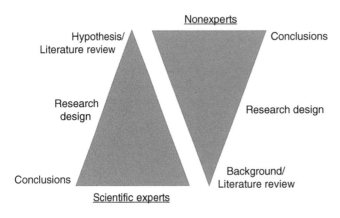

Figure 6.3 Comparison of How Scientific Experts and Non-Experts Typically Present Crisis and Risk Information.

boundaries of stakeholder engagement and about the fair and equitable distribution of scarce or limited resources (e.g., conflicts about why some individuals or groups are getting more help and assistance than others).[27] A key to negotiation is communicating the value of not fighting, acknowledging shared concerns, uncovering interests, engaging in interest-based bargaining, and finding win-win solutions.

Many of the principles of negotiation come from the larger field of conflict resolution and alternative dispute resolution. Negotiation can be especially challenging in risk and crisis situations because of high emotions. Fisher et al. (2011) and Ury (1993) provide detailed and research-based guidance on negotiation principles and techniques for emotionally charged issues.[28] Fisher and Ury emphasize that negotiation is a two-way process of exchanging promises and commitments to reach a satisfactory agreement. The primary purpose of a negotiation is to produce a fair and sustainable outcome within a reasonable amount of time. If negotiations fail, it is important to know in advance your desired "best alternative to the negotiated agreement," or BATNA. BATNA is defined in the negotiation literature as the most advantageous alternative course of action a party can take if negotiations fail and an agreement cannot be reached. BATNA requires listening for interests, interest-based bargaining, and seeking out win-win solutions.

At the center of negotiation is interaction with people who may have differing personalities, goals, objectives, and agendas. Thus, understanding and mastering interpersonal dynamics is a fundamental communication skill in a successful negotiation. Interpersonal dynamics and human emotions are manifested in many forms in a negotiation and can dominate the proceedings, especially when important decisions are being made. An effective negotiator should therefore be knowledgeable and aware of the presence and effect of these dynamics and emotions.

Trust Building Skills: Crisis spokespersons should be able to share messages that focus on the primary factors that determine trust. Most importantly, these include transparency and telling the truth. These also include listening, caring, empathy, competency, honesty, openness, and dedication.

Nonverbal Communication Skills: Nonverbal communication refers to communication effected by means other than words. When confronting a threat, risk, or high-stress issue, people are often highly attentive to nonverbal cues. In high concern situations, nonverbal cues or signals, such as how you look and how you sound, can be even more important than what you say. Nonverbal communications can provide over 75% of message content related to trust, with nonverbal information enhancing or diminishing the chances that information will be heard, understood, and trusted.

Holding constant other variables, the effectiveness of information shared by a risk communicator or spokesperson will be substantially affected by cues and signals such as facial expressions, body language, and vocal variety. Depending on the individual, situation, and context, nonverbal communication in high concern situations: (1) is quickly noticed, (2) can override verbal communication when inconsistencies are present (e.g., body language and tonality are often perceived as more accurate indicators of meaning and emotions than the words themselves); and (3) depending on the cue or signal, is often interpreted negatively. For example, sitting back in a chair can be interpreted as indifference; crossing ones arms can be interpreted as closed and defiant.

In a classic and widely cited set of studies, Mehrabian conducted research on nonverbal communication.[29] He found that 7% of a message is conveyed through words, 38% through vocal elements such as tone, and 55% through nonverbal body language elements such as eyes, facial expressions, gestures, and posture.

In Mehrabian's research, subjects were asked to listen to a recording of a speaker saying words such as *"maybe"* in different ways to convey liking, neutrality, and disliking. They were also shown photos of a woman's face conveying the same three emotions. They were then asked to speculate about the emotions heard in the recorded voice, seen in the photos, and both together. The subjects correctly identified the emotions 50% more often from the photos than from the voice. In a second study, subjects were asked to listen to nine recorded words, three meant to convey liking, three to convey neutrality, and three to convey disliking. Each word was pronounced in three different ways. When asked to speculate about the emotions being conveyed, subjects were more influenced by the tone of voice than by the words themselves.

Because of significant limitations of the research, it is unclear the extent to which Mehrahian's original formula – 7% words/38% tone of voice/55% body language formula – can be applied or extended to overall nonverbal communication. For example, Mehrabian used few people in his experiments; the research design did not take into account the extent to which the speakers could produce the required tone of voice; the experimental situation was artificial with no context; the communication model used for the experiment was oversimplified; little weight was attached to the characteristics of the observers making judgments; and the experimental subjects were aware of the purpose of the experiments. As a result, the research findings are directly applicable only to situations where (1) a speaker is using only a few words, (2) the tone of the spoken words is inconsistent with the meaning of the word, and (3) a judgment is being made about the feelings of the speaker. With these critiques in mind, Mehrabian's research nonetheless began a series of research projects confirming the power of speech and, more broadly, the power of nonverbal communication.[30]

The list of nonverbal communication skills presented in Table 6.11 summarizes research on nonverbal information derived from research on kinesics, i.e., gestures, body movements, facial expressions (such as expressions indicating anger, sadness, surprise, happiness, fear, or disgust), eye behavior, and posture, as well as vocal behavior, i.e., how something is said. One nonverbal cue alone typically means little. Clusters of nonverbal cues provide critical information about a spokesperson and the messages being presented.

The exact meaning of a nonverbal communication depends upon the context and culture in which it occurs. In Western culture, the most important nonverbal body language cues are eyes/facial expressions, posture, voice, hands, and overall appearance. Nonverbal cues that are interpreted highly negatively in Western European and American culture are shown in Table 6.11; nonverbal pitfalls are listed in Table 6.12. Cultural differences can especially create pitfalls in global communication. Cultural differences in interpreting verbal communication and nonverbal communication cues are discussed in detail in Chapter 8 of this book: Foundational Principles: Trust, Culture, and Worldviews.

6.4.1   Case Study: Lac-Mégantic Rail Tragedy

The Lac-Mégantic rail disaster occurred in the Canadian town of Lac-Mégantic in the eastern province of Quebec on 6 July 2013. An unattended freight train carrying crude oil rolled down a grade and derailed in the downtown area of the town, resulting in fire and the explosion of multiple tank cars. More than 40 people were killed. More than 30 buildings in the town's center, roughly half of the downtown area, were destroyed. The blast radius was over a half-mile. The tragedy was one of the deadliest rail accidents in Canadian history.

The rail company at the center of the Lac-Mégantic tragedy – Montreal, Maine, and Atlantic Railway – violated nearly every basic crisis communications principle. For example, executives from the company failed to: (1) arrive at the scene quickly; (2) express caring and empathy for those affected; and (3) acknowledge responsibility for the company's role in the tragedy.[31]

Effective crisis communication begins with anticipation, preparation, and practice. One of the most important recommendations in the crisis communication literature is to have a crisis

Table 6.11 Nonverbal signals and cues with possible negative interpretation.

Poor eye contact:	Dishonest, closed, unconcerned, nervous, lying
Sitting back in chair:	Not interested, unenthusiastic, unconcerned, withdrawn, distancing oneself, uncooperative
Arms crossed on chest:	Not interested, uncaring, not listening, arrogant, impatient, defensive, angry, stubborn, not accepting
Rocking movements:	Nervous, lack of self-confidence
Pacing back and forth:	Nervous, lack of self-confidence, cornered, angry, upset
Frequent hand-to-face contact:	Dishonest, deceitful, nervous
Hidden hands:	Deceptive, guilty, insincere
Touching and/or rubbing nose:	Doubt, disagreement, nervous, deceitful
Touching and/or rubbing eyes:	Doubt, disagreement, nervous, deceitful
Pencil chewing/hand pinching:	Lack of self-confidence, doubt
Jingling money in pockets:	Nervous, lack of self-confidence, lack of self-control, deceitful
Constant throat clearing:	Nervous, lack of self-confidence
Drumming on table, tapping feet:	Nervous, hostile, anxious, impatient, bored
Clenched hands:	Anger, hostile, uncooperative
Locked ankles/squeezed hands:	Deceitful, apprehensive, nervous, tense, aggressive
Palm to back of neck:	Frustration, anger, irritation, hostility
Tight-lipped:	Nervous, deceitful, angry, hostile
Licking lips:	Nervous, deceitful
Frequent blinking:	Nervous, deceitful, inattentive
Slumped posture:	Nervous, poor self-control
Raising voice/high-pitched tone:	Nervous, hostile, deceitful
Shrugging shoulders:	Unconcerned, indifferent

communication plan that anticipates crisis scenarios, stakeholders, and questions; identifies appropriate messages, messengers, and means, or communication; and practices what plan specifies. Such a plan allows for a timely and appropriate response to a crisis situation. Unfortunately, based on the response by the company, interviews, and media reports, it is not clear whether a crisis communication plan did not exist or if one was simply ignored.

Edward Burkhardt, president of Rail World, which owns the Montreal, Maine, and Atlantic railroad, arrived at Lac-Mégantic on 10 July 2013. His many failures as a crisis communicator included:

Late Arrival and Communications: Burkhardt was criticized by the media, citizens, and others for arriving at the scene of the fire and explosion four days after the tragic event.

Speculation: Burkhardt speculated about the cause of the tragedy instead of relying on confirmed facts. As reported in the *New York Times*, Burkhardt first blamed "tampering with the

Table 6.12 Key nonverbal pitfalls in Western European culture.

- Poor eye contact: Poor eye contact, combined with facial expressions, can communicate lack of listening, caring, concern, competence, expertise, honesty, and transparency
- Insufficient volume: Poor volume can communicate lack of confidence, competence, and openness
- Poor enunciation/pronunciation
- Poor vocal pacing and rhythm
- Repetitive use of distracting words such as "ok," "like," "uhm" or "uh"
- Poor posture: can communicate lack of confidence and authority
- Repetitive gestures and motions: gestures such as hand waving, throat-clearing, jingling keys or change, and pacing can be distracting and communicate lack of confidence
- Poor grooming
- Distracting attire or jewelry

train's locomotives."[32] He then shifted blame to the local volunteer fire department that helped extinguish the massive fire. Later, at the press conference in Lac-Mégantic where residents booed him, Burkhardt blamed a railroad engineer, asserting the engineer had not set the brakes properly. Burkhardt's speculation was challenged by the first responders and by the Canadian government agency, Transportation Safety Canada. Burkhardt's speculation later proved to be incorrect.

Poor Media Skills: Upon arrival at the town, Burkhardt held a spontaneous press conference in the middle of the street in the town of Lac-Mégantic. He was greeted with jeers and heckled by residents during the press conference. He answered questions only in English in a primarily French-speaking community. He failed to offer a French-speaking spokesperson in a Canadian province where French is an official language and the language of everyday conversation. At the press conference, Burkhardt criticized journalists for their poor "manners." While Burkhardt did express an "abject apology," at the end of the press conference he joked that his net worth had taken a hit from the tragedy.

6.4.2 Disaster and Emergency Warnings

Disaster and emergency warnings add a special dimension of urgency to crisis communication. They also involve dramatically increased stakes. It is almost impossible in today's world to go through a week without hearing news about the failure of warnings related to a terrorist attack, earthquake, flood, wildfire, a nuclear plant accident, food contamination, or an outbreak of an infectious disease.

Disaster and emergency warnings can create additional stress, anxiety, and fear. In response to a warning, people are often confused about where to turn and what to do. People may have to decide immediately what specific protective action, if any, to adopt.

Disasters and emergency warnings are typically one-way communications. Two-way communications take place primarily in the pre- and postphases of a disaster or emergency. In a disaster or emergency, the warning system focuses heavily, and sometimes exclusively, on messages alerting people to the disaster or emergency and informing them about protective actions. Messages are ideally presented to stakeholders by authorities who speak with "one voice" or at least in harmony with one another. This assumes disaster and emergency authorities have coordinated and collaborated with one another, which is unfortunately not always the case. Effective disaster and emergency warnings require risk managers and communicators to understand and anticipate the information needs of stakeholders and where stakeholders may disagree. This type of knowledge can be gained through effective stakeholder engagement.

6.4.2.1 Designing Effective Warnings

Each type of disaster or emergency requires a different type of warning. As shown in Table 6.13, warning messages are subject to a diverse set of challenges and potential failures.

Disaster and emergency warnings vary not only by the type of disaster or emergency but by the distinctive characteristics of the disaster or emergency. These include the probability, speed, magnitude, and consequences of an impact. The primary consequences of impact are harm and damage. However, secondary impacts are also important. In some cases, secondary impacts can be more devastating than the initial harm or damage. For example, earthquakes can produce fires, leaks of toxic chemicals, and explosions at industrial facilities. Hurricanes and floods can cause contamination of drinking water supplies and the shutdown of water and electric utilities.

6.4.2.2 Steps in the Disaster and Emergency Warning Process

Disaster and emergency warning systems are designed to get the immediate attention of people and encourage prompt action. Actions can be protective actions, such as evacuation, or a continued state of alertness.

Terminology related to disaster and emergency warnings can be confusing. For example, the US National Weather Service uses two terms for weather disasters based on the degree of certainty: *watch* and *warning*. The term watch is used when the risk of a weather event has increased significantly but its occurrence, location, and/or timing is still uncertain. A watch notice is issued to provide people enough lead time to set their disaster plans in motion. A watch notice lets a person know weather conditions are favorable for a hazardous event to occur. It literally means "be on guard!" By comparison, a warning notice is issued when a weather event poses an imminent threat to life or property or has a very high probability of occurring in a short period of time.

Table 6.14 lists four major steps in the disaster and emergency warning process. Attitudes and beliefs influence each step in the warning process. For example, attitudes and beliefs affect how people attend to the warning, perceive the warning, use existing knowledge and past experiences to assess the warning, trust the warning, intend to act in response to the warning, and actually behave in response to the warning. Behavior, the ultimate outcome of the decision process, is a function of expectations about outcomes resulting from the adopted protection

Table 6.13 Disaster and emergency warnings: sources of failure.

Failure to anticipate:

- the ability of people to help solve problems on their own
- the ability of people to cope and be resilient
- the willingness of people to help others
- the complexity of what motivates people to take recommended protective actions
- the complexity of what people have learned from previous disasters and emergencies
- the complexity of how people actually behave in disasters and emergencies
- the importance of social networks (e.g., family, neighbors, friends, and co-workers) for specific types of disaster and emergency behaviors, such as evacuation behavior
- the importance of understanding preferred sources of information in a disaster or emergency situation
- the complexity of creating the high levels of self- and group-efficacy needed to master the challenges created by a disaster or emergency

action and the actual or true abilities and capabilities of a person to engage in or perform the protective action.

As noted by Lindell and Perry and the National Research Council/National Academy of Sciences, each phase in the reaction to a warning is affected by a vast number of situational, psychological, economic, social, and cultural factors.[33] Examples of factors that determine if a warning will be effective are listed in Table 6.15.

Table 6.14 Phases in the process of absorbing and acting on disaster warnings.

---

1) Receiving the disaster or emergency warning
   - processing the information
   - verifying the information
   - assessing actions being taken by others, including friends, relatives, first responders and disaster management authorities
   - assessing information about sources of assistance
   - assessing information about sources of additional information
2) Understanding the disaster or emergency warning
   - believing the warning refers to a real threat
   - assessing recommended protective actions
   - believing the recommended protective actions are needed
3) Personalizing the warning
   - believing the recommended protective actions are personally needed
   - believing the warning applies to them or to the people or things they value
   - believing the recommended protective actions are the best available methods of personal protection
   - believing the recommended protective action can be done effectively and efficiently
   - believing the recommended protective action will protect what they value
   - believing in self- and group efficacy
4) Implementing the warning
   - taking or not taking action
   - moving from intention to actual behavior

Table 6.15 Factors that determine if a warning will be effective.

A warning is more likely to be effective if it:

- is offered by a credible source of information
- is presented through clear, plain language using a minimal number of technical terms, jargon, or acronyms
- contains specific recommended protective actions (including who, what, where, when, why, and how)
- identifies possible adverse outcomes if recommended protective actions are not taken
- is accurate, timely, relevant, and truthful
- is presented though authoritative language
- is targeted to a specific stakeholder group or audience
- is consistent with past messages or explains changes
- is repeated multiple time through multiple communication channels

---

6.5   Communicating Effectively about Blame, Accountability, and Responsibility

The first communication priority in the response phase of a crisis is to provide messages about what happened, what is being done in response, and what protective actions people can take. In some cases, these questions are often followed almost immediately by questions about who is to blame and who is responsible. Crisis communications experts have identified a diverse set of strategies for responding to questions about blame and responsibility. Many of these strategies are grounded in crisis communication theory, such as Coombs's *Situational Crisis Communication Theory* and Benoit's *Image Repair Theory* and *Image Restoration Theory*.[34] These theories examine the psychological, social, cultural, and situational variables that affect the choice of strategy and the potential for the reputational or image repair strategy to be effective. The theories also examine the ability of different strategies to generate sympathy, stakeholder emotions, and support.

Crisis communication strategies focused on addressing blame, accountability, and responsibility are typically based on several assumptions. First, crises are events that can severely damage an organization's reputation. Second, reputation matters because it is a highly valued resource and asset. Third, attributions of blame, accountability, and responsibility affect how people perceive a crisis and those involved, how they will react to the crisis, and how they will interact with individuals and organizations in a crisis. Finally, attributions of blame, accountability, and responsibility are based on a complex set of psychological, cultural, sociodemographic, political, and economic variables.

As a crisis unfolds, one of the first strategic steps regarding blame is to determine how stakeholders perceive responsibility. For example, if an individual or organization is perceived as a victim and not the cause of the crisis, as might be the case in a natural disaster, low or minimal responsibility is typically assigned. If an individual or organization is perceived as the cause of the crisis, as might be the case in an industrial accident, strong responsibility is typically assigned. Based on analysis of the situation and stakeholder perceptions, one or more crisis communication strategies are adopted. These are listed in Table 6.16. These are optional strategies, and many engender serious ethical issues. The list includes the kinds of strategies that have, both appropriately and inappropriately, been employed to restore reputation and repair image following a crisis.

Crisis communication strategies related to blame, accountability, and responsibility raise important ethical issues, as well as legal and policy issues. These include who makes the decisions about blame, accountability, and responsibility; what are the limits of blame, accountability, and responsibility; and when complete information should be shared, with whom complete information should be shared, how information should be shared, and how complete and transparent the information should be.

Because of fears about blame, accountability, and responsibility, many leaders and crisis managers do not fully disclose crisis-related information. They act under the assumption that others may never find out about the situation or problems. However, as Coombs points out:

> Research consistently demonstrates that a crisis does less reputation damage when the organization is first to report the crisis. The same exact crisis does less damage when the organization first reports it than when the news media or another source is the first to report the crisis.[35]

Table 6.16 Communication strategies employed to address blame and the loss of reputation in a crisis. (Note: While these are often used, many are ineffective, unethical, or inappropriate.)

Denial strategies and messages: claim there is no crisis.

Shifting blame/scapegoat strategies and messages: blame an outside entity for the crisis.

Attack the accuser strategies and messages: confront and attack the group or person claiming a crisis exists.

Bolstering/ingratiation strategies and messages: praise other organizations, remind people of past good works, or remind people about the positive track record and qualities of leaders and the organization.

Concern/compassion strategies and messages: express concern for victims and offer compensation in the form or money, goods, or services.

Regret strategies and messages: express how badly they feel about the crisis.

Defensive/excuse strategies and messages: claim lack of information about or control over the situation; claim lack of control or lack of intent to do harm.

Minimization/justification strategies and messages: claim little or no irreversible damage was caused by the crisis; minimize the important or perceived damage caused by the crisis.

Good intention strategies and messages: claim those in charge meant well or that those in charge did not mean for the crisis to happen.

Transcendence strategies and messages: ask for the crisis to be seen in a different context or perspective.

Differentiation strategies and messages: compare the actions of those in charge to the actions or misdeeds of others.

Provocation strategies and messages: claim those in charge were forced into the crisis by the actions of others.

Corrective action strategies and messages: promise or set a goal to change, restore the pre-crisis status quo, and prevent repetition.

Apology strategies and messages: accept full responsibility for the crisis, admit guilt, express remorse, and ask for forgiveness.

6.6 Communicating an Apology

An apology is a regretful acknowledgment of an offense, failure, or undesirable event. An apology, if sincere, authentic, carefully constructed, and delivered, can be a powerful and effective tool for restoring trust and restoring reputation. It helps address outrage and can reset or help heal a broken relationship.

Effective apologies enable those who have been offended to believe they have been listened to and recognized.[36] Ineffective apologies can create outrage. An example of an ineffective apology was the one offered by Oscar Munoz, chief executive officer of United Airlines, following the forceful removal of a passenger physically dragged off a plane due to overbooking a United Flight:

> This is an upsetting event to all of us here at United. I apologize for having to reaccommodate these customers.
>
> (Oscar Munoz, 10 April 2017)

In a crisis, a range of undesirable events can lead to the need for an apology. These include failing to meet expectations, poor performance, a disparaging remark, inappropriate use of humor, poor choice of language, poor behavior, errors, not fulfilling a promise, accidents, scandals, mishaps, and mistakes.

Table 6.17 Key factors in a successful apology.

Primary factors

1) Apologizes with sincerity and admits to the mistake, harm, or offence
2) Accepts ownership and responsibility without assigning blame elsewhere
3) Commits to a future that will not repeat the mistake, harm, or offence

Secondary factors

4) Provides evidence of regret, remorse, sorrow, penitence (e.g., compensation)
5) Is timely (a too late apology loses power, effectiveness, and impact)
6) Gives power to the offended party

Elements of a good apology are shown in Table 6.17.

An effective apology is more than accepting blame, expressing regret, or saying "I'm sorry." An individual or organization that apologizes must demonstrate an understanding of what and where things went wrong and accept responsibility. An effective apology rejects a "deny and defend" strategy or any attempts to divert attention. It replaces the "deny and defend" strategy with acknowledging responsibility and apologizing. It rejects blaming others and offering excuses. It rejects statements such as: "I'm sorry that I made you feel that way," or, worse, "It was not my intention to. . ."

The effectiveness of the apology will also depend on a wide variety of other factors, including the seriousness of the offense, the nature of the offense, the person or organization delivering the apology, the context, and the way the apology is delivered. Holding constant other variables, the more serious the offense, the greater the need to include all the elements in Table 6.17.

Discussions about the use of apologies often include concerns that apologies create legal liabilities for the apologist. However, a review of formal and common law indicates apologies rarely constitute evidence of guilt.[37] In addition, "I'm sorry" laws in more than half of the US states indicate apologies cannot be used in court as evidence of guilt.

Apologies also raise important ethical issues and questions. For example, if compensation is offered in a crisis event involving deaths, ethical issues are raised by putting a price on someone's life.

The central question is always whether the apology is sincere. Questions include: Was the apology offered solely to save the reputation of an individual or organization? Was the apology a direct or indirect means to duck blame or assign blame to others? Was the apology offered to avoid liability? Was the information shared in the apology truthful and complete?

6.6.1 Case Study: Maple Leaf Foods and the Listeria Food Contamination Crisis

Maple Leaf Foods is a major food processing company headquartered in Toronto, Canada. The company employs approximately 24,000 people across Canada, the United States, Europe, and Asia. In 2008, Maple Leaf Foods had sales over $5.2 billion. In August 2008, Canada had a crisis involving one of the worst cases of food contamination in the country's history. The crisis was caused by an outbreak of listeria due to contamination of meat products – specifically cold cuts – produced by Maple Leaf Foods. The contamination led to dozens of sick people and more than 20 deaths.

The bacterium Listeria monocytogenes, commonly referred to as listeria, is found in soil, vegetation, sewage, water, and the feces of animals and humans. Listeria bacteria can also be found in unpasteurized dairy products, raw vegetables and meats, and processed foods including deli meats and hot dogs. Eating foods spoiled with listeria monocytogenes can cause serious illness including brain and blood infections and, in extreme cases, death. People who eat foods contaminated with listeria may carry the bacteria and still not develop listeriosis. Those who are most vulnerable to developing the disease include the elderly, infants, and those with a compromised immune system. Foods that are tainted with listeria do not look, smell or taste off. Listeria, which can be killed through the cooking and pasteurization processes, will continue to grow in foods that are in the refrigerator.

Maple Leaf's response to the crisis and its apology stands as one of the most effective in the history of crisis communication.[38] Maple Leaf company officials responded in a timely manner with statements of caring and a compassionate apology. Maple Leaf instituted a voluntary recall before the outbreak was officially linked to their plant. Upon official confirmation of the link, Maple Leaf expanded the recall to all products from the plant.

In a press conference held at Maple Leaf Headquarters, president and CEO of Maple Leaf Foods Michael McCain responded to the Canadian Food Inspection Agency and Public Health Agency of Canada conclusion that the strain of listeria bacteria linked to the illness and death of several consumers matched the listeria strain identified in some Maple Leaf food products.

> To those people who are ill, and to the families who have lost loved ones, I offer my deepest and sincerest sympathies. Words cannot begin to express our sadness for their pain. . . This week our best efforts delivering the highest quality, safe food have failed us. For that we are deeply sorry. We know this has shaken consumer confidence in us. Our actions will continue to be guided by putting their interest first.[39]

In his testimony to a subcommittee of the Canadian Parliament, McCain admitted that his company failed in its efforts to protect consumers and was responsible for the deaths of 22 Canadians:

> [W]e did take responsibility and accountability for this, because it occurred in our plant, on our watch, with Canadian consumers eating our product. We have an obligation to produce a safe product, and it's an obligation we've held very close for 100 years. We had systems and protocols in place that we felt were best practice, and they failed us. So accountability and responsibility for that series of events does rest very squarely on our shoulders as an organization, and I'm personally accountable for that organization, so that rests very squarely on my shoulders.[40]

In contrast to the reputational strategy used by many other companies in crisis situations, such as blaming others, Maple Leaf Foods' response to the crisis was immediate and highly visible. Besides his apology at the press conference, McCain posted his apology in newspaper and television advertisements and on the corporate website. Maple Leaf sought no justification or excuse. It accepted accountability and responsibility. For example, in a *Globe and Mail* newspaper interview, McCain said:

> Collectively, we just tried to figure out what was the "right thing to do" in the middle of this terrible situation. It really isn't about a complicated strategy. We have a highly principled set of values in our company, and they guided us throughout, including putting consumers first and being clear and accepting responsibility.[41]

Media reports informed readers and viewers that total direct costs of the recall were estimated to be high, and class action lawsuits were being filed. In response, McCain announced that, knowing there was desire to assign blame, he stated firmly that the buck stopped with him. Rather than denying responsibility or blaming others, Maple Leaf accepted its guilt and sought forgiveness. McCain indicated that while the company's standards of practice and quality control had been breached, this was not representative of the corporation's identity or its standards of business practice.

McCain's apology was widely praised by mainstream broadcast and print media outlets.[39] It was frequently cited as a textbook example of crisis management and communication.

One hiccup in this case occurred on 17 September 2008. Canadian Agricultural Minister Gerry Ritz made national news as a result of comments he made about the Maple Leaf crisis during a conference call with government officials that were made public. Ritz was quoted as saying, "This is like a death by a thousand cuts. Or should I say cold cuts." Ritz apologized for his remarks. However, various groups called for his resignation. New Democratic Party Leader Jack Layton responded by saying, "Canadians are dying because of the mismanagement of our government. . . there should absolutely never be that kind of humor. . .. It illustrates the government is not taking this matter as seriously as they should." A spokesman for Prime Minister Harper released a statement saying Ritz's comments were tasteless and completely inappropriate.

By shutting down its plant in Toronto, initiating a nationwide recall, and taking precautionary measures necessary to minimize risks, Maple Leaf began to regain consumer confidence by late 2008. As part of this trust rebuilding effort, Maple Leaf assembled a group of leading food safety experts, including officials from the Canadian Food Inspection Agency and the Public Health Agency of Canada. Maple Leaf asked the group to advise the company on operational changes to prevent future contamination incidents. Based on recommendations from the group, Maple Leaf undertook several plant-wide measures to build and rebuild public confidence in its products. The company communicated these changes through various traditional and social media outlets.

Using social media, Maple Leaf Foods launched an external company blog to create a direct conversation with the public. In his first blog, McCain explained why the company was using social media. He indicated it was his belief it was Maple Leaf's responsibility to maintain a continuous, honest, and open line of communication with the public. The blog would also provide updates on the company's progress toward food safety, address topics and questions related to listeria management, and respond to new events and issues as they occurred.

Maple Leaf compensated victims. It also reached a settlement agreement with the principal groups that had launched class-action suits.

Following inspections conducted by the Canadian Food Inspection Agency, the Maple Leaf plant was reopened. Upon the reopening, McCain took groups of reporters on organized tours of the facility and communicated the company's improved safety protocols. When Maple Leaf announced its products would be returning to retail outlets and were safe and ready for purchase, it launched a new media campaign. The theme of the campaign was "Passionate People, Passionate about

Food." The campaign showed microbiologists working at the Toronto plant and then serving food to children in their homes. Maple Leaf also launched a new website dedicated to informing the public about steps it was taking to improve food safety. The company also launched another website containing a national listeria education and outreach program.

Maple Leaf survived the crisis. By January 2009, 70–75% of Maple Leaf sales had recovered.

6.6.2 Case Study: Southwest Airlines Apology

On Tuesday morning, 17 April 2018, an engine on Southwest Flight 1380 jet exploded over Pennsylvania and debris hit the plane. Jennifer Riordan, a 43-year-old bank executive from Albuquerque, New Mexico, was sucked partway out of the jet when a window shattered. She died later from her injuries. The Boeing 737, bound from New York to Dallas with 149 people aboard, made an emergency landing in Philadelphia. Part of the reason this incident was so widely reported was that passengers were immediately sharing their experiences and perspectives on social media. For example, one passenger paid $8 for inflight wi-fi while he thought the plane was going to crash so that he could broadcast what was happening on Facebook Live and say a farewell to friends and family.

Passengers who were on board Southwest Flight 1380 received letters of apology from the airline the next day. Gary Kelly, the chairman and chief executive officer of Southwest Airlines, signed the letter. The letter was one page and had five paragraphs. In the first paragraph, the company offered its apologies:

> "On behalf of the entire Southwest Airlines Family, please accept our sincere apologies for the circumstances surrounding Flight 1380 on Tuesday morning, April 17th."

The second paragraph of the letter identified what the company was doing. It read:

> "While the National Transportation Safety Board investigates the catastrophic engine failure, our primary focus and commitment is to assist you in every way possible."

In the third paragraph of the letter, the company detailed how customers could get help from Southwest Airlines staff.

In the fourth paragraph of the letter, Southwest took steps to repair relationships with the passengers. It reads:

> We value you as our Customer and hope you will allow us another opportunity to restore your confidence in Southwest as the airline you can count on for your travel needs. In this spirit, we are sending you a check in the amount of $5,000 to cover any of your immediate financial needs. As a tangible gesture of our heartfelt sincerity, we are also sending you a $1,000 travel voucher (in a separate e-mail), which can be used for future travel.

Such payments are not unusual in such situations. It puts money in the hands of people who may need it for expenses. Southwest Airlines did not do what other airline companies have done: offering compensation in exchange for people offering to drop all claims against the company.

In the fifth and final paragraph, the letter ended with the company repeating its apologies: "Again, please accept our heartfelt apologies."

6.7 Communications in the Postcrisis Recovery Phase

Communications in the *postcrisis recovery phase* are focused primarily on the knowledge gained in the days, weeks, months, and years following a crisis event. Communications in the postcrisis phase include messages focused on methods and means for recovery, and revolve around issues such as resilience, coping, recovery, failures, gaps, lessons learned, and postcrisis actions, along with secondary and tertiary crises and other impacts that resulted from the original event or from steps taken to address it.[42]

Recovery is best achieved when people, organizations, and communities are resilient. To accomplish resilience, strategies and messages are needed that identify psychological, social, cultural, economic, legal, and political incentives. Each incentive contributes to a desire for a return to normal. However, full recovery from a severe crisis is seldom achieved. Instead, a "new normal" is created under which people continue to live their lives.

Another communication challenge in the postcrisis recovery phase is stress. It is common for individuals involved in a crisis to experience extreme stress and anxiety. Many people can be expected to exhibit a wide range of stress-related thoughts, feelings, and behaviors. Extreme stress and anxiety can overwhelm an individual's ability to cope and process information and can severely impair communications. For example, stress can lead to poor problem-solving ability, lowered alertness, poor judgment, difficulty calculating, poor memory, ease of distraction, and inattention to detail. It can also lead to hypervigilance, withdrawal, and substance abuse. In a major crisis, significant numbers of people may experience long-term stress-related behavioral symptoms. These symptoms may lead to posttraumatic stress disorder (PTSD). Emotional reactions to stress that impair communications following a crisis include guilt, fear, shock, sadness, irritability, anger, disbelief, emotional numbness, panic attacks, and depressed mood. Emotions following a major crisis include:

- Anxiety and distress, e.g., people may be asking or thinking: Where can I turn for help? Will anything be left for me? What awful things are ahead? What do I do now?
- Anger, e.g., people may be asking or thinking: How could such awful things have happened to us? Why is no one helping? Doesn't anybody care about us anymore? Why have we been forgotten? Where are authorities when we need them? Why are we getting so little information? Why are we being treated so badly? Why are some people getting more than we are?
- Misery, depression, and empathy, e.g., people may be asking or thinking: Will things ever be the same? What can you possibly say to those who have lost everything?
- Disappointment and betrayal, e.g., people may be asking or thinking: Why do the authorities keep ignoring our wishes and demands? What have we done to justify this suffering?

Besides these stress-related reactions, individuals experiencing grief may go through stages, each of which can impair communications: trauma, shock, denial (e.g., by ignoring warnings or ignoring messages to take protective actions), anger (e.g., through emotional outbursts or assigning blame to others), bargaining (e.g., by trying to find things to do that will mitigate or solve the problem), depression, acceptance of loss, and forgiveness. Lack of information in the postcrisis phase promotes anxiety and encourages the spread of rumors and misinformation. Rumors and misinformation flourish in information vacuums.

Of the many individuals and groups that often experience extreme stress after crises, first responders and their families merit particular attention. Their stress levels are often exacerbated by demands for, and shortages of, resources to address the crisis. Failures to communicate effectively to first responders and their families can lead to additional stress, performance failures, and poor resilience. Resilience is created by the emotional support and communications from professional counselors and by a person's own social networks, including co-workers, family members, and faith-based ministers. Crisis managers need to facilitate these types of communication support. For example, many organizations assign a "care officer" to communicate with bereaved family members about financial and other issues following the death of a first responder.

An additional source of stress in the postcrisis stage is stigma. Stigma is a mark of shame or discredit, a strong feeling of disapproval. Stigma, from the Greek word of the same spelling meaning "mark, puncture," came into English through Latin to mean a mark burned into the skin to signify disgrace. Stigma can attach to people, places, and things. For example, following a radiological disaster, stigma can attach to the people exposed to radiation but who are not dangerous to others (e.g., people from the affected areas perceived as *polluted*, *contaminated*, and to be *shunned*), to the geographical areas affected (e.g., rumors and misinformation about those who live and travel to the areas affected), and to local products from the areas affected (e.g., avoidance of food, drink, and products from the areas affected). Stigma can linger long after a crisis. Stigma results in attitudes and behaviors ranging from avoidance and isolation to denials of assistance and services.

Key questions addressed in the postcrisis recovery stage include: What was done well or not well? Were there gaps and mistakes in precrisis preparedness and crisis response phases? Were precrisis preparedness efforts and crisis response efforts effective or ineffective? How can we improve our communication processes? The postcrisis phase offers opportunities to: (1) analyze and review crisis communication efforts for lessons learned; and (2) integrate lessons learned into the crisis communication plan to prevent recurrence and better prepare for future events. The postcrisis/recovery phase presents opportunities to deliver promised information, provide updates on recovery and prevention, and engage with stakeholders on remembrance and memorial events.

6.7.1 Case Study and Case Diary: New York City's Communication Trials by Fire, from West Nile to 9/11

I have a special affinity for New York City. I was born and raised in New York. Except for college in England, I went to school in New York City. For many years, I have worked with city officials on a diverse set of risk, high concern, and crisis communication issues.

The period from 1999 to 2001 was a particularly difficult time for New York City. The City faced many crises: an outbreak of the West Nile virus in 1999, the World Trade Center terrorist attack in September 2001, and anthrax terrorist attacks in September and October 2001. Characteristics of New York City – including its large size, diverse population, and location as a media, cultural, and financial center – increased the risk, high concern, and crisis communication challenges. Through a continual process of postcrisis evaluation, the City developed strong crisis management and communication preparedness strategies.

During this period of turmoil, I worked with various City departments on crisis communication. My primary counterpart in city government was Sandra Mullin, director of communications and associate commissioner of the City's Department of Health and Mental Hygiene. New York's crisis management and communication approach was founded on the following foundational principles.

First, carefully listen to and understand stakeholders' values and concerns.

Second, recognize and adapt to the fact that many people and groups use risks as proxies or surrogates for other more general social, economic, legal, political, or cultural concerns and agendas.

Third, be proactive, initiating risk and communication efforts quickly with adequate resources and clear objectives.

Fourth, adopt best practices for stakeholder engagement and dialogue.

Fifth, engage in outreach activities and develop messages based on empirically based information about who is perceived to be most trustworthy; who is best suited to communicate risk and crisis messages; which messages are most effective; which messages are most respectful of different values and worldviews; which messages raise moral or ethical issues; which messages are most respectful of established institutional processes and expectations; and where, when, and how the risk and information can best be communicated.

In her review of New York City's crisis management and communication strategy, Mullin cited lessons drawn from risk and crisis communication theories articulated by Sandman, Covello, and others.[43]

1) Go out with the news quickly, no matter how inconvenient.
2) People's perception of risk may differ from what we want their perception of risk to be.
3) Do not just recognize risk perception differences. Acknowledge and empathize with people's concerns.
4) Risk communication should be carefully considered and assessed as part of a broader public health goal.
5) Use communication strategies to help people get used to the scary and the unthinkable.
6) Overcommunicate – especially during the early phase of a crisis.
7) Press conference performance matters.
8) Make sure credible spokespeople are prepared to communicate with the media and make time to do so.
9) Develop relationships with community groups and communicate with them often.
10) Get ready for the next crisis: get to know reporters; write down a communications plan; create and continuously update contact lists; develop "what if" fact sheets, press releases, and message palettes; and build communication teams.

Based on these lessons learned, New York City undertook numerous risk and crisis communication activities in which I took part. These are listed in Table 6.18.

6.7.2 Case Study: Johnson & Johnson and the Tylenol Tampering Case

The Johnson & Johnson Tylenol case study is a widely cited example of effective crisis communication messaging during the crisis event and in the postcrisis stage. Leadership at Johnson & Johnson, the maker of the pain reliever Tylenol, sets an example for other organizations by making the protection of public health and safety the company's key message through both actions and words.

In 1982, a person or persons replaced capsules of the over-the-counter pain reliever Tylenol with cyanide-laced capsules. The poisoned Tylenol was purchased from pharmacies and food stores in the Chicago area. Seven people died. Before the crisis, Tylenol was the leader in the pain reliever field, accounting for over 35% of the market share for pain relievers. Tylenol outsold all four of the other leading painkillers. After the seven deaths, Johnson & Johnson's market share for pain relievers dropped to 7%.

Upon being informed that people had ingested poisoned Tylenol, James Burke, Johnson & Johnson's chairman, alerted consumers, via the media, not to consume any Tylenol product. Burke told consumers not to resume using the product until the extent of the tampering could be determined and the investigation was concluded.

In response to the crisis, Johnson & Johnson stopped the production and advertising of extra-strength Tylenol. It withdrew all Tylenol capsules from store shelves in Chicago and the

Table 6.18 Synopsis of New York City Crisis Communication Strategy and Activities: 1999-2001.

1) Establish a baseline of public and community leaders' knowledge, attitudes, information needs, and trust levels.
2) Develop a crisis communication plan with contact lists, templates, and protocols for what to do before, during, and after a crisis.
3) Improve internet and intranet sites including developing a "NYC Aware" section of the agency website for crisis-related information.
4) Develop an electronic centralized contact mailing list as a means of providing regular communications before a crisis event occurs and as a means of developing a system for rapid communication if and when a crisis event does occur.
5) Enhance cross-cultural and translation capacity to facilitate information dissemination and sharing with immigrant and non-English speaking populations and communities.
6) Prepare city leaders to better communicate in crisis situations by providing risk and crisis communication training to over 500 key city leaders, including those at city hall; the police, fire, emergency management, public health, environmental protection, and transportation departments; public and private hospitals; and the mental health community.
7) Provide media and crisis communication training for over 100 key communicators from different agencies.
8) Develop materials for likely crisis scenarios, such as crisis preparedness brochures, as well as fact sheets, press releases, and materials for those with low literacy levels.
9) Meet formally and informally with key media representatives to discuss crisis preparedness issues and needs and concerns on both sides.
10) Convene ethnic and community media roundtables focused on crisis issues.
11) Develop and implement crisis-related educational campaigns, such as the "*If you see something, say something*" public information campaign.
12) Work closely with emergency management officials to coordinate communication activities within the Incident Command Structure.
13) Develop drills based on the crisis communication plan.

surrounding area. Burke followed the Chicago withdrawal with a recall of all 31 million bottles of Tylenol with a retail value of over $100 million. The withdrawal of Tylenol from store shelves costs Johnson & Johnson millions of dollars.

By withdrawing all Tylenol from the marketplace, even before the results of the investigation were known, and even though company executives were convinced the company was not responsible for the tampering, Johnson & Johnson communicated one of the key attributes of trust: caring and concern. Burke was highly visible and accessible throughout the crisis, presenting highly personal messages at press conferences and on television.

In the first week of the crisis, Johnson & Johnson established a 1–800 hotline for consumers to call. They also held numerous press conferences, established a toll-free line for news organizations to call, and published daily messages with updated statements. The company provided counseling and financial assistance for families of victims, even though the company was not responsible for the product tampering. New information was released daily, consistent with the crisis communication dictum: "Tell the truth, tell it well, tell it early, tell it all, and tell it fast."

Johnson & Johnson's response to the tampering incident was not without errors and mistakes. For example, in response to a question about whether cyanide was used in any part of the manufacturing process, Johnson & Johnson company executives said definitively "no." However, on that same day, company executives learned that cyanide was in fact used in the quality assurance process. The company released this information and apologized. While the reversal embarrassed the company briefly, Johnson & Johnson's quick response, clear explanation, and openness help them recover trust.

Also, Johnson & Johnson was slow to recall the product. The poisonings occurred on 29 September 1982. Investigations by police discovered a common denominator: all the victims had recently taken Tylenol. Tylenol samples were examined and were found to be tainted with cyanide. Police issued urgent warnings via the media, and they also sent patrols through the Chicago metropolitan area issuing warnings over loudspeakers to discontinue the use of all Tylenol products. In parallel with the police activities, Johnson & Johnson distributed warnings to hospitals and distributors and halted Tylenol production and advertising. However, it was not until 5 October 1982 that Johnson & Johnson issued a nationwide recall of Tylenol.

Six months after the crisis occurred, Johnson & Johnson reintroduced Tylenol into the marketplace with tamper-proof packaging. The package was triple sealed and tamper-proof – a glued box, a plastic seal over the neck of the bottle, and a foil seal over the mouth of the bottle. Tylenol became the first product in the industry to use the new tamper-resistant packaging. In order to motivate consumers to buy the product, they offered various monetary incentives.

Johnson & Johnson was able to recover the market share it had lost during the crisis. The organization was able to reestablish the Tylenol brand name as one of the leading over-the-counter consumer products in the United States. In 1990, Fortune magazine inducted Burke into its National Business Hall of Fame, noting, "Few managers of corporate crises have survived an episode of the perfect crime – unsolved murder – in which their product was the murder weapon and their customers the innocent victims. . . He managed it so well that he not only restored Tylenol, his company's single most important profit-maker, to preeminence, but he also enhanced the company's fine reputation. During his tenure, Burke also took Johnson & Johnson's earnings and stock prices to new highs."

In an interview published in the book *Lasting Leadership*, Burke emphasized the value of Johnson & Johnson's credo in the crisis, which dated back to the company's founding in 1887. The credo stated the company is responsible first to its customers, then to its employees, the community, and the stockholders, in that order. Burke also noted: "Trust has been an operative word in my life. [It] embodies almost everything you can strive for that will help you to succeed. You tell me any human relationship that works without trust, whether it is a marriage or a friendship or a social interaction; in the long run, the same thing is true about business."

6.7.3 Case Study: Flint, Michigan and Contaminated Drinking Water

The drinking water contamination crisis in Flint, Michigan that began in 2014 represents a classic case of a water contamination crisis compounded by massive failures in risk and crisis communication and messaging. In 2014, as a cost-saving measure by a nearly bankrupt city, Flint changed its water source from water treated by the City of Detroit Water and Sewerage Department to the Flint River. Soon after the change, Flint residents complained about dark-colored, foul-tasting, smelly water and skin rashes, and hair loss.

In addition to these serious complaints, the drinking water also had developed serious lead contamination problems. The corrosive Flint River water caused lead from aging pipes to leach into the water supply, causing high levels of lead in the water and the blood of children. Lead leached out from aging pipes into thousands of homes. Flint failed to properly treat the water, and the state failed to properly test it. Drinking water testing conducted in February 2015 found that a significant proportion of samples of Flint's drinking water had lead levels well above the regulatory standards set by the US Environmental Protection Agency. Test samples showed lead levels were in some cases more than 100 times the regulatory standard.

After the city announced the water was contaminated with bacteria and, later, chemicals that cause cancer, state officials insisted nothing was seriously wrong. A leaked report from the federal Environmental Protection Agency warning of lead contamination was dismissed as the work of a "rogue employee." When pediatricians in Flint reported a spike in lead in children's blood, a state official dismissed the information in the report as simply "data." The state admitted something was wrong only after scientists from Virginia Tech went to Flint to test the water and found elevated lead levels in 40% of homes.

Flint pediatrician Dr. Mona Hanna-Attisha began to suspect something had gone terribly wrong with the public water supply. Dr. Hanna-Attisha worked as a physician at a state-funded medical center. She was stone-walled by officials when she sought research data on the water, then given falsified numbers. There were also accusations against her personally and that her speaking up could endanger the medical center's funding. But she found research that showed a startling and dangerous upswing in lead in the city's water. According to her book on the Flint crisis, she went public with her research and received significant backlash, including threats against her.[44]

As a result of the lead crisis, more than 8,000 children were exposed to contaminated water. Lead can damage human brains and nervous systems in irreversible ways, especially those of children. In parts of Flint, the percentage of children with high levels of lead in their blood doubled after the switch from Detroit city water to the polluted Flint River. Children exposed to lead have lower IQs and are more likely to have difficulty focusing and paying attention. They can have difficulties with learning, speaking, and language processing. They are more likely to be impulsive and aggressive and to be diagnosed with ADHD. Many of these problems are lifelong. According to the US Environmental Protection Agency, there is no demonstrated safe concentration of lead in blood.

Even before the lead crisis, Flint was struggling. About 40% of its residents live in poverty. Michigan Governor Rick Snyder had appointed an emergency manager, an unelected official with near-total control over the city's finances because Flint was near bankruptcy. In January 2016, the city was declared to be in a state of emergency by the governor of Michigan. Several government officials – including one from the City of Flint, two from the Michigan Department of Environmental Quality, and the regional administrator from the US Environmental Protection Agency – resigned over the mishandling of the crisis. In April 2016, criminal charges were filed against several government officials by the attorney general of Michigan.

The contamination of water and the declaration of a state of emergency raised many operational and legal issues. These included allegations of violations of the Safe Drinking Water Act, misconduct in office, tampering with evidence, cover-ups, and communicating false information. In January 2016, the Natural Resources Defense Council and its local partners sued the City of Flint and Michigan state officials, seeking to secure safe drinking water for Flint residents. Fourteen months later, the state and the local government agreed to replace lead service lines and institute a transparent and effective lead-monitoring system.

Compounding these operational and legal problems were massive risk and crisis communication failures by officials at virtually every level of government. Among these many failures were the failure by officials to provide accurate, caring, and timely messages in response to the hundreds of questions posed by the media, regulators, health care providers, and the public. For example, throughout the early stages of the crisis, city and state representatives continually told Flint citizens their water was safe to drink. A final report by a task force commissioned by the Michigan governor accused officials of "intransigence, unpreparedness, delay, and inaction." Throughout the crisis, officials failed to follow the first and most basic cardinal rule of risk and crisis communication: *Tell the truth, tell it well, tell it early, tell it all, and tell it fast.*

Risk and crisis communication messages were poorly prepared and delivered by Michigan officials in response to virtually every stakeholder question, representing all the "5Ws and 1H" (Who, What, Where, When, Why, and How). Even answers to basic questions were poorly prepared and delivered, including:

- Who sets the lead standards for drinking water in Flint?
- What is the lead standard for drinking water in Flint?
- Is the lead standard for drinking water in Flint the same as elsewhere?
- Where were the most serious lead problems found in Flint?
- When did Flint first discover lead problems in the drinking water?
- Why is it taking so long to fix the lead problems in Flint?
- How can the lead problems in Flint and elsewhere best be fixed?

Government officials in Michigan could have provided answers to these and other water contamination questions by referring to the more than 200 peer-reviewed water contamination message maps published in 2007 by the Environmental Protection Agency.[45]

As more and more citizens voiced their concerns, state officials were "callous and dismissive," according to a report by the independent Flint Task Force, established by Michigan's governor Rick Snyder in October 2015. Government officials acknowledged and communicated there was a serious problem only after residents elevated public awareness and garnered national attention. And in the end, they only acknowledged, communicated, and attempted to correct the problem only after being compelled to do so by the courts in response to a lawsuit and subsequent mediation.

The Flint case study, along with the case studies of New York City and Johnson & Johnson, highlights many of the crisis communication principles that guide all three phases of a crisis, with each phase informing the other. Effective crisis communication relies on a cycle of good planning and preparedness, responsiveness during the crisis, and learning from postcrisis evaluation.[46] While the crisis event typically gathers the most attention, the handling of each phase has the potential for an enormous impact on an organization and its stakeholders.

6.8 Chapter Resources

Below are additional resources to expand on the content presented in this chapter.

Angus, K., Cairns, G., Purves, R., Bryce, S., MacDonald, L., & Gordon, R. (2013). *Systematic Literature Review to Examine the Evidence for the Effectiveness of Interventions That Use Theories and Models of Behaviour Change: Towards the Prevention and Cntrol of Communicable Diseases.* Stockholm: European Centre for Disease Prevention and Control.

Arvai, J., Campbell-Arvai, V. (2014). Risk communication: Insights from the decision sciences, in *Effective Risk Communication*, eds. Arvai, J., Rivers, L. New York: Earthscan,

Baron, J., Hershey, J.C., Kunreuther H. (2000). "Determinants of priority for risk reduction: The role of worry." *Risk Analysis* 20(4):413–428.

Barton, L. (2001). *Crisis in Organizations.* Cincinnati, Ohio: College Divisions Southwestern.

Bennett, P., Caiman, K. (eds.) (1999). *Risk Communication and Public Health.* New York: Oxford University Press.

Benoit, W. L. (1995). *Accounts, Excuses, and Apologies: A Theory of Image Restoration*. Albany: State University of New York Press.

Benoit, W. L. (2005). "Image restoration theory," in *Encyclopedia of Public Relations*, vol 1, ed. R. L. Heath. Thousand Oaks, CA: Sage.

Benoit, W. L. and Pang, A. (2008). "Crisis communication and image repair discourse," in *Public Relations: From Theory to Practice*, eds. T. L. Hansen-Horn and B. D. Neff. New York: Pearson.

Berland G., Elliott, M., Morales, L., Algazy, J., and Kravitz, R. (2001). "Health information on the Internet: accessibility, quality, and readability in English and Spanish." *Journal of the American Medical Association* 285:2612–2621.

Bier, V.M. (2001). "On the state of the art: risk communication to the public." *Reliability Engineering & System Safety* 71(2):139–150.

California Department of Public Health (2015). *Crisis and Emergency Risk Communication Toolkit*. Sacramento, CA: California Department of Public Health. Accessed at: http://cdphready.org/wp-content/uploads/2015/03/crisis_and_emergency_risk_communication_toolkit_july_2011.pdf

Campbell, Joseph. (2008). *The Hero with a Thousand Faces*, 2nd edition. Novato, CA: New World Library.

CDC (US Centers for Disease Control and Prevention) (2019). *Crisis and Emergency Risk Communication*. Atlanta, GA: Centers for Disease Control and Prevention.

Chess, C., Hance, B.J., and Sandman, P.M. (1989). *Planning Dialogue with Communities: A Risk Communication Workbook*. New Brunswick: Rutgers University, Cook College, Environmental Media Communication Research Program.

Chung, I.J. (2011). "Social amplification of risk in the Internet environment." *Risk Analysis* 31(12):1883–1896.

Coombs, W.T. (1995). "Choosing the right words: The development of guidelines for the selection of the 'appropriate' crisis-response strategies." *Management Communication Quarterly* 8(4):447–476.

Coombs, W.T. (2004). "A theoretical frame for post-crisis communication: Situational crisis communication theory." in *Attribution Theory in the Organizational Sciences: Theoretical and Empirical Contributions*, ed. M.J. Martinko. Greenwich, CT: Information Age Publishing.

Coombs, W.T. (2006). *Code Red in the Boardroom: Crisis Management as Organizational DNA*. Westport, CT: Praeger.

Coombs, W.T. (2007). "Protecting organization reputations during a crisis: The development and application of situational crisis communication theory." *Corporate Reputation Review* 10:163–176.

Coombs, W.T. (2009a). "Conceptualizing crisis communication," in *Handbook of Crisis and Risk Communication*, eds. R. L. Heath and H. D. O'Hair. New York: Routledge.

Coombs, W.T. (2009b). "Crisis, crisis communication, reputation, and rhetoric," in *Rhetorical and Critical Approaches to Public Relations*, eds. R. L. Heath, E. L. Toth, and D. Waymer. New York: Routledge.

Coombs, W.T. (2019). *Ongoing Crisis Communication: Planning, Managing, and Reporting*. Thousand Oaks, CA: SAGE Publications.

Coombs, W.T., and Holladay, S.J. (2005). "Exploratory study of stakeholder emotions: Affect and crisis," in *Research on Emotion in Organizations, Volume 1: The Effect of Affect in Organizational Settings*, eds. N. M. Ashkanasy, W. J. Zerbe, and C. E. J. Hartel. New York: Elsevier.

Coombs, W.T., and Holladay, S.J. (2012). *Handbook of Crisis Communication*. Hoboken, NJ: Wiley.

Coombs, T.W. (2014). *Crisis Management and Communications –Updated*. Institute for Public Relations. Accessed at: https://instituteforpr.org/crisis-management-communications/

Coombs, T.W. (2019). *Ongoing Crisis Communication: Planning, Managing, and Responding*, 5th edition. Thousand Oaks, CA: Sage.

Coombs, W.T., and Holladay, S.J. (2012). *Handbook of Crisis Communication.* Hoboken, NJ: Wiley.

Covello, V.T. (1983). "The perception of technological risks: A literature review." *Technological Forecasting and Social Change* 23:285–297.

Covello. V.T. (1984). "Social and behavioral research on risk: Uses in risk management decisionmaking." *Environment International* 10(5–6): 541–545.

Covello, V.T. (1991). "Risk comparisons and risk communication: Issues and problems in comparing health and environmental risks," in *Communicating Risks to the Public*, eds. R.E. Kasperson and P.J.M. Stallen. Dordrecht, The Netherlands: Kluwer Academic Publishers.

Covello V.T. (1992). "Risk communication, trust, and credibility." *Health and Environmental Digest* 6(1):1–4.

Covello V.T. (1993). "Risk communication, trust, and credibility." *Journal of Occupational Medicine* 35:18–19.

Covello, V.T. (2003). "Best practice in public health risk and crisis communication." *Journal of Health Communication* 8:1–5.

Covello, V.T. (1998). *Risk Perception, Risk Communication, and EMF Exposure: Tools and Techniques for Communicating Risk Information.* Vienna, Austria: World Health Organization/ICNRP International.

Covello, V.T. (2006). "Risk communication and message mapping: A new tool for communicating effectively in public health emergencies and disasters." *Journal of Emergency Management* 4(3):25–40.

Covello, V.T. (2008). "Strategies for overcoming challenges to effective risk communication," in *Handbook of Risk and Crisis Communication*, eds. Heath, R. and H. O'Hair. New York: Routledge

Covello, V.T. (2011a). *Guidance on Developing Effective Radiological Risk Communication Messages: Effective Message Mapping and Risk Communication with the Public in Nuclear Plant Emergency Planning Zones. NUREG/CR-7033.* Washington, DC: Nuclear Regulatory Commission.

Covello, V.T. (2011b). *Developing an Emergency Risk Communication (ERC)/Joint Information Center (JIC) Plan for a Radiological Emergency. NUREG/CR-7032.* Washington, DC: Nuclear Regulatory Commission.

Covello, V.T. (2011c). "Risk communication, radiation, and radiological emergencies: strategies, tools, and techniques." *Health Physics,* November, 101(5):511–530.

Covello, V.T. (2014). "Risk communication," in *Wiley-Blackwell Encyclopedia of Health, Illness, Behavior, and Society*, eds. W. Cockerham, R. Dingwall, and S. Quah. Oxford: Blackwell.

Covello, V.T (2016). "Risk communication," in *Environmental Health: From Global to Local.* 3rd edition, ed. Frumkin, H. San Francisco: Jossey-Bass/Wiley. 988–1008.

Covello, V.T., Allen, F. (1988). *Seven Cardinal Rules of Risk Communication.* Washington, DC: US Environmental Protection Agency.

Covello, V., Becker, S., Palenchar, M., Renn, O., and Sellke, P. (2010). *Effective Risk Communications for the Counter Improvised Explosive Devices Threat: Communication Guidance for Local Leaders Responding to the Threat Posed by IEDs and Terrorism.* Washington, DC: Department of Homeland Security.

Covello, V., and Hyer R. (2020). *COVID-19: Simple Answers to Top Questions: Risk Communication Field Guide Questions and Key Messages.* Arlington, VA: Association of State and Territorial Health Officers. Accessed at: https://www.hsdl.org/?view&did=835774

Covello V.T., and Hyer R. (2014). *Top Questions On Ebola: Simple Answers.* Arlington, VA: Association of State and Territorial Health Officials.

Covello, V.T., McCallum, D.B., and Pavlova, M.T. eds. (1989). *Effective Risk Communication: The Role and Responsibility of Government and Nongovernment Organizations.* New York: Plenum.

Covello, V.T., Minamyer, S., and Clayton, K. (2007). *Effective Risk and Crisis Communication during Water Security Emergencies: Summary Report of EPA Sponsored Message Mapping Workshops.* Cincinnati, OH: US Environmental Protection Agency.

Covello, V.T., and Mumpower, J. (1985). "Risk assessment and risk management: an historical perspective. *Risk Analysis* 5(2):103–120.

Covello, V.T., Peters, R.G., and Wojtecki, J.G. (2001). "Risk communication, the West Nile virus epidemic, and bioterrorism: Responding to the communication challenges posed by the intentional or unintentional release of a pathogen in an urban setting." *Journal of Urban Health* 78:382–391.

Covello, V.T., and Sandman, P.M. (2001). "Risk communication: Evolution and revolution" in *Solutions to an Environment in Peril*, ed. A. Wolbarst. Baltimore: Johns Hopkins University Press.

Covello, V.T., von Winterfeldt, D., and Slovic P. (1987). "Communicating scientific information about health and environmental risks: Problems and opportunities from a social and behavioral perspective" in *Uncertainties in Risk Assessment and Risk Management*, eds. V.T. Covello, A. Moghissi, and V. Uppulori. New York: Plenum Press.

Crick, M. J. (2021). "The importance of trustworthy sources of scientific information in risk communication with the public." *Journal of Radiation Research* 62(S1):1–6.

Cummings, L. (2014). "The 'trust' heuristic: Arguments from authority in public health." *Health communication* 29(10):1043.

Davies, C.J., Covello, V.T., and Allen, F.W., eds. (1987). *Risk Communication: Proceedings of the National Conference.* Washington: The Conservation Foundation.

Dean, D.H. (2004). "Consumer reaction to negative publicity: Effects of corporate reputation, response, and responsibility for a crisis event." *Journal of Business Communication* 41:192–211.

Downs, J. (2014). "Video interventions for risk communication," in *Effective Risk Communication*, eds. Arvai, J. and Rivers, L. New York: Earthscan.

Durodié, B. (2006). *The Concept of Risk.* London: The Nuffield Trust and the UK Global Health Programme.

Dykes, B. (2020). *Effective Data Storytelling: How to Drive Change with Data, Narrative and Visuals.* Hoboken, NJ: Wiley.

Fang, D., Fang, C.L., Tsai, B.K., Lan, L.C., and Hsu, W.S. (2012). "Relationships among trust in messages, risk perception, and risk reduction preferences based upon avian influenza in Taiwan." *International Journal of Environmental Research and Public Health* 9(8)2742–2757.

Fazio, L.K., Brashier, N.M., Payne, B.K., and Marsh, E.J. (2015). "Knowledge not protection against illusory truth." *Journal of Experimental Psychology* 144(5):993–1002.

Fearn-Banks, K. (2016). *Crisis Communications: A Casebook Approach.* New York: Routledge.

Fink, S. (1986). *Crisis Management: Planning for the Inevitable.* New York, NY: American Management Association.

Fischhoff, B. (1989). "Helping the public make health risk decisions," in *Effective Risk Communication: The Role and Responsibility of Government and Nongovernment Organization*, eds. V.T. Covello, D.B. McCallum, and M.T. Pavlova. New York: Plenum Press.

Fischhoff, B. (1995). "Risk perception and communication unplugged: Twenty years of progress." *Risk Analysis* 15(2):137–145.

Fischhoff B. (2005). "Risk perception and communication," in *Handbook of Terrorism and Counter-Terrorism*, ed. D. Kamien. New York: McGraw-Hill.

Fischhoff, B., Slovic, P., and Lichtenstein, S. (1978). "How safe is safe enough? A psychometric study of attitudes towards technological risks and benefits." *Policy Sciences* 9(2):127–152.

Fischhoff, B., Slovic, P., and Lichtenstein, S. (1981). "Lay foibles and expert fables in judgments about risk," in *Progress in Resource Management and Environmental Planning*, eds. T.O'Riordan and R. K. Turner. New York: Wiley.

Fischhoff, B., Lichentenstein, S., and Slovic, P. (1981). *Acceptable Risk*. New York: Cambridge University Press.

Fischhoff, B., Bostrom, A., and Quadrel, M.J. (2002). "Risk perception and communication" in *Oxford Textbook of Public Health: The Methods ofPpublic Health*. 4th edition, eds. R. Detels, J. McEwan, R. Beaglehole, and J. Heinz. New York: Oxford University Press.

Fleishman-Mayer, L., Bruine de Bruin, W. (2014). "The 'Mental Models' methodology for developing communications: Adaptions for information public risk management decisions about emerging technologies," in *Effective Risk Communication*, eds. Arvai, J. and L. Rivers. New York: Earthscan.

Flynn, T. (2009). *Authentic Crisis Leadership and Reputation Management: Maple Leaf Foods and 2008 Listeriosis Crisis*. Hamilton, Ont.: DeGroote School of Business, McMaster University.

Fox-Glassman, K.T., and Weber, E.U. (2016). "What makes risk acceptable? Revisiting the 1978 psychological dimensions of perceptions of technological risks." *Journal of Mathematical Psychology* 75:157–169.

Frandsen, F., and Johansen, W. (2017). *Organizational Crisis Communication: A Multi-vocal Approach*. Thousand Oaks, CA: Sage Publications.

Germani, F., and Biller-Andorno, N. (2021). "The anti-vaccination infodemic on social media: A behavioral analysis." *PLOS One* March 2021.

Glik, D.C. (2007). "Risk communication for public health emergencies." *Annual Review of Public Health* 28:33–54.

Grunig, J.E. (1997). "A situational theory of publics: Conceptual history, recent challenges and new research," in *Public Relations Research: An International Perspective*, eds. D. Moss, T. MacManus, and D. Vercic. London: International Thomson Business Press.

Gustafson, P.E. (1998). "Gender differences in risk perception: theoretical and methodological perspectives." *Risk Analysis* 18(6):805–811.

Hampel, J. (2006). "Different concepts of risk: A challenge for risk communication." *International Journal Medical Microbiology* 296(40):5–10.

Health Protection Network. (2008). *Communicating with the Public about Health Risks*. Glasgow: Health Protection Scotland.

Heath, R.L,. and O'Hair, H.D. (eds.). (2020). *Handbook of Risk and Crisis Communication*. New York: Routledge.

Heath, R.L., Palenchar, M.J., Proutheau, S., and Hocke, T.M. (2007). "Nature, crisis, risk, science, and society: What is our ethical responsibility?" *Environmental Communication* 1(1):34–42.

Henwood, K., Pidgeon, N., Parkhill, K., and Simmons, P. (2010). "Researching risk: Narrative, biography, subjectivity." *Forum: Qualitative Social Research* 11(1):1–22.

Holmes, B.J. (2008). "Communicating about emerging infectious disease: the importance of research." *Health, Risk & Society* 10:349–360.

Huang, Y.H. (2006). "Crisis situations, communication strategies, and media coverage: A multi-case study revisiting the communicative response model." *Communication Research* 33:180–205.

Hunter, P.R., and L. Fewtrell (2001). "Acceptable Risk." in *Water Quality: Guidelines, Standards and Health (World Health Organization)*, eds. L. Fewtrell and J. Bartram, London: IWA Publishing. Accessed at: https://www.who.int/water_sanitation_health/dwq/iwachap10.pdf

Hyer, R.N., and Covello, V.T. (2007). *Effective Media Communication in Public Health Emergencies: A Field Guide*. Geneva: World Health Organization.

Hyer, R.N., and Covello, V.T. (2017). *Top Questions on Zika: Simple Answers*. Arlington, VA: Association of State and Territorial Health Officials.

Infanti, J., Sixsmith, J., Barry, M.M., Núñez-Córdoba, J.M., Oroviogoicoechea-Ortega, C., and Guillén-Grima, F.A. (2013). *A Literature Review on Effective Risk Communication for the Prevention and Control of Communicable Diseases in Europe*. Stockholm: European Centre for Disease Control and Prevention.

Janssen, A., Landry, S., and Warner, J. (2006). "'Why tell me now?' The public and healthcare providers weigh in on pandemic influenza messages." *Journal of Public Health Management Practice* 12:388–94.

Jamieson, K.H., Lammie, K., Warlde, G., et al. (2003). "Questions about hypotheticals and details in reporting on anthrax." *Journal of Health Communication* 8:121–131.

Johnson, B.B., and Covello, V. (1987). *The Social and Cultural Construction of Risk: Essays on Risk Selection and Perception*. Dordrecht: D. Reidel Publishing.

Johnson, B.B. (1999). "Ethical issues in risk communication." *Risk Analysis* 19(3):335–348.

Jong, W., and Dückers, M.L.A. (2019). "The perspective of the affected: What people confronted with disasters expect from government officials and public leaders." *Risk, Hazards & Crisis in Public Policy* 10(1):14–31.

Kasperson, R., Renn, O., Slovic, R, Brown, H., Emel, J., Goble, R., Kasperson, J., and Ratick, S. (1988). "The social amplification of risk: A conceptual framework." *Risk Analysis* 8(2):177–187.

Kahan, D, Braman, D., Cohen, G., Slovic, P., and Gastil, J. (2010), "Who fears the hpv vaccine, who doesn't, and why: an experimental study of the mechanisms of cultural cognition," *Law and Human Behavior* 34(6):501–51.

Kahan, D., Slovic, P., Braman, D., Cohen, G., & Gastil, J. (2009), "Cultural Cognition of the Risks and Benefits of Nanotechnology," *Nature Nanotechnology*, 4 (2): 87–90.

Leiss, W. (1995). "Down and dirty: the use and abuse of public trust in risk communication." *Risk Analysis* 15(6):685–692.

Lewis, M. (2021). *The Premonition: A Pandemic Story*. New York: Norton.

Lin, I.H., and Petersen, D. (2008). *Risk Communication in Action: The Tools of Message Mapping*. Washington, DC: US Environmental Protection Agency.

Lindell, M.K., and Perry, R.W. (2012). "The protective action decision model: theoretical modifications and additional evidence." *Risk Analysis* 32:616–632.

Linn, M.R., and Tinker, T.L. (1994). *A Primer on Health Risk Communication Principles and Practices*. Washington, DC: Agency for Toxic Substances and Disease Registry.

Lindell, M.K., and Perry, R.W. (2012). "The protective action decision model: theoretical modifications and additional evidence." *Risk Analysis* 32(4):616–632.

Lindenfeld, L., Smith, H.M., Norton, T., and Grecu, N.C. (2014). "Risk communication and sustainability science: lessons from the field." *Sustainability Science* 9(2):119–127.

Liu, T., Zhang, H., and Zhang, H. (2020). "The impact of social media on risk communication of disasters." *International Journal of Environmental Research and Public Health* 883:1–17.

Löfstedt, R.E. (2008) "What environmental and technological risk communication research and health risk research can learn from each other." *Journal of Risk Research* 11(1-2):141–167.

Lundgren, R.E., and McMakin, A.H. (2018). *Risk Communication: A Handbook for Communicating Environmental, Safety, and Health Risks*. Hoboken, NJ: Wiley.

Malecki, K.M.C., Keating, J.A., and Safdar, N. (2021). "Crisis communication and public perception of COVID-19 risk in the era of social media." *Clinical Infectious Diseases* 72(4):697–702.

Markwart, H., Vitera, J., Lemanski, S., Kietzmann, D., Brasch, M., and Schmidt, S. (2019). "Warning messages to modify safety behavior during crisis situations." *International Journal of Disaster Risk Reduction* 38:101235.

Maslow, A. (1970). *Motivation and Personality*. New York: Harper and Row.

McComas, K.A. (2006). "Defining moments in risk communication research: 1996-2005." *Journal of Health Communication* 11(1):75–91.

Mebane, F., Temin, S., and Parvanta, C.F. (2003). "Communicating anthrax in 2001: A comparison of CDC information and print media accounts." *Journal of Health Communication* 8:50–82.

Mileti, D.S. (1990). "Warning systems: a social science perspective," in *Preparing for Nuclear Power Plant Accidents*, eds. D. Golding, J.X. Kasperson, and R.E. Kasperson. Boulder, CO: Westview Press.

Mileti, D.S., and Fitzpatrick, C. (1991). "Communication of public risk: its theory and its application." *Sociological Practice Review* 2(1):20–8.

Mileti, D.S., and Peek, L. (2000). "The social psychology of public response to warnings of a nuclear power plant accident." *Journal of Hazardous Materials* 75:181–194.

Mileti, D.S., and Sorensen, J.H. (1988). "Planning and implementing warning systems," in *Mental Health Response to Mass Emergencies*, ed. M. Lystad. New York, NY: Brunner-Mazel.

Miller, G. (1956). "The magical number seven, plus or minus two: some limits on our capacity for processing information" *Psychological Review* 101(2):343–352.

Mitroff, I. (2004). *Crisis Leadership: Planning for the Unthinkable*. Hoboken, NJ: Wiley.

Morgan, M.G., Fischhoff, B., Bostrom, A., and Atman, C. (2001). *Risk Communication: The Mental Models Approach*. New York: Cambridge University Press.

Mousavi, S.Y., Low, R., and Sweller, J. (1995). "Reducing cognitive load by mixing auditory and visual presentation auditory and visual presentation modes." *Journal of Educational Psychology* 87(2):319–334.

Mullin, S. (2003), "The anthrax attacks in New York City: The "Giuliani press conference model" and other communication strategies that helped." *Journal of Health Communication* 8:15–16.

National Research Council (1989). *Improving Risk Communication. Committee on Risk Perception and Risk Communication*. Washington: National Academy Press.

National Research Council. (1996). *Understanding Risk: Informing Decisions in a Democratic Society*. Washington: National Academy Press.

Neeley, L. (2014). "Risk communication and social media," in *Effective Risk Communication*, eds. Arvai, J. and Rivers, L. New York: Earthscan.

Person, B., Sy, F., Holton, K., Govert, B., and Liang, A.(2004). "Fear and stigma: the epidemic within the SARS outbreak." *Emerging Infectious Diseases* 10(2):358–363.

Peters, R.G., Covello, V.T., and McCallum, D.B. (1997). "The determinants of trust and credibility in environmental risk communication: An empirical study." *Risk Analysis* 17(1):43–54.

Pew Research Center (2021). *Social Media Use in 2021*. Accessed at: https://www.pewresearch.org/internet/2021/04/07/social-media-use-in-2021/

Pidgeon, N., Kasperson, R., Slovic, P., eds. (2003). *The Social Amplification of Risk*. New York: Cambridge University Press.

Plough, A., Sheldon, K. (1987). "The emergence of risk communication studies: social and political context." *Science, Technology, & Human Values* 12(3/4): 4–10.

Powell, D., Leiss, W. (1997). *Mad cows and Mother's Milk: The Perils of Poor Risk Communication*. Montreal: McGill- Queen's University Press.

Prasad, A. (2021). "Anti-science Misinformation and Conspiracies: COVID–19, Post-truth, and Science & Technology Studies (STS)." *Science, Technology and Society*. April 2021.

Rader, M.H. (1981). "Dealing with information overload." *Personnel Journal* 60(5):373–375.

Rahn, M., Tomczyk, S., Schopp, N. and Schmidt, S. (2021). "Warning messages in crisis communication: risk appraisal and warning compliance in severe weather, violent acts, and the COVID-19 pandemic." *Frontiers in Psychology* 12:1–10.

Ratzan, S.C., and Moritsugu, K.P. (2014). "Ebola crisis-communication chaos we can avoid." *Journal of Health Communication* 19(11):1213–1215.

Rayner, S. (1992). "Cultural Theory and Risk Analysis," in *Social Theories of Risk*, eds. Krimsky, S. and Golding, D. Westport, CT: Praeger.

Renn, O. (2008). *Risk Governance: Coping with Uncertainty in a Complex World*. London, UK: Earthscan.

Renn, O., and Levine, D. (1991). "Credibility and trust in risk communication," in *Communicating Risks to the Public*, eds. R. Kasperson and P. Stallen. Dordrecht, Netherlands: Kluwer Academic Publishers.

Reynolds, B., and Seeger, M.W. (2005). "Crisis and emergency risk communication as an integrative model." *Journal of Health Communication* 10(1):43–55.

Sandman, P.M. (1998). "Hazard versus outrage in the public perception of risk," in *Effective Risk Communication: The Role and Responsibility of Government and Nongovernment Organization*, eds. V.T. Covello, D.B. McCallum, and M.T. Pavlova. New York: Plenum Press.

Sandman, P. and Lanard, J. (2012). *Crisis Communication: Guideline for Action*. Fairfax, VA: American Industrial Hygiene Association. Accessed at: https://www.psandman.com/media.htm#AIHAvid

Santos, S., Covello, V.T., and McCallum, D. (1996). "Industry response to SARA Title III: pollution prevention, risk reduction, and risk communication." *Risk Analysis* 16(1):57–65.

Schultz, F., Utz, S., and Goritz, A. (2011)." Is the medium the message? Perceptions of and reactions to crisis communication via twitter, blogs, and traditional media." *Public Relations Review* 37(3):430–437.

Seeger, M. W. (2006). "Best practices in crisis communication: An expert panel process." *Journal of Applied Communication Research* 34(3):232–244.

Seeger, M.W. (2006). "Best practices in crisis communication: an expert panel process." *Journal of Applied Communication Research* 34(3):232–244.

Seeger, M.W., Sellnow, T.L., and Ulmer, R.R. (2003*). Communication and Organizational Crisis*. Westport, CT: Praeger.

Sellnow, T.L. and M.W. Seeger (2021). *Theorizing Crisis Communication*. 2nd edition. New Jersey: Hoboken.

Sellnow, T.L., and Seeger, M.W. (2013). *Theorizing Crisis Communication*. Malden (MA): Wiley-Blackwell.

Sellnow, T.L., Ulmer, R.R., Seeger, M.W., and Littlefield, R.S. (2009). *Effective Risk Communication: A Message-Centered Approach*. New York: Springer.

Sheppard, B., Janoske, M., and Liu, B. (2012). *Understanding Risk Communication Theory: A Guide for Emergency Managers and Communicators*. Report to Human Factors/Behavioral Sciences Division, Science and Technology Directorate, US Department of Homeland Security. College Park, MD.

Silverman, C. (2020). *Verification Handbook for Disinformation and Media Manipulation*, 3rd edition. Brussels, Belgium and Maastricht, the Netherlands: European Journalism Centre. Accessed at: https://datajournalism.com/read/handbook/verification-3

Slovic, P. (1986). "Informing and educating the public about risk." *Risk Analysis* 6(4):403–415.

Slovic, P., Fischhoff, B., and Lichtenstein, S. (1986). "The psychometric study of risk perception," *Risk Evaluation and Management*, eds. V. Covello, J. Menkes, and J. Mumpower. Boston: Springer, pp. 3–24.

Slovic, P. (1987). "Perception of risk." *Science* 236: 280–285.

Slovic, P. (1999). "Trust, emotion, sex, politics, and science: surveying the risk-assessment battlefield." *Risk Analysis* 19(4):689–701.

Slovic, P., Finucane, M., Peters, E., and MacGregor, D. (2004). "Risk as analysis and risk as feelings: some thoughts about affect, reason, risk and rationality." *Risk Analysis* 24(2):1–2.

Sohn, Y.J., and Lariscy, R.W. (2014). "Understanding reputational crisis: definition, properties." *Journal of Public Relations* 26 (1):23–43.

Stallen, P.J.M (ed.) (1991). *Communicating Risks to the Public: International Perspectives.* Dordrecht: Kluwer.

Starr, C. (1969). "Social benefits versus technological risks." *Science* 165(3899)(Sep. 19, 1969):1232–1238.

Swire-Thompson, B., and D. Lazer (2020). "Public health and online misinformation: challenges and recommendations." *Annual Review of Public Health* 41(1):433–451.

Thomas, C.W., Vanderford, M.L., and Crouse Quinn, S. (2008). "Evaluating emergency risk communications: A dialogue with the experts." *Health Promotion Practice* 9(4):5–12.

Tinker, T.L., and Silberberg, P.G. (1997). *An Evaluation Primer on Health Risk Communication Programs and Outcomes.* Washington, DC: Department of Health and Human Services.

Tinker, T.L, Covello, V.T., Vanderford, M.L., Rutz, D., Frost, M., Li, R., Aihua, H., Chen, X., Xie, R., and Kan, J. (2012). "Disaster Risk Communication" Chapter 141, in *Textbook in Disaster Medicine*, ed. S. David. Dordrecht, Netherlands: Wolters Kluwer.

Ulmer, R. R., Sellnow, T. L., and Seeger, M. W. (2009). "Post-crisis communication and renewal," in *Handbook of crisis and Risk Communication*, eds. R. L. Heath and H. D. O'Hair. New York: Routledge.

Veil, S., Reynolds, B., Sellnow T.L., and Seeger, M.W. (2008). "CERC as a theoretical framework for research and practice." *Health Promotion Practice* 9(4):26–34.

Walaski, P.F. (2011) *Risk and Crisis Communication.* Hoboken, NJ: Wiley.

World Health Organization (2020). *Call for Action: Managing the Infodemic--A Global Movement to Promote Access to Health Information and Harm from Health Misinformation Offline Communities.* Accessed at: https://www.who.int/news/item/11-12-2020-call-for-action-managing-the-infodemic

World Health Organization. (2004). *Communication Guidelines for Disease Outbreaks.* WHO Expert Consultation on Outbreak Communications, Singapore, September 21-23, 2004. Geneva, Switzerland: World Health Organization.

World Health Organization. (2005). *WHO Outbreak Communication Guidelines.* Geneva: World Health Organization.

Wouter Jong (2020). "Evaluating Crisis Communication. A 30-item Checklist for Assessing Performance during COVID-19 and Other Pandemics," *Journal of Health Communication* 25 (12): 962–970.

Endnotes

1 Centers for Disease Control and Prevention. *Crisis and Emergency Risk Communication.* (2014). p. 31

2 Covello, V., Merkhofer, M. (1993). *Risk Assessment Methods: Approaches for Assessing Health and Environmental Risks.* New York: Plenum Press.

3 Mitroff, I. (2004). *Crisis Leadership: Planning for the Unthinkable.* Hoboken, NJ: Wiley.

4 See, e.g., Sandman, P.M. (1998). *"Hazard versus outrage in the public perception of risk" in Effective Risk Communication: The Role and Responsibility of Government and Nongovernment Organization*, eds. V.T. Covello, D.B. McCallum, and M.T. Pavlova. New York: Plenum Press. See also: Covello, V.T., Sandman, P.M. (2001). *"Risk communication: Evolution and revolution" in Solutions to an Environment in Peril*, ed. A. Wolbarst. Baltimore: Johns Hopkins University Press.

5 In the United States, Great Britain, and other countries, a one in a million risk is the level of acceptable or tolerable risk set by many government agencies for members of the public. It designates a level of risk for which no further improvements in safety are needed. As noted by Hunter and Fewtrell (2002, p. 209), the one in a million level was set by the British Health and Safety Executive "after considering risks in other contexts, with a risk of 1 in 1,000,000 being roughly the same as the risk of being electrocuted at home and a hundredth that of dying in a road traffic accident." The Environmental Protection Agency's National Contingency Plan "provides for an acceptable individual risk range, below which no response activity is necessary, anywhere within the range of 10 to the minus 4 (1 in 10,000) to 10 to the minus 6 (1 in 1,000,000)." See also: McClure, D.G. (2014). "All that one in a million talk." *Michigan Journal of Environmental and Administrative Law*. January 25, 2014. Accessed at http://www.mjeal-online.org/632/ "McClure, D.G." Accessed at: http://www.mjeal-online.org/632/; See also Graham, J. (1993). "The legacy of one in million." *Risk in Perspective* 1(1)/ Accessed at: https://cdn1.sph.harvard.edu/wp-content/uploads/sites/1273/2013/06/The-Legacy-of-One-in-a-Million-March-1993. https://cdn1.sph.harvard.edu/wp-content/uploads/sites/1273/2013/06/The-Legacy-of-One-in-a-Million-March-1993.pdf

6 See for example, Lewis, M. (2021) The *Premonition: A Pandemic Story,* NY: Norton, for a description of early warning signs missed for the COVID-19 pandemic.

7 Report to the President by the Presidential Commission of the Space Shuttle Challenger Accident (1986): https://spaceflight.nasa.gov/outreach/SignificantIncidents/assets/rogers_commission_report.pdf

8 Mitroff, I. (2004). *Crisis Leadership: Planning for the Unthinkable*. Hoboken, NJ: Wiley.

9 https://www.psandman.com/col/deepwater4.htm

10 The Guardian. *"Gulf oil disaster: BP admits missing warning signs hours before blast."* Accessed at: https://www.theguardian.com/environment/2010/sep/08/deepwater-horizon-rig-bp-report

11 https://www.theguardian.com/environment/2010/sep/08/deepwater-horizon-rig-bp-report

12 https://www.theguardian.com/environment/2010/sep/08/deepwater-horizon-rig-bp-report

13 https://www.psandman.com/col/deepwater4.htm

14 https://www.newsweek.com/what-not-say-when-your-company-ruining-world-7350; http://www.bbc.co.uk/news/mobile/10360084

15 http://science.time.com/2010/07/25/oil-spill-goodbye-mr-hayward/

16 https://www.theguardian.com/business/2010/jul/26/tony-hayward-bp-russia-gulf-oil-spill

17 http://science.time.com/2010/07/25/oil-spill-goodbye-mr-hayward/

18 Wessely, S. (2005). "Short and Long Term Psychological Responses to the New Terrorism" in eds. S. Wessely and V.N. Krasnov, *Psychological Responses to the New Terrorism: a NATO-Russia Dialogue*. Amsterdam; Oxford: IOS Press. p.185

19 Mileti, D.S., Peek, L. (2000). "The social psychology of public response to warnings of a nuclear power plant accident." *Journal of Hazardous Materials* 75:90; See also Mileti, D.S., Sorensen, J.H. (1988). "Planning and implementing warning systems," in *Mental Health Response to Mass Emergencies*, ed. M. Lystad. New York, NY: Brunner-Mazel; Mileti, D.S. (1990) "Warning Systems: A Social Science Perspective," in *Preparing for Nuclear Power Plant Accidents*, eds. D. Golding, J. X. Kasperson, and R. E. Kasperson. Boulder, CO: Westview Press.

20 See, e.g., https://www.washington.edu/uwem/files/2017/10/Crisis-Communications-Plan-Final-Oct-2017.pdf; http://cdphready.org/wp-content/uploads/2015/03/crisis_and_emergency_risk_communication_toolkit_july_2011.pdf; http://new.paho.org/hq/dmdocuments/2010/PAHO_CommStrategy_Eng.pdf

21 Covello, V.T. (2011b). Developing an Emergency Risk Communication (ERC)/Joint Information Center (JIC) Plan for a Radiological Emergency. NUREG/CR-7032. Washington, DC: Nuclear

Regulatory Commission; National Mining Association (2010.) Media and Community Crisis Communication Planning Template. Accessed at: https://www.partnersglobal.org/wp-content/uploads/wpallimport/files/NMAs-Media-and-Community-Crisis-Communication-Planning-Template-.pdf

22 Covello, V.T. (2011). Developing an Emergency Risk Communication (ERC)/Joint Information Center (JIC) Plan for a Radiological Emergency. Washington, DC: US Nuclear Regulatory Commission. Office of Nuclear Security and Incident Response. Accessed at: https://www.hsdl.org/?view&did=4482

23 Covello, V.T. (2011). Guidance on Developing Effective Radiological Risk Communication Messages: Effective Message Mapping and Risk Communication with the Public in Nuclear Plant Emergency Planning Zones. (NUREG/CR-7033). Washington, DC: US Nuclear Regulatory Commission. Office of Nuclear Security and Incident Response. Accessed at: https://www.nrc.gov/reading-rm/doc-collections/nuregs/contract/cr7033/

24 http://drexel.edu/dornsife/research/centers-programs-projects/center-for-public-health-readiness-communication/social-media-library/

25 See, e.g.,: https://www.fema.gov/incident-command-system-resources; See also http://www.icscanada.ca/

26 Covello, V.T. (2011). *Developing an Emergency Risk Communication (ERC)/Joint Information Center (JIC) Plan for a Radiological Emergency. NUREG/CR- 7032*. Office of Nuclear Security and Incident Response, US Nuclear Regulatory Agency Accessed at: https://www.hsdl.org/?abstract&did=4482

27 McMains, M.J. and Mullins, W.C. (2015). *Crisis Negotiations*. New York: Routledge.

28 Fisher, R., Ury, W., and Patton, B. (2011). *Getting to Yes: Negotiating Agreement Without Giving In*. New York: Penguin Books; Ury, W. (1993). *Getting Past No: Negotiating in Difficult Situations*. New York: Bantam Books.

29 Mehrabian, A. (1972). *Nonverbal Communication*. New York: Transaction Publications.

30 See, e.g.,: Archer D., & Akert R. M. (1977) "Words and everything else: verbal and nonverbal cues in social interpretation." *Journal of Personality and Social Psychology*, 35, 443–449.Mehrabian A., & Wiener M. (1967) "Decoding of inconsistent communications." *Journal of Personality and Social Psychology*, 6, 109–114; Ekman, P., Friesen, W. V., O"Sullivan, M., &Scherer, K. (1980). "Relative importance of face, body, and speech in judgements of personality and affect." *Journal of Personality and Social Psychology*, 38, 270–277.Mehrabian A., & Ferris S. R. (1967) "Inference of attitudes from nonverbal communication in two channels." *Journal of Consulting Psychology*, 31, 248–452; Jones E. J., LeBaron C. D. (2002) "Research on the relationship between verbal and nonverbal communication: emerging integrations." *Journal of Communication*. Special Issue, 52, 499-521; Krauss, R. M., Apple, W., Morency, N. Wenzel, C., &Winton, W. (1981) "Verbal, vocal and visible factors in judgements of another's affect." *Journal of Personality and Social Psychology*, 40, 312–320; Trimboli, A., Walker, Michael B. (1987) "Nonverbal dominance in the communication of affect: A myth?" *Journal of Nonverbal Behavior*. 11, 180–190; Wallbott H. G., & Scherer K. R. (1986) "Cues and channels in emotion recognition." *Journal of Personality and Social Psychology*, 51, 690–699.

31 https://www.forbes.com/sites/johnbaldoni/2013/07/15/how-edward-burkhardt-is-making-the-lac-megantic-accident-even-worse/#3b4868a635c4

32 https://www.nytimes.com/2013/07/13/business/before-blast-hauling-oil-revived-a-tiny-railroad.html

33 Lindell,M.K., Perry, R.W. (1987). "Warning mechanisms in emergency response systems." *International Journal of Mass Emergencies and Disasters* 5(2) 137–153; National Research Council/National Academy of Sciences (1991). *Real-Time Earthquake Monitoring: Early Warning and Rapid*

Response. Washington, DC : National Academies Press. See also Lindell and Perry (1992). *Behavioral Foundations of Community Emergency Planning.* Washington, DC : Hemisphere Publishing Corporation. See also Dynes, R.R., Quarantelli, E.L. (1976). "The family and community context of individual reactions to disaster" in *Emergency and Disaster Management: A Mental Health Sourcebook,* eds. H. Parad, H.L. Resnik, and L. Parad. Bowie, MD: Charles Press; Mileti, D.S. (1975). *Natural Hazard Warning Systems in the United States: A Research Assessment.* Boulder, CO: Institute of Behavioral Science, University of Colorado; Mileti, D.S., Fitzpatrick, C. (1991). "Communication of public risk: Its theory and its application." *Sociological Practice Review* 2 (1):20–28; Mileti, D.S., Peek, L. (2000). "The social psychology of public response to warnings of a nuclear power plant accident." *Journal of Hazardous Materials* 75: 181–194; Mileti, D.S., Sorensen, J.H. (1988). "Planning and implementing warning systems," in *Mental Health Response to Mass Emergencies,* ed. M. Lystad. New York, NY: Brunner-Mazel; Sorensen, J.H. (2000). "Hazard warning systems: Review of 20 years of progress." *Natural Hazards Review* 119-125;AU: Please provide the volume number for author "Sorensen, J.H." Quarantelli, E.L. (1990). *The Warning Process and Evacuation Behavior: The Research Evidence.* Newark, DE: Disaster Research Center, University of Delaware.

34 See, e.g., Coombs, W. T. (2004). "A theoretical frame for post-crisis communication: Situational crisis communication theory." in ed. M.J. Martinko, *Attribution Theory in the Organizational Sciences: Theoretical and Empirical Contributions.* Greenwich, CT: Information Age Publishing.

35 Coombs, T.W. (2014). "Crisis Management and Communications –Updated." Institute for Public Relations. https://instituteforpr.org/crisis-management-communications/ p. 4

36 See, e.g., Battistella, E. (2014). *Sorry About That: The Language of Public Apology.* Oxford. Oxford University Press; Benoit, W. (1995). *Accounts, Excuses, and Apologies: A Theory of Image Restoration.* Albany: State University of New York Press; Fehr, R., & Gelfand, M. J. (2010). "When apologies work." *Organizational Behavior and Human Decision Processes* 113(1), 37–50; Fernandez, C. and Fernandez, R. (2014) "Know how to make a powerful apology." in eds: Fernandez, C. and Fernandez, R. *It-Factor Leadership.* Carrboro, NC, Fast-Track Leadership, Inc.; Frantz, C. M., & Bennigson, C. (2005). "Better late than early: The influence of timing on apology effectiveness." *Journal of Experimental Social Psychology* 41(2), 201–207; Kim, P.H., Ferrin, D.L., Cooper, C.D., & Dirks, K.T. (2004). "Removing the shadow of suspicion: The effects of apology versus denial for repairing competence- versus integrity-based trust violations." *Journal of Applied Psychology* 89(1), 104–118; Lazare, A. (2004). *On Apology.* New York: Oxford University Press; Schweitzer, M.E., Brooks, A.W., Galinsky, W. (2015). "The organizational apology." *Harvard Business Review* 93 (9); 44–52.

37 Patel, A. & Reinsch, L. (2003). "Companies can apologize: Corporate apologies and legal liability." *Business and Professional Communication Quarterly* 66 (1) 9–25.

38 Tattrie, J. (2009, February 20). "Maple Leaf's handling of listeria crisis set 'the gold standard', experts say." Metro Halifax. Accessed at: http://www.metronews.ca/; See also Howell, G., Miller, R. (2010). "Maple Leaf Food: Crisis and Containment Case Study." *Public Communication Review* 1(1):47–55.

39 Maple Leaf Foods (2008) "Maple Leaf CEO Michael H. McCain Responds to Determination of Link to Plant," Accessed at https://www.mapleleaffoods.com/news/maple-leaf-ceo-michael-h-mccain-responds-to-determination-of-link-to-plant/

40 Parliament of Canada, Subcommittee on Food Safety of the Standing Committee on Agriculture and Agri-Food, Evidence, No 3, 16:25, 2nd Session, 40th Parliament, Ottawa, April 20, 2009. Accessed at https://www.ourcommons.ca/DocumentViewer/en/40-2/AGRI/report-3/page-27

41 "Maple Leaf CEO McCain Took Your Questions," Globe and Mail, Dec. 2, 2008. Accessed at: https://www.theglobeandmail.com/report-on-business/maple-leaf-ceo-mccain-took-your-questions/article22502266/

42 For example, accidents at nuclear power plants, such as Three-Mile Island, Chernobyl, and Fukashima had impacts on long term mental health, as well as issues such as public support for nuclear power, increased public concerns about new technologies such as genetic engineering and nanotechnology.

43 Mullin, S. (2003). "New York City's communication trials by fire, from West Nile to SARS." *Biosecurity and Bioterrorism: Biodefense Strategy, Practice, and Science* 1(4): 267–272 "Mullin, S." Published online in 2004 and accessed at: https://www.liebertpub.com/doi/abs/10.108 9/153871303771861478?journalCode=bsp

44 Hanna-Attisha, M. (2018). *What the Eyes Don't See: A Story of Crisis, Resistance, and Hope in an American City*. New York: Penguin Random House.

45 See Covello, Minamyer, and Clayton (2007). *Effective Risk and Crisis Communication During Water Security Emergencies*. Washington, D.C. US Environmental Protection Agency. Accessed at: https:// nepis.epa.gov/Exe/ZyNET.exe/60000CES.txt?ZyActionD=ZyDocument&Client=EPA&In dex=2006%20Thru%202010&Docs=&Query=&Time=&EndTime=&SearchMethod=1&TocRestric t=n&Toc=&TocEntry=&QField=&QFieldYear=&QFieldMonth=&QFieldDay=&UseQField=&Int QFieldOp=0&ExtQFieldOp=0&XmlQuery=&File=D%3A%5CZYFILES%5CINDEX%20DATA%5C 06THRU10%5CTXT%5C00000000%5C60000CES.txt&User=ANONYMOUS&Password=anonymo us&SortMethod=h%7C-&MaximumDocuments=1&FuzzyDegree=0&ImageQuality=r75g8/r75g8/ x150y150g16/i425&Display=hpfr&DefSeekPage=x&SearchBack=ZyActionL&Back=ZyActionS&B ackDesc=Results%20page&MaximumPages=1&ZyEntry=3

46 See, e.g., Wouter Jong (2020). "Evaluating crisis communication: A 30-item checklist for assessing performance during COVID-19 and other pandemics," *Journal of Health Communication*, 25 (12): 962–970.

7

Foundational Principles: Perceptions, Biases, and Information Filters

---

CHAPTER OBJECTIVES

This chapter covers the evidence- and science-based foundational principles that explain how people perceive, filter, and react to information when they are under stress. These principles are at the core of what differentiates risk, high concern, and crisis communication from other communication fields.

At the end of this chapter, you will be able to:

- Describe why perceptions equal or become reality and why perceptions are as relevant to communication effectiveness as facts.
- Summarize the range and nature of factors that affect message perception under stress, from psychological and social to neurological.
- Demonstrate the concept of "acceptable risk" and present approaches to gaining agreement about what is acceptable.
- Determine how to account for filters, biases, and other factors in crafting messages.

> The major public health challenges since
> 9/11 were not just clinical, epidemiological,
> technical issues. The major challenges
> were communications. In fact, as we move
> into the twenty-first century, communication may
> well become the central science of public
> health practice.
>
> *- Dr. Edward Baker. US Assistant Surgeon General, 2001*

7.1   Case Diary: "A" Is for "Apples"

It was a cold day in February 1989. I was sitting in my office at Columbia University, working late. I took a break from my work and walked over to a colleague's office to watch an episode of CBS's *60 Minutes*. We had agreed earlier in the day to watch the program together. We were joined by several graduate students.

Communicating in Risk, Crisis, and High Stress Situations: Evidence-Based Strategies and Practice, First Edition.
Vincent T. Covello.
© 2022 by The Institute of Electrical and Electronics Engineers, Inc. Published 2022 by John Wiley & Sons, Inc.

The television guide showed that *60 Minutes*, a high-rated television news program with more than 40 million viewers, would present results from their investigation of the risks to children of eating apples and drinking apple juice made from apples treated with Alar. Alar was the trade name for *daminozide*, a chemical made by Uniroyal Chemical Company. Alar was sprayed on apple trees and absorbed by the tree to regulate growth; this made harvesting easier and kept apples from falling off the trees before ripening, ensuring they would be red and firm for transport and storage. Since residues of Alar would be inside the apple, it could not be washed off. Alar in apples is odorless and tasteless.

Parents were told by *60 Minutes* that children are at greater risk of cancer than adults from *Alar* because they drink so much apple juice. The average preschooler drinks 18 times more apple juice than their parents.

Alar was listed with the US Environmental Protection Agency (EPA) as a pesticide and widely used by apple growers. However, it was not a pesticide. It was a plant-growth chemical. Alar was first approved for the United States in 1963. Many modern scientists still disagree about the safety of Alar, including its imminent danger and its potential long-term cumulative threat to health. It was used on apples until 1989 when the manufacturer voluntarily withdrew it from the market.[1]

I was particularly interested in the program because it raised many complex risk communication challenges. As a researcher, I knew an attack against apples would be perceived, on both a conscious and subconscious level, as more than just an attack against apples as a safe and healthy food. At the psychological level, an attack against apples would be perceived as a threat and attack on values. Apples are a symbol for, or associated with, over 15 valued things, including health (e.g., "an apple a day keeps the doctor away"), love (e.g., "you are the apple of my eye"), home and hearth (e.g., apple pie and motherhood), innocence (Snow White), success ("the Big Apple"), education (an apple for a teacher), fertility (Johnny Appleseed), modern science (e.g., the apple that fell on the head of the young Isaac Newton, leading to the "aha moment" that prompted him to come up with his law of gravity), gold (the golden apple as appears in various Greek and Roman myths, including the 12 labors of Hercules), and, perhaps most important, knowledge of all good and evil, immortality, sin, and temptation (e.g., Adam and Eve and the expulsion from paradise for eating an apple from the forbidden tree). For a risk communicator, an attack against apples was a communication challenge at the highest level.

Second, I had a personal stake in what would happen. The year before, I had published a guide and policy document for the EPA titled "The Seven Cardinal Rules of Risk Communication."[2] One section of the document discussed the importance of addressing risk perceptions. I was curious to see if the EPA spokesperson would follow the guidance in the EPA document.

Third, I knew that the lead investigator would be Ed Bradley, a senior reporter for *60 Minutes*. I had seen several television interviews done by Bradley. I knew Bradley had a reputation for asking the same question over and over again. His interviews often appeared to me to be closer to a police interrogation than to a media interview. I predicted that Bradley would likely raise questions about each of the factors and filters that determine public perceptions of risk, such as trust, caring, empathy, honesty, trustworthiness, competence, voluntariness, fairness, benefits, effects on children, catastrophic potential, and personal control. For example, the use of Alar unfairly gave minimal benefits to the consumer. Most of the benefits went to the farmers using Alar so that they could keep apples on the tree longer. Alar residues could not be washed or peeled off an apple. It would be pointed out that this takes control away from consumers. The danger that might be posed by Alar was a dreaded, catastrophic, adverse outcome (cancer) for thousands of children.

Fourth, I was curious to see how the EPA spokesperson would communicate the complex science of risk assessment to lay consumers. I was familiar with the EPA official Bradley would be interviewing: Dr. Jack Moore, the acting administrator of the EPA and a highly respected technical professional and toxicologist. I was especially interested because I had just published a layperson guide to risk assessment for the White House Council on Environmental Quality.[3] The document discussed a complex scientific controversy in risk assessment and one that was central to the assessment of the safety of Alar: the appropriateness and value of drawing conclusions about the risks to humans from studies of mice exposed to amounts of a chemical well beyond what humans consume. I knew that the EPA had previously concluded, mostly from studies of mice, that Alar was safe at the current levels of human exposure. But could the EPA spokesperson explain this clearly and effectively in an intense broadcast media format?

Finally, I knew the interview would be a challenge, regardless of who the lead reporter or spokesperson was. The story had all the elements of powerful storytelling: villains, victims, and heroes.[4] The EPA would be presented as the villain. Children would be presented as the victims. And *60 Minutes* would be presented as the hero exposing the outrage.

I turned on the program and was stunned by what I was seeing and hearing. The program began, as usual, with the *60 Minutes* image and sound of a ticking clock. The clock disappeared. The screen showed a consumer in the supermarket buying apples, and a disembodied voice said: "The most potent cancer-causing chemical in our food supply is a pesticide sprayed on apples to keep them on the trees longer and make them look better." The voice offering this information was stating it as a fact, not an allegation. The voice was easily recognizable: it was Ed Bradley, the *60 Minutes* reporter.

The picture on the TV changed to a man wearing a white shirt and red tie, sitting far back in his chair, nodding his head with a crowded bookcase behind him. As expected, Bradley was interviewing Dr. Jack Moore. In a voiceover, Bradley said: "The EPA's acting administrator, Dr. Jack Moore, acknowledged that the EPA has known about the cancer risk for sixteen years."

The TV picture shifted to Bradley. He was sitting forward in his chair and looking down over the rims of his glasses at Dr. Moore. His hands were waving, and he appeared angry and frustrated. Bradley said to Dr. Moore: "A lot of these chemicals got on the market when we did not know they were cancer-causing agents, and they are on the market now, and we know that they do cause cancer." Dr. Moore replied: "That's correct." Bradley continued: "But you say we can't take them off because they are already on the market and they went on the market when we didn't know they cause cancer." Dr. Moore replied: "That's the paradox of the statute." Bradley hit back with: "That's crazy."

The picture on the television shifted to Bradley, sitting in front of a picture of a large, shiny red apple marked with skull and crossbones. Bradley concluded: "And who is most at risk? Children who may someday develop cancer from this one chemical called daminozide."

Watching the program, I attempted to figure out what Dr. Moore was trying to say when he said: "That's correct. That's the paradox of the statute." It certainly was not plain English, a key best practice in effective risk, high concern, and crisis communications. I thought he might have been trying to say that the agency did not see a need to go beyond restrictions on Alar already in place; second, that the agency's hands were tied by environmental regulations and laws; and third, that, based on years of research, the EPA had concluded that the short-term risks of eating apples treated with Alar are not significant and do not pose an imminent or major threat to the health of children.

Within a few minutes, the head of the EPA had violated virtually all the cardinal rules of risk communication, including the most damaging risk communication failure: failing to address perceptions of risk. Through his verbal and nonverbal communication, he communicated a message that could be interpreted in a way that suggested he and his agency did not care about the health of children or assign any priority to this issue.

The scare that started with the *60 Minutes* Alar segment was followed the next day with congressional testimony by the actress Meryl Streep criticizing the EPA for its handling of Alar. Schools yanked apples off their menus; parents threw out their apple juice; the company that made Alar stopped selling it; thousands of apple growers lost their businesses or had to be bailed out. The Alar scare "had great economic and other repercussions that still continue."[5]

The high levels of fear, economic damage, and lost trust in the EPA resulting from the Alar scare could have been avoided if the EPA spokesperson had recognized the fundamental principles of risk perception described in this chapter. This case demonstrates the importance – and necessity – of knowledge and understanding of public perceptions of risk and how these perceptions play on the media stage.

7.2 Message Perception and Reception in High Concern Situations

Foundational principles for communicating effectively in risk, high concern, and crisis situations are described in a large body of scientific research. Over the past five decades, hundreds of articles have been published in peer-reviewed scientific journals on these foundational principles. The focus of many of these articles is how individuals and groups perceive, understand, interpret, and react to risk, high concern, and crisis information. (A selection of valuable resources is shown in the Resource section of this chapter.)

The design and delivery of a message in a risk, high concern, and crisis situation are built on three foundational principles derived from the behavioral and neuroscience literature:

1) People under stress typically want to know you care before they care what you know.
2) People under stress typically have difficulty processing information – hearing, understanding, and remembering.
3) People under stress typically focus most on the negative information they hear.

Each of these principles points to difficulties people have processing information in risk, high concern, and crisis situations. One of the most exciting early scientific discoveries related to the difficulties people have processing information, even in low-stress situations, was made in 1956 when Professor George Miller from Princeton University wrote:

> My problem is that I have been persecuted by an integer. For seven years this number has followed me around, has intruded in my most private data, and has assaulted me from the pages of our most public journals.[6]

These are the first words in a seminal paper in the history of psychology and neuroscience. The number that plagued Professor Miller was "seven." First published in a leading psychological journal, "The Magical Number Seven, Plus or Minus Two: Some Limits on Our Capacity for Processing Information," the paper established Miller as a leading scholar in cognitive psychology and neuroscience. It also helped lay the foundation for the entire field of cognitive psychology and neuroscience. Before Miller's work, researchers and practitioners had long known the human brain had a limited attention span and capacity for processing information, including short-term and long-term memory. What they did not know were the limits of processing and memory. Miller quantified the limits of short-term memory for low-stress situations. He tested short-term memory via a variety of tasks, including asking research subjects to repeat a set of digits. In all research designs, Miller found the average limit to be seven items.

Bell Laboratories supported the research of Professor Miller. It is why local phone numbers originally became seven digits long. Miller found that seven-digit numbers were as long as people could remember without forgetting or making errors. More than seven to nine items received at a time created information overload for short-term memory.

Table 7.1 The rule of three for message development

---

1) Three key messages

   (e.g., You can do "x." You can do "y." We recommend you do "z"; The problem is "x." The solution is "y." The challenge is "z."; We know "x." We don't know "y." We are doing "z" to gain more information.)

2) Three supporting pieces of information for each key message

   (e.g., facts, proofs, third party endorsements, links, visuals)

3) Three repetitions of the three key messages

Miller's "magical number seven" as a message-development tool supplemented the much better-known message development tool, the "magical number three." The number three, or the Rule of Three, is one of the most powerful best practices in the toolbox of the risk, high concern, and crisis communicator.[7] The Rule of Three, which has three parts, is shown in Table 7.1.

The Rule of Three has its roots in ancient Greece. Aristotle posited that a story should have a beginning, middle, and end. The anthropologist Joseph Campbell said a great story should have villains, victims, and heroes.[8] In literature and the movies, there are three amigos, three little pigs, three musketeers, three stooges, and three bears. The Declaration of Independence guarantees three inalienable rights: life, liberty, and the pursuit of happiness. Safety warnings are typically most effective when expressed in threes. For example, the three steps a fire victim should follow to minimize injury are "stop, drop, and roll." The public health message widely communicated during the 2009–2010 pandemic influenza outbreaks was: (1) wash your hands; (2) cover your cough or sneeze; and (3) stay home if you are sick. Holding constant other variables, the ability of the human brain to process information – hear, understand, and remember – declines when stressed. Under stress, people have trouble retaining seven messages, and it is better to present content in groups of three.

Building on the communications work of Miller and others, behavioral science and neuroscience researchers have explored how people process information in both low-stress and high-stress situations. In both low- and high-stress situations, the human brain typically goes through a five-part process that shapes perceptions, actions, and behaviors: (1) receiving or not receiving the information; (2) understanding or not understanding the information, (3) believing or not believing the information, (4) deciding the personal relevance or lack of relevance of the information, and (5) deciding what to think, what not to think, what to do, and what not to do. Each step involves a complex set of neurological processes and decisions by different parts of the brain and is affected by the mental filters described in Section 7.3.

Message development is especially challenging when people are extremely fearful, stressed, worried, anxious, or highly concerned about a threat. They typically struggle with processing incoming information – hearing, understanding, and remembering. A central goal of risk, high concern, and crisis communication message development is to overcome the communication barriers created by fear, stress, worry, and anxiety.

7.3 Message Filter Theory: A Set of Principles Drawn from the Behavioral and Neuroscience Literature

Effective message development is based on a deep understanding of how the human brain processes messages about risks and threats. Before information about a risk or threat is acted upon, the information goes through multiple filters. These filters include (1) social amplification filters, (2) mental shortcut filters, (3) knowledge and belief filters, (4) personality filters, (5) negative dominance filters, (6) trust determination filters, and (7) cultural filters. (See Table 7.2). This chapter discusses the first five filters; the last two are discussed in Chapter 8. These filters, and the factors that determine how the filters operate, are the intellectual bedrocks upon which effective message development for risk, high concern, and crisis communication situations are built. Message filter theory helps explain the success or failure of a risk, high concern, and crisis communication message.

A central theme in the behavioral and neuroscience literature on message development is how the human brain forms perceptions and decides about risks and threats. Perceptions and decisions about risks and threats encompass all the ways the brain realizes risks and threats.

An important paradox identified in the risk, high concern, and crisis message-development literature is that the risks and threats that endanger or harm people, and which should make people fearful, are often different from the risks and threats that actually make people fearful. There is a low correlation between the ranking of the risks and threats that make the public fearful compared to what technical and scientific experts believe are harmful.[9]

There are risks and threats that make people fearful but may actually cause little harm. There are also risks and threats that harm people but do not make people fearful. For example, a person may worry about getting sick from urban air pollution while they chain-smoke cigarettes. The paradox of what people fear, and what may appear on the surface to be irrational thinking, is partly explained by the factors that cause the fear response of the brain – fight, freeze, or flight. Besides values, experience, and emotion, among several of the most important fear factors are trust, voluntariness, personal controllability, familiarity, fairness, benefits, dread, and uncertainty. These psychological, subjective, and emotional factors – together with available technical facts about a risk or threat, as well as a wide variety of dynamic situational, biological, personality, experiential, networking, social, demographic, economic, and cultural factors – are filters through which the brain determines how to respond to information about risks and threats. These filters profoundly affect fear. Technical facts alone are seldom sufficient to address fear. In many situations, only a small proportion of fear is driven by technical facts; a large proportion of fear is driven by psychological, socioeconomic, and cultural fear factors.

Differences between how technical risk experts and non-risk experts perceive risks are among the greatest challenges in risk, high concern, and crisis communication. For example, in one of the most cited studies in the risk perception literature, technical risk experts and nonexperts were asked to rank 30 sources of risk.[10] The rank orders of the lists were considerably different. A central conclusion of the research was that risk communication differs from risk education. Effective risk communication requires an understanding of the perceptions, beliefs, interpretations, and value systems of the target audience.

The brain's perceptions of risks and threats are a function of two key sets of factors: (1) those reflecting the degree to which the risk or threat is unknown, and (2) those reflecting the degree to which the risk or threat is dreaded. Unknown risks and threats include those that are perceived to

Table 7.2 Message filters

---

1) Social amplification filters
2) Mental shortcut filters
3) Knowledge and belief filters
4) Personality filters
5) Negative dominance filters
6) Trust determination filters
7) Cultural filters

---

be uncertain, not well understood, unobservable, new, and have delayed consequences. Dreaded risks and threats include those that are perceived to be potentially catastrophic, certain to be harmful, involuntary, uncontrollable, affecting future generations, and unfair. Controlling for other variables, the unknown and dreaded characteristics of a risk or threat are among the most important levers that determine and predict fear.

One of the first scientific and research-based observations of the power of fear factors and filters on decision-making and the brain was made by Starr.[11] Starr's analysis, and the studies that followed, indicated fear factors profoundly affect levels not only of fear but also of worry, anxiety, anger, outrage, and acceptability. People will often accept risks or threats as much as 1000 times greater if they are voluntary and perceived to be under their personal control. Similarly, people will often accept risks as much as 1000 times greater if they perceive the activity that generates the risk or threat to be clearly beneficial.

Psychological, socioeconomic, and cultural filtering of messages counters the conventional notion that "facts speak for themselves." There are risky behaviors that people commonly engage in even though these behaviors have been determined by science to be high in risk. Common examples include smoking or driving under the influence of alcohol. At the same time, people can become concerned and outraged over risks and threats that are determined by scientists and experts to be low.

7.4 Case Study: COVID-19 and Risk Perception Factors

Pandemics such as COVID-19 trigger neurological and emotional risk perception and fear factors that can exponentially elevate the level of fear from a perceived risk. Neuroscience and behavioral science research indicate that these risk perceptions and fear factors function consciously and unconsciously within the brain. As shown in Table 3.6 of Chapter 3, there are 20 primary perception and fear factors that can raise the level of worry and anxiety (i.e. that function in the brain like "hot buttons" and can hijack a person's decision-making abilities). The COVID-19 pandemic brought most of these factors into play.

The following are several of the most important hot buttons pushed by COVID-19 during 2020.

- *Trust in responsible authorities and institutions*: People are often more concerned about activities or actions where the responsible assessor or manager is perceived to be untrustworthy than they are about activities or actions where the responsible assessor or manager is perceived to be trustworthy. *COVID-19 hot button*: Messages from authorities were inconsistent, politicized, and often contradictory, leaving the public confused about whom to trust.
- *Voluntariness:* Risks from activities considered involuntary or imposed are judged to be greater and are therefore less readily accepted, than risks from activities that are seen to be voluntary. *COVID-19 hot button*: Exposure to the virus that causes COVID-19 is perceived to be involuntary. Except for those who sign up for a drug clinical trial, people do not choose to be infected with COVID-19.
- *Scale/catastrophic potential:* Risks from activities viewed as having the potential to cause a significant number of deaths and injuries grouped in time and space are judged to be greater than risks from activities that cause deaths and injuries scattered or random in time and space. *COVID-19 hot button*: The effects of COVID-19 are particularly catastrophic for vulnerable populations such as the elderly, people with underlying health conditions, and those with limited access to health care. Hospitalizations and deaths occurred in spikes grouped in time and place.
- *Familiarity/exotic*: Risks from activities viewed as unfamiliar or exotic are judged to be greater than risks from activities viewed as familiar. *COVID-19 hot button*: COVID-19 was unfamiliar. COVID-19 did not exist in humans before the Wuhan outbreak in China in late 2019.

- *Understanding/visibility*: People are often more concerned about activities or actions perceived to be characterized by poorly understood and invisible exposure mechanisms or processes than about activities or actions perceived to be characterized by visible and apparently well-understood exposure mechanisms or processes. *COVID-19 hot button*: The virus that causes COVID-19 is not capable of being seen by the naked eye. Exposure mechanisms, such as whether the virus is airborne and dangerous to humans in small droplets, were poorly understood for several months after the first discovery of the disease, and thus the source of danger was not seen or understood.

- *Uncertainty*: Risks from activities that pose highly uncertain risks are judged to be greater than risks from activities that appear to be relatively well known to science. *COVID-19 hot button*: COVID-19 is novel and characterized by huge uncertainties. Because of the evolving nature of scientific understanding, scientists and public health officials had to frequently change their answers to questions about COVID-19, including questions about causes, symptoms, and the effectiveness of control measures.

- *Controllability*: Risks from activities viewed as under the control of others are judged to be greater and are less readily accepted than those from activities that appear to be under the control of the individual. *COVID-19 hot button*: Individuals have only limited ability to control their protection from COVID-19. Individuals can influence the spread of COVID-19 through protective behaviors such as frequent handwashing, use of face masks, self-quarantine, and social distancing, but total protective isolation is difficult.

- *Effects on children:* People are often more concerned about activities or actions that are perceived to adversely affect children or specifically put children in the way of harm or risk (e.g., asbestos in school buildings, milk contaminated with radiation or toxic chemicals) than about activities or actions engaged in by adults. *COVID-19 hot button*: While COVID-19 causes more severe illness in adults than children, the risks, controversies, potential impacts, and uncertainties about school closing and remote learning caused enormous concern.

- *Personal stake*: Risks from activities in which people have a personal stake are judged to be greater than risks from activities that appear to pose no direct or personal threat. *COVID-19 hot button*: COVID-19 protection requires changes in personal behavior. Individuals are advised to distance from others, including relatives, and engage in actions that are very personal (such as mask wearing and sanitizing) and restrictive of their personal freedoms (e.g., not traveling, attending events, and socializing indoors). Many individuals also have enormous personal stakes in the economic effects of shutdowns and restrictions.

- *Victim identity*: People are often more concerned about activities or actions that are perceived to cause harm to identifiable victims or named persons than about risks that are statistical and affect persons that are nameless or faceless. *COVID-19 hot button*: Many people know or know about individuals who have become ill with COVID-19 because the disease is so widespread. While the media presents many "nameless" statistics, they also present individual stories.

- *Pleasurable/dread*: Risks from activities that evoke fear, terror, or anxiety are judged to be greater than risks from activities that do not arouse such feelings or emotions. *COVID-19 hot button:* Individuals fear contracting COVID-19 and dread the health, economic, and social consequences of the COVID-19 pandemic. These include lockdowns, debilitating illness, hospitalization, death, closed businesses, bankruptcies, loss of livelihood, financial volatility, overwhelmed hospitals, and self-quarantine.

- *Awareness/media attention*: Risks from activities that receive considerable media coverage are judged to be greater than risks from activities that receive little. *COVID-19 hot button*: For most of 2020, traditional and social media outlets ran lead media stories about COVID-19 almost every day. Through websites, news outlets, and social media platforms, people were continually reminded of dangers and threats to their personal welfare and society. This also extended to the COVID -19 vaccines. For example, mainstream and social media posted dramatic stories about individuals experiencing a very rare type of blood clot after receiving a vaccine. Although the risk was extremely low and the incidence rare, the negative information was amplified because negative content drowns out other information. For anyone already anxious about vaccination, these stories increased their concern and vaccine hesitancy.

- *Fairness*: People are often more concerned about activities or actions that are perceived to be characterized by an inequitable or unfair distribution of risks, costs, and benefits than about activities or actions perceived to be characterized by an equitable or fair distribution of risks, costs, or benefits. *COVID-19 hot button*: Data have shown that COVID-19 has greater spread and more serious consequences in less affluent and minority communities. Workers who have front-line jobs, such as in restaurants, public transportation, and grocery stores, cannot work from home, while many professionals can, creating an inequity in the threat and consequences. The illness also poses significantly more dangerous symptoms and the threat of death to the elderly and those with underlying health conditions.

- *Benefits*: Risks from activities that seem to have unclear, questionable, or diffused personal or economic benefits are judged to be greater than risks from activities that have clear and tangible benefits. *COVID-19 hot button*: COVID-19 carries no benefits for people infected with the disease, other than the hope of possible future immunity.

- *Reversibility*: People are often more concerned about activities or actions that are perceived to have potentially irreversible, permanent adverse outcomes than about activities or actions perceived to have potentially reversible adverse outcomes or effects. *COVID-19 hot button*: In more serious cases, COVID-19 can cause long-term health issues or death.

- *Nature of evidence*: People are often more concerned about activities or actions that are based on risk assessments from human studies than about activities or actions based on risk assessments from nonhuman studies. *COVID-19 hot button*: COVID-19 clearly affects humans, with extensive evidence regarding impact on humans.

7.4.1 Social Amplification Filters

Psychological filters can combine either to intensify or to diminish perceptions and fears of risks or threats. Amplification processes are described and explored by theorists of social amplification.[12] Social amplification researchers have focused much of their work on the paradox described above: that relatively minor risks, threats, and concerns, as assessed by experts and scientific professionals, are often perceived by the public as major risks, threats, and concerns.

The main thesis of social amplification is that information about perceived risks and threats interacts with psychological, social, economic, institutional, and cultural processes and factors that amplify or diminish how people respond to such information. According to social amplification theory, information about a risk is communicated forward by information brokers who occupy social amplification "stations." Amplification brokers include scientists, leaders, risk managers, cultural groups, activist groups, opinion leaders, social networks, traditional broadcast media, traditional print media, and social media, such as blogger sites. Only a fraction of all incoming information is actually used by these information brokers. Each broker changes the original risk or threat message by amplifying or diminishing the message, focusing on secondary economic, social, legal, and political impacts. For example, the adverse impacts of an industrial accident resulting in few injuries, deaths, or property damage may nonetheless be amplified by those pointing to the secondary adverse impacts, including loss of trust in risk and crisis management institutions, additional insurance costs, and lawsuits.

7.4.2 Mental Shortcut Filters

The human brain uses numerous mental shortcuts to calculate the probability that an adverse outcome will occur, making receiving or not receiving information more complicated than it may first appear. Receiving or not receiving information is influenced by a complex set of mental shortcuts scientifically called *heuristics*. These shortcuts help the brain gather and remember information selectively.

Because of these shortcuts, the brain typically filters out a large amount of information and uses only a small subset of available information to decide about risks and threats. For example, individuals often assign a greater probability to events that are easy to recall or about which they are frequently reminded, such as by the news media or in discussions with friends or colleagues. The brain also overestimates the occurrences of rare and dramatic causes of fatalities while underestimating the frequency of more common causes of death. Deaths from botulism, for example, are overestimated and considered more common than they actually are. Deaths from diabetes and stroke are estimated to be less frequent than they actually are.

One explanation for why the brain acts this way is that it is an efficient way to process information. The human brain is continually subjected to information about benefits, costs, and risks. The brain cannot possibly take all the time needed to carefully assess each piece of information – such as all the benefits, costs, and risks of getting up from a chair and taking a walk. People process information quickly to avoid threats and prevent harm. Quick and efficient processing of information about a threat is a basic survival skill; it is done instinctively and as an automatic reflex.

Several of the most important mental shortcuts the brain uses to estimate and judge risks and threats are listed in Table 7.3 and described below.

1) **Information availability**: The availability of information about an event (i.e. information that is accessible or easily remembered) often leads to an overestimation of its frequency. The brain assigns a greater probability to events about which it is frequently reminded (e.g., in the news media or in conversations) or to events easy to recall or imagine because of concrete examples or dramatic images.
2) **Conformity and relativity**: Conformity and relativity are the tendencies to adopt attitudes and behaviors relative to the attitudes or behavior of others, because others are doing something, or because others believe it is true.
3) **Anchoring and adjustment**: Anchoring and adjustment are tendencies on the part of the brain to give undue weight to the first piece or one piece of information in assessing probabilities. The first piece of information functions as the reference point or anchor. It sets the tone for what will follow.

Table 7.3 Mental short-cuts the brain uses to judge and estimate risks and threats

1) Information availability
2) Conformity and relatively
3) Anchoring and adjustment
4) Framing
5) Affect/feelings/emotions
6) Overconfidence
7) Confirmation bias

4) **Framing:** Framing is the tendency for the brain to reach conclusions based on how information is framed and presented. Different frames, such as the words used, have a profound effect on perceptions of risks and threats. Framing is often important in the discussion of perceived losses and gains. In general, a loss is perceived as more significant, and therefore more important to avoid, than an equivalent gain. Responses to a message are highly affected by whether the message highlights the positive or negative aspects of an event, situation, or action.

5) **Affect/feelings/emotions**: Affect is the tendency for perceptions and judgments of risk to be strongly influenced by feelings and emotions.[13] An "affect" is the instinctive emotional and subjective response. A common example is the feeling of dread. It is a mental shortcut whereby the brain relies on instinct, emotions, and "gut" feelings in forming perceptions about a risk or threat. *Feelings are facts for the person experiencing them.* The feeling can be positive or negative and can guide risk and threat judgments. Positive feelings and emotions typically result in lower concerns about a risk or threat. Negative feelings and emotions typically result in higher concerns about a risk or threat.[14] For example, research consistently finds that feelings and emotions – such as fear and guilt – are among the strongest predictors of risk-related attitudes and behaviors toward issues such as climate change.

6) **Overconfidence**: Overconfidence is the tendency of the brain to overestimate abilities to avoid harm. It is most evident when people perceive they are in control of a situation. Unfounded overconfidence leads to reduced feelings of susceptibility. A majority of people, for example, consider themselves less likely than average to get cancer, get fired from their job, or get mugged. Many people resist using seat belts because of the unfounded belief that they are better or safer than the average driver. In a similar vein, many teenagers engage in high-risk behaviors (e.g., texting and driving, smoking, and unprotected sex) because of their perceptions of invincibility.

7) **Confirmation bias**: Confirmation bias is the tendency of the brain to search for, interpret, favor, and recall information that confirms existing perceptions, attitudes, beliefs, and values.[15] Confirmation bias is more likely to be observed in high stress, high concern, high-stakes, low-trust, and emotionally charged situations or when deeply entrenched beliefs are challenged. Confirmation bias leads to the tendency to accept information that is consistent with existing beliefs and to ignore information that is not. Once a belief about a risk is formed, new evidence is frequently made to fit the belief, contrary information is filtered out, ambiguous data are interpreted as confirmation, and consistent information is seen as "proof." Strongly held beliefs about risks, once formed, change slowly. When supported by multiple sources of trusted information, beliefs can be extraordinarily persistent even in the face of contrary evidence.

This last bias – confirmation bias – is among the powerful biases affecting the human brain under conditions of high uncertainty. Confirmation bias is closely related to the concept of cognitive dissonance. Cognitive dissonance can be defined as the stress experienced by a person who holds two or more contradictory beliefs related to an attitude or behavior. To relieve the stress caused by cognitive dissonance, the brain will frequently change where it searches for information, how hard it will search, how it interprets what it finds, and what beliefs will be formed. Change tactics used by the brain include ignoring or denying contradictory information.

Confirmation bias is largely performed unconsciously. It often results in discounting, ignoring, or avoiding information that is inconsistent with one's existing beliefs. Existing beliefs include a person's (1) assessment or expectation of the existence and likelihood of an event (e.g., the existence and likelihood of climate change), (2) beliefs about the causes of an event (e.g., human activity is a, or the, major contributor to climate change), and (3) beliefs about the consequences of an event (e.g., climate change will cause global warming and irreversible and catastrophic environmental impacts). A person is more likely to engage in confirmation bias when the brain – especially the brain's amygdala/threat-detection circuit and the parallel cognitive systems in the neocortex – evaluates the situation as an important threat or the person has a significant vested interest.[16]

A classic case of confirmation bias is illustrated by the concerns and fears people have regarding climate change.[17] People seek information that supports their existing beliefs about (1) if global warming exists, (2) the degree to which it is caused by human activities (if it exists), (3) expert opinion on climate change, global warming, and its causes, and (4) the potential for climate change, if it exists, to cause adverse impacts on the economy. People listen to their own biases and ignore or discount facts and opinions that are contrary to their biases.

7.4.3 Knowledge and Belief Filters

An important topic in the message development literature concerns representations of what the brain knows, believes, and expects about a risk or threat. These representations are often referred to as *mental models*. Researchers have used mental models to analyze a wide variety of perceptions of risk-related issues, including electromagnetic fields, climate change, radon, sexually transmitted diseases, and emerging technologies.[18] Based on interviews with laypeople and experts, researchers have produced numerous useful insights. In studies of radon, for example, researchers found major misconceptions and gaps in public knowledge and beliefs. These misconceptions and gaps substantially affected perceptions, attitudes, and behaviors. The researchers noted: "Few knew that radon and its decay products are relatively short-lived, leading some to conclude that once a home is contaminated, nothing can be done to clean it up."[19]

Mental models are a key element in *game theory*. Game theory provides a mathematical framework for analyzing the knowledge, beliefs, expectations, and choices made by individuals. According to game theory, individuals have knowledge and preferences for outcomes and probabilities for each outcome, and each player chooses strategies based on those preferences, i.e., on their mental models. Choices include, for example, the number of game plays, the sequential structure of the game, the possibility of forming coalitions with other players, and other players' preferences over outcomes. Like game strategists, communicators also need to consider how stakeholders make choices that stem from their mental models.[20]

The mental models that individuals construct from their prior knowledge and beliefs affect message development in the following ways:

1) Messages should be designed to address how the brain understands a risk or threat, including the receiver's knowledge, beliefs, information gaps, and misperceptions. For example, for a health, safety, and environmental issue, it is important to first gather information about stakeholder knowledge and beliefs about exposure, adverse consequences, mitigation procedures, and protective actions.
2) What a person knows and believes about a risk or threat will significantly affect how that person interprets and uses existing and new information.
3) Empirical research can uncover discrepancies between how nonexperts understand a risk or threat (i.e., their mental models) and how experts, such as technical, engineering, and scientific professionals, understand the same risks or threats (their mental models).

4) One of the primary goals of risk, high concern, and crisis communication message development is to increase knowledge and reduce or remove discrepancies between how nonexperts (e.g., the public or a specific stakeholder) perceive and understand a risk or threat, and how experts (e.g., technical, engineering, and scientific professionals) perceive and understand the same risks or threats.

7.4.4 Personality Filters

Responses to messages are strongly influenced by personality. Although people seldom can be described by one discrete personality type, there are many personality metrics that measure and identify personality types. One of the most common is the Myers-Briggs personality metric. This metric is based in part on the theories of psychological types described by the psychoanalyst C. G. Jung. The typology identifies and describes 16 personality types. Questions for the metric include: Do you prefer to focus on the outer world or on your own inner world (the extraversion [E] vs. introversion [I] personality type)? Do you prefer to focus on the basic information you take in, or do you prefer to interpret and add meaning (the sensing [S] vs. intuition [N] personality type)? When making decisions, do you prefer to first look at logic and consistency or first look at the people and special circumstances (the thinking [T] vs. feeling [F] personality type)? In dealing with the outside world, do you prefer to get things decided, or do you prefer to keep options open (the judging [J] vs. perceiving [P] personality type)?

According to Myers-Briggs and other personality theories, distinct personality types process and communicate information differently. Personality affects how a person's brain prefers to process information and how open a person's brain is to new information. Personality type is a major determinant of (1) how a person is, or prefers to be, perceived by others, and (2) how a person relates, or prefers to relate, to others. For example, a person identified as having the personality characteristic extroversion [E] may prefer to talk out loud about a risk issue and hear aloud what others have to say. They may therefore prefer a town hall meeting format for the sharing and exchange of information with experts and others about a risk-related issue. They may get angry and upset when open forums for discussion, such as a town hall meeting, are not made available. By comparison, a person identified as having the personality characteristic introversion [I] may prefer to read what experts and others have to say about a risk issue and submit their comments in writing. They may therefore prefer to view what is posted on an organization's website or read handout materials provided at an open house or information forum. As another example, a person identified as having the personality characteristic thinking [T] may give greater weight to logic and technical facts while the personality type feeling [F] may give more weight to what others feel and express when making judgments and decisions about a risk.

7.4.5 Negative Dominance/Loss Aversion Filters

Holding constant other variables, the human brain puts more weight on negative messages than on positive messages. Negative messages can be about events, objects, activities, actions, and personal traits.[21] The human brain puts an especially heavy weight on one particular type of negative message: messages about losses. As the Nobel Prize winner Professor Daniel Kahneman observed, most people are averse to losses. People do not like to lose that which they already have.[22]

Kahneman, together with his colleague, Amos Tversky, formally introduced into the academic literature the term "loss aversion."[23] The work of Kahneman, Tversky, and others indicates that the brain's rules of engagement for negative and loss information differ greatly from the rules of engagement for positive and gain information. These different rules of engagement are listed in Table 7.4.

As a result of these and other cognitive processes, the weight of negative messages in relation to positive messages is often two or three times greater or more.[24] In practical terms, *it often takes on average at least three positive messages to offset a single negative one. Similarly, it often takes on average at least three benefit messages to offset one negative message about costs or losses.*

Research focused on specific high-stress scenarios indicates that the impact of negative messages versus positive messages can be five times as great. For example, Professor John Gottman at the University of Washington conducted groundbreaking research on the likelihood of a wedded couple getting divorced or remaining married.[25] The stressors leading to divorce included (1) finances, (2) communications, (3) work, (4) in-laws, (5) children, (6) division of labor, and (7) intimacy.

Based on research related to divorce, the single biggest determinant of divorce is communications and, more specifically, the ratio of positive to negative comments that the partners offer to each other. Gottman, for example, found the optimal ratio was five positive messages for every negative one. For those couples who ended up divorced, the ratio was approximately three positive comments for every four negative comments. Gottman found that the cause of divorce was not only a disproportionate number of negative verbal messages but also a disproportionate number of negative nonverbal messages, especially facial expressions.[26] For reading facial expressions, Gottman relied heavily on the research of Paul Ekman and his colleagues.[27]

Gottman's research and that of others on negatives also revealed that negative messages should not entirely be shunned. For example, they often provide useful information and help identify things that are not working, such as miscommunications. It is the disproportionate number of negative to positive messages that creates problems.[28]

Negative dominance and loss aversion often lead to a marked preference by people for reassuring messages that they will not face losses. While the assurance of safety may not be possible, messages about risk need to be constructed recognizing that people will focus more on the negative statements than on the positive statement.

Table 7.4 Rules for engagement by the brain with negative vs positive information

Controlling for other variables, the human brain typically:

1) is more sensitive to messages about negatives and losses than to messages about positives and gains of equal value;

2) is willing to exert more energy, effort, and resources *to avert* a negative or loss than they are for an equivalent gain;

3) is willing to exert more energy, effort, and resources *to avoid* losing something it already has than to possess something of equal value;

4) processes and filters negative messages faster and more thoroughly than positive messages;

5) stores and recalls unpleasant memories and experiences more than it stores and recalls positive memories and experiences;

6) considers negative aspects of an experience before other considerations;

7) pays closer attention to and remembers longer messages that contain negatives, e.g., words such as "no," "not," "never," "nothing," "none," and, more generally, words, phrases, or images with strong negative associations or connotations.

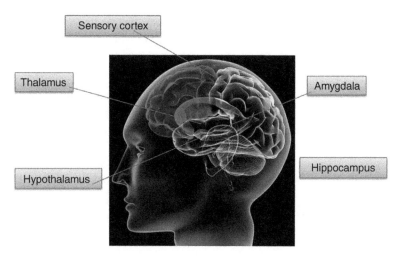

Figure 7.1 Critical Components in Human Brain's Risk and Threat Protection System

7.5 Message Filters and the Brain

The human brain filters information about risks and threats through several brain structures, including the amygdala. The amygdala is commonly associated with the concepts of fight, freeze, or flight. The amygdala is a very primitive part of the brain. It is located deep in the brain and is about the size of an almond. Because of its primitive nature, it is sometimes called the reptilian or lizard brain.

The brain's risk and the threat-detection systems are essential for processing emotions, including processing messages that induce fear about risks or threats. These other brain structures include the prefrontal cortex (where rational thinking primarily takes place), the sensory cortex, the brain stem, the septum, the hippocampus, the thalamus, and the hypothalamus (See Figure 7.1.). Together with the amygdala, it is in these parts of the brain where emotions such as fear are primarily processed.

The role of the amygdala is frequently misunderstood. Initially, the amygdala was considered to be "the organ of fear."[29] More recent neuroscience studies have expanded the role of the amygdala, including recognition of unpleasant negative emotions, recognition of both pleasant and unpleasant stimuli, memory, attention, and alertness.[30] Multiple connections of the amygdala indicate it may be more than a simple risk, threat, or danger detector.[31]

It is now understood that no single brain area is totally responsible for any single brain function.[32] The notion of exclusive brain areas is an intellectual residue from when most evidence about the brain was obtained via observations of damage to a particular area of the brain. Neuroscientists now understand information processing as a function of complex neurological systems that begin with the activation of particular parts of the brain that take the lead, such as the amygdala and hypothalamus, and then function as part of an interconnected system.[33]

7.6 Message Filters, Perceptions, and Models of Human Behavior

Messages about risks and threats influence perceptions and behaviors. However, there is a less-than-perfect correlation between perceptions and behaviors. Reasons for this are complex and are discussed in the research literature on behavior and health, especially the research literature associated with the Health Belief Model.[34] The Health Belief Model assumes people want to stay healthy and avoid illness and will therefore adopt behaviors they believe will protect them from illness. Controlling for other variables, the model identifies messages that can narrow the gap between perceptions, attitudes, beliefs, intentions, and behaviors, described as follows.

Perceived susceptibility messages: These are messages that influence a person's beliefs about how vulnerable and susceptible they are to an adverse outcome arising from a risk or threat.

Perceived severity messages: These are messages that influence a person's beliefs about the seriousness and magnitude of an adverse outcome arising from a risk or threat.

Perceived benefits messages: These are messages that influence a person's beliefs about the advantages of taking specific actions to reduce, eliminate, or manage an adverse outcome arising from a risk or threat.

Perceived barriers: These are messages that influence a person's beliefs about the negative aspects and costs of adopting specific behaviors to reduce, eliminate, or manage an adverse outcome arising from a risk or threat.

Perceived self-efficacy: These are messages that influence a person's beliefs about their ability to carry out and to successfully perform a behavior to reduce, eliminate, or manage an adverse outcome arising from risk or threat.

Perception and belief models have been applied successfully to messages related to a wide variety of health-related behaviors. Most prominently, they have been applied successfully to smoking behavior.

7.7 Message Filters, Perceptions, and Persuasion

Science- and evidence-based research focused on the links between message filters, message development, and persuasion began in the middle of the last century and is ongoing.[35] Systematic observations on messages designed to persuade can be traced back to the recommendations made by Aristotle over 2000 years ago. Aristotle argued that for virtually all effective purposes, communication is persuasion. We persuade for seven reasons: (1) to share information, ideas, and feelings; (2) to influence one another; (3) to coordinate activity; (4) to build relationships; (5) to acquire goods and services; (6) to entertain and express ourselves; and (7) to create and sustain culture-based values and beliefs.

In his classic work *Rhetoric: The Art of Persuasion*, Aristotle hypothesized that messages designed to persuade are of three types, which he called *appeals* or *arguments*: [36]

1) *appeals* to authority (in Greek, *ethos*)
2) *appeals* to logic (in Greek, *logos*)
3) *appeals* to emotion (in Greek, *pathos*)

Appeals to authority are messages designed to persuade via characteristics of the source of information. The authoritativeness of a source of information, in turn, is based on perceived attributes, including expertise, experience, track record, character, qualifications, credentials, reputation, familiarity, dedication, value similarity, and attractiveness. Appeals to authority also include messages designed to persuade via reference to rules, regulations, standards, and promises. Appeals to logic are messages designed to persuade through proofs, apparent proofs, and reason. Proof- and reason-based messages are often based on facts and statistics. Appeals to emotion are messages designed to persuade by putting the audience into a particular emotional frame of mind.

Communication techniques that have an especially strong impact on emotional responses include visuals and storytelling.

Among factors that enhance the persuasiveness of a message are:[37]

1) attractiveness and likability of the person or organization supplying the information;
2) similarity of values, positions, attitudes, preferences, background, and worldviews;
3) reciprocity and commitment (e.g., if you give me something, I must give you something back of equal or higher value; reciprocally, if I give you something, you must give me something back of equal or higher value);
4) claims about the scarcity or rarity of the desired thing, such as an object, status, idea, or experience;
5) expertise and authority;
6) conformity with what trusted others believe or are doing;
7) consistency and resonance with a person's desires and self-image;
8) trust in the source of information;
9) status or power of the information source.

A special and polarizing topic in the literature on persuasion is the effectiveness of messages using fear to persuade. Some researchers have argued vigorously for the effectiveness of fear appeals. Other researchers have argued equally vigorously that fear appeals are counterproductive and can backfire. In 2015, results were published from a comprehensive meta-analysis investigating the effectiveness of fear appeals for influencing attitudes, intentions, and behaviors.[38] The investigation tested predictions from numerous theories. The results showed:

1) fear appeals can positively influence attitudes, intentions, and behaviors;
2) fear appeals are most effective when they consider the ability of a person to carry out and perform a recommended behavior;
3) fear appeals are most effective when focus is on a onetime only vs. repeated behavior;
4) fear appeals are most effective when they are focused on a specific target audience.

Fear appeals were used successfully when the US government required television and radio stations to give millions of dollars of free airtime for antismoking ads. Many of these ads were graphic, disturbing, and highly frightening. Smoking rates plummeted.[39]

The places where fear messages have been least effective are messages to teenagers about health and safety risks (e.g., about everything from smoking to speeding to unprotected sex). This is due in part to peer pressure, mood management, distrust of adults, image maintenance, addiction, persuasive advertising, social encouragement, rebellious behavior, and the perception of many teenagers that they are invulnerable or have the extra protection or shield provided by youthfulness.[40]

In summary, people can respond usefully to messages that use fear but only in particular kinds of circumstances and when the messages address a person's beliefs about their ability to carry out and successfully reduce, eliminate, or manage an adverse outcome.

7.8 Message Filters, Perceptions, and Ethics

A polarizing topic in the literature on perceptions, message filters, and persuasion is ethics. Questions about ethics, truthfulness, propaganda, and manipulation have long been at the center of discussions about messages designed to persuade. In her classic treatise *Lying*, Bok argues that few if any human groups, organizations, institutions, or states can succeed in the long run when serious questions are raised about the truthfulness of leaders.[41] Bok also distinguishes between intentionally misleading others and unknowingly uttering a falsehood.

As a guide to ethical risk, high concern, and crisis communication, virtually every professional association has a "code of ethics."[42] These codes are typically a statement of abiding principles supported by explanations and position papers that address specific issues. They are also typically presented not as a set of rules but rather as a guide that encourages all who engage in the profession to take responsibility for the information they provide, regardless of communication format or medium. While these various codes have differences, most share common elements, including the principles of truthfulness, accuracy, objectivity, impartiality, fairness, and public accountability. For example, the Society for Risk Analysis holds its members to the following codes:[43]

Members of the Society for Risk Analysis shall:

- Hold paramount the truth in all matters associated with risk analysis.
- Observe the laws, regulations, and ethical standards regarding the conduct of risk research and practice, including guidelines for human and animal studies.
- Give due consideration to the ethical, legal, social, and policy implications of their research, advice, and communications.
- Conduct their work with objectivity and themselves with integrity, being honest and truthful in reporting and communicating their research and assessments.
- Employ sound analytic methods in the effort to identify, characterize, and assess methods for addressing risks.
- Practice high standards of workplace, occupational, and environmental health and safety for the benefit of themselves, their coworkers, their families, their communities, and society.
- Abstain from professional judgments influenced by an undisclosed conflict of interest and disclose any material or professional conflicts of interest.
- Conduct themselves honorably, responsibly, ethically, and lawfully so as to enhance the honor, reputation, and usefulness of the risk analysis profession.

Like many broader ethical systems, "codes of ethics" typically discuss the principle of "limitation of harm." An example of the "limitation of harm" principle would be withholding information that might result in harm to someone if shared.

7.9 Message Filters and the Issue of Acceptable Risk

One of the most intense and contentious issues faced by societies is what level of risk or threat is acceptable, what level of risk or threat is insignificant, what level of risk or threat can be ignored, how safe is safe enough, and whom to trust in determining what is acceptable.[44] For example, some people will view an individual lifetime risk of one in a million as insignificant, and therefore acceptable. Others will perceive the same risk number to be large, significant, and therefore unacceptable.

Questions about acceptability are at the heart of many risk-related controversies, such as climate change, nuclear power, genetic engineering, and nanotechnology. In the year 2020, questions about risk acceptability were at the forefront in discussions about public health measures to stop the spread of the virus that causes COVID-19.[45] These measures included lockdowns, community stay-at-home orders, requirements to wear face masks, social distancing, and restrictions on social gatherings. Individuals and groups opposed to these measures argued they violated their personal freedoms and interfered with their right to make their own decisions about acceptable risks. They also argued that the costs of these measures – especially to the economy – were greater than the benefits. This was in contrast to public health officials and the many people who gave primacy to stopping the spread of the disease.

At an individual level, answers to these questions often come down to a personal decision based on personality and context, such as choosing whether to eat only organic foods and whether to avoid all foods known to have been treated with pesticides or that contain additives. On a group, community, or societal level, where values conflict and decisions often result in winners and losers, decisions about what level of risk is acceptable and how safe is safe enough are much more complex and difficult. Before a decision can be made, information about the risk or threat must go through several filters.

Since nearly all activities and actions carry some level of risk or threat, a zero-risk or threat goal would require an absolute ban on the activity or action of concern. However, when an activity or action is banned, there are usually risks and threats associated with the substitutes. In practice, zero-risk goals are difficult, if not impossible, to achieve.

Many regulations and standards explicitly or implicitly recognize that there are levels of risk or threats that are so small that they are not significant or worthy of attention. These small or insignificant risks and threats are called *de minimis* risks or threats. The term *de minimis* is derived from the legal doctrine *de minimis non curat*: the law does not concern itself with trifles. Proponents of a *de minimis* approach argue that regulations and standards should establish *de minimis* levels and should set regulations or standards only for those risks or threats greater than the *de minimis* level. There is as yet no full consensus within societies on what constitutes a *de minimis* or acceptable level of risk. For example, in the United States, many federal government agencies consider a risk or threat level of one in a million an acceptable level for the public, but that is not a universal standard.

Proponents of an acceptable risk approach to managing risks and threats argue that setting an acceptable risk level helps organizations and agencies (1) decide when an action or agent poses a significant risk or threat; (2) set consistent levels of risk or threat for regulatory or management action; and (3) focus their attention on truly risky and threatening activities and avoid spending scarce resources on trivial risks and threats.

By comparison, critics of an acceptable risk approach argue that the use of this approach is problematic at best and impossible at worst. This is because of fundamental, deeply based, and unresolvable perceptual differences among individuals and groups about what level of risk or threat is acceptable, what level of risk or threat is insignificant, what level of risk or threat can be ignored, and how safe is safe enough.

7.9.1 Factors in Determining Acceptable Risk

One criticism of the acceptable risk approach is that acceptable risk levels are typically defined in relation to a probability of experiencing an adverse outcome. An example would be setting an acceptable risk level to one in a million increased chance of contracting a disease in a lifetime. Critics argue the probability of experiencing an adverse outcome should not, and is not, the sole factor that should define an acceptable risk or threat level. Several other factors need to be considered.

First, psychological and cultural factors need to be considered to determine an acceptable risk or threat. As indicated elsewhere in this book, more than a dozen perception/fear/high concern factors need to be considered when confronting an emotionally charged issue. The most important factor is trust. Other factors include perceived benefits, personal controllability, voluntariness, fairness of the distribution of costs and benefits, reversibility, catastrophic potential, and factual certainty.

Second, a risk or threat may be perceived as acceptable and trivial if only a few people are affected. However, this same risk or threat may be perceived as unacceptable and significant if numerous people are affected.

Third, adverse outcomes need to be considered in what is acceptable, and not everyone will agree on what is an adverse outcome and what is the probability of that adverse outcome. For example, some will advance arguments about risks and threats to future generations and ecosystems, while others will give more weight to the present.

Fourth, a risk or threat may be perceived as acceptable and insignificant if viewed alone. However, that same risk or threat may not be perceived as acceptable and insignificant if considered as part of a cumulative burden of risk or threat. Cumulative effects therefore need to be considered in setting an acceptable risk level.

Fifth, setting an acceptable risk level is difficult and sometimes impossible because groups, organizations, communities, and societies are continually changing. A risk or threat that at one time was deemed by a group, organization, community, or society to be acceptable and insignificant may not be seen as such in the future. For example, perceptions have radically changed regarding risks and threats to the environment caused by industrial activities.

Unfortunately, there is no numerical level of risk – other than zero – that will ever be likely to receive universal acceptance. Eliminating all risks or threats is not only impossible but would force populations to give up many of the things they value. Thus, except where prohibited by law, decision makers and risk managers face the task and difficulty of identifying levels of risk or threat that are greater than zero but that are also acceptable and "safe enough."

7.9.2 Strategies for Addressing Acceptable Risk

In making decisions about acceptable risks and threats, risk managers typically address the question through one or more communication strategies, any of which may engender trust or distrust among interested and affected stakeholders. A list of these strategies is shown in Table 7.5; they are discussed below.

The first strategy for addressing acceptable risk is to conduct a formal analysis and communicate the results to stakeholders. The analysis would identify the economic, social, and political costs of risk or threat reduction, the benefits gained from the activity or action that poses the risk or threat, and the availability of substitutes or alternatives.

A second strategy for addressing acceptable risk is to let the existing political and legal system decide what is acceptable. If there are no laws or regulations, or if the laws are ambiguous, the strategy would be to let the legal system and courts determine what is "reasonable." The term

Table 7.5 Communication strategies for addressing acceptable risk issues.

1) Conducting formal analysis.
2) Letting the existing political and legal system decide.
3) Using a comparative and precedent-based strategy.
4) Invoking the "precautionary principle."
5) Engaging trusted individuals and/or respected experts.
6) Encouraging public participation.

"reasonable" – or, more commonly, "unreasonable" – appears in several laws and regulations as a criterion for making decisions about acceptability. However, "reasonableness" is often left undefined, requiring lawyers, litigants, judges, and juries to determine what constitutes a reasonable risk or threat.

A third strategy for addressing acceptable risk is to use a comparative and precedent-based approach and communicate the results to stakeholders. According to this approach, acceptable risk decisions should be guided by comparisons with other risks or threats that people have already chosen to accept (i.e., levels of risk or threat accepted implicitly or explicitly in prior societal decisions). As discussed elsewhere in this book, the comparative approach is fraught with challenges and difficulties.

A fourth strategy for addressing acceptable risk is to invoke the "precautionary principle" and communicate the results to stakeholders. The precautionary principle, also called the "prudent avoidance approach" and the "precautionary approach," argues that all reasonable efforts should be taken to minimize potential risks when the magnitude of the risks is unknown or highly uncertain. Many researchers and observers have explored this approach.[46] In its most rigid form, the precautionary principle indicates that the lack of full scientific certainty should not be a reason for postponing measures to reduce or eliminate a risk or threat when the adverse consequences of a risk or threat are great or irreversible. Advocates of this approach argue the following: When an activity or action may pose a significant risk or threat with serious irreversible outcomes, precautionary measures should be taken even if cause-and-effect relationships have not been fully established.

The precautionary principle is frequently invoked in communications when there are significant perceived uncertainties or gaps in scientific understanding. The precautionary principle is founded in part on traditional and conventional communications, such as "it is better to be safe than sorry." There are several potential difficulties associated with the full implementation of the precautionary principle. For example, if applied inappropriately or over-rigidly, full implementation of the precautionary principle may (1) hinder or paralyze innovation; (2) give limited guidance to risk managers; (3) fail to recognize short- and long-term benefits of an activity or action; (4) fail to fully consider alternative courses of action; (5) result in arbitrary, *ad hoc*, and inconsistent risk management decisions; and (6) result in unintended adverse outcomes.

A fifth strategy for addressing acceptable risk is to engage trusted individuals and/or respected experts and let them decide. Results from these deliberations, such as by appointing commissions, committees, working groups, or expert panels, can then be documented and set down as policies, laws, protocols, standards, or regulations defining what is acceptable and how safe is safe enough.

The sixth strategy used to resolve the acceptable risk question is to determine what, if any, agreement can be achieved through public participation, stakeholder engagement, political processes, negotiation, mediation, and conflict-resolution methods. Stakeholder engagement processes are described in Chapter 4 of this book. Results from these processes can then be classified as decisions, policies, standards, laws, protocols, and regulations defining what is acceptable and how safe is safe enough.

The stakeholder engagement approach has many attractions, including that it is consistent with the first of the Seven Cardinal Rules of Risk Communication: Accept and involve the public as a legitimate partner.[47] However, this approach is plagued by many dilemmas, difficulties, challenges, and practical hurdles. For example, for the stakeholder-based approach to work and gain acceptance, stakeholders need to have access to all relevant risk-related information and the skills to interpret that information. Moreover, there needs to be agreement about the five *W*s (who, what, where, when, and why) and the one *H* (how). For example, *Who* should be engaged? *What* will stakeholders be asked to discuss or decide? *What* will be the objective of stakeholder engagement? *What* strength of evidence will be required? *What* assumptions will need to be made? *What* resources will be needed? *Where* will stakeholder engagements take place? *When* will stakeholder engagements take place in the decision-making process? *Why* will some stakeholders be included and others be left out? *How* will uncertainties and incomplete information be handled? *How* will perceptual biases that can distort risk judgments be managed? *How* will stakeholder opinions be canvased? *How* will stakeholders that are typically not active in public decision-making be engaged? *How* will vulnerable or underserved stakeholders be engaged? *How* will the engagement process be protected from "group think" or manipulation by special interests? *How* will experts, consultants, and advisers be selected and used? *How* will stakeholder agreement and consensus be determined? At a deeper level, *How* will the engagement process overcome subtle challenges to decision-making, such as "framing effects"?[48]

7.10 The Message is in the Mind of the Receiver

Many managers and communicators are frustrated by what they perceive as irrational attitudes and behaviors when people respond to risks and threats. This chapter challenges this notion of sheer irrationality and considers the beliefs and perceptions that may underpin what appear to be irrational and illogical responses. Years of research indicate that fighting risk perceptions with facts and logic alone is often ineffective. To be effective in risk, high concern, and crisis situations, managers and communicators need to work along with the brain's models and perceptions, rather than simply opposing them. To be effective, managers and communicators need to consider the many complex factors and filters through which information passes.

A key principle of risk, high concern, and crisis communication is that people's perceptions of a risk or threat depend less on factual evidence and logic than on complex perceptual factors and filters ranging from psychological, personality, and social-demographic factors to group affiliations and cultural and neurobehavioral factors. These factors and filters determine how people judge risk acceptability and how they react to information in a high-stress, high-stakes, or risk-related situation. Because these perceptual factors and filters come actively and prominently into play under stress, they are at the core of what differentiates risk, high concern, and crisis communication from other communication fields.

7.11 Chapter Resources

Below are additional resources to expand on the content presented in this chapter.

Árvai, J., Rivers, L. III., eds. (2014). *Effective Risk Communication*. London: Earthscan.

Arvai, J., Campbell-Arvai, V. (2014). "Risk communication: insights from the decision sciences," in *Effective Risk Communication*, eds. J. Arvai and L. Rivers III. London: Taylor and Francis.

Bavel, J.J.V., Baicker, K., Boggio, P.S. (2020). "Using social and behavioral science to support COVID-19 pandemic response." *Nature Human Behavior* 4:460–471.

Beck, M., Kewell, B. (2014). *Risk A Study of Its Origins, History and Politics*. Hackensack, NJ: World Scientific Publishing Company.

Bennett, P., Calman, K., eds. (1999). *Risk Communication and Public Health*. New York: Oxford University Press.

Bennett, P., Coles, D., McDonald, A. (1999). "Risk communication as a decision process," in *Risk Communication and Public Health*, eds. P. Bennett and K. Calman. New York: Oxford University Press.

Bier, V.M. (2001). "On the state of the art: risk communication to the public." *Reliability Engineering and System Safety* 71(2):139–150.

Bohnenblust, H., Slovic, P. (1998). "Integrating technical analysis and public values in risk based decision making." *Reliability Engineering & System Safety* 59:151–159.

Bostrom, A. (2003). "Future Risk Communication." *Futures* 35:553–573.

Bostrom, A., Atman, C., Fischhoff, B., Morgan, M.G. (1994). "Evaluating risk communications: completing and correcting mental models of hazardous processes, Part II." *Risk Analysis* 14(5):789–797.

Bourrier, M., Bieder, C., eds. (2018). *Risk Communication for the Future: Towards Smart Risk Governance and Safety Management*: Cham, Switzerland: Springer.

Breakwell, G.M. (2007). *The Psychology of Risk*. Cambridge, UK: Cambridge University Press.

Brosch, T. (2021). "Affect and emotions as drivers of climate change perception and action: A review." *Current Opinion in Behavioral Sciences* 42:15–21.

Centers for Disease Control and Prevention (CDC) (2012). *Emergency, Crisis, and Risk Communication*. Atlanta, GA: Centers for Disease Control and Prevention.

Chess, C., Hance, B.J., Sandman, P.M. (1986). *Planning Dialogue with Communities: A Risk Communication Workbook*. New Brunswick, NJ: Rutgers University, Cook College, Environmental Media Communication Research Program.

Chess, C., Salomone, K.L., Hance, B.J. (1995). "Improving risk communication in government: research priorities." *Risk Analysis* 15(2):127–135.

Chess, C., Salomone, K.L., Hance, B.J., Saville, A. (1995). "Results of a national symposium on risk communication: next steps for government agencies." *Risk Analysis* 15(2):115–120.

Chung, I.J. (2011). "Social amplification of risk in the Internet environment." *Risk Analysis* 31(12):1883–1896.

Coombs, W.T. (1998). "An analytic framework for crisis situations: Better responses from a better understanding of the situation." *Journal of Public Relations Research* 10(3):177–192.

Coombs, W.T. (2007). "Protecting organization reputations during a crisis: The development and application of situational crisis communication theory." *Corporate Reputation Review* 10(3):163–176.

Coombs, W.T. (2019). *Ongoing Crisis Communications: Planning, Managing, and Responding*. Thousand Oaks, CA: Sage Publications, Inc.

Covello V. (1983). "The perception of technological risks." *Technology Forecasting and Social Change: An International Journal* 23:285–297.

Covello V. (1989). "Issues and problems in using risk comparisons for communicating right-to-know information on chemical risks." *Environmental Science and Technology* 23(12):1444–1449.

Covello V. (1992). "Risk communication, trust, and credibility." *Health and Environmental Digest* 6(1):1–4.

Covello V. (1993). "Risk communication and occupational medicine." *Journal of Occupational Medicine* 35:18–19.

Covello, V. (2003). "Best practices in public health risk and crisis communication." *Journal of Health Communication* 8 (Suppl. 1):5–8; 148–151.

Covello, V. (2006). "Risk communication and message mapping: a new tool for communicating effectively in public health emergencies and disasters." *Journal of Emergency Management* 4(3):25–40.

Covello, V. (2014). "Risk communication," in *Environmental health: From Global to Local*, ed. H. Frumkin. San Francisco: Jossey-Bass/Wiley.

Covello, V., Allen, F. (1988). *Seven Cardinal Rules of Risk Communication*. Washington, D.C.: US Environmental Protection Agency.

Covello V., McCallum D., Pavlova M. (1989). *Effective Risk Communication: The Role and Responsibility of Government and Nongovernment Organizations*. New York: Plenum Press.

Covello, V., Merkhofer, M. (1993). *Risk Assessment Methods: Approaches for Assessing Health and Environmental Risks*. New York: Plenum Press.

Covello, V., Minamyer, S., Clayton, K. (2007). *Effective Risk and Crisis Communication during Water Security Emergencies*. *EPA Policy Report*; EPA 600-R07-027. Washington, D.C.: US Environmental Protection Agency.

Covello, V., Peters, R., Wojtecki, J., Hyde, R. (2001). "Risk communication, the West Nile virus epidemic, and bio-terrorism: Responding to the communication challenges posed by the intentional or unintentional release of a pathogen in an urban setting." *Journal of Urban Health* 78(2):382–391.

Covello, V., Sandman, P., Slovic, P. (1988). *Risk Communication, Risk Statistics, and Risk Numbers*. Washington, DC: Chemical Manufacturers Association.

Covello, V., Sandman, P. (2001). "Risk communication: Evolution and revolution," in *Solutions to an Environment in Peril*, ed. A. Wolbarst. Baltimore, MD: Johns Hopkins University Press.

Covello, V., Slovic, P., von Winterfeld, D. (1986). "Risk communication: A review of the literature." *Risk Abstracts* 3(4):171–182.

Covello, V., Slovic, P., von Winterfeld, D. (1987). *Risk Communication: A Review of the Literature*. Washington, DC: National Science Foundation.

Cox, Jr., A.L. (2012). "Confronting deep uncertainties in risk analysis." *Risk Analysis* 32(10):1607–1629.

Cummings, L. (2014). "The 'trust' heuristic: arguments from authority in public health." *Health Communication* 29(10):1043.

Cvetkovich, G., Vlek, C.A., Earle, T.C. (1989). "Designing technological hazard information programs: Towards a model of risk-adaptive decision making," in *Social Decision Methodology for Technical Projects*, eds. C.A.J. Vlek and G. Cvetkovich. Dordrecht: Kluwer Academic.

Cvetkovich, G., Siegrist, M., Murray R., Tragesser, S. (2002). "New information and social trust asymmetry and perseverance of attributions about hazard managers." *Risk Analysis* 22(2):359–367.

Davies, C.J., Covello, V.T., Allen, F.W., eds. (1987). *Risk Communication: Proceedings of the National Conference*. Washington: The Conservation Foundation.

Dietz, T. (2013). "Bringing values and deliberation to science communication." *Proceedings of the National Academy of Sciences* 110:14081–14087.

Dunwoody, S. (2014). "Science journalism," in *Handbook of Public Communication of Science and Technology*, eds. M. Bucchi and B. Trench. New York: Routledge.

Earle, T.C. (2010). "Trust in risk management: A model-based review of empirical research." *Risk Analysis* 30(4):541–574.

Earle, T.C., Siegrist, M. (2008). "On the relation between trust and fairness in environmental risk management." *Risk Analysis* 28(5):1395–1414.

Falk, E., Scholz, C. (2018). "Persuasion, influence, and value: Perspectives from communication and social neuroscience." *Annual Review of Psychology* 69:329–356.

Fearn-Banks, K. (2019). *Crisis Communications: A Casebook Approach*, 5th edition. New York: Routledge.

Fischhoff, B. (1995a). "Risk perception and communication unplugged: twenty years of process." *Risk Analysis* 15(2):137–145.

Fischhoff, B. (1995b). "Strategies for risk communication. Appendix C." in *National Research Council: Improving Risk Communication*. Washington, DC: National Academies Press.

Fischhoff, B. (2012). *Judgment and Decision Making*. New York: Earthscan.

Fischhoff, B. (2013). "The sciences of science communication," in *Proceedings of the National Academy of Sciences* 110 (Supplement 3):14033–14039.

Fischhoff, B., Brewer, N.T., Downs, J.S., eds. (2011). *Communicating Risks and Benefits: An Evidence-Based User's Guide*. Washington, DC: Food and Drug Administration.

Fischhoff, B., Davis, A. L. (2014). "Communicating scientific uncertainty," in *Proceedings of the National Academy of Sciences of the United States of America* 111(Suppl. 4):13664–13671.

Fischhoff, B., Kadvany, J. (2011). *Risk: A Very Short Introduction*. New York: Oxford University Press.

Fischhoff B, Lichtenstein S, Slovic P, Keeney D. (1983). *Acceptable Risk*. Cambridge, Massachusetts: Cambridge University Press.

Flynn, J., Slovic, P., Mertz, C.K. (1994). "Gender, race, and perception of environmental health risks." *Risk Analysis* 14(6):1101–1108.

Glik D.C. (2007). "Risk communication for public health emergencies." *Annual Review of Public Health* 28(1):33–54.

Hance, B.J., Chess, C., Sandman, P.M. (1990). *Industry Risk Communication Manual*. Boca Raton, FL: CRC Press/Lewis Publishers.

Halvorsen, P.A. (2010). "What information do patients need to make a medical decision?" *Medical Decision Making* 30 (5 Suppl):11S–13S.

Haight, J.M., ed. (2008). *The Safety Professionals Handbook: Technical Applications*. Des Plaines, IL: The American Society of Safety Engineers.

Heath, R. O'Hair, D., eds. (2009). *Handbook of Risk and Crisis Communication*. New York: Routledge.

Hess, R., Visschers, V.H.M., Siegrist, M., Keller, C. (2011). "How do people perceive graphical risk communication? The role of subjective numeracy." *Journal of Risk Research* 14(1):47–61.

Hyer, R.N., Covello, V.T. (2005). *Effective Media Communication During Public Health Emergencies: A World Health Organization Handbook*. Geneva: United Nations World Health Organization Publications.

Hyer, R., Covello, V.T. (2017). *Top Questions on Zika: Simple Answers*. Arlington, VA: Association of State and Territorial Health Officials.

Infanti, J., Sixsmith, J., Barry, M.M., Núñez-Córdoba, J.M., Oroviogoicoechea-Ortega, C. & Guillén-Grima, F.A. (2013). *A Literature Review on Effective Risk Communication for the Prevention and Control of Communicable Diseases in Europe*. Stockholm: European Centre for Disease Control and Prevention.

Jardine, C.G., Driedger, S.M. (2014). "Risk communication for empowerment: An ultimate or elusive goal?" in *Effective Risk Communication*, eds. J. Arvai and L. Rivers III. London: Earthscan.

Joslyn, S., LeClerc, J. (2012). "Uncertainty forecasts improve weather related decisions and attenuate the effects of forecast error." *Journal of Experimental Psychology* 18:126–140.

Joslyn, S., Nadav-Greenberg, L., Taing, M.U., Nichols, R.M. (2009). "The effects of wording on the understanding and use of uncertainty information in a threshold forecasting decision." *Applied Cognitive Psychology* 23:55–72.

Heath, R. O'Hair, D., eds. (2009). *Handbook of Risk and Crisis Communication*. New York: Routledge.

Kahneman, D. (2011). *Thinking, Fast and Slow*. New York: Macmillan Publishers.

Kahneman, D., Slovic, P., Tversky, A., eds. (1982). *Judgment Under Uncertainty: Heuristics and Biases*. New York: Cambridge University Press.

Kahneman, D., Tversky, A. (1979). "Prospect theory: An analysis of decision under risk." *Econometrica*, 47(2):263–291.

Kasperson, R.E. (1986). "Six propositions on public participation and their relevance for risk communication." *Risk Analysis* 6(3):275–281.

Kasperson, R.E. (2014). "Four questions for risk communication." *Journal of Risk Research* 17(10):1233–1239.

Kasperson, R.E., Golding, D., Tuler, S. (1992). "Social distrust as a factor in sitting hazardous facilities and communicating risks." *Journal of Social Issues* 48(4):161–187.

Kasperson, R.E., Renn, O., Slovic, P., Brown, H.S., Emel, J., Goble, R., Kasperson, J. X., Ratick, S. (1987). "Social amplification of risk: A conceptual framework." *Risk Analysis* 8(2):177–187.

Kasperson, R., Kasperson, J.X., Golding, D. (1999). "Risk, trust, and democratic theory," in *Social Trust and the Management of Risk*, eds. G. Cvetkovich and R. Löfstedt. London: Earthscan.

Kasperson, R. E., Stallen, P.J.M. eds. (1991). *Communicating Risks to the Public: International Perspectives*. Dordrecht, Netherlands: Kluwer.

Klampitt, P.G., DeKoch, R.J., Cashman, T. (2000). "A strategy for communicating about uncertainty." *The Academy of Management Executive* 14(4):41–57.

Lerner, J.S., Li, Y., Valdesolo, P., Kassam, K.S. (2015). "Emotion and decision making." *Annual Review of Psychology* 66:799–823

Lindenfeld, L., Smith, H., Norton, T., Grecu, N. (2014). "Risk communication and sustainability science: lessons from the field." *Sustainability Science* 9(2):119–127.

Löftstedt, R. (2003). "Risk communication. Pitfalls and promises." *European Review* 11(3):417–435.

Löfstedt, R. (2005). *Risk Management in Post-Trust Societies*. New York: Palgrave Macmillan.

Löfstedt, R.E., Bouder, F. (2014). "New transparency policies: Risk communication's doom?" in *Effective Risk Communication*, eds. J. Àrvai and L. Rivers III. London: Earthscan.

Lundgren, R.E., McMakin A.H. (2018). *Risk Communication: A Handbook for Communicating Environmental, Safety, and Health Risks*. Hoboken, NJ: Wiley-IEEE Press.

McComas, K.A. (2006). "Defining moments in risk communication research: 1996–2005." *Journal of Health Communication* 11(1):75–91.

McComas, K.A., Arvai, J., Besley, J.C. (2009). "Linking public participation and decision making through risk communication," in *Handbook of Risk and Crisis Communication*, eds. R. Heath and D. O'Hair. New York: Routledge.

Mileti, D., Nathe, S., Gori, P., Greene, M., Lemersal, E. (2004). *Public Hazards Communication and Education: The State of the Art*. Boulder, Colo.: Natural Hazards Center.

Miller, M., Solomon, G. (2003). "Environmental risk communication for the clinician." *Pediatrics* 112(1):211–217. Accessed at: https://pdfs.semanticscholar.org/33c9/8bd7f526d727e8249632df23c537db89cd4f.pdf

Morgan, M.G., Fischhoff, B., Bostrom, A., Atman, C.J. (2002). *Risk Communication: A Mental Models Approach*. Cambridge, UK: Cambridge University Press.

National Research Council (1989). *Improving Risk Communication*. Washington, D.C.: National Academies Press.

National Research Council (1996). *Understanding Risk: Informing Decisions in a Democratic Society*. Washington, DC: National Academies Press.

National Research Council) (2008). *Public Participation in Environmental Assessment and Decision Making*. Washington, DC: The National Academies Press.

National Academy of Sciences (2014). *The Science of Science Communication II: Summary of a Colloquium*. Washington, DC: The National Academies Press.

National Academy of Sciences (2017). *Communicating Science Effectively*. Washington, D.C.: The National Academies Press.

National Oceanic and Atmospheric Administration, US Department of Commerce. (2016). *Risk Communication and Behavior. Best Practices and Research*. Washington, DC: National Oceanic and Atmospheric Administration.

Peters, E. (2012). "Beyond comprehension: The role of numeracy in judgments and decisions." *Current Directions in Psychological Science* 21(1):31–35.

Peters, R., McCallum, D., Covello, V.T. (1997). "The determinants of trust and credibility in environmental risk communication: An empirical study." *Risk Analysis* 17(1):43–54.

Pidgeon, N., Kasperson, R., Slovic, P. (2003). *The Social Amplification of Risk.* Cambridge, UK: Cambridge University Press.

Renn, O. (1992). "Risk communication: Towards a rational discourse with the public." *Journal of Hazardous Materials* 29:465–579.

Renn, O. (2008). *Risk Governance: Coping with Uncertainty in a Complex World.* London, UK: Earthscan.

Renn, O., Levin, D. (1991). "Credibility and trust in risk communication," in *Communicating Risks to the Public*, eds. R. Kasperson and P. Stallen. Dordrecht, The Netherlands: Kluwer Academic Publishers.

Reynolds, B. (2019). *Crisis and Emergency Risk Communication.* Atlanta, GA: US Centers for Disease Control and Prevention.

Reynolds B., Seeger, M.W. (2005). "Crisis and emergency risk communication as an integrative model." *Journal of Health Communication* 10:43–55.

Rodrıguez, H., Dıaz, W., Santos, J., Aguirre, B. (2007). "Communicating risk and uncertainty: Science, technology, and disasters at the crossroads," in *Handbook of Disaster Research*, eds. H. Rodríguez, E.L. Quarantelli, and R.R. Dynes. New York: Springer.

Sandman, P.M. (1989). "Hazard versus outrage in the public perception of risk," in *Effective Risk Communication: The Role and Responsibility of Government and Non-Government Organizations*, eds. V.T. Covello, D.B. McCallum, M.T. Pavlova. New York: Plenum Press.

Seeger, M.W. (2006). "Best practices in crisis communication: An expert panel process." *Journal of Applied Communication Research* 34(3):232–44.

Sheppard, B., Janoske, M., Liu, B. (2012). "Understanding risk communication theory: A guide for emergency managers and communicators." *Report to Human Factors/Behavioral Sciences Division, Science and Technology Directorate, US Department of Homeland Security.* College Park, MD: US Department of Homeland Security.

Slovic, P. (1987). "Perception of risk." *Science* 236(4799):280–285.

Slovic, P. (1993). "Perceived risk, trust, and democracy." *Risk Analysis* 13(6):675–682.

Slovic, P. (2000). *The Perception of Risk.* London, UK: Earthscan.

Slovic, P. (2016). "Understanding perceived risk: 1978-2015." *Environment: Science and Policy for Sustainable Development* 58(1):25–29.

Slovic, P., M. Finucane, L., Peters, E., MacGregor, D.G. (2004). "Risk as analysis and risk as feelings: Some thoughts about affect, reason, risk, and rationality." *Risk Analysis* 24(2):311–322.

Slovic, P. Finucane, M.L., Peters, E., MacGregor, D.J. (2007). "The affect heuristic." *European Journal of Operational Research* 177(3):1333–1352.

Stallen, P.J.M., Tomas, A. (1988). "Public concerns about industrial hazards." *Risk Analysis* 8(2):235–245.

Steelman, T. A., McCaffrey, S. (2013). "Best practices in risk and crisis communication: Implications for natural hazards management." *Natural Hazards* 65(1):683–705.

Thompson, K.M., Bloom, D.L. (2000). "Communication of risk assessment information to risk managers." *Journal of Risk Research* 3:333–352.

Tuler, S.P., Kasperson, R.E. (2014). "Social distrust and its implications for risk communication: An example of high level radioactive waste management," in *Effective Risk Communication*, eds. J. Arvai and L. Rivers III. London: Earthscan.

Tversky, A., Kahneman, D. (1974). "Judgment under uncertainty: Heuristics and biases." *Science* 185(4157):1124–1131.

US Environmental Protection Agency. (2005). *Superfund Community Involvement Handbook*. EPA 540-K-05-003. Washington, D.C.: US Environmental Protection Agency.

US Department of Health and Human Services. (2006). *Communicating in a Crisis: Risk Communication Guidelines for Public Officials*. Washington, DC: US Department of Health and Human Services.

Walaski, P. (2011). *Risk and Crisis Communications: Methods and Messages*. Hoboken, NJ: Wiley.

Weinstein, N. D. (1987). *Taking care: Understanding and Encouraging Self-Protective Behavior*. New York: Cambridge University Press.

Wojtecki, J., Peters, R. (2007). in J. Gordon (2007). *The Pfeiffer Book of Successful Communication Skill-Building Tools*. Hoboken, NJ: Wiley.

van der Linden, S. (2015). "The social-psychological determinants of climate change risk perceptions: towards a comprehensive model. *Journal of Environmental Psychology* 41:112–124.

van Valkengoed, A.M., Steg, L. (2019). "Meta-analyses of factors motivating climate change adaptation behaviour." *Nature Climate Change* 9:158–163.

Xie, B., Brewer, M.B., Hayes, B.K., McDonald, R.I., Newell, B.R (2019). "Predicting climate change risk perception and willingness to act." *Journal of Environmental Psychology* 65:101331.

Xie, X.F., Wang, M., Zhang, R.G., Li, J., Yu, Q.Y. (2011). "The role of emotions in risk communication." *Risk Analysis* 31(3):450–465.

Endnotes

1 For readers interested in the complex science and issues associated with determining the safety of Alar, I recommend reading one of my books on risk analysis and assessment methods and then reading the hundreds of articles published about the Alar scare. See Cohrssen, J.J., and Covello, V.T. (1989). *Risk Analysis: A Guide to Principles and Methods for Analyzing Health and Environmental Risks*. Washington, DC: White House Council on Environmental Quality. See also Covello, V.T., and Merkhofer, M.W. (1993). *Risk Assessment Methods*. Boston: Springer Publishing Co.

2 Covello, V., and Allen, F. (1988). *The Seven Cardinal Rules of Risk Communication*. EPA Policy Document OPM-87-020. Washington, DC. US Environmental Protection Agency. Accessed at: https://archive.epa.gov/care/web/pdf/7_cardinal_rules.pdf.

3 Cohrssen, J.J., and Covello, V.T. (1989). *Risk Analysis: A Guide to Principles and Methods for Analyzing Health and Environmental Risks*. Washington, DC: White House Council on Environmental Quality.

4 See Campbell, J. (2008). *The Hero with a Thousand Faces: The Collected Works of Joseph Campbell*. Novato, CA: New World Library.

5 Friedman, S.M., Villamil, K., Suriano, R.A., Brenda, P., and Egolf, B.P. (1996). "Alar and apples: Newspapers, risk and media responsibility." *Public Understanding of Science*. 5(1):1–20, p. 1.

6 Miller, G. (1956). "The magical number seven, plus or minus two: Some limits on our capacity for processing information." *Psychological Review* 63(2) p. 81–97.

7 The "Rule of Three" is identified as "R3" in Figure 3.4, "Communication Templates and Tools for Risk, Crisis, and High Stress Communication" found in Chapter 3 of this book.

8 Campbell, J. (2008). *The Hero with a Thousand Faces*, 3rd edition. Novato, CA: New World Library.

9 See Chapter 2 in this book for a discussion of this issue.

10 See Slovic, P. (1987). "Perception of risk." *Science* 236(4799):280–285.

11 Starr, C. (1969). "Social benefits versus technological risks," *Science* 165(3899):1232–1238.

12 See, e.g., Pidgeon, N., Kasperson, R., and Slovic, P. (2003). *The Social Amplification of Risk.* Cambridge, UK: Cambridge University Press.

13 See, e.g., Xie, X.F., Wang, M., Zhang, R.G., Li, J., and Yu, Q.Y. (2011). "The role of emotions in risk communication." *Risk Analysis* 31(3):450–65; Lerner, J.S., Li, Y., Valdesolo, P., and Kassam, K.S. (2015). "Emotion and decision making." *Annual Review of Psychology*, 66:799–823; Slovic, P. Finucane, M.L., Peters, E., and MacGregor, D.J. (2007). "The affect heuristic." *European Journal of Operational Research*, 177(3):1333–1352,

14 See, e.g., Brosch, T. (2021). "Affect and emotions as drivers of climate change perception and action: A review." *Current Opinion in Behavioral Sciences*, 42:15–21; van der Linden, S. (2015). "The social-psychological determinants of climate change risk perceptions: Towards a comprehensive model. *Journal of Environmental Psychology*, 41:112–124; Xie, B., Brewer, M.B., Hayes, B.K., McDonald, R.I., and Newell, B.R. (2019). "Predicting climate change risk perception and willingness to act." *Journal of Environmental Psychology* 65:101331; van Valkengoed, A.M. and Steg, L. (2019). "Meta-analyses of factors motivating climate change adaptation behaviour." *Nature Climate Change*, 9:158–163.

15 See, e.g., Nickerson, R.S. (1998), "Confirmation bias: A ubiquitous phenomenon in many guises." *Review of General Psychology* 2(2):175–220.

16 See, e.g., LeDoux, J.E. (2003). "The emotional brain, fear, and the amygdala." *Cellular and Molecular Neurobiology* 23:727–738. See also LeDoux, J.E. (1996). *The Emotional Brain* (2002). New York: Simon & Schuster.

17 See,Drews, S., Van den Bergh, J.C.J.M. (2016): "What explains public support for climate policies? A review of empirical and experimental studies." *Climate Policy*, 16:855–876; Sullivan, A., and White, D. (2019). "An assessment of public perceptions of climate change risk in three Western US cities." *Weather, Climate, and Society.* April, 11:446–463. Accessed at https://journals.ametsoc.org/doi/10.1175/WCAS-D-18-0068.1;Leiserowitz, A.A. (2005). "American risk perceptions: Is climate change dangerous?" *Risk Analysis* 25:1433–1442; Leiserowitz, A.A., Maibach, E., Roser-Renouf, C., Rosenthal, S., and Cutler, M. (2017). *Politics and Global Warming.* Yale Program on Climate Change Communication/George Mason University Center for Climate Change Communication Report. Accessed at http://climatecommunication.yale.edu/wp-content/uploads/2017/07/Global-Warming-Policy-Politics-May-2017.pdf.; National Aeronautics and Space Administration (2019). US Global Change Research Program (2018). *Fourth National Climate Assessment Vol I + II.* Accessed at: https://climate.nasa.gov/evidence/. See also Sommerville, R.C., and Hassol, S.J. (2011). "Communicating the science of climate change." *Physics Today* 64 (10): 48–53: "Sommerville, R.C." Weber, E.U. (2016) "What shapes perceptions of climate change?" *WIREs* 7(1):125–134. Accessed at: https://onlinelibrary.wiley.com/doi/pdf/10.1002/wcc.377.

18 See, e.g., Fleishman-Mayer, L., Bruine de Bruin, W. (2014). The 'Mental Models' methodology for developing communications: Adaptions for information public risk management decisions about emerging technologies," in *Effective Risk Communication*, eds. Arvai, J., Rivers, L. London and New York: Earthscan.

19 Atman, C.J., Bostrom, A., Fischhoff, B., and Morgan, M.G. (1994). "Designing risk communications: Completing and correcting mental models of hazardous processes." *Risk Analysis* 14(5):779–88

20 The film *The Princess Bride* shows an entertaining example of game theory. Two of the main characters in the film engage in a game whose outcome is death for one of the players. The character in the film known only as the *Masked Man* poisons one of two goblets of wine with an odorless, tasteless poison that will cause death. He places one goblet in front of himself and the other goblet in front of his rival, Vizzini. As part of the game, Vizzini must then decide whether to drink the wine from the goblet in front of himself or drink the wine from the goblet in front of the *Masked Man*. After going through a long list of options and possible outcomes, Vizzini drinks from the cup he had chosen and dies from the poison. In the end (spoiler alert), it is revealed that the game was rigged. The *Masked Man* had poisoned both the cups. The *Masked Man* had spent years developing immunity to the poison. He had placed the poison in both goblets rather than just one. As the *Masked Man* predicted, Vizzini's mental model had not allowed for this unexpected option.

21 See, e.g., Rozin, P., and Royzman, E.B. (2001). "Negativity bias, negativity dominance, and contagion." *Personality and Social Psychology Review* 5(4):296–320. Rozin and Royzman point out in their article that negativity bias is manifested in four ways: (a) negative potency (negative entities are stronger than the equivalent positive entities), (b) steeper negative gradients (the negativity of negative events grows more rapidly with approach to them in space or time than does the positivity of positive events, (c) negativity dominance (combinations of negative and positive entities yield evaluations that are more negative than the algebraic sum of individual subjective valences would predict), and (d) negative differentiation (negative entities are more varied, yield more complex conceptual representations, and engage a wider response repertoire).

22 Kahneman, D. (2011). *Thinking, Fast and Slow*. New York: Macmillan Publishers.

23 Kahneman, D., Slovic, P., and Tversky, A., eds. (1982). *Judgment Under Uncertainty: Heuristics and Biases*. New York: Cambridge University Press; Kahneman, D., and Tversky, A. (1979). "Prospect theory: An analysis of decision under risk." *Econometrica*, 47(2):263–291.

24 See, e.g., Covello, V. T. (2009). "Strategies for overcoming challenges to effective risk communication," in *Handbook of Risk and Crisis Communication*, eds. R. Heath and D. O'Hare. New York: Routledge/Taylor & Francis.

25 See, e.g., Gottman, J. (1994). *What Predicts Divorce? The Relationship Between Marital Processes and Marital Outcomes*. Hillsdale, NJ: Lawrence Erlbaum Associates. See also Gottman, J. (2011). *The Science of Trust*. New York: W.W. Norton & Co.

26 Gottman's research resulted in a formula to predict divorce with an accuracy rate of as high as 90%. He evaluated specific negative communication patterns, including the ratio of positive to negative words and expressions exchanged between the partners, including gestures and body language. The most destructive negative communication patterns were criticism, contempt, defensiveness, and stonewalling.

27 Ekman developed a method for identifying facial expressions called the Facial Coding System (FACS), an anatomically based system for describing visual changes in groups of facial muscles.See, e.g.,Ekman, P. (2003). *Emotions Revealed: Recognizing Faces and Feeling to Improve Communication and Emotional Life*. New York: Owl Books/Henry Holt and Co. See also Ekman, J. (2009). *Telling Lies: Clues to Deceit*. New York: W.W. Norton and Co. See also, Goleman, D. (1995). *Emotional Intelligence: Why It Can Matter More Than IQ*. New York: Bantam Books.

28 See, e.g., Larrick, R., Wu, G. (1999). "Goals as reference points." *Cognitive Psychology* 38:79–109.

29 See, e.g., LeDoux, J.E. (2003). "The emotional brain, fear, and the amygdala." *Cellular and Molecular Neurobiology* 23:727–738.

30 See, e.g., Morrison, S.E., and Salzman, C.D. (2010). "Re-valuing the amygdala." *Current Opinion in Neurobiology* 20(2):221–230; see also Pessoa, L., and Adolphs, R. (2010). Emotion processing and the amygdala: From a 'low road' to 'many roads' of evaluating biological significance." *Nature Reviews Neuroscience* 11:773–783.

31 Kim, M.J., Loucks, R.A., Palmer, A.L., Brown, A.C., Solomon, K.M., and Marchante, A.N. (2011). "The structural and functional connectivity of the amygdala: From normal emotion to pathological anxiety." *Behavioral Brain Research* 223:403–410; see also Sander, D., Grafman, J., and Zalla,

T. (2003). "The human amygdala: An evolved system for relevance detection." *Reviews in the Neurosciences* 14:303–316; Sergerie, K., Chochol, C., and Armony, J. L. (2008). "The role of the amygdala in emotional processing: A quantitative meta-analysis of functional neuroimaging studies." *Neuroscience Biobehavioral Review* 32:811–830.

32 See, e.g., Purves, D., Jeannerod, M., and Coquery, J. M. (2005). *Neurosciences*, 3rd edition. Sunderland, MA: De Boeck Superieur; Sinauer Associates Inc.; see also Kandel, E. R., Schwartz, J.H., and Jessell, T.M. (2000). *Principles of Neural Science*, 4th edition. New York: McGraw-Hill.

33 See, e.g., Bonnet, L., Comte, A., Tatu, L., Millot, J.L., Moulin, T., Medeiros de Bustos, E. (2015). "The role of the amygdala in the perception of positive emotions: An "intensity detector." *Frontiers in Behavioral Neuroscience* 9: 178–183; Berntson, G.G., Cacioppo, J.T., and Tassinary, L.G., eds. (2007). *Handbook of Psychophysiology*. Cambridge, UK: Cambridge University Press; Kreibig, S.D. (2010). "Autonomic nervous system activity in emotion: A review." *Biological Psychology* 84:394–421.

34 See, e.g., Carpenter, C.J. (2010). "A meta-analysis of the effectiveness of health belief model variables in predicting behavior." *Health Communication* 25(8):661–669; Conner, M. & Norman, P. (1996). *Predicting Health Behavior. Search and Practice with Social Cognition Models*. Buckingham: Open University Press; Glanz, K., Rimer, B.K., and Lewis, F.M. (2002). *Health Behavior and Health Education: Theory, Research and Practice*. San Francisco, CA/Hoboken, NJ: Wiley; Glanz, K., and Bishop, D.B. (2010). "The role of behavioral science theory in development and implementation of public health interventions." *Annual Review of Public Health* 31:399–418; Janz, N.K., and Becker, M.H. (1984). "The health belief model: A decade later." *Health Education & Behavior* 11(1):1–47; Rosenstock, I. (1974). "Historical origins of the health belief model." *Health Education & Behavior* 2(4):328–335; Stretcher, V.J., and Rosenstock, R.M. (1997). "The health belief model." in *Cambridge Handbook of Psychology, Health and Medicine*, ed. A. Baum, Cambridge: Cambridge University Press.

35 See, e.g., Falk, E., and Scholz, C. (2018). "Persuasion, influence, and value: Perspectives from communication and social neuroscience. *Annual Review of Psychology* 69:329–356; Fisher, R., Ury, W., and Patton, B. (1991). *Getting to Yes*. New York: Penguin Books;Ury, W., Hovland, C.I., and Weiss, W. (1967). "The influence of source credibility on communication effectiveness," in, *Experiments in Persuasion*, eds. R.L. Rosnowand and E.J. Robinson, New York: Academic Press; McGuire, W.J. (1985). "Attitude and attitude change" in eds: G. Lindzey and E. Aronson, *Handbook of Social Psychology*. New York: Random House; Petty, R.E., and Cacioppo, J.T. (1986). "The elaboration likelihood model of persuasion." *Advances in Experimental Social Psychology* 19:123–205.

36 Aristotle (ND). *The Art of Rhetoric*. New York: Collins Classic Books.
Oxford, UK: Oxford University Press, 2018.

37 See, e.g., Cialdini, R.B. (2016). *Influence: The Psychology of Persuasion*, Revised Edition. New York: William Morrow and Co.; Cialdini, R.B. (2016). *Pre-suasion*, New York: Simon & Schuster.

38 Tannenbaum, M.B., Hepler, J., Zimmerman, R.S., Saul, L., Jacobs, S., Wilson, K., and Albarracín, D. (2015). "Appealing to fear: A meta-analysis of fear appeal effectiveness and theories." *Psychological Bulletin* 141(6):1178–1204.

39 For a review of the impacts of antismoking campaigns, see., e.g., Levy, D.T., Chaloupka, F., and Gitchell, J. (2004). "The effects of tobacco control policies on smoking rates." *Journal of Public Health Management and Practice*, 10(4): 338–353. The review looked at the impact of antitobacco media campaigns, which originally focused on educating consumers on the health risks of smoking and made use of television, radio, billboards, and print. The review reported that that adult and youth tobacco use in the United States declined following initiation of national media campaigns launched under the Fairness Doctrine in 1966, with reductions in per capita consumption as high as 4% per year. Researchers also found that antismoking campaigns are only one cornerstone in strategies to reduce smoking. Cigarette taxes and the passage of comprehensive clean air laws each have the potential to reduce smoking prevalence by 10% or more.

40 Gruber, J., Zinman, J. (2001). "Youth smoking in the United States: Evidence and implications," in *Risky Behavior among Youths*, ed. J. Gruber, Chicago: University of Chicago Press; p. 69–120. see also O'Donoghue, T., Rabin, M. (2001). "Risky behavior among youths: Some issues from behavioral economics" in *Risky Behavior among Youths*, ed. J. Gruber, Chicago: University of Chicago Press; p. 29–68. Pfeffer, D., Wigginton, B., Gartner, C., and Morphett, K. (2017). "Smokers' understandings of addiction to nicotine and tobacco: A systematic review and interpretive synthesis of quantitative and qualitative research." *Nicotine & Tobacco Research*, 9:1038–1046.

41 Bok, S. (1979). *Lying: Moral Choice in Public and Private Life*. New York: Vintage Books: A Division of Random House.

42 See, e.g., Society of Professional Journalists (2020). *Code of Ethics*. Accessed at: https://www.spj.org/ethicscode.asp; Public Relations Society of America (2020). Code of Ethics. Accessed at: https://www.prsa.org/about/ethics/prsa-code-of-ethics.

43 Society for Risk Analysis (2020). *Code of Ethics*. Accessed at: https://sra.org/our-ethics.

44 See, e.g., Fishhoff, B., Lichtenstein, S., Slovic, P., Derby, S., and Keeney, R., eds. (1981). Acceptable risk. Cambridge: Cambridge University Press. See also Hunter, P.R., and Fewtrell, L. (2001). "Acceptable risk," in *Water Quality: Guidelines, Standards and Health. Risk Assessment and Management for Water-Related Infectious Disease*, eds. Fewtrell, L., Bartram, J., London: IWA Publishing. pp. 207–227.

45 See, e.g., Bavel, J.J.V., Baicker, K., Boggio, P.S. et al. (2020). Using social and behavioral science to support COVID-19 pandemic response. *Nature Human Behavior*, 4:460–471.

46 See, e.g., Grandjean, P. (2004). "Implications of the precautionary principle for primary prevention and research." *Annual Review of Public Health* 25:199–223; Morris, J. (2000), "Defining the precautionary principle," in *Rethinking Risk and the Precautionary Principle*. Oxford, UK: Butterworth-Heinenmann; Sandin, P. (1999). "Dimensions of the precautionary principle." *Human and Ecological Risk Assessment* 5(5):889–907.

47 US Environmental Protection Agency (1988). *The Seven Cardinal Rules of Risk Communication. Policy Document OPA-87-20*. Washington, DC: Environmental Protection Agency. Accessed at: https://archive.epa.gov/publicinvolvement/web/pdf/risk.pdf

48 See Covello, V.T. (1998). "Risk communication." in *Handbook of Environmental Risk Assessment and Management*, ed. P. Callow, Oxford: Blackwell Science. PP. 520–541.

8

Foundational Principles: Trust, Culture, and Worldviews

CHAPTER OBJECTIVES

The chapter covers the foundational, evidence-based principles that explain how people determine their trust in communications in the context of risk, high concern, and crisis situations. The chapter also covers principles that explain how culture and worldviews affect the way individuals and groups judge trust and interpret information about risks and threats.

At the end of this chapter, you will be able to:

- Explain why gaining trust is the essential prerequisite to successful risk, high concern, and crisis communication.
- Recognize that there is an asymmetry in achieving and losing trust – it is hard to achieve but easy to lose.
- Recognize that there has been a precipitous drop in trust in major social institutions in the past half century.
- Describe how trust is determined by a large number of attributes, including perceptions of caring, competence, and honesty.
- Predict the range of factors that determine trust.
- Use appropriate factors in building trust.
- Describe why communication effectiveness lies in respecting differences, not in advocating similarities.
- Interpret how understanding the culture and worldviews of an audience is critical for developing communications that gain trust and are effective.
- Recognize the different aspects of culture and worldview that affect interpretation of, and trust in, messages about high concern issues.

The crucial differences which distinguish human societies and human beings are not biological. They are cultural.

—*Ruth Benedict (anthropologist).*

Difference is of the essence of humanity . . . The answer to difference is to respect it.
—*John Hume (Northen Irish politician and co-recipient of the 1998 Nobel Peace Prize)*

Communicating in Risk, Crisis, and High Stress Situations: Evidence-Based Strategies and Practice, First Edition.
Vincent T. Covello.

8.1 Case Diary: A Disease Outbreak in Africa

The single biggest threat to man's continued dominance on the planet is the virus.
—Joshua Lederberg, PhD, Nobel laureate

The international conference call on an Ebola virus outbreak I had organized did not start well. There were twelve people on the call. The first person to speak, a traditional health practitioner from West Africa, expressed his view that all Western medicine practitioners are elitists, racists, and neocolonialists, and that he did not wish to work with neocolonialists on the Ebola outbreak. The next person to speak, a health practitioner from a European nation, characterized West African traditional health practitioners as witch doctors and voodoo fanatics, and did not see why he and his colleagues should work with them on the Ebola outbreak. Following a few minutes of name-calling and raised voices, I intervened and called for a time-out. I announced I would speak with individuals on the call separately, and I would continue the conference call the next day.

It was the summer of 2014. I had been appointed by Dr. Margaret Chan, the director-general of the World Health Organization (WHO), to serve as senior adviser to the WHO Emergency Committee on Ebola. On 8 August 2014, Dr. Chan had declared the West Africa Ebola crisis to be a "Public Health Emergency of International Concern." In her emergency announcement, she pointed out that Ebola virus could kill millions of people if the current trajectory of the disease was not reversed.

I had previously worked with Dr. Chan and her colleagues on other disease outbreaks. These included the H1N1 swine flu outbreak and the H5N1 bird flu outbreak. Through my work for the CDC in the United States on severe acute respiratory syndrome (SARS), I had also interacted with public health officials in Hong Kong when Dr. Chan was serving as Hong Kong's director of health. In 2004, I went to Hong Kong and, together with colleagues, developed a lessons-learned document[1] and a comprehensive crisis communication plan to prepare for another possible viral outbreak in Hong Kong.

The purpose of the conference call was to discuss the Ebola outbreak in three West African countries: Sierra Leone, Liberia, and Guinea. The specific goal of the conference call was to discuss cooperation between traditional health practitioners and public health agencies in stopping a widespread practice that was spreading Ebola rapidly: touching, washing, and dressing the bodies of those who had died from Ebola. During the Ebola outbreak, these behaviors were a major source of contagion and death. Ebola victims were highly infectious even after death. If a person became infected with Ebola by touching the dead, their chances of dying from the disease were as high as 90 percent. Thousands of lives could be saved if we stopped mourners from touching the dead. But how? Traditional funeral practices that included touching, washing, and dressing the bodies of the dead were deeply embedded in West African culture.

The 2014–2016 outbreak of Ebola in West Africa was the largest, most severe, and most complex Ebola epidemic in history. Over 28,000 people were infected; over 11,000 people died. Ebola deaths were disproportionately concentrated among healthcare workers. Five hundred doctors, nurses, paramedics, and other healthcare workers died from Ebola. Liberia lost over 8 percent of its doctors, nurses, and midwives to Ebola.[2] During the 2014–2015 Ebola outbreak, no cures or vaccines were available. The best medical workers could do for a patient was provide "supportive care" through rest, extra oxygen, hydration fluids, and blood transfusions.

Once the Ebola virus finds a human host, it spreads rapidly to others through contact with body fluids such as blood, feces, vomit, sweat, semen, saliva, and tears. The Ebola virus also can cause a victim to bleed from all parts of their body. The virus can survive for several days after a person dies. Because a person who dies from Ebola is still highly infectious, touching, washing, and dressing the body of a dead person create a serious and deadly threat to health.

I had one primary communication goal for the Ebola conference call: to build trust and set the groundwork for constructive dialogue. Based on the first few minutes of the call, we were clearly off to a rough start.

During the time-out, I called a subset of those on the call. My strategy was to focus on "influencers" – persons who have the most power to affect the attitudes and decisions of their colleagues because of their authority, knowledge, position, or relationships. My goal was to uncover points of agreement that could be pursued through "interest-based bargaining" – a negotiation strategy developed by the Harvard University Program on Negotiation in which parties seek "win-win" solutions to their disagreements or disputes.[3]

My first call was to a leader of the West African health practitioner group. We had spoken several times previously and had gotten along very well. He had shared with me that he had studied medicine as a foreign student at Oxford University, but dropped out to become a traditional health practitioner in his own West African country. He told me that his views on "alternative medicine" and treating the whole person and not just symptoms were not received well by many of his teachers and fellow students at Oxford. I told him I had also been a foreign student in Great Britain, but at Oxford's rival, Cambridge University. We talked and shared stories about the challenges of being a foreign student in Great Britain. I told him that some of my views as a "Yankee in King Arthur's Court" were also not well received by some of my teachers and fellow students at Cambridge. We quickly became friends.

I asked my West African friend and colleague if he would join me on a conference call the following day with one of the leading members of the health practitioner group from international agencies and Western nations. I told him I wanted to focus the call on areas of agreement. I would take careful notes and write up a summary. If he and the other person on the call agreed with my summary, I would present it at the beginning of the next conference call with the larger group.

The most important communication challenge the three of us identified was the lack of trust of West African health practitioners by practitioners of Western medicine, and vice versa. My West African colleague reiterated several things he had said to me the day before. First, he said there was a prevailing belief among many people in West Africa that the practitioners of Western medicine in their country are outsiders and don't understand or respect West African culture. Second, there was a widely circulating rumor, believed by many people, that some practitioners of Western medicine deployed to West Africa are knowingly or unknowingly part of a genocidal plan to kill black Africans and might even be involved in harvesting human organs for sale in Europe and elsewhere. Third, some people in West Africa believe that if you go to a hospital or clinic for Ebola run by a practitioner of Western medicine, you will never return home. Fourth, many people in West Africa believe diseases are not caused by germs or viruses, as claimed by practitioners of Western medicine, but by evil spirits, and therefore the appropriate and effective way to treat a disease is to find the evil spirit that caused the illness and dispel it. Fifth, many people in West Africa believe that practitioners of Western medicine do not handle the dead respectfully. Many people believe the transition from life to death needs to be facilitated by touching, washing, kissing, and dressing the body of the deceased. These funeral and death rituals ensure the spirit of the deceased is at peace and will not haunt the living.

I began the conference call the next day with the larger group by presenting a summary of the conference call the previous day, our diagnosis of the problem, and our recommended actions. I said our subgroup agreed that each group – West African health practitioners and the practitioners of Western medicine – had something of value to offer the other. Among other things, practitioners of Western medicine could offer medicines, personnel, and medical technologies for treating Ebola. Practitioners of traditional medicine in West Africa could offer entrance to communities based in a centuries old, culturally based approach to healing that treated the whole person and not just symptoms. We stated that one of the most important skills a health practitioner could possess – regardless of whether they were a health practitioner based in local West African culture or a health practitioner based in Western medicine – is the ability to listen and to communicate genuine concern, empathy, and interest to a patient, the skills known as "a good bedside manner." A good bedside manner can create rapport with the patient, establish trust, and promote healing.

My subgroup of three made five specific recommendations to the larger group. First, we would agree to be civil and stop using derogative terms such as "witch doctor" and "colonialist" about each other. Second, we would agree to respect each other's beliefs about the causes and treatment of illness. Third, we would agree to cooperate with each other in locating sick individuals and in conducting contact tracing – identifying persons who may have come in contact with an infected person and collecting information from these contacts. Fourth, we would agree Ebola is powerful and evil, regardless of whether we call it a spirit or a germ. Fifth, we would agree to explore alternative ways of dealing with the body of a person who had died from Ebola. After much discussion, the method of dealing with the body we selected was based on the traditional belief that some evil spirits are so powerful they could kill or injure anyone who came near to them. This was why the body of a dead West African king could not be touched by mourners. Instead, a wooden effigy was made that mourners could touch. Adopting this practice to Ebola, if a person died of the powerful Ebola spirit, the local health practitioner would make a wooden effigy of the person who had died. International agencies would pay the cost for the making of the effigy. The local health practitioner would then lead mourners in touching, washing, and dressing the effigy. All those on the conference call agreed to promote this solution. It was a win-win.

This case study demonstrates the importance – and necessity – of building trust and adapting risk, high concern, and crisis communications to a specific cultural context. It demonstrates why trust, respect for differences, and constructive dialogue are prerequisites for successful risk, high concern, and crisis communication. One size does not fit all. Culture matters.

8.2 Trust Determination

A central theme in the behavioral and neuroscience literature on risk, high concern, and crisis situations is how the brain makes decisions about trust. Trust is the single most important filter through which information about risks and threats passes. Trust plays a crucial role in determining attitudes and behaviors. Trust is a fundamental prerequisite for informed decision-making. This is especially true as societies and technologies become more complex, when individuals lack knowledge to make decisions, and when scientific information is constantly changing or not available. Trust is a means for reducing complexity and uncertainty. Only when trust has been established can other communication goals, such as informed decision-making and constructive dialogue, be achieved.

Trust is a firm belief in the character, ability, strength, or truth of someone or something. It is a person or thing in which confidence is placed. Confidence only becomes less important when a person believes they have sufficient knowledge to judge a risk themselves and do not have to rely on the knowledge of others.

Renn and Levine (2014) defined trust as "the generalized expectancy that a message received is true and reliable and that the communicator demonstrates competence and honesty by conveying accurate, objective, and complete information."[4] Although trust, confidence, credibility, and believability are often used interchangeably, trust is a more enduring perception. These terms imply a judgment about the message or a source of information. Trust is achieved through a variety of means, including sustained actions, listening, and communications.[5]

8.3 Characteristics and Attributes of Trust

Trust is typically built over long periods of time. It is easily lost and, once lost, difficult to regain. Trust is determined by actions, verbal messages, and nonverbal messages. Trust can be attributed to a person, group, organization, or institution, or to an object (such as an inspected car), activity, or event associated with that person, group, organization, or institution.

The attributes that determine trust vary greatly from situation to situation and context to context. For example, in some situations, trust is earned by displaying or demonstrating compassion, caring, listening, and empathy. In other situations, trust is earned by acknowledging the legitimacy of fears, concerns, and emotions as facts. In yet other situations, trust is earned by citing compliance with widely accepted scientific or technical standards, procedures, and protocols.

As an illustration, after the Chernobyl accident, studies indicated that people had more trust in foreign scientists than in their own national experts.[6] With this in mind, crisis management authorities recognized very soon after the 2011 Fukushima accident in Japan that more trust would likely be given to an assessment of the accident by foreign independent experts than by Japanese experts. The expert group assigned this task was the United Nations Scientific Committee on the Effects of Atomic Radiation (UNSCEAR). UNSCEAR had a long-standing widely accepted high reputation in the radiation field. In conducting the assessment, the UNSCEAR expert group committed themselves to the following guiding principles: recording all decisions and data shared; ensuring all experts are scrutinized for possible conflicts of interest and met requirements for expertise; selecting collaborating experts based on balance and different perspectives; documenting and checking calculations so that they could stand up to scrutiny and criticism; candidly acknowledging and communicating unknowns and uncertainties; avoiding speculation; and clearly separating science from policy matters.[7] These guiding principles were specifically aimed at addressing concerns about the trustworthiness of the assessment.

At its most fundamental, trust is built on confidence and the belief that leaders, managers, and communicators will not ask of others that which they would not be willing to do themselves. Leaders, managers, and communicators are deemed trustworthy when they act in ways that show their concern for others and put the needs of others before their own needs.

A diverse set of factors enter into trust determinations. These include perceptions of caring and empathy; perceptions of competence and expertise; perceptions of honesty and openness; and perceptions of dedication and commitment.[8] Contextual factors that enter into trust determination include the personalities of those interested or affected, the immediacy of the situation, sociodemographic factors, economic factors, and cultural factors. For example, a person may apply different trust attributes or different weights to these attributes depending on how familiar they are with the risk or threat.

Table 8.1 lists several of the most important characteristics and attributes that determine trust in a person. Many of these characteristics and attributes align with the factors that determine risk perceptions (see Chapter 7 of this book). The close relationship between trust and perceptions points to the fact that trust is bound up tightly with more general and fundamental attitudes toward risks and threats. This tight binding of trust and perceptions helps explain why trust is so

Table 8.1 Personal characteristics that influence trust by others.

Characteristics that create trust	Characteristics that create distrust
Good listener	Inattentive
Caring	Unconcerned
Competent	Lack of expertise
Honest	Dishonest
Transparent	Evasive
Truthful	Deceitful
Consistent	Inconsistent
Fair	Unfair
Objective	Biased
Shares values, attitudes, and behaviors Mirrors attitudes and values	Dissimilar values, attitudes, and behaviors
Charismatic	Uninspiring
Altruistic	Selfish/narcissistic/arrogant
Sincere/authentic	Insincere/inauthentic
Provides clear and well-organized information	Provides unclear and disorganized information
Admits uncertainties	Claims certainty; uses absolutes and guarantees
Cites trusted sources	Cites questionable sources
Provides timely information	Does not provide information in a timely manner
Provides relevant information	Provides non-relevant information
Knowledgeable	Uninformed
Uses personal stories	Uses only abstract information
Provides relevant information	Does not provide needed information
Has excellent reputation or track record	Has poor reputation or track record
Meets expectations	Disappoints expectations
Confident	Lacks confidence
Experienced	Lacks experience
Ethical	Unethical
Acknowledges feelings and emotions	Disregards or minimizes feelings and emotions

dependent on perceptions and why it is so important to link research on trust to research on culture and worldviews.

Many of these same characteristics and attributes of trust were observed over two thousand years ago by Aristotle in *Rhetoric: The Art of Persuasion*. Aristotle postulated that trust in a person's character – *ethos* in Greek, which translates roughly to moral character – was one of the most important tools for persuasion. According to Aristotle, the major components of *ethos* are intelligence, wisdom, aptitude, ability, character, moral virtue, goodwill, shared values, shared preconceptions, and rapport with the target audience. These characteristics will be uppermost in the mind of the beholder in determining trust and are a prerequisite for effective persuasive communications.[9]

In one of the first empirical studies of trust in risk, high concern, and crisis situations, Peters, Covello, and McCallum found that the most important trust determination factors grouped around four major attributes of an individual or organization.[10] The following were attributes of trust:[11]

1) Perceptions of listening, caring, empathy, and compassion (e.g., words, gestures, and actions that communicated altruism, goodwill, fairness, objectivity, lack of bias, and value similarity)
2) Perceptions of competence and expertise (e.g., words, gestures, and actions that communicated knowledge, performance, wisdom, preparedness, dedication, experience, predictability, consistency, and track record)
3) Perceptions of honesty and openness (e.g., words, gestures, and actions that communicated transparency, truthfulness, candidness, fairness, accountability, responsibility, integrity, and moral principles)
4) Other situation-based factors (e.g., words, gestures, and actions that communicated dedication, commitment, consistency, perseverance, dependability, and purpose)

Failures by leaders, risk managers, and technical professionals to display these attributes of trust through words, gestures, and actions often result in distrust. Such failures undermine the ability of leaders to gain acceptance of policies, decisions, and recommendations. For example, in the initial hours and days following the nuclear power plant accidents at Three Mile Island in the United States, Chernobyl in Russia, and Fukushima in Japan, distrust was created by the failure of many leaders, risk managers, and technical professionals to communicate caring, empathy, competence, honesty, and transparency.[12]

The absence of one trust attribute can often be compensated for by an abundance of other trust attributes. For example, if the perception of competence and expertise of a communicator is rated low by listeners because of a perceived bias, prejudice, hearsay, rumors, or misinformation, this mistrust can be partially countered by indicators and cues that establish the communicator as a trusted source of information. These indicators could include facts – or cues to facts – that identify the person's expertise, such as their position within an organization, their achievements, their track record, endorsements, and citations to the scholarly or technical literature. The communicator can also demonstrate competence and expertise with deep knowledge of facts and data, anecdotes and stories that illustrate their extensive experience and skills. Additional indicators and cues for competence and expertise include references by the communicator to (1) awards and other symbols of recognition; (2) positions held in professional or technical organizations; (3) publications and postings on reputable professional or technical sites; and (4) being an invited speaker or lecturer at a respected institution or organization.

8.3.1 Trust and First Impressions

Initial impressions about the presence or absence of the key attributes of trust are often lasting impressions. This is vividly illustrated by Gladwell in his book *Talking to Strangers: What We Should Know About the People We Don't Know*.[13] Gladwell presents research on how people determine trust, especially strangers. His book was driven in part by his interest in a police stop that ended in tragedy. In 2015 a young African-American woman was pulled over by a white police officer for failure to signal a lane change. She became indignant; he became angry; she was arrested and locked up in a jail cell. She killed herself three days later. The audio recording of the encounter was widely distributed through social media channels. In his book, Gladwell reproduces the entire exchange and analyzes what contributed to the lack of trust on the part of both parties and the impact of first impressions.

The impact of first impressions on trust is also vividly illustrated by research done at the University of Glasgow. According to this research, people often make judgments about trust and personality within as little time as one second, especially if that person is a stranger.[14] Dr. Philip McAleer, the lead researcher, noted that from the first word a person hears they make judgments about the speaker that are difficult to change.[15] In one study, McAleer recorded sixty-four people, both men and women, from Glasgow, reading a paragraph that included the word "hello."[16] He then extracted all the "hellos" and had 320 participants listen to the different voices and rate them on ten unique personality traits, such as trustworthiness and confidence. What McAleer found was that the participants largely agreed on which voice matched which personality trait. One male voice was overwhelmingly voted the least trustworthy. The pitch of the untrustworthy voice was much lower than the male deemed most trustworthy. McAleer noted it didn't really matter whether the ratings of personality accurately reflected a speaker's true personality traits. What mattered was that there was virtually total agreement among respondents about the personality traits of a stranger. In less than a second, respondents made a virtually instantaneous judgment about the trustworthiness of another human being.

8.3.2 Loss of Trust

One of the major challenges related to trust faced by those communicating risk-related information is the current lack of trust for many social institutions. As Turner indicates, there has been a steady loss of trust in experts.[17] Survey results, such as the Pew Research Center results shown in Figure 8.1, indicate this is especially true for elected officials, business leaders, and journalists.[18]

This lack of confidence in social institutions is due to a variety of factors. For example, trust has been eroded by changes in the complexity, ubiquity, and scale of technology. The public is continually reminded of the limits of technology by technological failures.[19] These include historical disasters associated with a mechanical or engineering failure, such as the sinking of the *Titanic* – deemed to be virtually unsinkable – and the dramatic explosion of the *Hindenburg* blimp. These also include modern disasters that live vividly in the public memory, such as the NASA *Challenger* space shuttle accident, the *Exxon Valdez* oil spill, the BP/Deepwater Horizon oil spill, the Union Carbide chemical facility explosion in Bhopal, India, and the nuclear power plant accidents at Three Mile Island, Chernobyl, and Fukushima. Such accidents continue to undermine public trust in technology and industry around the world, for example, the 2020 warehouse explosion in Beirut, Lebanon, the 2019 Xiangshui chemical plant explosion, and a string of explosions at chemical facilities on the Texas Gulf Coast in 2020. Fears associated with these highly publicized disasters are compounded by lesser-known technological disasters associated with buildings, bridges, dams, and transportation vehicles where mechanical or engineering failure played a significant role.[20]

According to polls on trust conducted by one of the world's largest communication firms,

> . . . the past two decades have seen a progressive destruction of trust in societal institutions . . . Traditional power elite figures, such as CEOs and heads of state, have been discredited. The growth of social media platforms fully shifted people's trust from a top-down orientation to a horizontal one in favor of peers or experts. Now we are seeing a further reordering of trust to more local sources, with "My Employer" emerging as the most trusted entity, because the relationships that are closest to us feel more controllable.[21]

Americans' trust in medical scientists, scientists, military relatively high; fewer trust journalists, business leaders, elected officials

% of U.S adults who say they have a_____ (of) confidence in each of the following groups to act in the best interests of the public

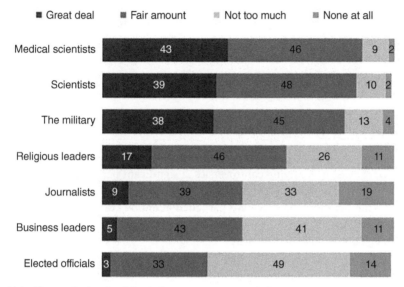

Note: Respondents who did not give an answer are not shown.
Source: Survey conducted April 20–26, 2020.

PEW RESEARCH CENTER

Figure 8.1 Trust in Institutions (2021). Source: Pew Research Center.[18]

The researchers reported the following:

> The informed public—wealthier, more educated, and frequent consumers of news—remain far more trusting of every institution than the mass population. In a majority of markets, less than half of the mass population trust their institutions to do what is right . . . Distrust is being driven by a growing sense of inequity and unfairness in the system. The perception is that institutions increasingly serve the interests of the few over everyone. Government, more than any institution, is seen as least fair; 57 percent of the general population say government serves the interest of only the few, while 30 percent say government serves the interests of everyone.

The 2020 Trust Barometer[22] indicated that none of the four major institutions in society – business, government, nongovernmental organizations (NGOs), and media – are seen by the public as both competent (delivering promises) and ethical. People today grant their trust based on two distinct attributes: competence (e.g., delivering on promises) and ethical behavior (e.g., doing the right thing and working to improve society). "Business ranks highest in competence, holding a massive

54-point edge over government as an institution that is good at what it does (64 percent vs.10 percent). NGOs led on ethical behavior over government (a 31-point gap) and business (a 25-point gap).

At the community level, experience with risk management authorities, such as public health and emergency management officials, can significantly affect perceptions of trust. Individuals who have had poor experiences with risk management authorities may be set in their negative perceptions of these authorities and refuse to alter their beliefs. All messages from risk management authorities will be treated with distrust, cynicism, and skepticism.

A common thread in virtually all risk, high concern, and crisis message development strategies is to build trust and, if needed, rebuild trust. Only when trust has been established can other goals, such as stakeholder engagement, constructive dialogue, and consensus building, be successfully achieved. Messages that portray risk management decisions as the result of stakeholder engagement and participation typically are judged more positively than those made using the "DAD" (decide, announce, defend) model. Participation helps to engender trust, and successful results require trust.

8.3.3 Gaining Trust

Because of the importance of trust in resolving controversies, a significant part of the risk, high concern, and crisis communication literature is focused on methods for gaining trust beyond what was discussed above. Three methods that have received particular attention are discussed below.

8.3.3.1 Gaining Trust through Stakeholder Engagement

As discussed in Chapter 5 of this book, stakeholder engagement processes, especially face-to-face meetings, open houses, information exchanges, and workshops, can enhance trust. They allow participants to engage in dialogue and an interactive exchange of information. Positive relationships developed through stakeholder engagement help increase confidence, reduce uncertainty, and speed up problem-solving and decision-making. They also allow participants to use nonverbal cues to identify potential misunderstandings.

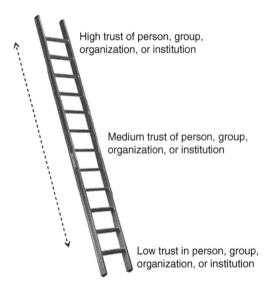

High trust of person, group, organization, or institution

Medium trust of person, group, organization, or institution

Low trust in person, group, organization, or institution

Figure 8.2 Trust ladder.

8.3.3.2 Gaining Trust through Trust Transference

Because of the importance of trust in risk, high concern, and crisis communication, a significant part of the research literature focuses on who is trusted and who is not. For example, research indicates messages from the following groups receive high ratings on trust for scenarios involving chemical and radiation risks: citizen advisory groups, health professionals, safety professionals, and educators.[23]

Knowing who is and who is not trusted is important for *trust transference*. Holding constant other variables, the principle of trust transference is that messages from a lower-trusted source of information can take on the trust held by a higher-trusted source. The higher source functions to validate or endorse

Figure 8.3 Trust ladder: Pharmaceutical Drug Safety (US, 2018). *Source:* Center for Risk Communication.

- Pharmacist
- Professor (medical research)
- Physician/Nurse/PhD
- Health official
- Friend/family member (with personal experience)
- Drug manufacturer
- Expert/consultant hired by manufacturer

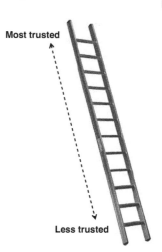

Most trusted

Less trusted

the message of the less trusted source. In other words, trust in messages can be substantially enhanced by validation and endorsements from trustworthy sources. Conversely, the principle of *trust reversal* states the following: holding constant other variables, a message from a lower-trusted source of information that attacks messages from a higher-trusted source loses trust. Trust ladders, such as that found in Figure 8.2, can be constructed for specific issues. Figure 8.3 illustrates the trust ladder for sources of messages about pharmaceutical drug safety and effectiveness.

8.3.3.3   Gaining Trust through Actions and Behavior

"I've learned that people will forget what you said, people will forget what you did, but people will never forget how you made them feel."

—*Maya Angelou (American author and poet)*

Leaders and managers are regularly evaluated by whether their articulated words are reflected in their actual behavior and actions. Behaviors and actions inconsistent with articulated words and values undercut trust. Many studies indicate that trust is gained when behaviors and actions are consistent with words.[24] The studies mirror the meaning attached to adages such as "actions speak louder than words," "practice what you preach," and "walk the talk." These phrases mean that a person's actions and behaviors match what they are saying and demonstrate what is true by example. These phrases also mean that a person's actions and behaviors are consistent with reciprocity and Golden Rule principles of treating others as you want to be treated.

8.4   Case Study: Trust and the Chernobyl Nuclear Power Plant Accident

In 1986, a major explosion at the Chernobyl Nuclear Power Plant in the Soviet Union resulted in many deaths and injuries. The effects of the accident continue to be felt.

The explosions that destroyed the Chernobyl reactor vessel and the ensuing fire resulted in a large release of radioactive materials to the environment. The cloud from the burning reactor spread radioactive materials over much of Europe. The greatest concentrations of contamination occurred over areas of the Soviet Union surrounding the reactor.

An estimated 200,000 emergency and recovery operation workers from the army and volunteers, power plant staff, local police, and fire services were initially involved in responding to the accident. The highest doses of radiation were received by these emergency workers and the on-site personnel. Millions of people living in areas contaminated with radioactive material had to be evacuated.

According to the United Nations International Atomic Energy Agency (IAEA), reliable information about radioactive contamination was unavailable initially to people in Chernobyl or the adjoining geographical areas and nations.[25] Access to reliable information remained inadequate for nearly two years following the accident. This failure and delay led to widespread distrust of official information and the mistaken attribution of many ill-health conditions to radiation exposure. The IAEA also noted that the Soviet government delayed the public announcement that the accident had even occurred. Information provision was selective and restrictive, particularly in the immediate aftermath of the accident. This approach left a legacy of mistrust surrounding official statements on radiation.

According to a range of opinion polls and sociological studies conducted in recent years, nearly two decades after the Chernobyl accident, residents of affected areas still lack the information they need to lead healthy, productive lives, in significant part due to distrust. Although accurate information is now accessible and governments have made many attempts at dissemination, many misconceptions and myths about the threat of radiation persist, promoting a paralyzing fatalism among residents.[26] Overcoming mistrust of information about Chernobyl remains a major challenge to this day. This is due in part to the early secrecy with which Soviet authorities treated the accident. It is also due in part to the complex scientific language often used by scientists and engineers in presenting risk information. The moral of this case is that communication about a risk or threat will fail when there is a lack of transparency.

8.5 Case Diary: The Fukushima Japan Nuclear Power Plant Accident[27]

In 2011, I was privileged to be part of an international group of experts enlisted to assist the crisis communications related to the nuclear power plant accident in Fukushima, Japan. All involved recognized that the disaster could have global implications.

The disaster occurred at the Fukushima Daiichi Nuclear Power Plant in Fukushima Prefecture, Japan. The disaster was started by an earthquake greater than 9.1 on the Richter scale, which in turn generated a tsunami of over thirty feet. The tsunami swept over the plant's seawall, flooded the plant's reactor buildings with seawater, and destroyed the plant's emergency generators. This led to multiple nuclear meltdowns, multiple explosions, and the release of radioactive contamination into the air and water.

All involved knew communications would be difficult. We recognized our most difficult task would be to build and restore trust in the messages from risk management authorities. Radiation is one of the most feared of all health risks. When radiation is released as the result of a radiological accident, it typically generates great fear and distrust.

Our first trust challenge was to address what appeared to be a growing distrust of risk management authorities by those evacuated from the Fukushima area, by residents of areas downwind from the Fukushima nuclear power plant (which encompassed a large area of Japan), and by plant workers and first responders. We knew distrust would significantly reduce the ability of those affected to process risk information – to hear, understand, and remember. Virtually all the characteristics associated with distrust and heightened perceptions of risk are present in a major radiological incident. These characteristics include, but are not limited to, perceptions that the risks posed by the incident are involuntary, highly uncertain, inescapable, caused or aggravated by human error, unfamiliar, potentially the cause of death or dreaded illnesses such as cancer, likely to cause harm to children and pregnant women, and being managed by untrustworthy risk management authorities.

Our second trust challenge was the need to coordinate the communication efforts and crisis communication plans of multiple organizations.[28] For the Fukushima disaster, these organizations included, among others, TEPCO (the utility company that operated the nuclear power plants); the Japanese Government; the UN IAEA (which sets international radiation standards); the UN Scientific Committee on the Effects of Atomic Radiation (UNSCEAR); the US Department of Defense (which operates military bases throughout Japan); nuclear regulatory agencies around the world; and concerned environmental and public health agencies from around the world. Many of these organizations were issuing vastly different and confusing messages about the accident.

Our third trust challenge was to overcome messages promoting distrust in risk management authorities posted on social media platforms. This was a relatively new challenge for many of us involved in crisis communications. Japanese social media channels were highly active and were overshadowing mainstream media channels. They were spreading rumors and misinformation. Most seriously, the sites were unwisely encouraging people not to evacuate and to take potassium iodide (KI) instead – misinformation that had potentially deadly consequences, since evacuation was a paramount concern.

Our fourth trust communication challenge was to translate technical language and jargon used by scientists and engineering professionals to communicate about radiation into information that the lay public could understand. Jargon and technical language can breed distrust among nonexperts. Massive confusion was being caused by statistics and scientific notation. Different countries and organizations were using different terms to describe radiation levels. For example, scientists and engineering professionals from Japan and international organizations were using radiation terms such as *sieverts*, *grays*, *becquerels*, and *coulombs per kilogram* to describe radiation levels. Scientists from the United States were using radiation terms such as *rems*, *rads*, *curies*, and *roentgens* to describe radiation levels. Additional technical terms were being bandied about, including *millisieverts*, *microsieverts*, *micrograys*, *beta radiation*, *gamma radiation*, *alpha radiation*, Bq/m^3, Bq/cm^3, and *joules per kilogram*. One press release actually included all these terms. The rationale for the plethora of technical terms being used was the desire by scientists and engineering professionals to be precise and accurate in their descriptions of levels of radiation. However, it unfortunately complicated the messages and was a major cause of public distrust.

We knew we needed to overcome the many confusing and conflicting communications put out by so many and varied individuals and organizations – ranging from official and unofficial communications to accurate but often confusing or incomprehensible communications to bizarre. And we needed to advise the frightened and highly stressed population about the urgent actions they needed to take to protect themselves from danger, including evacuating their homes. The most intense and immediate issues were related to evacuation, which was in turn directly linked to trust. No one wants to leave their home, especially not without advance notice and for an unknown amount of time. Such evacuation is undertaken only if individuals trust the authorities asking them to embark on such an unwanted activity. Because of this, we needed to be sure our strategies were working to build trust, inform decision making, and constructively influence attitudes and behaviors. We especially wanted to know how we were doing with the social media sites that were posting misinformation. We quickly instituted a process of testing and evaluation, including conducting focus groups and interviews, and monitoring media channels. Results of these efforts demonstrated that our work was helping to build trust, restore trust, and increase the knowledge needed by people to make informed decisions about their health and safety.

An enormous crisis like the Fukushima disaster demonstrates the importance – and necessity – of effective risk, high concern, and crisis communication. Messages that are heard, trusted, and understood can combat powerful and widespread misinformation. The right information delivered at the right time by the right person can save thousands of lives. The case study illustrates the importance of coordinating the risk and crisis communication plans from the responding organizations, agreeing on the clarity and content of key messages across the organizations, and holding joint press conferences.

8.6 Gaining Trust in High-Stakes Negotiations

In 1979, a highly respected, research-based Program on Negotiation was founded at Harvard University. The program was focused on negotiations in low-trust, high concern, and high-stakes situations. The program mandate was to improve the theory, teaching, and practice of negotiation through evidence-based research. In 1981, Professor Roger Fisher and Professor William Ury – cofounders of the program – published two seminal works: *Getting to Yes* and *Getting Past No*. These two books radically changed the way researchers and practitioners thought about negotiation. The basic approach advocated by Fisher and Ury was to negotiate in such a way that participants developed trusting relationships and worked together to achieve a mutually beneficial outcome.

Studies done by the researchers at the Harvard program and researchers inspired by the program point to seven basic tenets for building trust in high-stakes negotiations. Many of these tenets are derived from the larger field of risk, high concern, and crisis communication.

The first tenet for building trust in high-stakes negotiations is to separate the person from the problem. Participants in high-stakes negotiations may behave irrationally, impulsively, and emotionally, with such behavior rooted in conscious or unconscious fears, needs, and perceptions. Participants in high-stakes negotiations may also offer viewpoints based on mental shortcuts and cognitive biases – unconscious brain processes referred to as "heuristics" that distort decision making under conditions of uncertainty. Active listening skills are typically needed to uncover these fears, needs, perceptions, and biases.

The second tenet for building trust in high-stakes negotiations is to separate positions from interests. Interests are basic needs, wants, and motivations. Interests are the fundamental drivers of high-stakes negotiations. Although often hidden and unspoken, interests nonetheless guide what is done and said in high-stakes negotiations. The successful separation of positions from interests leads to "interest-based" bargaining. Interest-based bargaining means avoiding the natural tendency of focusing on positions (what the other side says they want) and focusing instead on their interests (why the other side is saying what they want). Active listening skills are typically needed to uncover interests. Active listening skills include asking open-ended questions, paraphrasing, reflecting back without judgment, making eye contact, taking notes, and showing empathy. People who feel they are being heard often become less oppositional and willing to listen to other points of views.

The third tenet for building trust in high-stakes negotiations is to focus on building trusting relationships. Successful negotiators recognize that those with whom they are negotiating are not opponents but counterparts. Treating counterparts as opponents typically leads to an adversarial relationship.

The fourth tenet for building trust in high-stakes negotiations is to work cooperatively to discover win-win solutions. Win-win solutions typically require a commitment to mutual gains. They require each party acknowledge, respect, and recognize the concerns, needs, and desires of the other party and make a commitment to address these concerns, needs, and desires. They also require that each party be transparent and respect different perceptions of what is a loss and gain.

The fifth tenet for building trust in high-stakes negotiations is preparation. Successful negotiators can predict the behavior of the other side and anticipate their actions. They can predict the objectives and negotiation styles of their counterparts. They can know the supervisory chain of their counterparts and predict who will be needed to ratify a negotiated outcome. They can predict what their counterparts will perceive to be fair and equitable. They can predict the consequences for their counterparts if the negotiation fails. They can predict which style of negotiation – ranging from very aggressive and assertive to passive and accommodating – is likely to be most successful for the particular negotiation.

The sixth tenet for building trust in high-stakes negotiations is to know your best alternative to a negotiated agreement, often referred to as the BATNA. The BATNA is the best outcome that will result if the negotiation fails. The BATNA needs to be identified in the preparation phase. It enables the negotiator to walk away from the negotiating table. It also enables the negotiator to avoid traps and pitfalls, such as inappropriately revealing a weak BATNA.

The seventh tenet for building trust in high-stakes negotiations is self-awareness. Successful negotiators are aware that they have a personal style of communicating that affects their interactions with others. They are emotionally intelligent, aware of their emotional state, and aware of how their emotional state might influence the negotiation dynamic. They recognize the importance of controlling their emotions and temper. They expect conflict but can keep calm and avoid responding too quickly.

8.7   Case Diary: Gaining Trust and the SARS Outbreak in Hong Kong

SARS is a viral respiratory illness first identified in mainland China in 2002. SARS reached Hong Kong in March 2003.[29] From March to June 2003, 1750 cases had been identified in Hong Kong. During that same period, more than 275 people died of the disease.

The SARS outbreak created an unprecedented challenge for those in Hong Kong and around the world charged with communicating to the public and the media. Communication efforts were hampered by the failure of risk management authorities in mainland China to be honest and transparent about SARS.[30] Before the advent of SARS in Hong Kong, mainland China had experienced outbreaks of SARS in November 2002 that reached their peak in February 2003. Lack of transparency by risk management authorities on intense outbreaks of SARS in mainland China led to heightened anxieties and fears in Hong Kong.

The Hong Kong SARS Expert Committee asked my colleague, Dr. Timothy Tinker, and me at the Consortium for Risk and Crisis Communication (the Consortium) to work closely with Hong Kong health care professionals, public health professionals, and government officials in Hong Kong to identify lessons learned about trust from Hong Kong's experience with SARS.[31] Our report was followed by the development of a comprehensive strategic and operational plan for effective health risk communication in both crisis and noncrisis situations.

Our report listed the following lessons learned about gaining trust through actions and communications:

1) At the start of a crisis, it is important to be able to quickly turn to a prepared media crisis communication plan. A sound media communication plan helps an organization to proceed confidently and to quickly establish leadership and credentials.

2) To instill trust, crisis management authorities must be the first to deliver bad news.

3) Information shared by crisis management authorities should be accurate (always) and truthful (always).

4) For the crisis spokespersons to be trusted, they must be willing to admit to uncertainties, acknowledge emotions, communicate caring and empathy, admit mistakes, quickly correct misinformation and rumors, convey clear messages, provide complex information in layers that gradually increase in complexity, make extensive use of visuals (graphics, drawings, maps, charts, flowcharts, photographs, videos, personal anecdotes, and stories), and tailor messages to specific groups.

5) For the crisis management authorities to be trusted, they must form alliances and partnerships with other trusted sources of information, be transparent, serve as role models for recommended behaviors, avoid sharing conflicting messages, provide the right amount of information to medical practitioners who may feel overloaded or underserved, avoid unnecessary delays in informing the public, break through a noisy and cluttered media environment through daily media appearances, and understand public knowledge and opinion through stakeholder engagement and monitoring.

6) For the crisis management system to be trusted, ensure that those providing information in an official capacity do not speak or act beyond their expertise, roles, or responsibilities, and ensure that structures are in place that allow easy data sharing of risk- and crisis-related information between government and nongovernmental organizations.

8.8 Trust and Culture

Culture plays a key role in trust determination. Culture can be defined as the web of meaning shared by members of a particular society or group. Culture comprises shared elements, such as beliefs, values, customs, assumptions, attitudes, behaviors, possessions, expectations, and symbols. It is reflected in behaviors such as communication style, relationships, habits, language, reward systems, teamwork, work, recruitment, appearance, and gestures. Conscious or unconscious cultural bias and discrimination occurs when behavioral elements of culture are negatively labeled and when these labels are applied in a broad-brush fashion to entire groups. The term *culture* applies to more than nationalities and ethnic groups. It also applies to organizations and groups. For example, research has identified significant differences in the way men and women perceive risks and threats. Men and women often have different levels of concern about the same risks, trust different sources of information, and often perceive different risks altogether.[32]

Culture is a filter through which information passes. Culture-based beliefs largely determine what is perceived as true or false. Culture-based beliefs largely determine what is perceived as likable or unlikable. Culture-based beliefs largely determine what is perceived as good or bad.

8.9 Cultural Competency

Navigating cultural differences in message development presents significant challenges. Cultural competency for effective message development requires risk, high concern, and crisis communicators to:

1) Recognize their own cultural biases or prejudices
2) Develop an in-depth understanding of the culture of interest
3) Value and respect diversity.

Figure 8.4 The iceberg model of cultural diversity.

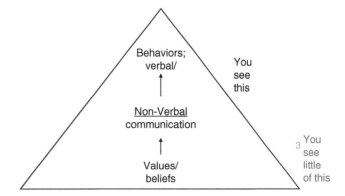

As shown in Figure 8.4 – the Iceberg Model of Cultural Diversity – cultural competency also requires knowledge and understanding of the values and core beliefs that lie beneath observed behaviors.

In Japan, for example, the behaviors surrounding the exchange of business cards are complex, highly formalized, and reflect important underlying values and beliefs about status, identity, and relationships. In contrast to the often casual exchange in the West, the value attached to business cards in Japan leads to a process for exchanging cards that involves multiple steps, including handing over the card with two hands, bowing an appropriate amount, turning the card toward the other person, and placing the card in a special holder separate from money and credit cards. Omitting or shortcutting these steps would imply a lack of respect for the card owner's position and status.

As another example, in Afghanistan, high cultural value is placed on honor, bravery, family, justice, religion, protection of women, elders, sanctuary, revenge, and loyalty. High cultural value is particularly placed on hospitality, especially among the Pashtuns, a major group in Afghanistan. Hospitality is viewed by the Pashtuns as more than a social obligation. It is also a religious obligation and a point of honor. It is captured by the Afghan proverb: "Honor the guest. Even though he be an infidel, open the door." Known as *Melmastyā,* hospitality is required for all visitors and strangers, regardless of their race, religion, national affiliation, or economic status. *Melmastyā* takes precedence over other things. Once under the roof of the host, a guest can neither be harmed nor surrendered, even if the person is an archenemy. A guest must eat slowly because the plate will be constantly filled. *Melmastyā* means offering tea to a guest. Sharing tea together indicates a relationship and puts the responsibilities of friendship on both the guest and the host.

Several of the most important cultural differences affecting trust in risk, high concern, and crisis communication are listed in Table 8.2. Failure to recognize these differences can lead to misunderstandings, stress, and distrust.

8.9.1 Different Communication Styles

The way people communicate the same message varies widely between and within cultures. Words and phrases are frequently used in different ways. For example, the meaning and interpretation of the English word "yes" vary greatly from culture to culture, depending on context, intonation, and where it is positioned in a sentence. The word translated in English from another language to mean "yes" may not signify agreement or disagreement. In Hindi, for example, the word for "yes," "*achha,*" is complex. Its meanings include "yes," "good," "okay," "oh," "agreement," "really," and

Table 8.2 Cultural differences affecting trust in risk, high concern, and crisis situations.

1) Different communication styles
2) Different attitudes and approaches toward conflict
3) Different nonverbal communication
4) Different attitudes and approaches to decision making
5) Different attitudes and approaches toward information disclosure
6) Different attitudes and approaches to knowing
7) Different attitudes and approaches toward conversation and discourse
8) Different attitudes and approaches toward the use of humor

"awesome." It may simply mean: "I heard what you said" or "I understand what you said." In the same way, the word "*hai*" in Japanese is usually translated as "yes." However, its actual meaning is very similar to the Hindi word "*achha*."

Similarly, the meaning and interpretation given to the English word "*no*" vary from culture to culture. In many cultures, the use of the word "no" is discouraged because it may cause the person delivering or receiving the word to lose face. In Japan, for example, the word "no" is often avoided as a direct answer or response to a question. It is often replaced with expressions such as: "That would be a difficult;" "Please let me think about it;" "I believe it would be very hard to do this;" or "I would like to study this more." These phrases express different degrees of "no-ness." These and similar expressions become a stronger, firmer, and more emphatic way of saying "no" if the Japanese person adds specific forms of nonverbal communication, such as sucking in the air between clenched teeth and then releasing the air with a long, sighing "Saa . . ."

8.9.2 Different Attitudes and Approaches toward Conflict

In many cultures, open conflict and disagreements are viewed as embarrassing and demeaning. Communications with the potential to cause open disagreement or conflict are avoided. Differences are worked out quietly behind the scenes.

8.9.3 Different Nonverbal Communication

Different cultural norms exist for nonverbal communication through the eyes, facial expressions, posture, hand gestures, body gestures, personal distance, dress, jewelry, general appearance, and vocal characteristics (e.g., loudness or speed of delivery). For example, different cultural norms exist regarding the degree of assertiveness in communicating a message. Some cultures consider raised voices to be threatening or a sign of anger, whereas other cultures consider an increase in volume as a sign of enthusiasm and commitment. Another example is the American "okay" sign: touching the index figure to the thumb to make a circle, a positive gesture to Americans with a clear affirmative meaning. In other contexts, this hand gesture has multiple, varied culture-based meanings, including a sign for an insult, the evil eye, sex, the number zero, and a coin or money.

8.9.4 Different Attitudes and Approaches to Decision Making

The roles individuals play in decision making vary widely from culture to culture. For example, in some cultures, decision-making responsibilities are shared, while in other cultures a strong value is placed on reserving decision-making power solely to those in positions of power. In some cultures, individual recognition and initiative in decision-making are encouraged, while in others consulting those in power or achieving consensus is essential.

8.9.5 Different Attitudes and Approaches toward Information Disclosure

In some cultures, transparency and sharing of emotional or personal information is encouraged, while in other cultures such transparency and sharing is considered inappropriate and discouraged.

8.9.6 Different Attitudes and Approaches to Knowing

Significant differences occur among cultures in the way people acquire knowledge. For example, some cultures place a high value on information acquired through objective means (such as counting and measuring), while other cultures appreciate less tangible ways of knowing, such as intuition.

8.9.7 Different Attitudes and Approaches toward Conversation and Discourse

Cultures vary in the assumptions and rules governing conversations and discourse. These assumptions and rules cover such diverse areas as the following:

1) Opening or closing a conversation
2) Interrupting and taking turns during a conversation
3) Using silence
4) Shifting between topics
5) Attending to what is being said
6) Using gestures to make or emphasize a point
7) Sequencing and timing of elements in a speech, presentation, or conversation
8) Appropriate and inappropriate conversational topics and words
9) Using storytelling and anecdotes.

8.9.8 Different Attitudes and Approaches toward the Use of Humor

Cultures vary greatly in the assumptions and rules governing the use of humor in conversation and discourse. Humor is extremely difficult to use effectively cross-culturally, as it often does not translate well. What is perceived to be humorous and funny in one culture is often not perceived as humorous in another. In high-stress situations, a remark intended as humorous is often interpreted as a sign of not caring, not taking the issue seriously, or trivializing the importance of an issue.

8.10 Risk Perceptions, Trust, and Cultural Theory

Research focused on culture, risks, and threats typically has its academic roots in the behavioral and social sciences. Researchers from the discipline of psychology typically focus their work on the subjective and emotional factors that influence messages about risks and threats. Researchers from the disciplines of anthropology, economics, sociology, social psychology, and political science typically focus their work on the social, organizational, ethnic, social network, and group factors that influence communications about risks and threats.

One of the major contributions of research on culture for risk, high concern, and crisis communications is that it emphasizes the important filtering role of social and cultural factors in setting risk agendas and in determining which risks or threats will be emphasized or de-emphasized. Cultural researchers argue that cultural factors and vested interests are central to decisions about trust, understanding risk perceptions, and understanding why some facts about risks and threats are accepted and others are rejected.[33]

Cultural theory expands on explanations of human behavior proposed in *rational choice theory*. Rational choice theory, rooted in economic theory, treats human attitudes, intentions, behaviors, and judgments of trust as largely reflections of the implicit or explicit weighing of costs and benefits. Cultural theory also expands on explanations of human attitudes, intentions, and behaviors found in *psychometric theory*. Psychometric theory, rooted in psychology, social psychology, and behavioral economics, treats human attitudes, intentions, behaviors, and judgments of trust as largely shaped by emotions and often distorted by biases.

Cultural theory is based on the following fundamental assumption: since risks and threats are virtually infinite in number, effectively and efficiently, societies and groups with a society need to select which risks and threats are important and acceptable. Risk selection is necessary because it is impossible to be aware of, and attend to, all risks and threats. There is only so much a group or society can worry about. It would be overwhelmingly stressful for people in societies to attend to the thousands of risks and threats people could be exposed to in their daily lives. Since groups and societies cannot look in all directions at once, they must decide which risks to fear most, which risks are worthy of attention and concern, which risks are worth taking, and which risks can safely be taken off the radar screen.

Cultural theory is also based on several additional premises. First, to function effectively and efficiently, societies, and groups within a society, need to decide what actions to take to control, reduce, or eliminate risks and threats identified as important or not acceptable. Second, societal and group consensus is often lacking about which risks and threats are important and acceptable. For example, some societies, or groups within a society, may decide certain risks and threats – such as the risks and threats associated with nuclear power, genetically engineered organisms, fracking, childhood vaccination, climate change, and illnesses from a disease outbreak such as COVID-19 – are of high concern and must be addressed, mitigated, reduced, or eliminated. Other societies, or groups within a society, may decide that these same risks and threats are of low concern and can be safely ignored. Third, the lack of societal and group consensus on what risks and threats are important and acceptable can cause strong emotional reactions and conflict. Fourth, what is perceived as a risk and threat, what is perceived as important, and what is perceived as acceptable are largely social constructs stemming from social-demographic and cultural factors. Perceived risks and threats reflect deeper cultural norms, values, policies, or institutions. Fifth, cultural attitudes toward risks and threats are deeply embedded in the conscious and unconscious mind. Sixth, culture profoundly shapes risk selection. Seventh, the risks and threats that are finally selected for attention by a society, or groups within a society, are not necessarily chosen because of facts and data. The risks and threats selected for attention may have little relation to actual threats and dangers as determined by technical experts, and may be among the least likely to adversely affect people or the things they value.

The primary goal of cultural theory research is to uncover (1) how particular risks and threats come to be selected for societal attention and action; (2) how groups, communities, and societies decide what should be done in response to risks and threats; (3) how to reduce conflict; and (4) how trust in risk management authorities and policies is determined by groups, communities, and societies. According to cultural theory, every culture has its own distinctive portfolio of risks and threats believed to be worthy of concern. Every culture highlights some risks and threats and downplays others. Every culture adopts methods for reducing, controlling, or eliminating some risks and threats but not others. Every culture has a hierarchy of trusted sources of information about risks and threats. Every culture exaggerates or minimizes risks and threats according to the social and cultural acceptability of the underlying activities.

Cultural theory, as it relates to risks and threats, has its intellectual origins in a landmark publication by the political scientist Aaron Wildavsky and the anthropologist Mary Douglas titled, *Risk and Culture: An Essay on the Selection of Technological and Environmental Dangers.*[34] The book expands on an earlier book by Douglas titled *Purity and Danger,*[35] which represented a paradigm shift in thinking about risks. For example, it challenged conventional ideas about pollution. Douglas argued forcefully and passionately in her book for the importance of culture and context in understanding trust and risk perceptions. For example, based on her research, Douglas determined it is often a mistake to view food safety communications – such as the injunction in the Old Testament against eating pork – as simply forerunners of modern food safety communications. According to Douglas, food safety communications often serve a variety of purposes in a society other than health and safety. These include preventing cultural intermixing and marriage, setting cultural boundaries, separating one cultural group from another, declaring cultural solidarity, and symbolically bringing order into the world through culture-based classifications, rules, and regulations.

8.11 Risk Perceptions, Trust, and Worldviews

Perhaps the deepest, most pervasive, and most central influence on risk perceptions and trust are worldviews. Worldviews are how people frame and interpret their world. They are important because they profoundly influence how individuals, groups, and organizations behave and whom they trust.

Worldviews influence perceptions and thinking on a wide variety of issues, operating at a deep, often unconscious level. If information about a risk, threat, or high-concern issue is consistent with an individual's worldviews, they will typically pay attention to and trust the information. If the information is in conflict with their worldviews, people will often ignore or discount the information being shared. Understanding worldviews is an essential element in crafting effective communications in risk, high concern, and crisis situations.

There are at least five major types of worldviews that influence perceptions of risk and trust: (1) individualism, (2) egalitarianism, (3) communitarianism, (4) hierarchy, and (5) fatalism. Communications are filtered through these perspectives. There is a large overlap of types and significant variance within each type.

1) **Individualism**: Individualism is characterized by beliefs that (1) groups and organizations function best when there are few rules, regulations, and restrictions and (2) people should be free to make their own choices and decisions without interference from outside bodies. People who believe in individualism often use pronouns such as "I," "my," and "me." Communications directed at individualists are typically most trusted and effective when they focus on individual choice, options, control, benefits, and preparedness.

2) **Egalitarianism**: Egalitarianism is characterized in beliefs that (1) priority should be given to equality and fairness; (2) people should reject hierarchal ranks; and (3) people should be free to

decide without interference. Communications directed at egalitarians are typically most trusted and effective when they focus on freedom, equality, and fairness.

3) **Communitarianism**: Communitarianism is characterized by beliefs that individuals, groups, and organizations function best when people work together toward a common goal and good. People who believe in communitarianism often use pronouns such as "we" and "our." Communications directed at communitarians are typically most trusted and effective when they focus on teamwork and group preparedness.

4) **Hierarchy**: Hierarchy is characterized by beliefs that (1) priority should be given to rules, regulations, and restrictions set by established and traditional authorities; (2) groups and organizations function best when there are clearly defined and hierarchically ordered ranks; (3) groups and organizations function best when people agree to behave under clearly defined social roles and established social norms; and (4) if it is good for the group, it is good for the individual. Communications directed at those who value hierarchy are typically most trusted and effective when they focus on rules, regulations, standards, and accepted processes or procedures.

5) **Fatalism**: Fatalism is characterized by beliefs that (1) future events are determined by destiny; (2) people have no power to influence the future; (3) people should be apathetic or resigned to what will happen; and (4) because of destiny and inevitability, acceptance is more appropriate than resistance. Communications directed at fatalists are typically most trusted and effective when they include concrete, personal, and positive narratives demonstrating that change is possible (e.g., a personal story of risk avoidance or recovery, ideally relating to a person or group known to the stakeholder).

The five types of worldviews described above play an important role in the who, what, where, when, why, and how of communications about risks, threats, and trust. However, people seldom can be described by one discrete worldview. Also, which, if any, worldview is applied depends heavily on context and social psychological factors. Worldviews, together with context and social psychological factors, help explain why communications about a particular risk, threat, or source of information can cause radically different perceptions and judgments.

Research on worldviews has been criticized for a paucity of empirical studies and for failing to recognize the complexity of decisions about risks, threats, and trust. For example, many people from specific cultures and ethnic groups clearly hold fatalistic worldviews. However, the explanatory value of fatalistic worldviews for delineating differences between people in their judgments about risks, threats, and trust often disappear after controlling for other explanatory factors such as age, personality (e.g., pessimism), socioeconomic status, fear, cynicism, and superstition.[36] Moreover, many people who hold fatalistic worldviews can simultaneously hold beliefs about personal initiative, responsibility, and the power to change things, (e.g., "God helps those who help themselves").

As this chapter and the previous chapter on perceptions have shown, it is typically not a straightforward process to get people to comprehend, believe, and follow preventive and protective safety measures. Understanding perceptions, culture, trust, and beliefs is essential to communicating information that can be heard.

8.12 Case Diary: Fame, Family, and Fear in Public Health Communications

Public health measures, including well-executed communication plans and a vaccine, successfully averted a serious flu pandemic in 2009 when the H1N1 virus spread, but getting the messages across was a difficult challenge. The communication plan had to fit the population.

I was invited to work with public health experts in a Latin American country to help them convince the population to get vaccinated. There was considerable resistance, and the entreaties of public health officials were being ignored. I realized we had to delve into an understanding of the belief systems that were creating the barriers to getting vaccinated. Testing insights with my colleagues, I developed a model of cultural values in that country and their implications for risk communication, mapping beliefs to perceptions, and perceptions to messages. The model is shown in Figure 8.5. It has since been used in many circumstances, including by CDC and other public health organizations for vaccine communications about childhood diseases and about disease outbreaks such as H1N1, Ebola, and COVID-19. The model considers the key role of values such as the centrality of family, spirituality, independence, and machismo in perceptions, motivations, and decision making.

The model led us to a variety of strategies. A key one was to employ the ladder of trust and identify spokespersons who could capture the support and attention of the public. We turned to sports figures and to the church.

We had overwhelming success in the realm of sports. The public health team convinced a leading athlete of the day to pose for a photo promoting vaccination. Surrounded by his family – a key message – and baring his muscular chest and biceps as he extended his arm for a vaccination, the photo became a sensation.

Unexpectedly, we had a more challenging time persuading church officials to support the vaccine campaign. We met with officials from the Catholic Church and asked them to help us promote our three key public health measures: wash your hands, cover your cough or sneeze, and stay home if you are sick. We also sought their help in overcoming hesitancy by the population to getting the H1N1 vaccine. The church fathers listened to us sympathetically and gravely told us they saw their primary mission as one of healing the spirit; healing the body was our problem.

We left the meeting surprised and stumped. How might we persuade the church fathers to see that they should not separate body and spirit? I turned to the research-based negotiation techniques in my repertoire. Where would be the "win" for the church in helping us prevent bodily illness? As we probed that question, my colleagues in the government realized that if the pandemic continued to spread, they would have to institute strict public health measures, including shutting down public gatherings such as church services. We immediately communicated this imminent possibility to church officials. The officials conferred among themselves and called us the next day. They reported they would be happy to help us get our public health messages across.

A few days later, I was sitting with my colleagues at at a Sunday service, a translator whispering into my ear as the priest spoke. The priest had especially invited our team to attend and hear how clearly and firmly he would communicate to the hundreds of parishioners at his service. At the end of the service, I was delighted to hear the expected messages. Then my eyes opened wider and wider as the translator went on. The priest announced that the vaccination was so important that he had stationed nuns who were nurses at the back of the church; they were armed with shots for H1N1 vaccine. Parishioners should roll up their sleeves as they exited.

I was pleased and amazed at the action orientation of the priest, which well exceeded our expectations for effective, proactive communication. I was even more delighted as he told the congregation that they had to protect their mothers by getting vaccinated. Then his next words made my jaw drop. He pronounced that anyone who made their mother sick by not taking appropriate preventative measures would burn in hell for eternity. I knew fear messages are often successful, but this was certainly one I had not expected. At the end of the service, there was a long line of parishioners for the nurse nuns to vaccinate.

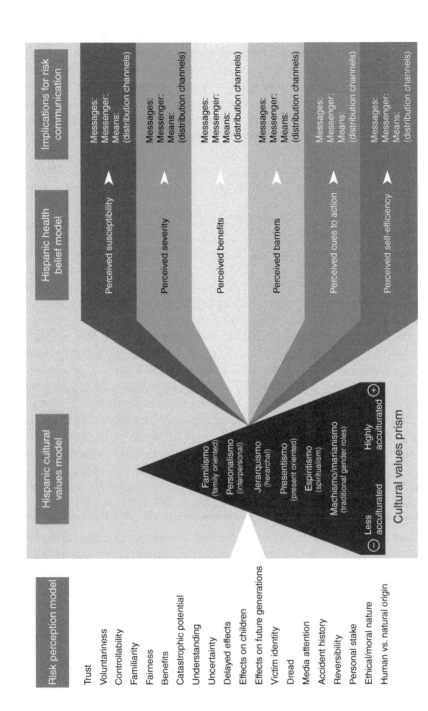

Figure 8.5 Cultural diversity: Sample model of cultural values.

The foundational principles for risk, high concern, and crisis communication are an essential basis for developing communication strategies. Successful strategy design requires analysis and creative problem-solving that draws deeply into understanding and application of these principles.

8.13   Chapter Resources

Below are additional resources to expand on the content presented in this chapter. "Aristotle" (Original date: 350 B.C.E). *Rhetoric: The Art of Persuasion (2015)*. New York: Mockingbird Publishing.

Árvai, J., Rivers, L. III., eds. (2014). *Effective Risk Communication*. London: Earthscan.

Arvai, J., Campbell-Arvai, V. (2014). "Risk communication: Insights from the decision sciences," in *Effective Risk Communication*, eds. J. Arvai and L. Rivers III. London: Taylor and Francis.

Beck, M., Kewell, B. (2014). *Risk A Study of Its Origins, History and Politics*. Hackensack, NJ: World Scientific Publishing Company.

Bennett, P., Calman, K., eds. (1999). *Risk communication and Public Health*. New York: Oxford University Press.

Bennett, P., Coles, D., McDonald, A. (1999). "Risk communication as a decision process," in *Risk Communication and Public Health*, eds. P. Bennett and K. Calman. New York: Oxford University Press.

Bier, V.M. (2001). "On the state of the art: risk communication to the public." *Reliability Engineering and System Safety* 71(2):139–150.

Bohnenblust, H., Slovic, P. (1998). "Integrating technical analysis and public values in risk based decision making." *Reliability Engineering & System Safety* 59: 151–159.

Bostrom, A. (2003). "Future risk communication." *Futures* 35:553–573.

Bostrom, A., Atman, C., Fischhoff, B., Morgan, M.G. (1994). "Evaluating risk communications: Completing and correcting mental models of hazardous processes, part II." *Risk Analysis* 14 (5):789–797.

Bourrier, M., Bieder, C., eds. (2018). *Risk Communication for the Future: Towards Smart Risk Governance and Safety Management*: Cham, Switzerland: Springer.

Breakwell, G.M. (2007). *The Psychology of Risk*. Cambridge, UK: Cambridge University Press.

Centers for Disease Control and Prevention (CDC) (2012). *Emergency, Crisis, and Risk Communication*. Atlanta, GA: Centers for Disease Control and Prevention.

Chess, C., Hance, B.J., Sandman, P.M. (1986). *Planning Dialogue with Communities: A Risk Communication Workbook*. New Brunswick, NJ: Rutgers University, Cook College, Environmental Media Communication Research Program.

Chess, C., Salomone, K.L., Hance, B.J. (1995). "Improving risk communication in government: Research priorities." *Risk Analysis* 15(2):127–135.

Chess, C., Salomone, K.L., Hance, B.J., Saville, A. (1995). "Results of a national symposium on risk communication: Next steps for government agencies." *Risk Analysis* 15(2):115–120.

Chung, I. J. (2011). "Social amplification of risk in the Internet environment." *Risk Analysis* 31(12):1883–1896.

Coombs, W.T. (1998). "An analytic framework for crisis situations: Better responses from a better understanding of the situation." *Journal of Public Relations Research* 10(3):177–192.

Coombs, W.T. (2007). "Protecting organization reputations during a crisis: The development and application of situational crisis communication theory." *Corporate Reputation Review* 10(3):163–176.

Coombs, W.T. (2019). *Ongoing Crisis Communications: Planning, Managing, and Responding.* Thousand Oaks, CA: Sage Publications, Inc.

Covello V. (1983). "The perception of technological risks." *Technology Forecasting and Social Change: An International Journal* 23:285–297.

Covello V. (1989). "Issues and problems in using risk comparisons for communicating right-to-know information on chemical risks." *Environmental Science and Technology* 23(12):1444–1449.

Covello V. (1992). "Risk communication, trust, and credibility." *Health and Environmental Digest* 6(1):1–4.

Covello V. (1993). "Risk communication and occupational medicine." *Journal of Occupational Medicine* 35:18–19. Accessed at: https://journals.lww.com/joem/citation/1993/01000/risk_communication_and_occupational_medicine.11.aspx

Covello, V. (2003). "Best practices in public health risk and crisis communication." *Journal of Health Communication* 8 (Suppl. 1):5–8;148–151.

Covello, V. (2006). "Risk communication and message mapping: A new tool for communicating effectively in public health emergencies and disasters." *Journal of Emergency Management* 4(3):25–40.

Covello, V. (2014). "Risk communication," in *Environmental health: From Global to Local*, ed. H. Frumkin. San Francisco: Jossey-Bass/Wiley.

Covello, V., Allen, F. (1988). *Seven Cardinal Rules of Risk Communication.* Washington, D.C.: US Environmental Protection Agency.

Covello V., McCallum D., Pavlova M. (1989). *Effective Risk Communication: The Role and Responsibility of Government and Nongovernment Organizations.* New York: Plenum Press.

Covello, V., Merkhofer, M. (1993). *Risk Assessment Methods: Approaches for Assessing Health and Environmental Risks.* New York: Plenum Press.

Covello, V., Minamyer, S., Clayton, K. (2007). *Effective Risk and Crisis Communication during Water Security Emergencies. EPA Policy Report*; EPA 600-R07–027. Washington, D.C.: US Environmental Protection Agency.

Covello, V., Peters, R., Wojtecki, J., Hyde, R. (2001). "Risk communication, the West Nile virus epidemic, and bio-terrorism: Responding to the communication challenges posed by the intentional or unintentional release of a pathogen in an urban setting." *Journal of Urban Health* 78(2):382–391.

Covello, V., Sandman, P., Slovic, P. (1988). *Risk Communication, Risk Statistics, and Risk Numbers.* Washington, DC: Chemical Manufacturers Association.

Covello, V., Sandman, P. (2001). "Risk communication: Evolution and revolution," in *Solutions to an environment in peril*, ed. A. Wolbarst. Baltimore, MD: Johns Hopkins University Press.

Covello, V., Slovic, P., von Winterfeldt, D. (1986). "Risk communication: A review of the literature." *Risk Abstracts* 3(4):171–182.

Covello, V., Slovic, P., von Winterfeld, D. (1987). *Risk Communication: A Review of the Literature.* Washington, DC: National Science Foundation.

Cox, L.A. Jr. (2012). "Confronting deep uncertainties in risk analysis." *Risk Analysis* 32(10):1607–1629.

Crick, M.J. (2021. "The importance of trustworthy sources of scientific information in risk communication with the public." *Journal of Radiation Research*, 62(S1):i1–i6.

Cummings, L. (2014). "The 'trust' heuristic: arguments from authority in public health." *Health communication* 29(10):1043.

Cvetkovich, G., Vlek, C.A., Earle, T.C. (1989). "Designing technological hazard information programs: towards a model of risk-adaptive decision making," in *Social Decision Methodology for Technical Projects*, eds. C.A.J. Vlek and G. Cvetkovich. Dordrecht: Kluwer Academic.

Cvetkovich, G., Siegrist, M., Murray R., Tragesser, S. (2002). "New information and social trust asymmetry and perseverance of attributions about hazard managers." *Risk Analysis* 22(2):359–367.

Davies, C.J., Covello, V.T., Allen, F.W., eds. (1987). *Risk Communication: Proceedings of the National Conference*. Washington: The Conservation Foundation.

Dietz, T. (2013). "Bringing values and deliberation to science communication." *Proceedings of the National Academy of Sciences* 110:14081–14087.

Dunwoody, S. (2014). "Science journalism," in *Handbook of Public Communication of Science and Technology*, eds. M. Bucchi and B. Trench. New York: Routledge.

Earle, T.C. (2010). "Trust in risk management: A model-based review of empirical research." *Risk Analysis* 30(4):541–574.

Earle, T.C., Siegrist, M. (2008). "On the relation between trust and fairness in environmental risk management." *Risk Analysis* 28(5):1395–1414.

Engdahl, E., Lidskog R. (2014). "Risk, communication and trust: Towards an emotional understanding of trust. *Public Understanding of Science*, 23(6):703–717.

Fearn-Banks, K. (2019). *Crisis Communications: A Casebook Approach*, 5th edition. New York: Routledge.

Fischhoff, B. (1995a). "Risk perception and communication unplugged: Twenty years of process." *Risk Analysis* 15(2):137–145.

Fischhoff, B. (1995b). "Strategies for risk communication. Appendix C." in *National Research Council: Improving Risk Communication*. Washington, DC: National Academies Press.

Fischhoff, B. (2012). *Judgment and Decision Making*. New York: Earthscan.

Fischhoff, B. (2013). "The sciences of science communication," in *Proceedings of the National Academy of Sciences* 110 (Supplement 3):14033–14039.

Fischhoff, B., Brewer, N.T., Downs, J.S., eds. (2011). *Communicating Risks and Benefits: An Evidence-based User's Guide*. Washington, DC: Food and Drug Administration.

Fischhoff, B., Davis, A. L. (2014). "Communicating scientific uncertainty," in *Proceedings of the National Academy of Sciences of the United States of America* 111(Suppl. 4): 13664–13671.

Fischhoff, B., Kadvany, J. (2011). *Risk: A Very Short Introduction*. New York: Oxford University Press.

Fischhoff B, Lichtenstein S, Slovic P, Keeney D. (1983). *Acceptable Risk*. Cambridge, Massachusetts: Cambridge University Press.

Fjaeran, L., Aven, T. (2021) "Creating conditions for critical trust – How an uncertainty-based risk perspective relates to dimensions and types of trust." *Safety Science* 133:105008.

Dohle, S., Wingen, T., Schreiber, M. (2020). "Acceptance and adoption of protective measures during the COVID-19 pandemic: The role of trust in politics and trust in science." *Social Psychological Bulletin* 15(4):1–23.

Dominic Balog-Way, D., McComas, K., Besley, J. (2020). "The evolving field of risk communication." *Risk Analysis*, 10.1111/risa.13615, 40, S1, (2240-2262), (2020).

Flynn, J., Slovic, P., Mertz, C.K. (1994). "Gender, race, and perception of environmental health risks." *Risk Analysis* 14(6):1101–1108.

Glik D.C. (2007). "Risk communication for public health emergencies." *Annual Review of Public Health* 28(1):33–54.

Hance, B.J., Chess, C., Sandman, P.M. (1990). *Industry Risk Communication Manual*. Boca Raton, FL: CRC Press/Lewis Publishers.

Halvorsen, P.A. (2010). "What information do patients need to make a medical decision?" *Medical Decision Making* 30(5 Suppl):11S–13S.

Haight, J.M., ed. (2008). *The Safety Professionals Handbook: Technical Applications*. Des Plaines, IL: The American Society of Safety Engineers.

Heath, R. O'Hair, D., eds. (2009). *Handbook of Risk and Crisis Communication*. New York: Routledge.

Hess, R., Visschers, V.H.M., Siegrist, M., Keller, C. (2011). "How do people perceive graphical risk communication? The role of subjective numeracy." *Journal of Risk Research* 14(1):47–61.

Hyer, R.N., Covello, V.T. (2005). *Effective Media Communication During Public Health Emergencies: A World Health Organization Handbook*. Geneva: United Nations World Health Organization Publications.

Hyer, R., Covello, V.T. (2017). *Top Questions on Zika: Simple Answers*. Arlington, VA: Association of State and Territorial Health Officials.

Infanti, J., Sixsmith, J., Barry, M.M., Núñez-Córdoba, J.M., Oroviogoicoechea-Ortega, C. & Guillén-Grima, F.A. (2013). *A Literature Review on Effective Risk Communication for the Prevention and Control of Communicable Diseases in Europe*. Stockholm: European Centre for Disease Control and Prevention.

International Atomic Energy Agency. (2011), *Stakeholder Involvement Throughout the Life Cycle of Nuclear Facilities*. IAEA Nuclear Energy Series No. Ng-T-1.4, IAEA, Vienna: International Atomic Energy Agency.

International Atomic Energy Agency. (2006). *The Chernobyl Forum:2003–2005. Chernobyl's Legacy*. Vienna: International Atomic Energy Agency.

Jardine, C. G., Driedger, S. M. (2014). "Risk communication for empowerment: An ultimate or elusive goal?" in *Effective Risk Communication*, eds J. Arvai and L. Rivers III. London: Earthscan.

Joslyn, S., LeClerc, J. (2012). "Uncertainty forecasts improve weather related decisions and attenuate the effects of forecast error." *Journal of Experimental Psychology* 18:126–140.

Joslyn, S., Nadav-Greenberg, L., Taing, M.U., Nichols, R.M. (2009). "The effects of wording on the understanding and use of uncertainty information in a threshold forecasting decision." *Applied Cognitive Psychology* 23:55–72.

Heath, R. O'Hair, D., eds. (2009). *Handbook of Risk and Crisis Communication*. New York: Routledge.

Kahneman, D. (2011). *Thinking, Fast and Slow*. New York: Macmillan Publishers.

Kahneman, D., Slovic, P., Tversky, A., eds. (1982). *Judgment Under Uncertainty: Heuristics and Biases*. New York: Cambridge University Press.

Kahneman, D. Tversky, A. (1979). "Prospect theory: An analysis of decision under risk." *Econometrica*, 47(2):263–291.

Kasperson, R.E. (1986). "Six propositions on public participation and their relevance for risk communication." *Risk Analysis* 6(3):275–281.

Kasperson, R.E. (2014). "Four questions for risk communication." *Journal of Risk Research* 17 (10):1233–1239.

Kasperson, R.E., Golding, D., Tuler, S. (1992). "Social distrust as a factor in sitting hazardous facilities and communicating risks." *Journal of Social Issues* 48(4):161–187.

Kasperson, R. E., Renn, O., Slovic, P., Brown, H. S., Emel, J., Goble, R., Kasperson, J. X., Ratick, S. (1987). "Social amplification of risk: A conceptual framework." *Risk Analysis* 8(2):177–187.

Kasperson, R., Kasperson, J.X., Golding, D. (1999). "Risk, trust, and democratic theory," in *Social Trust and the Management of Risk*, eds. G. Cvetkovich and R. Löfstedt. London: Earthscan.

Kasperson, R. E., Stallen, P.J.M. eds. (1991). *Communicating Risks to the Public: International Perspectives*. Dordrecht, Netherlands: Kluwer.

Klampitt, P.G., DeKoch, R.J., Cashman, T., "A strategy for communicating about uncertainty." *The Academy of Management Executive* 14(4):41–57.

Lindenfeld, L., Smith, H., Norton, T., Grecu, N. (2014). "Risk communication and sustainability science: lessons from the field." *Sustainability Science* 9 (2):119–127.

Löftstedt, R. (2003). "Risk communication. Pitfalls and promises." *European Review* 11(3):417–435.

Löftstedt, R. (2005). *Risk Management in Post-Trust Societies*. New York: Palgrave Macmillan.

Löftstedt, R E., Bouder, F. (2014). "New transparency policies: Risk communication's doom?" in *Effective Risk Communication*, eds J. Àrvai and L. Rivers III. London: Earthscan.

Lundgren, R.E., McMakin A.H. (2018). *Risk Communication: A Handbook for Communicating Environmental, Safety, and Health Risks*. Hoboken, NJ: Wiley-IEEE Press.

McComas, K.A. (2006). "Defining moments in risk communication research: 1996–2005." *Journal of Health Communication* 11(1):75–91.

McComas, K.A., Arvai, J., Besley, J.C. (2009). "Linking public participation and decision making through risk communication," in *Handbook of Risk and Crisis Communication*, eds. R. Heath and D. O'Hair. New York: Routledge.

Mileti, D., Nathe, S., Gori, P., Greene, M., Lemersal, E. (2004). *Public Hazards Communication and Education: The State of the Art*. Boulder, Colo.: Natural Hazards Center.

Miller, M., Solomon, G. (2003). "Environmental risk communication for the clinician." *Pediatrics* 112 (1):211–217. Accessed at: https://pdfs.semanticscholar.org/33c9/8bd7f526d727e8249632df23c537 db89cd4f.pdf

Morgan, M.G., Fischhoff, B., Bostrom, A., Atman, C. J. (2002). *Risk Communication: A Mental Models Approach*. Cambridge, UK: Cambridge University Press.

National Research Council (1989). *Improving Risk Communication*. Washington, D.C.: National Academies Press.

National Research Council (1996). *Understanding Risk: Informing Decisions in a Democratic Society*. Washington, DC: National Academies Press.

National Research Council) (2008). *Public Participation in Environmental Assessment and Decision Making*. Washington, DC: The National Academies Press.

National Academy of Sciences (2014). *The Science of Science Communication II: Summary of a Colloquium*. Washington, DC: The National Academies Press.

National Academy of Sciences (2017). *Communicating Science Effectively*. Washington, D.C.: The National Academies Press.

National Oceanic and Atmospheric Administration, US Department of Commerce. (2016). *Risk Communication and Behavior. Best Practices and Research*. Washington, DC: National Oceanic and Atmospheric Administration.

Peters, E. (2012). "Beyond comprehension: The role of numeracy in judgments and decisions." *Current Directions in Psychological Science* 21(1):31–35.

Peters, R., McCallum, D., Covello, V.T. (1997). "The determinants of trust and credibility in environmental risk communication: An empirical study." *Risk Analysis* 17(1):43–54.

Pidgeon, N., Kasperson, R., Slovic, P. (2003). *The Social Amplification of Risk*. Cambridge, UK: Cambridge University Press.

Power-Hays, A., McGann, P. (2020). "When actions speak louder than words." *New England Journal of Medicine* 383:1902–1903.

Renn, O. (1992). "Risk communication: Towards a rational discourse with the public." *Journal of Hazardous Materials* 29:465–579.

Renn, O. (2008). *Risk Governance: Coping with Uncertainty in a Complex World*. London, UK: Earthscan.

Renn, O., Levine, D. (1991). "Credibility and trust in risk communication," in *Communicating Risks to the Public*, eds. R. Kasperson and P. Stallen. Dordrecht, The Netherlands: Kluwer Academic Publishers.

Reynolds, B. (2019). *Crisis and Emergency Risk Communication*. Atlanta, GA: US Centers for Disease Control and Prevention.

Reynolds B., Seeger, M.W. (2005). "Crisis and emergency risk communication as an integrative model." *Journal of Health Communication* 10:43–55.

Rodrıguez, H., Dıaz, W., Santos, J., Aguirre, B. (2007). "Communicating risk and uncertainty: Science, technology, and disasters at the crossroads," in *Handbook of Disaster Research*. New York: Springer.

Rogers, M., Richard Amlôt, R., James Rubin, G., Wessely, S., Krieger, K. (2007). "Mediating the social and psychological impacts of terrorist attacks: The role of risk perception and risk communication", *International Review of Psychiatry* 19(3):279–288.

Sandman, P.M. (1989). "Hazard versus outrage in the public perception of risk," in *Effective Risk Communication: The Role and Responsibility of Government and Non-Government Organizations*, eds. V.T. Covello, D.B. McCallum, and M.T. Pavlova. New York: Plenum Press.

Seeger, M.W. (2006). "Best practices in crisis communication: An expert panel process." *Journal of Applied Communication Research* 34(3):232–244.

Siegrist, M. (2021). "Trust and risk perception: A critical review of the literature." *Risk Analysis* 41(3):480–490.

Sheppard, B., Janoske, M., Liu, B. (2012). "Understanding risk communication theory: A guide for emergency managers and communicators." *Report to Human Factors/Behavioral Sciences Division, Science and Technology Directorate, US Department of Homeland Security*. College Park, MD: US Department of Homeland Security.

Slovic, P. (1987). "Perception of risk." *Science* 236 (4799): 280–85.

Slovic, P. (1993). "Perceived risk, trust, and democracy." *Risk Analysis* 13(6): 675–682.

Slovic, P. (1999). "Trust, emotion, sex, politics, and science: Surveying the risk-assessment battlefield." *Risk Analysis* 19(4):689–701.

Slovic, P. (2000). *The Perception of Risk*. London, UK: Earthscan.

Slovic, P. (2016). "Understanding perceived risk: 1978–2015." *Environment: Science and Policy for Sustainable Development* 58(1):25–29.

Slovic, P., M. Finucane, L., Peters, E., MacGregor, D.G. (2004). "Risk as analysis and risk as feelings: Some thoughts about affect, reason, risk, and rationality." *Risk Analysis* 24(2):311–322.

Siegrist, M., Cvetkovich, G. (2000). "Perception of hazards: The role of social trust and knowledge." *Risk Analysis* 20(5):713–719.

Siegrist, M. and Zingg, A. (2014). "The role of public trust during pandemics: Implications for crisis communication." *European Psychologist* 19(1):23–32.

Song, H., McComas, K.A., Schuler, K.L. (2018). "Source effects on psychological reactance to regulatory policies: The role of trust and similarity." *Science Communication* 40(5):591–620.

Stallen, P.J.M., Tomas, A. (1988). "Public concerns about industrial hazards." *Risk Analysis* 8(2):235–245.

Steelman, T.A., McCaffrey, S. (2013). "Best practices in risk and crisis communication: Implications for natural hazards management." *Natural Hazards* 65(1):683–705.

Thompson, K.M., Bloom, D.L. (2000). "Communication of risk assessment information to risk managers." *Journal of Risk Research* 3:333–352.

Tuler, S. P., Kasperson, R.E. (2014). "Social distrust and its implications for risk communication: An example of high level radioactive waste management," in *Effective Risk Communication*, eds. J. Arvai and L. Rivers III. London: Earthscan.

Tversky, A., Kahneman, D. (1974). "Judgment under uncertainty: heuristics and biases." *Science* 185(4157):1124–1131.

US Environmental Protection Agency. (2005). *Superfund Community Involvement Handbook*. EPA 540-K-05-003. Washington, D.C.: US Environmental Protection Agency.

US Department of Health and Human Services. (2006). *Risk Communication Guidelines for Public Officials*. Washington, DC: US Department of Health and Human Services.

Varghese, N.E., Sabat, I., Neumann-Böhme, S., Schreyögg, J., Stargardt, T., Torbica, A. (2021). "Risk communication during COVID-19: A descriptive study on familiarity with, adherence to and trust in the WHO preventive measures." *PLoS ONE* 16(4):e0250872.

Walaski, P. (2011). *Risk and Crisis Communications: Methods and Messages*. Hoboken, NJ: Wiley.

Weinstein, N.D. (1987). *Taking Care: Understanding and Encouraging Self-Protective Behavior.* New York: Cambridge University Press.

Wingen, T., Berkessel, J.B., Englich, B. (2020). "No replication, no trust? How low replicability influences trust in psychology." *Social Psychological & Personality Science* 11(4):454–463.

Wojtecki, J., Peters, R., Gordon J. (2007). *The Pfeiffer Book of Successful Communication Skill-Building Tools.* Hoboken, NJ: John Wiley & Sons.

Endnotes

1 See the case study in this chapter on SARS and Hong Kong.

2 Evansa, D. K., Goldstein, M., Popova, A. (2015). "Health-care worker mortality and the legacy of the Ebola epidemic." *The Lancet Global Health* 3(8):439–440. Accessed at: doi:10.1016/ S2214-109X(15)00065-0.

3 See Fisher, R., Ury, W., Patton, B. (2011). *Getting to Yes: Negotiating Agreement Without Giving In.* New York: Penguin Books; Ury, W. (1993). *Getting Past No: Negotiating in Difficult Situations.* New York: Bantam Books.

4 Renn, O., and Levine, D. (1991). "Credibility and Trust in Risk Communication." *Communicating Risks to the Public.* Dordrecht, Netherlands: Kluwer Academic Publishers, p. 176.

5 See., e.g., Covello V. (1992). "Risk communication, trust, and credibility." *Health and Environmental Digest* 6(1):1–4; Covello V. (1993). "Risk communication and occupational medicine." *Journal of Occupational Medicine* 35:18–19; Covello, V. (2003). "Best practices in public health risk and crisis communication." *Journal of Health Communication* 8 (Suppl. 1): 5–8; Diermeier, D. (2013). "Building Trust During Times of Crisis." *Public Gaming International* June. 51–54; Diermeier, D. (2011). *Reputation Rules: Strategies for Building Your Company's Most Valuable Asset.* New York: McGraw-Hill.

6 See, e.g., Drottz-Sjoeberg, B.M., Rumyantseva, G.M., Martyvshov, A.N. (1993). *Perceived Risks and Risk Attitudes in Southern Russia in the Aftermath of the Chernobyl Accident.* Report No.13. Stockholm: Centre for Risk Research, Stockholm School of Economics.

7 See Crick, M.J. (2021. "The importance of trustworthy sources of scientific information in risk communication with the public." *Journal of Radiation Research* 62(S1):i1–i6.

8 See, e.g., Covello V.T. (1993). "Risk communication and occupational medicine." *Journal of Occupational Medicine* 35:18–19. Accessed at: https://journals.lww.com/joem/ citation/1993/01000/risk_communication_and_occupational_medicine.11.aspx

9 In the resources section of this document, see the discussion of trust in the book by Lundgren and McMakin (2014) and in the article by Peters, Covello, and McCallum (1997).

10 Peters, R. G., Covello, V. T., and McCallum, D. B. (1997). "The Determinants of Trust and Credibility in Environmental Risk Communication: An empirical study." *Risk Analysis* 17(1): 43–54; See also Covello, V., Peters, R., Wojtecki, J., and Hyde, R. (2001). "Risk communication, the West Nile virus epidemic, and bioterrorism: Responding to the communication challenges posed by the intentional or unintentional release of a pathogen in an urban setting." *Journal of Urban Health* 78: 382–391.

11 See Figure 2, Trust Determination Factors, in Chapter 2.

12 See, e.g., Covello, V. T. (2011)."Risk Communication, Radiation, And Radiological Emergencies: Strategies, Tools, And Techniques." *Health Physics.* (Nov)101(5):511–30.

13 Gladwell, M. (2019) Talking to Strangers: What We Should Know About the People We Don't Know. New York: Little, Brown, and Co.

14 McAleer, P., Alexander Todorov, A., and Belin, P. (2014). "How Do You Say 'Hello'? Personality Impressions from Brief Novel Voices" PLOS/ONE. Accessed at: https://journals.plos.org/plosone/article?id=10.1371/journal.pone.0090779. See also Park, S. J., Yeung, G., Vesselinova, N., Kreiman, J., Keating, P. A., and Alwan, A. (2018). "Towards understanding speaker discrimination abilities in humans and machines for text-independent short utterances of different speech styles." *Journal of the Acoustical Society of America* 144(1): 375–386.

15 http://www.npr.org/blogs/health/2014/05/05/308349318/you-had-me-at-hello-the-science-behind-first-impressions.

16 The study was inspired in part by the Hollywood film *Jerry Maguire*, where the character played by Renée Zellweger tells the character played by Tom Cruise, "You had me at hello."

17 Nichols, T. (2017). *The Death of Expertise. The Campaign against Established Knowledge and Why It Matters.* Oxford, UK and New York: Oxford University Press.

18 Pew Research Institute (2020). "Americans' trust in military, scientists relatively high; fewer trust media, business leaders, elected officials." Accessed at: https://www.pewresearch.org/ft_19-03-21_scienceconfidence_americans-trust-in-military/

19 People are continually reminded about technological failures by the telling and retelling – some accurate and some fantasy -- of technological failures in books, articles, and films such as *The China Syndrome* for nuclear power plant accidents, *The Andromeda Strain* for microbes, and *Contagion and Outbreak* for disease outbreaks.

20 For a regularly updated list and description of technological failures in the United States and globally, see the website Disasters published by the libraries of Pennsylvania State University. Accessed at https://guides.libraries.psu.edu/c.php?g=423498&p=2892860. The website notes that separating engineering and machine failures and from human error and negligent maintenance is often difficult.

21 Edelman Polling Co (2019). *2019 Edelman Trust Barometer: Executive Summary*. Accessed at: https://www.edelman.com/sites/g/files/aatuss191/files/2019-01/2019_Edelman_Trust_Barometer_Executive_Summary.pdf.

22 Edelman (2020). *2020 Edelman Trust Barometer: Executive Summary*. Accessed at https://www.edelman.com/trustbarometer; See also Gallup Organization (2020). *Confidence in Institutions: 1975 to 2020*: Accessed at: https://news.gallup.com/poll/1597/confidence-institutions.aspx.

23 See, e.g., McCallum, D. B., Hammond, S. L., Morris, L. A., and Covello, V. T. (1990). *Public Knowledge and Perceptions of Chemical Risks in Six Communities: Analysis of a Baseline Survey.* (EPA 230-01-90-074). Washington, DC: US Environmental Protection Agency.

24 See, e.g., Brown, M. E., and Treviño, L. K. (2006). Ethical leadership: A review and future directions. *The Leadership Quarterly* 17:595–616; Brown, M.E., Treviño, L.K., and Harrison, D. (2005). Ethical leadership: A social learning perspective for construct development and testing. *Organizational Behavior and Human Decision Processes* 97:117–134; Carmen Tanner, C., Brügger, A., van Schie, S., and Lebherz, C. (2015). Actions speak louder than words. *Zeitschrift für Psychologie / Journal of Psychology* 218 (4):225–233; Harter, S. (2002). Authenticity, in *Handbook of Positive Psychology*, eds. C.R. Snyder, S. Lopez (pp. 382–394). Oxford, UK: Oxford University Press.; Kanungo, R. N. (2001). Ethical values of transactional and transformational leaders. *Canadian Journal of Administrative Sciences* 18:257–265; Jones, E. E., Kanouse, D. E., Kelley, H. H., Nisbett, R. E., Valins, S., and Weiner, B. (eds.) (1972). *Attribution: Perceiving the Causes of Behavior.* Morristown, NJ: General Learning Press; May, D. R., Chan, A.Y.L., Hodges, T.D., and Avolio, B.J. (2003). Developing the moral component of authentic leadership. *Organizational Dynamics* 32:247–260; Rokeach, M. (1973). *The Nature of Human Values.* New York, NY: Free Press; Schmidt, W.H., and Posner, B.Z. (1982). *Managerial Values and Expectations.* New York, NY: Amacom; Sosik, J.J. (2005). The role of personal values in the charismatic leadership of corporate managers: A model and preliminary field study. *The Leadership Quarterly* 16:221–244; Treviño, L.K., Brown, M., and Hartman, L.P. (2003). A qualitative investigation of perceived executive ethical leadership: Perceptions from inside and outside the executive suite. Human Relations 55:5–37; Walumbwa, F.O., Avolio, B.J., Gardner, W.L., Wernsing, T.S., and Peterson, S.J. (2008). Authentic leadership: Development and validation of a theory-based measure. *Journal of Management* 34:89–126.

25 International Atomic Energy Agency (2006). *The Chernobyl Forum: Communication with the Public in a Nuclear or Radiological Emergency.* Vienna, Austria: International Atomic Energy Agency.

26 International Atomic Energy Agency (2006). *The Chernobyl Forum: Communication with the Public in a Nuclear or Radiological Emergency.* Vienna, Austria: International Atomic Energy Agency. Page 34.

27 See, for example: Funabashi, Y., and Kitazawa, K. (2012). "Fukushima in review: A complex disaster, a disastrous response." *Bulletin of the Atomic Scientists* 68(2):9–21; Investigation Committee on the Accident at the Fukushima Nuclear Power Stations of Tokyo Electric Power Company. (2012, July). Final Report. Accessed at: http://www.cas.go.jp/jp/seisaku/icanps/eng/nal-report.html; Kushida, K.E. (2014). "The Fukushima nuclear disaster and the democratic party of Japan: Leadership, structures, and information challenges during the crisis." *Japanese Political Economy* 40(1):29–68; National Diet of Japan, Fukushima Nuclear Accident Independent Investigation Commission. (2012). The Official Report of the Fukushima Nuclear Accident Independent Investigation Commission. Accessed at: http://www.nirs.org/fukushima/naiic_report.pdf; Acton, J.M., Hibbs, M. (2012). Why Fukushima Was Preventable. Carnegie Endowment for International Peace. Accessed at: http://carnegieendowment.org/les/fukushima.pdf; Svendsen, E.R. (2013). "A new perspective on radiation risk communication in Fukushima, Japan." *Journal of the National Institute of Public Health* 62(2):196–203.

28 See, e.g., Covello, V. T. (2011). *Guidance on Developing Effective Radiological Risk Communication Messages.* (NUREG/CR-7033). Washington D.C.: US Nuclear Regulatory Commission.

29 See, e.g., Hung L. S. (2003). "The SARS epidemic in Hong Kong: What lessons have we learned?" *Journal of the Royal Society of Medicine August* 96(8):374–378.

30 See, e.g., Huang Y. (2004). The SARS Epidemic and Its Aftermath in China: A Political Perspective. In Knobler, S., Mahmoud, A., and Lemon, S., eds., *Learning from SARS: Preparing for the Next Disease Outbreak.* Washington (DC): National Academies Press (US). Accessed at: https://www.ncbi.nlm.nih.gov/books/NBK92479/.

31 See Tinker, T. L, Covello, V. T., Vanderford, M. L., Rutz, D., Frost, M., Li, R., Aihua, H., Chen, X., Xie, R., and Kan, J. (2012). "Disaster risk communication." Chapter 141, in *Textbook in Disaster Medicine*, ed. S. David. Dordrecht, Netherlands: Wolters Kluwer. Accessed at: http://www.procommunicator.com/wp-content/upLoads/2012/04/a.article.risk-communication.Chapter-141.pdf.

32 See, e.g., Gustafson, P. (2008) "Gender differences in risk perception: Theoretical and methodological perspectives." *Risk Analysis* 18(6):805–811; Flynn, J., Slovic, P., and Mertz, C. (1994) "Gender, race, and perception of environmental health risks." *Risk Analysis* 14(6):1101–1108.

33 See, e.g., Covello, V., Johnson, B. (1987). "The social and cultural construction of risk: Issues, methods, and case studies," in *The Social and Cultural Construction of Risk*, eds. Johnson, B., and Covello, V. New York: Springer; Rayner, S. (1992) "Cultural theory and risk analysis," in *Social Theories of Risk*. Westport, CT: Praeger; Wildavsky, A., and Dake, K. (1990). "Theories of risk perception: Who fears what and why?" *Daedalus* 4:41–60; Douglas, M., and Wildavsky, A. (1983). *Risk and Culture: An Essay on Selection of Technological and Environmental Dangers.* Berkeley: University of California Press; Douglas, M. (1966). *Purity and Danger: An Analysis of Concepts of*

Pollution and Taboo. London: Routledge & Kegan Paul; Gross, R., and Rayner, S. (1985). *Measuring Culture: A paradigm for the analysis of social organization*. New York: Columbia University Press; Sederberg, P. (1984). *The Politics of Meaning: Power and Explanation in the Construction of Social Reality*. Tucson: University of Arizona Press; Dake, K. (1991). "Orienting dispositions in the perception of risk: An analysis of contemporary worldviews and cultural biases," *Journal of Cross-Cultural Psychology* 22:61–82; Dake, K. (1992). "Myths of nature—culture and the social construction of risk," *Journal of Social Issues* 4:21–37; Coughlin, R., and Lockhart, C. (1998). "Grid-group theory and political ideology. A consideration of their relative strengths and weaknesses for explaining the structure of mass belief systems." *Journal of Theoretical Politics* 10:33–58; Rippl, D. (2002). "Cultural theory and risk perception: A proposal for a better measurement." *Journal of Risk Research* 5(2):147–165.

34 M. Douglas, M., and Wildavsky, A. (1983). *Risk and Culture: An Essay on Selection of Technological and Environmental Dangers*. Berkeley: University of California Press. Douglas, M. (1966). *Purity and Danger: An Analysis of Concepts of Pollution and Taboo*. London: Routledge & Kegan Paul.

35 Douglas, M. (1966). *Purity and Danger: An Analysis of Concepts of Pollution and Taboo*. London: Routledge & Kegan Paul.

36 Flórez, K. R., Aguirre, A. N., Viladrich, A., Céspedes, A., Alicia De La Cruz, A., and Abraído-Lanza, A.F. (2009). "Fatalism or destiny? A qualitative study and interpretative framework on Dominican women's breast cancer beliefs." *Journal of Immigrant Minority Health* 11 (4):291–301.

9

Best Practices for Message Development in High Concern Situations

---

CHAPTER OBJECTIVES

This chapter describes best practices and tools for developing messages that can be heard, understood, and believed by individuals experiencing the stress of a risky, high concern, or crisis situation.

At the end of this chapter, you will be able to:

- Explain why messages need to be crafted with the recognition that they may be heard through stress.
- Recognize when to employ tools to assist in addressing stress factors.
- Identify key tools for crafting messages that can be effectively received in the context of stress.
- Describe the steps in the process of message map creation.

9.1   Case Diary: Mapping Through a Maze of COVID Confusion

It was February 2020 and public health officials across the US – and the globe – were caught in a whirlwind. In January, there had been a report of a few cases of an entirely new, novel, and sometimes deadly virus. By February, cases had exploded into a pandemic. Across the world, people were in a state of confusion and panic. Everyone wanted answers, sometimes to life-and-death questions: What is COVID-19? Are my family and I safe? How can I protect myself? How can we stop this? What are our government officials doing about this?

There were no simple answers; evidence and research were sparse – and sometimes conflicting – and the picture was changing every day. The medical and public health issues were complex, yet the anxious public wanted clear, concise information. State public health officials, who were madly trying to cope with devising policies and protective actions, were deluged with requests for information.

A colleague and I had been working for several years with the Association of State and Territorial Health Officials (ASTHO) on issues related to crisis communication. Among other projects, we had created for them FAQs on Ebola and Zika that were highly successful in answering questions posed frequently by the public. ASTHO turned to us again and asked us to produce a publication that would become *COVID-19: Simple Answers to Top Questions*, an online publication for state and territorial health officials and directors that went through six updated editions between March and November 2020.

Normally, in such a high-impact situation, ASTHO would rely heavily on information produced by the Centers for Disease Control and Prevention, the country's authoritative agency for disease control and prevention. Unfortunately, the pandemic had quickly become a politicized issue, and CDC's expert information was being reviewed and often redacted by non-experts in the White House. (President Donald Trump later revealed, in an interview conducted 19 March 2020, with journalist Robert Woodward, that, "I wanted to always play it [the pandemic] down. I still like playing it down because I don't want to create panic.")

Delivering information about COVID-19 in a comprehensible manner was a huge challenge – a challenge that could not have been met without the application of message mapping. Message maps are risk communication tools used to help organize complex information and make it easier to express current knowledge. The development process distills information into clear and easily understood messages. Answers to questions are presented initially in no more than three to five short sentences that convey key messages, ideally in the least number of words possible. The approach is based on surveys showing that lead or front-page media and broadcast stories usually convey only a soundbite: three to five messages, usually in less than nine seconds for broadcast media or 27 words for print.[1] Each primary message normally has three to five supporting messages that can be used when and where appropriate to provide context for the issue being mapped.

The health officials eagerly made use of the messages. The maps gave them three layers of messaging: (1) soundbites to use in the press, (2) clear, brief explanatory messages that are easily comprehended by a worried reader/listener, and (3) citations to the scientific research so that the expert health officials could trace the sources of information.

By the end of 2020, and after six published revisions, the ASTHO message mapping document contained over 75 questions and answers about COVID-19.[2] The document was downloaded by over 300,000 persons. Readers of the document commended the state and territorial health officials for providing a document with accurate, timely, user-friendly, and evidence-based information about COVID-19. Readers also expressed appreciation for the frequent updates of the document. They said that the message maps contributed greatly to their ability to make informed decisions about COVID-19 and that they aided their ability to communicate the information to others.

9.2 Introduction

Planning and preparation for communicating in risk, high concern, and crisis situations greatly increase the likelihood that the resulting messages will contribute positively to informed decision-making – a critical objective. Messages that are well constructed, well-practiced, and well delivered will enhance understanding, reduce misinformation, and provide a solid foundation for constructive engagement, dialogue, and the exchange of information among stakeholders. This chapter will describe the tools and best practices for developing messages that can be heard, believed, and understood by individuals facing risk, high concern, and crisis situations. Shown in Figure 9.1 are the three key messages of this chapter.

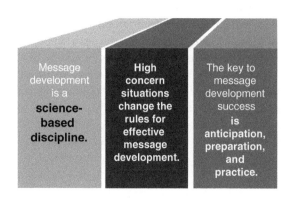

Figure 9.1 Message development for high-concern situations: three key messages.

The first part of the chapter discusses how to develop more effective messages in the context of knowledge about the brain's decision-making processes under high stress. The second part of the chapter provides detailed guidance for constructing clear messages through the essential tool of message mapping.

9.3 Crafting Messages in the Context of Stress and High Concern Decision-Making

In risk, high concern, and crisis situations, and because of the stress, fear, and emotions aroused by such situations, decisions about risks and threats are continuously made in the brain but are made differently than in less stressful daily life.[3] The following decision-making factors come into play in high-stress situations and relate most directly to best practices in message development: *trust determination, impaired comprehension, negative dominance*, and *the emotional impact of stress*.

9.3.1 Trust Determination and Messaging in High-Stress Situations

At times of stress, fear, and emotional arousal, the brain's processes for *trust determination* are significantly affected. More than usual, people will want to know you are listening, caring, competent, and honest. Most important, they will want to know that you care before they care what you know, and they will use this to decide whether to trust the messenger. Trusting the messenger is a prerequisite to hearing your message. As a result of activation of the brain's risk/threat detection system, a practical consequence for message development is that additional efforts must be made in high-stress situations to communicate messages that demonstrate, through actions and authentic words, trust determination attributes such as listening, caring, empathy, and compassion

9.3.1.1 The CCO Best Practice

A powerful communication tool for creating trust is the *CCO* template. In fact, it is the single most powerful best practice in the toolbox of risk, high concern, and crisis communicator because it directly addresses the issue of trust, which is the cornerstone for effective communication. As shown in Table 9.1, the *CCO* best practice calls for three messages: messages of compassion, conviction, and optimism.

The first "C" in the best practice calls for the strongest type of caring: authentic compassion. Caring is also communicated by active listening ("I hear what you are saying") and genuine empathy ("I can only imagine what you are feeling"). Less powerful, but still effective, are messages that communicate that you believe the issue being discussed is important or that you have been listening carefully to what is being said. ("What you are saying is important.")

A frequently quoted example of a message that conveyed compassion were the words spoken by New York City's then mayor, Rudolf Giuliani, on 9/11 when he said: "The number of casualties will be more than any of us can bear ultimately." The words of General Norman Schwarzkopf about the Gulf War are also a strong example of empathy: "It doesn't take a hero to order men into battle; it takes a hero to be one of those men who goes into battle."

Table 9.1 CCO template/best practice.

- Compassion (**C**) (listening, caring, or empathy)
- Conviction (**C**) (firm belief, confidence, or commitment)
- Optimism (**O**) (hope)

The second message – the second "C" message in the best practice – calls for a message of conviction, commitment, or firm belief. Examples include messages from public health leaders, such as Dr. Anthony Fauci of the US National Institutes of Health, during the COVID-19 crisis that conveyed faith and confidence in the safety and effectiveness of new vaccines to bring the pandemic to an end.

The third message – the "O" message in the best practice – calls for a message of optimism, hope, self-efficacy, group efficacy, resilience, calmness, support, and connectedness.[4] In developing messages for a high-stress situation, effective communicators consider: What message of evidence-based hope and optimism will inform, connect with, and/or inspire the target audience? Especially during periods of uncertainty, staying positive with learned or active optimism is critical to mental health. It is a central component of "Psychological First Aid."[5] It requires a conscious focus on what is going right or going well.

9.3.2 Impaired Comprehension and Messaging in High-Stress Situations

People in high-stress situations often have difficulty processing information. As a result of activation of the brain's risk/threat detection, additional efforts must be made in high-stress situations to communicate simple, clear messages that are repeated often by a variety of trusted sources.

One of the most powerful of practices for developing messages that are easily grasped is the *27/9/3* best practice. This is a practice for structuring messages. Other templates, such as the CCO, identify important content and tone. The 27/9/3 best practice most directly addresses one of the most difficult challenges in risk, high concern, and crisis communication: that people in high-stress situations typically have great difficulty hearing, understanding, and remembering information. The *27/9/3* best practice advises communicators to serve up information about a high concern issue in buckets or chunks of information made up of no more than 27 words, nine seconds of material, and three messages, whichever comes first. It also requires communicators to share their main messages first: telling people clearly and concisely at the beginning what they want them to know or what the target audience wants to know, then telling them more about these top-line messages, then telling them again what they told them first. This approach delivers the bottom line first.

9.3.3 Negative Dominance and Messaging in High-Stress Situations

In circumstances of stress, fear, and emotional arousal, the way the brain processes negative information is significantly affected. People in high-stress situations often put more weight on negative information than on positive information. Research has found that having three or more positive messages to counterbalance each negative message (such as bad news) helps mental health in high-stress and uncertain situations.[6]

As a result of *negative dominance*, those who are crafting messages in high-stress situations should make additional efforts to communicate positive, constructive messages. Several template tools described in Figure 5 in Chapter 3 help achieve this communication goal and overcome the hardwiring in the brain. These include the *One Negative Equals Three Positives (1N=3P)* best practice, the *I Don't Know (IDK)* best practice, and the *Know/Don't Know (KDK)* best practice.

The most powerful of these tools is the *1N=3P* best practice. It is sometimes called the "bad news" best practice. The *1N=3P* best practice calls for four messages, ideally preceded by an authentic and sincere message of listening, caring, empathy, or compassion. This preamble helps create a soft landing for the upcoming negative or bad news. The first message in the *1N=3P* best practice – the "N" in the formula – is the negative message or bad news. The second, third, and fourth messages – the "3P" in the formula – are the positive, constructive, or solution-oriented messages.

9.3.4 Emotional Impact and Messaging in High-Stress Situations

In high-stress situations, the brain will often rely more on emotions and instinct than on technical facts to make decisions about risks and threats.[7] As a result, those who are crafting messages in high-stress situations need to make additional efforts to communicate messages about issues that are important to emotional responses, such as trust, benefits, personal control, voluntariness, self-efficacy, familiarity, and fairness. Several best practices listed in Figure 5 of Chapter 3 help achieve this communication goal and overcome the hardwiring in the brain.

One of the most powerful of communication tools to address emotions in high-stress decision-making is the *TBC* best practice. As shown in Table 9.2, the first message – the "T" message – is a message that communicates attributes of trust, such as caring, honesty, or competence. The second message – the "B" message – is a message that communicates personal, group, organizational, community, and societal benefits. The third message – the "C" message – is a message that communicates personal control, such as a voice in addressing and resolving the issue, a choice among options and alternatives, and/or easy access to clear and accurate knowledge and information about the risk, threat, or issue.

Several practical guideposts for message development can be drawn from the literature on the impact of emotions on decision-making in stressful situations. To address emotional factors, the five most important perceptions that should be considered in message development are *voluntariness*, *personal control*, *familiarity*, *fairness*, and *trust*. Tips for constructing messages that address these factors in a positive, constructive, effective way are shown below.

Voluntariness. To communicate voluntariness, deliver messages that:

1) make the risk more voluntary;
2) encourage dialogue by using two-way communication channels;
3) ask permission;
4) ask for informed consent.

Table 9.2   The TBC tool (trust, benefits, control).

Trust (T)
1) **Caring**
2) **Honesty**
3) **Competence**
Benefits (B)
1) **Personal**
2) **Group/Organizational**
3) **Societal**
Control (Personal) (C)
1) **Voice**
2) **Choice**
3) **Knowledge**

Personal Control. To communicate personal control, deliver messages that:

1) give stakeholders a voice in the decision-making process;
2) give stakeholders choices, options, and alternatives for reducing or eliminating risks and threats;
3) share knowledge in a clear, understandable way, such as through hierarchical layers moving from simple to more complex;
4) identify things for people to do (for example, precautions, preventative actions, protective actions, treatments, and memorials);
5) indicate your willingness to cooperate and share authority and responsibility with others;
6) delegate and give important roles and responsibilities to others;
7) tell people how to recognize problems or symptoms;
8) tell people how and where to get further information;
9) encourage people to connect with others for support. (The single best predictor of human resilience during periods of high uncertainty is support from other people).

Familiarity. To communicate familiarity, deliver messages that:

1) use analogies to make the unfamiliar familiar;
2) have strong visual content or story-telling;
3) use appropriate comparisons;
4) describe means for exploring issues in greater depth.

Fairness. To communicate fairness, deliver messages that:

1) acknowledge possible inequities;
2) address inequities;
3) discuss dilemmas, options, and trade-offs.

Trust. To communicate trust, deliver messages that indicate:

1) voluntariness, controllability, familiarity, honesty, openness, transparency, disclosure, knowledge, competence, objectivity, and fairness;
2) caring, concern, and empathy;
3) similarities between the sender and receiver, or at least a common personal stake in the issue;
4) alliances and partnerships with trusted partners and independent third-party sources of information;
5) willingness to be held responsible and accountable for errors and omissions.

9.3.4.1 Case Study: Hoarding Toilet Paper at the Outset of the 2020 COVID-19 Pandemic

The hoarding of toilet paper at the beginning of the 2020 COVID-19 pandemic provides an illuminating case study of the ways in which emotions affect the processing of information in high-stress situations. In March 2020 toilet paper suddenly became a precious commodity. Supermarket shelves were emptied. As supplies on the shelves. emptied, people fought over the last remaining rolls and stockpiled any toilet paper they could find.

Emotion-based factors help explain the hoarding of toilet paper and the resulting shortages. The following are nine of the most prevalent factors:

1. Personal control.
Perceptions of personal control are among the most important emotional hot buttons affecting attitudes and judgments about a risk or threat. Hoarding toilet paper was a way for individuals to take back personal control in a world that seemed to be spinning out of control.

2. Perceptions of scarcity.

The hoarding of toilet paper was rooted in an emotional response to the perception of scarcity of something many felt was essential to personal hygiene and their sense of self. When people saw supermarkets limiting how much they could buy of something perceived as essential, they bought as much toilet paper as they could as a precaution for themselves and their families. Having appropriate supplies is one way to feel safe and secure from the enemy – in this case, the virus that causes COVID-19.

3. Primal fear.

The hoarding of toilet paper has roots in primal fears and instinct. Toilet paper is associated with the Freudian unconscious mind, toilet training, lack of cleanliness, unpleasant smells, filth, death, disease, and being presentable to others. Toilet paper became a symbol of health and a source of comfort.

4. The unfamiliar.

The hoarding of toilet paper was a natural response to a threat that was perceived as new, unfamiliar, and exotic. In many natural disaster situations, there is a general consensus about what is needed. People generally know what to do and what others are likely to do if a hurricane or tornado is approaching. COVID-19 was a previously unknown threat and there was a high uncertainty. Uncertainty can breed hoarding and panic buying.

5. Social media activity.

Social media has become one of the dominant ways people learn about new risks and threats. The COVID-19 pandemic was the proximate cause of concerns about shortages of toilet paper, and social media kicked these concerns into high gear. On social media, people were inundated with descriptions and dramatic images of overloaded shopping carts and people fighting over the last roll of toilet paper on a supermarket shelf.

6. Options.

When preparing for a natural disaster, one can stock up on different types of foods. People find lots of options. For toilet paper, for many people, there were no alternative products. There were no substitutes and people felt they had no options other than to hoard.

7. Guidance from others.

The hoarding of toilet paper was rooted in the idea that human beings are fundamentally social animals and seek guidance about new risks by looking at how others are behaving. From an evolutionary perspective, guidance from others can be helpful. If people don't know how to react to a new risk or threat, they can adopt the behavior of others. Seeing videos of panic buying may send a signal to the brain to do the same.

8. Distrust created by inconsistent messaging.

Messages about toilet paper supply were often inconsistent, and not trusted. This was part of a picture of a larger environment of distrust caused by inconsistent and conflicting messages about COVID-19. While many traditional sources of trusted information were saying that the situation was dire, other traditional sources were saying that there was nothing to worry about, that COVID-19 was no worse than the seasonal flu, or that there were more important things to worry about, such as the economy. Shoppers watching shelves empty of toilet paper lacked the trust to believe messages about supply or to follow requests not to stockpile.

9. Worry budgets.

People have worry budgets. A person can worry about only so much before the stress becomes unbearable. A ready supply of toilet paper takes this item off the worry list.

9.4 Message Mapping

One of the most powerful strategic tools for developing clear and concise risk, high concern, and crisis messages is the *message map*. A message map comprises layered and hierarchically organized information that can aid communicators in their responses to anticipated questions or concerns. It is a visual aid that provides, at a glance, messages developed for anticipated questions. Message maps are useful for quickly answering questions in a way that respond to people's cognitive difficulties (difficulty hearing, understanding, and remembering information) during periods of stress and high emotion.

Message maps tell communicators what questions to expect and the recommended answers to these questions. They allow multiple communicators to work from a consistent set of messages across diverse platforms. They enable communicators to "check off" their key points and minimize "communicator's regret" (i.e. regretting saying something or regretting not saying something).

As illustrated in the templates in Tables 9.3 and 9.4, message maps can be presented in a box or bullet format. Both formats enable communicators to respond quickly with timely, accurate, clear, concise, consistent, trusted, and relevant information.

Examples of actual message maps are shown in Table 9.5, for the box format, and Table 9.6, for the bullet format. The text of these examples is message maps created for the Centers for Disease Control and Prevention to communicate about pandemic influenza.[8]

Table 9.7 displays a list of questions reflecting intense public concern about the Ebola outbreak in West Africa in 2014. The full list of questions and mapped answers that were developed for the Association of State and Territorial Health Officials is available in Covello and Hyer (2014)[9].

Message maps are like the pre-scripted templates and talking points used by many organizations for media interviews. However, message maps differ in that they (1) have layered information, starting with clear, simple messages and moving in hierarchical layers to more complex information, (2) are constructed to adhere strictly to the principles of risk, high concern, and crisis communication, (3) can be repurposed for virtually any communication format, and (4) can be developed in advance to address what stakeholders want to know and what the communicator would like them to know.

9.4.1 Benefits of Message Maps

Message maps provide the foundation for well-constructed, practiced, and well-delivered messages that reduce misinformation and support informed decision-making. Once drafted, the effectiveness of the message maps can be evaluated through focus groups, interviews, and other evaluation techniques (see Chapter 11).

Once developed, message maps can be brought together in one place to produce a message playbook. Messages in the playbook can then be converted or repurposed individually or collectively into virtually any type of communication. These include fact sheets, news releases, media

Table 9.3 Message Map Template (Box Format)

ISSUE: Stakeholder Question or Concern:		
Key Message 1	**Key Message 2**	**Key Message 3**
Supporting Info. 1.1	Supporting Info. 2.1	Supporting Info. 3.1
Supporting Info. 1.2	Supporting Info. 2.2	Supporting Info. 3.2
Supporting Info. 1.3	Supporting Info. 2.3	Supporting Info. 3.3

Table 9.4 Message map template (bullet format).

Issue: . . .

Stakeholder Question or Concern: . . .

1) **Shorter Answer:** (Key Messages/Bullets in Boldface)

   • . . .

   • . . .

   • . . .

2) **Longer Answer:** (Key Messages/Bullets in Boldface; Two to Three Supporting Points/Bullets under Each Key Message/Bullet)

Key Message 1: . . .

Supporting Information:

   • . . .

   • . . .

   • . . .

Key Message 2: . . .

Supporting Information:

   • . . .

   • . . .

   • . . .

Key Message 3: . . .

Supporting Information:

   • . . .

   • . . .

   • . . .

---

advisories, text messages, emails, blogs, billboards, social media postings (e.g., Twitter, Facebook, Instagram, TikTok, and YouTube), website content, pamphlets, mail inserts, flyers, media kits, direct mailings, door hangers, brochures, scripts for news conferences, media interviews, videos, and scripts for hotlines. Message maps can also be used as content for educational materials (e.g.,

Table 9.5 Message map for pandemic influenza (box format).

Issue: Pandemic Influenza

Stakeholder Question or Concern: How is pandemic influenza different from seasonal flu?

Key Message 1:	Key Message 2:	Key Message 3:
Pandemic influenza is caused by an influenza virus that is new to people.	**The timing of an influenza pandemic is difficult to predict.**	**A pandemic influenza outbreak is likely to have more severe health effects than seasonal flu.**
Supporting Info. 1.1: Seasonal flu is caused by viruses that are already among people.	**Supporting Info. 2.1:** Seasonal flu occurs every year, usually during winter.	**Supporting Info. 3.1:** Pandemic influenza is likely to affect more people than seasonal flu.
Supporting Info. 1.2: Pandemic influenza may begin with an existing influenza virus that has changed.	**Supporting Info. 2–2:** Pandemic influenza has happened nearly 30 times in recorded history.	**Supporting Info. 3.2:** Pandemic influenza could severely affect a broader set of the population, including young adults.
Supporting Info. 1.3: Few people would be immune to a new influenza virus.	**Supporting Info. 2.3:** An influenza pandemic could last longer than the typical flu season.	**Supporting Info. 3.3:** A severe pandemic outbreak could change daily life, including limitations on travel and public gatherings.

Table 9.6 Message map for pandemic influenza (bullet format).

Issue: Pandemic Influenza

Stakeholder Question or Concern: What is pandemic influenza flu?

1) Shorter Answer:

- **Pandemic flu is a worldwide flu outbreak.**
- **Pandemic influenza is different from seasonal flu.**
- **Public health departments are prepared to respond to a pandemic flu outbreak.**

2) Longer Answer:

- **Pandemic flu is a worldwide flu outbreak.**
  - A pandemic flu outbreak is caused by a new influenza virus.
  - The flu spreads from person to person and is highly contagious.
  - Pandemic flu is expected to have a high death rate.
- **Pandemic influenza is different from seasonal flu.**
  - Seasonal outbreaks of flu are caused by viruses that have already spread among people.
  - Pandemic influenza is the development of a new virus that most people in the world have never been exposed to and for which they have no immunity.
  - Vaccines for pandemic influenza will not be available initially.
- **Public health departments are prepared to respond to a pandemic flu outbreak.**
  - Through monitoring systems, public health departments will track a pandemic influenza outbreak.
  - In the event of a pandemic influenza outbreak, trained staff and partners will distribute medical support and medications.
  - Public health departments will coordinate with hospitals and emergency care facilities to meet public needs.

Source: A full list of pandemic influenza questions and mapped answers can be found in the following publication: US Department of Health and Human Services (2006). *Pre-Event Message Maps*. Washington, DC: Department of Health and Human Services. Accessed at: https://www.calhospitalprepare.org/sites/main/files/file-attachments/pandemic_pre_event_maps.pdf

Table 9.7 Questions about Ebola.

101) What is Ebola?
102) Why is the Ebola virus called "Ebola"?
103) What should people know about the current Ebola situation?
104) How is Ebola different from other diseases that pass from one person to another?
105) Is Ebola the most dangerous disease that humans have ever encountered?
106) What should people be doing about Ebola now?
107) What are the symptoms of Ebola?
108) How long does the Ebola virus live on contaminated surfaces like bed sheets, pillows, stethoscopes?
109) Can a blood test show if a person has the Ebola virus before they have symptoms?
110) How do you confirm infection with the Ebola virus?
111) Can you catch Ebola by touching the skin of someone who has symptoms?
112) What is the risk of exposure to Ebola?
113) How contagious is the Ebola virus?
114) How is Ebola spread?
115) Are all body fluids (blood, mucus, tears, saliva, sweat, vomit, mucus, semen, feces, urine) dangerous and for how long?
116) Who can spread the Ebola virus to others?
117) How long can the Ebola virus survive on surfaces like table-tops and doorknobs?
118) Can Ebola spread by airborne means?
119) What exactly do you mean when you say that "Ebola is not airborne"?
120) Can Ebola mutate to airborne transmission?
121) How quickly can an Ebola test be done?
122) How can people be exposed to Ebola?
123) Can a dog or cat get Ebola?
124) . . .

instructional films), information forums, community meetings, open houses, radio and television talk shows, chat rooms, and face-to-face meetings.

One major benefit of message maps is that they can be used as part of virtually any stakeholder-engagement activity. They can provide the starting place for dialogue and conversations with stakeholders at any stage in the stakeholder-engagement process. Message maps are particularly useful because they distill complex scientific and engineering information into easily understood and accessible messages presented in a layered format with multiple tiers.

Message maps are especially helpful when prepared in advance for potential crises. Although actual crises are difficult to predict, a large percentage of messages can be created in advance for potential crises. In a crisis, speed is critical. Messages must be timely, clear, accurate, frequent, and relevant. These goals can be accomplished by having messages prepared, pre-vetted, and readily available. Since message maps can be prepared before a crisis, they also can be approved in advance, saving valuable time in an urgent situation. They allow crisis managers and communicators to establish themselves as the go-to source for information and focus on issues at hand. In the post-crisis phase, they can help stakeholders better understand risks and threats, build resilience, build self- and group efficacy, build hope, act constructively, recover more quickly, and gain or regain trust in leaders.

As shown in Table 9.8 message maps provide multiple strategic and tactical communication benefits.

9.4.2 Message Maps and the Brain

Message maps are consistent with how the brain makes complex decisions. According to the "SCARF" model of brain functioning (Status, Certainty, Autonomy, Relatedness, Fairness), the brain processes information considering five issues before making a decision in a high-stress situation:[10]

Status: information about caring, empathy, listening, importance to others, and rewards.

Certainty: information about the probability and magnitude of a future occurrence with adverse consequences (i.e., risks).

Autonomy: information about the personal control of risks.

Relatedness: information about connectedness, relationships, support from others, belonging, and safety.

Fairness: information about equity and the distribution of resources, benefits, costs, and risks.

Table 9.8 Benefits of message maps.

Managers and communicators can use message mapping to:

1) Decide what questions you want to answer or that need to be answered.
2) Decide what questions can or should be answered by others.
3) Function as a repository and archive of user-friendly and vetted responses to anticipated stakeholder questions in a structured, clear, concise, accessible, and organized format.
4) Promote dialogue about messages both inside and outside the organization.
5) Ensure the organization and its partners have consistent messages.
6) Ensure the organization speaks with one voice or multiple voices in harmony with one another.
7) Serve as "a port in a storm" when answers and information are lacking.
8) Prevent misstatements and omissions of key facts.
9) Provide information in a format that is easily understood and processed by those experiencing stress.

Information about these five issues is processed in two systems of the brain, called *System 1 Thinking* and *System 2 Thinking*. *System 1 Thinking* and *System Thinking 2* are two distinct models of decision-making by the brain.[11] *System 1 Thinking* is located primarily in the limbic system of the brain, which includes the amygdala (the fight-freeze-flight part of the brain), the hypothalamus, the thalamus, and the hippocampus. *System 2 Thinking* is located primarily in the prefrontal cortex, which is located in the front of the frontal lobe of the brain.

System 1 Thinking is typically instinctive, automatic, emotional, fast, and unconscious. It takes place in the emotional part of the brain that examines threats. It is efficient and requires little energy or attention. When the brain spots a threat or risk, it may send a message to the prefrontal cortex, which is the reasoning and logical thinking part of the brain. However, it often overcome by instinct and emotion. Moreover, to simplify the decision-making process in response to a threat, it often uses mental shortcuts that can result in bias and errors. In response to a threat, instinct and emotion can hijack the prefrontal systems of the brain and make bad choices and decisions.

System 2 Thinking is a slow, rational, thoughtful, and controlled way of thinking. It requires more energy than *System 1 Thinking*. Once activated, *System 2* has the ability to filter the instincts of *System 1*. It links memories and experiences from the past to the current situation. It considers the costs and benefits of choices and actions, examining the consequences using the prefrontal cortex part of the brain.

An example of *System 1 Thinking* is the decision to swat a fly that has landed on your nose or even the decision to get up from a chair. It is a gut feeling, the way the brain makes hundreds of decisions each day. An example of *System 2 Thinking* is the decision to wear a face mask to protect yourself from COVID-19, or the decision to buy a house. The more complex a decision is, the more likely the brain will engage in *System 2 Thinking*.

The SCARF model, the *System 1 Thinking / System 2 Thinking* model, and other neuroscience models indicate that complex decision-making is a brain process resulting in the selection of a suitable course of action from several alternative possibilities. When making a complex decision, the brain typically breaks the problem down into a hierarchically ordered series of smaller decisions. For example, when deciding how to solve a problem, the brain collects information about the problem, diagnoses the problem, and selects preferred solutions to the problem. Making an informed decision is relatively straightforward when there is little uncertainty. However, when there are significant uncertainties, the decision-making process gets more complicated and the brain requires additional information.

Neuroscience research suggests that decision-making under stress takes place through a hierarchy of steps.[12] The first step is to compute confidence in the outcomes of each decision or option. The next steps are computations designed to gain more and more confidence. In other words, the brain creates an information hierarchy and navigates through that hierarchy. In high-stress situations, such as when confronting the threats associated with a major change, the human brain is most receptive when information is presented in multiple tiers.[13] Messages presented in multiple tiers of increasing complexity allow the brain to slow down and help the brain remember information better.

The clear, concise messages contained in a message map help break through a brain phenomenon called *mental noise*. Mental noise, cognitive overload, and mental stress caused by real or perceived risks and threats can reduce a person's ability to process information by more than 80 percent.[14] *Mental noise* is "the constant chatter in the mind, comprising illogical and non-sequential thoughts, anxiety, fear and several other variables. It is the constant inner dialogue that goes on almost invariably at times, and plays with our inner focus, causing disturbances to our thought patterns and resulting in distraction of our much needed attention."[15]

In the traditional low-stress model of communication, information moves more readily to the receiver without encountering enough noise to inhibit comprehension. In high-stress situations, mental noise interferes with each stage of the communication process and creates a new process at each stage. According to neuroscience and mental noise research, communicators are most effective in high-stress situations when they avoid mental noise and cognitive overload by (1) limiting the number and length of messages that need to be processed, (2) limiting the length of the communication, (3) repeating the message several times, and (4) engaging in active listening and encouraging feedback.

Table 9.9 compares the processing of information in low-stress versus high-stress situations.

9.4.3 The Development of Message Mapping

Message mapping was first adopted by government agencies as a strategic risk and crisis communication tool in the aftermath of 9/11 and the anthrax attacks in the fall of 2001. Early in 2002, for example, the Centers for Disease Control and Prevention conducted intensive message mapping workshops focused on the communication challenges posed by a potential smallpox attack. These workshops produced dozens of message maps with useful content about smallpox and a bioterrorist attack. The CDC smallpox effort was followed by message mapping workshops focused on virtually every kind of bioterrorism weapon of mass destruction, including anthrax, botulism, plague, ricin, tularemia, sarin, and viral hemorrhagic fever.

Since 2002, US agencies at the federal, regional, state, county, and local levels have conducted dozens of message mapping workshops addressing a wide variety of risks. For example, in 2009, the CDC produced message maps for pandemic influenza.[16] The CDC noted the following:

> Message maps are risk communication tools used to help organize complex information and make it easier to express current knowledge. The development process distills information into easily understood messages written at a 6th grade reading level. Messages are presented in three short sentences that convey three key messages in 27 words. The approach is based on surveys showing that lead or front-page media and broadcast stories usually convey only three key messages usually in less than 9 seconds for broadcast media or 27 words for print. [17]

Table 9.9 Information processing in low concern/low risk/low stress situations versus information processing in high concern/high-risk/high-stress situations

The Human Brain Under Low Stress (Holding Constant Other Variables)	The Human Brain Under High Stress (Holding Constant Other Variables)
• can process an average of seven items of information	• can process an average of three items of information.
• will process information in a linear order (e.g., as in one, two, three).	• will process information according to what it perceives to be first and last.
• will process information at its educational and intelligence level.	• will process information well below its educational or intelligence level.
• will determine trust based largely on perceptions of the sender's competence and expertise.	• will determine trust based largely on perceptions of the sender's listening, caring, empathy, and compassion.
• will determine trust based largely on multiple pieces of information communicated over an extended period of time.	• will determine trust in as little as one to thirty seconds.

Message maps produced through these and related efforts became the basis for various risk, high concern, and crisis communication products. These included facts sheets, websites, hotlines, presentations, content for social media, and media interviews.

Several foundational lessons have resulted from the hundreds of message mapping efforts conducted to date.[18] These efforts established the importance of (1) comprehensively identifying and describing key and nontraditional stakeholders early in the communication process, (2) anticipating stakeholder questions and concerns before they are raised, (3) partnering with internal and external stakeholders in developing, reviewing, and distributing message maps, and (4) establishing easy access for leaders, managers, and communicators to vetted central repositories of message maps.

Another lesson learned from early message mapping efforts was that the process of generating message maps can be as important as the end product itself. Message mapping exercises give risk and crisis managers and communicators an opportunity to resolve differences of opinion early in the management and communication process. Management teams included people with subject-matter expertise in communications, policy, law, science, and technology. The process also gave management teams early warnings when the information database was incomplete and when existing policies or procedures were inadequate. Based on the knowledge obtained through the message mapping process, issue-management teams were able to identify and fill gaps in information and policy.

Message mapping is now widely accepted as a method for preparing, ahead of time, responses to questions frequently asked by interested or affected persons (stakeholders) during crises. Many government agencies and private sector organizations have sponsored hundreds of message mapping exercises focused on different types of risks and crises. At the federal level, these agencies have included the US Department of Energy, the US Department of Health and Human Resources, the US Department of Defense, the US National Institutes of Health, the US Food and Drug Administration, the US Department of Agriculture, the US Environmental Protection Agency, and the US Centers for Disease Control and Prevention.[19] As a result of these efforts, thousands of message maps exist for a diverse set of risk and crisis scenarios. These include message maps for biological agents, chemical agents, energy practices, radiological events, terrorism, water contamination, and disease outbreaks such as SARS, pandemic influenza, Ebola, Zika, and COVID-19.[20]

An important challenge in message mapping is updating. For example, the Association for State and Territorial Health Officers had to update their COVID-19 message map document six times during 2020 to keep up with the science. Also, when message map documents sit on the shelf for long periods of time, new staff may not be aware of them.

9.4.4   Case Study: Message Maps and Asbestos

The Agency for Toxic Substances and Disease Registry (ATSDR), part of the Centers for Disease Control and Prevention, developed a series of messages, based on principles of message mapping, for people concerned about exposure to asbestos.[21]

Asbestos is a class of minerals that naturally form long, thin, very strong fibers. Because of its sturdy and heat-resistant properties, asbestos was mined and used in making many products, including insulation, fireproofing and acoustic materials, wallboard, plaster, cement, floor tiles, brake linings, and roofing shingles. Asbestos is still present in older buildings and older materials. It is also used in products such as automobile brakes and roofing materials.

Asbestos is a potentially toxic substance. It can cause several diseases, including asbestosis, pleural disease, lung cancer, and mesothelioma. People exposed to asbestos do not necessarily develop diseases. The risk of disease depends on many factors, including, but not limited to, how much asbestos is in the air, how often exposure occurred, the duration of exposure, and whether the person exposed smoked tobacco.

Asbestos has been mined worldwide, mostly during the 20th century. In the United States, mining asbestos has ended. However, during mine operations, millions of tons of asbestos ore were excavated. The ore-excavation process generated tons of waste rock and ore. Runoff from the waste entered the surrounding surface water and streams. Affected streams often exhibit a milky color, a direct result of suspended asbestos fibers in the runoff.

After the mines were closed, many of these areas became popular places for recreation, such as hiking and camping. Asbestos exposure is typically not a problem if solid asbestos is left alone and not disturbed. However, people who visited the site could have been exposed to asbestos by breathing asbestos fibers released into the air from disturbed rocks or soil. These fibers can be inhaled into the lungs and remain there for a lifetime.

Because asbestos occurs in the environment, both naturally and from the breakdown or disposal of old asbestos products, ATSDR developed the following three key messages:

1) Disturbing asbestos materials can release tiny asbestos fibers, too small to see, into the air.
2) Some people who breathed asbestos fibers over many years have developed asbestos-related diseases.
3) Some of these diseases can be serious or even fatal.

Each key message was expanded with supporting facts and information, including visuals. The ATSDR messaging document also contained key messages and supporting facts noting that being exposed to asbestos does not mean a person will develop health problems. As noted by Covello and Merkhofer,[22] many things need to be considered and understood, including: (1) how long and how frequently the person was exposed to asbestos, (2) how long it had been since the exposures to asbestos had occurred, (3) how much asbestos was in the exposures, (4) whether the person smoked cigarettes (cigarette smoking combined with asbestos exposure greatly increases the chances of getting lung cancer), (5) the size and type of asbestos to which the person was exposed, and (6) other preexisting lung conditions that might exacerbate or accelerate the risk of lung cancer.

9.4.5 Steps in Developing a Message Map

As shown in Table 9.10, there are six steps in message mapping. Each step is described below.

9.4.5.1 Step 1: Identify, Profile, and Prioritize Key Stakeholders
The first step in message mapping is to identify and prioritize key stakeholders for a selected risk, high concern, or crisis issue, topic, or scenario. A stakeholder is a person, group, or organization

Table 9.10 Steps in message mapping

---

1) Identify, profile, and prioritize key stakeholders.
2) Develop lists of stakeholder questions and concerns.
3) Develop key messages.
4) Develop supporting information.
5) Test, practice, and deliver the message maps through the most appropriate communication channel.
6) Repurpose the message maps for other communication channels.

---

that has a vested interest in an issue, is impacted or affected by the issue, cares about how the issue turns out, and/or can influence the views and behaviors of others. For example, for an accident scenario at an industrial facility, key stakeholders would typically include the management and employees at the facility, victims of the accident, families and friends of victims of the accident, people living near the site of the accident, first responders, emergency managers, government officials, regulators, mainstream broadcast and print media, bloggers, subject-matter experts, investors, shareholders, contractors, customers, industry partners, suppliers, educators, and the public at large.

A major crisis can involve tens to tens of thousands of stakeholders. Every risk, high-concern, and crisis issue involves a distinctive set of stakeholders. Each stakeholder may have a distinctive set of characteristics. Knowledge about stakeholder characteristics is typically acquired through brainstorming and through what is known as *formative research*. Formative research identifies the characteristics of stakeholders, including their interests, attitudes, practices, behaviors, and informational needs. It is typically conducted at the beginning of the communication effort and continues throughout. It provides the benefits listed in Table 9.11.

Formative research includes literature reviews, qualitative research, and quantitative research. Qualitative research includes in-depth interviews, focus group discussions, and observational studies (e.g., research on what people actually do instead of what they say they do). Quantitative research includes surveys and other data gathering that allow for statistical analysis.

Based on formative research, potential key stakeholders can be described based on the factors listed in Tables 9.12 and 9.13.

Once key potential stakeholders have been identified, the next step is to determine communication objectives for each key stakeholder group and what obstacles will exist in achieving these objectives. For example, what do stakeholders want from you? What do you want from the stakeholder? How will the stakeholder use the information you provide? How will you use the information provided by the stakeholder? Who else is sharing information about the issue with the stakeholder?

An effective strategy for determining objectives is through the popular acronym SMART objectives: Specific, Measurable, Achievable/Assignable, Realistic/Reasonable, and Time Bound. An example of a time-bound objective would be the percentage of the target audience that by a set time has heard a crisis communication message or has taken a recommended protective action. Message mapping exercises that begin with SMART objectives are more likely to be successful because there is a clear understanding of what they are trying to achieve and how they will do it. SMART objectives can be set for each stakeholder. Potential objectives are listed in Table 9.14. Objectives such as those listed in the table are influenced by a number of factors, including resource

Table 9.11 Benefits of formative research.

Formative research:
1) helps ensure communication efforts are culturally appropriate;
2) begins the stakeholder engagement process and promotes constructive dialogue and partnerships;
3) provides vital information about what stakeholders know and want to know and what they don't know and don't want to know;
4) helps communicators decide who are the most important target audiences;
5) helps the communicator understand the factors that influence the stakeholder's behavior and determine the best ways to reach the stakeholder.

Table 9.12 Stakeholder characteristics.

1) *Sociodemographic Characteristics*: What are the gender, race, ethnicity, average income, age, education, occupation, family status, residential status, and organizational affiliations of members of the stakeholder group?

2) *Size*: How many members are in the stakeholder group? How many members of the stakeholder group normally attend functions?

3) *Preferred channels for communications*: What communication channels do they use and prefer for accessing information about the risk, threat, or issues, such as mainstream broadcast media, mainstream print media, the internet, social media, social networks, face-to-face meetings, public meetings, or open houses?

4) *Influence*: How able is the stakeholder to influence, change, or put pressure on the views or behaviors of others regarding the risk, threat, or issue?

5) *Interest level*: How concerned, worried, upset, anxious, fearful, or outraged are they about the risk, threat, or issue? How eager are they to learn more about the risk, threat, or issue? How much attention and time are they currently giving to, or would be willing to give to, issues surrounding the risk, threat, or issue?

6) *Experiences*: What is their prior experience with the risk, threat, or issue?

7) *Intentions*: What actions are they intending to take regarding the risk, threat, or issue?

8) *Trust*: How much do they trust different sources of information about the risk, threat, or issue?

9) *Competition*: What other risks, threats, or issues are vying or competing for their attention? What other organizations do they compete with for support or funding?

10) *Literacy and numeracy skills*: What are the literacy and numeracy skills of the average member of the stakeholder group? How diverse are the literacy and numeracy skills of members of the stakeholder group? What is the average level of education of members of the stakeholder group? What is the range of levels of education completed?

11) *Attitudes*: What are their opinions, views, vested interests, ways of thinking, feelings, and emotions? How much variability is there in attitudes, opinions, and feelings among members of the stakeholder group?

12) *Knowledge*: How knowledgeable are they about the risk, threat, or issue? How knowledgeable are they about where to get information about the risk, threat, or issue?

13) *Cultural biases*: What are the dominant worldviews of members of the stakeholder group?

14) *Leadership*: Who are the leaders of the stakeholder group? Who has led the stakeholder group in the past? How much influence do past and present leaders have on members of the stakeholder group? How much clearance and approval is needed for leaders to take action?

15) *Resources*: What internal and external resources, such as money, space, and personnel, are available to the stakeholder group?

16) *Partnerships*: Who are the allies of the stakeholder group? To what extent can the stakeholder group enlist the support of allies?

Table 9.13 Stakeholder types.

1) *Allies/Partners/Supporters/Friends*. Stakeholders who are allies, partners, supporters, and friends fall along a spectrum of support levels, including those who share, are aligned with, and/or support your goals and objectives; will trust the information you provide; will want you to succeed in a way they can support; will treat you as a co–team member; will provide input; and will provide resources, help, information, and expertise.

2) *Adversaries/Opponents/Enemies*. Stakeholders who are adversaries, opponents, and enemies fall along a spectrum of the opposition, including those who have different goals and objectives than you; will use different methods and means than you to achieve their goals and objectives; will distrust the information you provide; will undermine your efforts; will not trust you; will have significant and pointed questions and concerns; and will deny you help, resources, information, and expertise.

3) *Neutrals/Undecided*. Stakeholders who are neutrals and undecided fall along a spectrum of interest including those who are apathetic, wary, noncommitted, unaware, and cautious.

Table 9.14 Potential risk, high-concern, and crisis communication objectives.

---

1) **Informed Decision-Making**: to improve understanding among stakeholders.
2) **Right-to-Know**: to be transparent and disclose information about a risk or threat.
3) **Perceptions and Judgments**: to influence or change perceptions about levels of acceptable or tolerable risk or threat.
4) **Trust Building**: to build, maintain, or rebuild trust in a specific source of information or in the management process.
5) **Risk Reduction or Elimination**: to encourage adoption of specific attitudes, intentions, behaviors, or guidelines aimed at reducing or eliminating a risk or threat.
6) **Participation**: to encourage those interested or affected to become more involved, engage in constructive dialogue, and exchange information.

---

issues, time issues, legal issues, organizational requirements, the risk itself, and stakeholder requirements.[23]

Given these different objectives, it is important to have one single, overriding communication objective, sometimes called a SOCO (Single Overriding Communication Objective), specified before proceeding to the next step. At the most general, the overall goal of risk, high concern, and crisis communication should not be to defuse public concerns. It should be to produce informed stakeholders that are involved, interested, reasonable, thoughtful, solution-oriented, and collaborative.

A key consideration of risk communication is that the target will rarely be a single audience, but will usually entail a variety of audiences, and messages must be tailored to consider the different audiences that are likely to have different interests, values, and levels of intelligence, education, and understanding. Prioritizing which groups are most important to address is typically based on several criteria. These include (1) level of interest the stakeholder has in the issue, (2) power of the stakeholder to influence actions and decisions, (3) power of the stakeholder to block or advance operational and communication objectives, and (4) legal or moral requirements to consult with the stakeholder. A grid can be formed based on stakeholder interest and power (e.g., high power, high interest; low power, low interest).

9.4.5.2 Step 2: Develop Lists of Stakeholder Questions and Concerns

The second step in message mapping is to develop lists of potential questions and concerns for each key stakeholder group. The lists should be constructed as comprehensively as possible based on available resources. The lists should include what stakeholders want or need to know to make informed decisions. The appendices to this chapter contain several examples of such lists.

Questions and concerns in the lists can be grouped into categories based on various criteria. For example, questions and concerns can be grouped based on (1) who has raised the question or concern, (2) who will or should address the question or concern, (3) levels of stakeholder concern, such as whether the issue is of high concern, medium concern, or low concern to a particular stakeholder, (4) when these questions or concerns are likely to be raised, and (5) broad categories of concern such as trust concerns, health concerns, safety concerns, environmental concerns, quality of life concerns, political concerns, economic concerns, ethical and moral concerns, social concerns, vulnerable population concerns, historical concerns, and cultural concerns. Stakeholder questions and concerns may fall into tens or hundreds of categories and subcategories.

Lists of specific stakeholder questions and concerns can be generated through a variety of research methods. These include reviews of media stories (e.g., mainstream broadcast and published stories and social media blogs), results from searches using search engines such as Google, on websites, public meeting records, legislative transcripts, complaint logs, hotline logs, toll-free number logs, media logs, and social media content (for example, Facebook, Twitter, Instagram, and YouTube postings). When stakeholder questions and concerns are accumulated through a thorough and systematic process, a realistic goal is to anticipate identifying at least 95 percent of the questions that will actually be asked.[24]

Lists of stakeholder questions and concerns can be expanded and validated through (1) interviews with subject-matter experts, (2) workshops or discussion sessions with stakeholders and special interest groups, (3) interviews with individuals with experiences related to the risk, high concern, or crisis scenario, (4) consultations and conversations with individuals or organizations who represent, or who are members of, the target audience, and (5) consultations and conversations with partners and communication experts who have successfully developed communication products for the particular target audience.

Surveys, focus groups, and workshops are useful for uncovering questions and concerns. For example, workshops sponsored by the US Environmental Protection Agency provided valuable insights into the questions likely to be asked by stakeholders in a water contamination crisis. The working groups uncovered more than 200 questions for six water contamination scenarios.[25]

Lists of questions and concerns can number in the hundreds. For example, in a study conducted by the US Nuclear Regulatory Commission focused on the nuclear power plant accidents at Three Mile Island in the US, Chernobyl in Russia, and Fukushima in Japan, Covello (2011) identified more than 400 questions.[26]

9.4.5.3 Case Study: Stakeholder Questions, Terrorism, and Disasters

On 26 February 1993, terrorists exploded a bomb in the parking garage of the World Trade Center in New York City. If the bomb had been placed in the lobby of the building, it would have killed hundreds. The purpose of setting off the bomb in the parking garage was to bring down the building itself.

The 1993 World Trade Center terrorist attack was the first indication that international terrorism posed a major threat to the United States. Intelligence gathered by the FBI, the New York City Police Department, and other security forces indicated that terrorists had targeted New York City, had identified major vulnerable structures in the city and knew how to build the type of bomb that could bring down a major structure. One of the first risk and crisis communication documents produced in response to the 1993 bombing was titled *The Most Frequently Asked Questions by Journalists and the Public Following a Major Crisis, Emergency, or Disaster*. An updated version of this document can be found in Appendix 9.1 of this chapter. The document was used by emergency response organizations in New York City to anticipate questions that might arise as a result of another terrorist attack, develop draft answers, and practice delivery of these answers.

9.4.5.4 Step 3: Develop Key Messages

> *What are the proper proportions of a maxim? A minimum of sound to a maximum of sense.*
> —Mark Twain.

The third step in message mapping is to develop key or core messages for all the anticipated questions and concerns or for a prioritized subset of the questions and concerns of key stakeholders. The ideal number of key messages is three from the classical Latin phrase "*omne trium perfectum*" ("everything that comes in threes is perfect"). If well-constructed, three messages have the benefit of combining both brevity and rhythm with a small amount of information, creating an easily identifiable pattern. Key messages structured in threes create a triangle, one of the strongest structures in the physical universe. It functions as a table with three legs – structured with the minimum number of legs to make it stable.

Two generalized examples follow:

> Example 1:
> Key/Core Message 1: Many people have had difficulty with "X."
> Key/Core Message 2: Those who had difficulty with "X" can now do "Y."
> Key/Core Message 3: "Y" will provide people with "Z."
>
> Example 2:
> Key/Core Message 1: Many people want to know how to reduce the risk of "X."
> Key/Core Message 2: The best way to reduce the risk of "X" is "Y."
> Key/Core Message 3: More information about "Y" can be obtained from "Z."

Key messages are intended to address the information needs of the target audience. They are what leaders and communicators want people to take away from the exchange. They are what leaders and communicators want the target audience to think, feel, or do after they receive the key messages. Key messages should be based on what the target audience most needs to know or most wants to know. Key messages function individually or collectively to provide facts, updates, and recommended actions.

Key messages are developed most effectively through brainstorming sessions with a message mapping team. The message mapping team typically consists of subject-matter experts, risk communication experts, policy experts, legal experts, management experts, and a facilitator. The brainstorming sessions produce message scripts, usually as complete sentences or keywords. These sentences or keywords are then entered as key messages onto the message map template.

Tier One of a message map – the top line in the box format or the boldfaced key messages in the bullet format – is sometimes referred to as "the elevator speech." It is the amount of information that can be shared with another person about an important issue in a quick elevator ride. It is the first few slides in a slide presentation. It is the "Tell me what you want me to know" part of the Triple T communication model (Tell me what is important to know, Tell me more, Tell me again).

The second tier of a message map adds detail and supporting information to the elevator speech. Tier Two information functions as a transition from the general to more specific. The level of technical detail and use of technical language in Tier 2 may change depending on the audience.

Tier One messages – the key or top-line messages in a message map – are particularly useful as tweets for Twitter or as sound bites for media outlets. A *sound bite* is a short phrase or statement extracted from a longer statement that captures the essence of the longer statement.

Sound bites are frequently used in news broadcasts. They have become increasingly short. For example, an "analysis of all weekday evening network newscasts (over 280) from Labor Day to Election Day in 1968 and 1988 revealed that the average 'sound bite' fell from 42.3 seconds in 1968 to only 9.8 seconds in 1988. Meanwhile, the time the networks devoted to visuals of the candidates, unaccompanied by their words, increased by over 300 percent."[27] Research by the Center for Risk Communication indicates that the average sound bite on evening network newscasts in the year 2020 was nine seconds.

Sound bites are an essential element in effective media communication. When presented as short, memorable, and quotable messages – e.g., less than nine seconds – they will often be repeated through traditional and social media channels. Sound bites help ensure key messages are carried in news stories. Pre-scripted sound bites also reduce the risk of inaccurate paraphrasing by journalists.

One method for assuring key messages are delivered and heard is bridging. Bridging is a communication technique aimed at focusing and directing the receiver to key messages. Examples of bridging statements include "And what this all means is . . ." and "And what's most important to remember is . . ." Forty examples of bridging statements are provided in Chapter 12, in Table 12.7.

One of the most important rules for constructing key messages is to ensure they are clear and understandable to the target audience. For the public at large in the United States, clear and understandable messages typically fall all within the sixth- to ninth-grade reading level. Several organizations have developed guidelines for preparing such messages. For example, the Centers for Disease Control and Prevention developed a "Clear Communication Index." An index is a research-based tool designed to help communicators develop and assess public communication materials. The index has four introductory questions: Who is your primary audience? What do you know about the audience? What is your primary communication objective? What is the main message statement in the material? The index then evaluates the clarity of the messages based on twenty scored items drawn from the communications science literature. The index draws on various "readability tools," such as the "readability" utility found in Microsoft Word. It also draws on several thesauri and glossaries containing plain English words or phrases for technical terms.

Guidelines for developing clear and simplified messages can be found in Tables 9.15, 9.16, and 9.17.

Table 9.15 Best practices for creating clear messages.

1) **Simplify.** The higher the level of stress, fear, or anxiety, the greater the need to simplify the language and carefully structure messages from simple to more complex.

2) **Test for Readability/Comprehension**. Use the *Readability* utility included with most word-processing software to measure the readability level of the information.

3) **Use Clear Language**. *Readability* formulas are only one tool used to achieve clarity. Readability formulas are mechanical in nature and look at readability and comprehension characteristics such as sentence length, vocabulary, and position of nouns and verbs. However, they do not address the many communication characteristics that contribute to clarity. Clear language means:

   a) Use simple and correct grammar.

   b) Use short, simple words and sentences.

   c) Be careful when providing numbers expressed with decimal points as they can easily be misinterpreted or misunderstood.

   d) Avoid the use of jargon, acronyms, and abbreviations.

   e) Define new or unfamiliar terms so the target audience can understand them.

   f) Use short sentences to define new terms.

   g) Provide a glossary.

   h) Introduce the concept before introducing a new term or explain the new term soon after using it.

   i) If possible, ask the audience to identify terms that are not understood.

   j) Check frequently for understanding and misunderstanding.

   k) Use new or highly technical terms only if they are important for the target audience to know and remember.

   l) Avoid new or technical terms that have a different meaning from their common usage.

Table 9.16 The structure of a clear message.

1) Contains no more than three to five message points.
2) States the most important messages first and last.
3) Focuses on what the target audience wants to know.
4) Is recalled easily.
5) Provides advance warning when complex or difficult material will be presented.
6) Breaks down complex topics into smaller parts.
7) Uses the "Triple T Model" for presenting complex information: **t**ell your audience briefly what you are going to tell them; **t**ell them more about each point; **t**ell them again briefly what you told them.
8) Is tested in advance for content that could be misunderstood.
9) Encourages those interested or affected to provide feedback.
10) Uses the active voice for writing and speaking.
11) Provides complex information in tiers or layers of information that increase gradually in complexity.

9.4.5.5 Step 4: Develop Supporting Information

The fourth step in message map construction is to develop supporting information for the key messages. Once the attention and interest of the audience have been gained by the three key/core messages offered in the message map, the key/core messages can be expanded and amplified with additional information. Supporting information opens up the key messages like a fan. Supporting points match the interest of the stakeholder and fit the length and time constraints of the chosen communication format.

Supporting information includes facts, evidence, proofs, or explanations that elaborate or expand on each of the key messages. For example, if a water advisory has been issued with the key message "Don't drink the tap water," supporting information could include three but no more than five of the following messages: (1) Boil tap water before use, (2) Do not use water that has not been boiled for preparing food, cleaning your teeth, or washing wounds, (3) Remember your pets – they should not drink tap water that has not been boiled, (4) You can still use tap water for washing and bathing without having to boil it, and (5) You can still use tap water for general household purposes and toilet flushing.

Table 9.17 Methods for enhancing message clarity.

1) Tell stories that illustrate the point.
2) Use metaphors and analogies.
3) Use visuals (for example, graphics, drawings, maps, charts, flowcharts, infographics, paintings, photographs, video, and highlighted text) to support the message.
4) Use visual cues, such as boldface, color, shapes, lines and arrows, font size, spacing, alignment, and headings.
5) Use simple graphics.
6) Use colors to enhance meaning, but do not depend on colors to convey a message.
7) Beware of colors that are difficult to distinguish from surrounding colors.
8) Determine if the message content is consistent with culturally accepted ways of presenting or accessing information.
9) Respect and allow for the diverse nature of the target audience (for example, enlarge the type face and font size for audiences who are elderly or sight-impaired).

If one of the key messages in a message map about a disease outbreak is "Wash your hands," supporting information could include (1) Washing your hands is the number one preventative way to control the spread of disease-causing viruses and bacteria, (2) Wash your hands using warm water and soap, (3) Wash your hands for at least 20–30 seconds or as long as it takes to sing "Happy Birthday," (4) Use an alcohol-based hand sanitizer when there is no water available to wash your hands, and (5) Wash your hands more frequently after contact with a person who is sick.

Supporting information can also take the form of visuals, graphics, analogies, personal stories, or citations to trusted sources of information. The same principles that guide the construction of key messages also guide the development of supporting information.

Supporting information is critical to effective message mapping. It adds depth, which is often missed when information is presented only through sound bites. A core characteristic of a message map is the layering of information. Information is presented in tiers, from simple to more complex. This layering strategy addresses, at least in part, the problem that people who are stressed, worried, and fearful often have difficulty processing information. Layering information gives control to the individual. It allows the person to access information at whatever level they feel able to absorb.

The layered structure of a message map also responds to different information needs. For some people, simple, short messages are sufficient. Others want more facts and detail. Layered information and additional detail also help build trust. Audiences for message maps are constantly assessing trust. Supporting information can be documented and referenced in an electronically archived message map. Every item of additional information provides additional evidence to believe what is being communicated. Trust arises from delivering messages and arguments backed by evidence, not just by further explanation.

The layered structure of a message map can be challenging. Risk and crisis information is often complex. The complexity of such information is exacerbated because scientists are trained and positively reinforced for providing precise, thorough, and appropriately nuanced information. As a result, the initial simple and clear messages of a message map are often equated with "dumbing down" information. While jargon and technical language increase the efficiency and effectiveness of communications with technically trained peers, jargon and technical language decrease the effectiveness of communications with audiences who do not share the same technical training and background.

9.4.5.6   Step 5: Testing the Message Map

The fifth step in message mapping is conducting a systematic evaluation of the message map using standardized and accepted testing and evaluation procedures (see Chapter 11, "Evaluation"). Message testing and evaluation should begin by asking subject-matter experts who are not directly involved in the original message mapping process to validate the accuracy of the information contained in the maps. Message testing should then be done with individuals or groups who are, or who can serve as surrogates for, key internal and external target audiences. Among the most common methods for testing and evaluating message maps are interviews and focus groups. A focus group is a group of people assembled to participate in a facilitated discussion about a particular issue. A focus group can demonstrate whether the content of the layers in message maps is comprehensive, convincing, and effective. Chapter 11, "Evaluation," contains a detailed case study on mosquito control where message maps were tested using focus groups.

Table 9.18 Template for a PowerPoint presentation or letter based on message mapping.

(Note: Each slide in the PowerPoint deck should be viewed by the audience for two minutes or less.)

Slide 1/Paragraph 1: Background: includes messages on why the issue is important.

Slide 2/Paragraph 2: Key Messages: contains the three key messages from the message map.

Slides 3–6/Paragraphs 3–6: Supporting Information: Provides the supporting information for each of the three key messages drawn from the message map. May include statistics, citations to research, anecdotes, visuals, and/or validation of the key messages by the third-party sources.

Slide 6/Paragraph 7: **Key Messages Repeated**: Provides the same key messages as in Slide 2/Paragraph 2.

Slide 7/Paragraph 7: Next Steps: Provides three or fewer messages addressing the following questions: What's next? Where do we go from here?

Extra Credit: Strategically placed visuals, such as graphics, photographs, charts, analogies, metaphors, and stories, can be used to support and reinforce the key messages or supporting information.[1]

---

[1]See, e.g., Tufte, E. (1983). *The Visual Display of Quantitative Information.* Cheshire, CT: Graphics Press; Tufte, E. (1990) *Envisioning Information.* Cheshire, CT: Graphics Press; Tufte, E. (1996). *Visual Explanations.* Cheshire, CT: Graphics Press; Roam, D. (2009). *The Back of the Napkin: Solving Problems and Selling Ideas with Pictures.* London, UK: Penguin Books.

A critical step in message map testing is to share the message maps with partners and have them vetted. Sharing and vetting of message maps enhance message consistency and coordination across communication partners.

9.4.5.7 Step 6: Repurpose Maps through Appropriate Information Channels

The sixth and final step in message mapping is to repurpose the message maps into the communication format or channel that will most effectively reach and engage the target audience. (Communication formats and channels are discussed in Chapter 4, "Stakeholder Engagement.") Formats include fact sheets, brochures, websites, blogs, telephone hotline scripts, posters for an open house or information forum infographics, podcasts, videos, briefings, PowerPoint presentations, letters or emails sent in response to a complaint or criticism, and TED- or TEDx-style talks.[28] For example, Table 9.18 provides a template for a slide deck for a message map repurposed as a presentation or for a letter sent in response to a complaint.

9.5 Summary

Table 9.19 contains principles for communication in high-stress environments. The principles listed in the table are reminders that requirements for successful high-stress communication are often radically different than the rules for low-stress communication. These principles are critical not only for risk, high concern, and crisis communicators but also for leaders and for risk and crisis managers. Awareness of the psychological, behavioral, and neurological impact of stress on information processing will shape message development and communication strategy.

In conclusion, message development for risk, high concern, and crisis situations is a complex task requiring extensive planning, preparation, and practice. Practice alone does not make perfect; deliberate and evaluated practice makes perfect. Messages need to be conveyed with caring, conviction, and optimism. Until you know what people will think or behave when they are stressed, you do not know how to communicate with that person or group.

Table 9.19 Messaging principles for risk, high-concern, and crisis situations.

Messages should be structured recognizing that the brain processes information differently in high-concern, high-risk, and high-stress environments. When under stress, people will typically:

1) have difficulty hearing, understanding, and remembering information.
2) want to know that you care before they care what you know.
3) focus more on negative information than positive information.
4) determine trust based to a large extent on perceptions of listening, caring, and empathy.
5) resist changing their initial perceptions of trust.
6) determine trust in as little as one to thirty seconds (once determined, perceptions and judgments are highly resistant to change).
7) determine trust by as much as 75 percent through information communicated nonverbally.
8) have most trust for messages endorsed by trustworthy independent third-party sources of information.
9) respond best to messages that contain less than 27 words, nine seconds, and three points.
10) are quick to assign a negative meaning to nonverbal messages and not give the benefit of the doubt.
11) remember best messages presented through visuals.
12) remember most whatever they hear or see first and last.
13) process information only well below their educational and intelligence level.
14) actively look for visual information to support verbal messages—the visual part of the brain becomes a highly active player in processing high-stress information.
15) require at least three to four positive or constructive messages to offset bad news.
16) challenge the truthfulness of messages containing unmodified absolutes (e.g., "all," "always," "every," "nothing," and "none").
17) more vividly remember messages with strong negative imagery.
18) respond negatively to sentences beginning with the expression "as you know" or the equivalent (e.g., "Obviously;" "Clearly"). People often find such expressions insulting and condescending.
19) respond best to messages that address important perception, outrage, and fear factors such as benefits, control, voluntariness, dread, fairness, reversibility, morality, origin, and familiarity.
20) focus more on emotions and feelings than on technical or scientific facts.

9.6 Chapter Resources

Below are additional resources to expand on the content presented in this chapter.

Angus, K., Cairns, G., Purves, R., Bryce, S., MacDonald, L., & Gordon, R. (2013). *Systematic Literature Review to Examine the Evidence for the Effectiveness of Interventions That Use Theories and Models of Behavior Change: Towards the Prevention and Control of Communicable Diseases*. Stockholm: European Centre for Disease Prevention and Control.

Baird, B. (1986). "Tolerance for environmental health risks: The influence of knowledge, benefits, voluntariness, and environmental attitudes." *Risk Analysis* 6(4):425–435.

Baron, J., Hershey, J.C., Kunreuther H. (2000). "Determinants of priority for risk reduction: the role of worry." *Risk Analysis* 20(4):413–428.

Becker, S.M. (2004). "Emergency communication and information issues in terrorist events involving radioactive materials." *Biosecurity and Bioterrorism* 2(3):195–207.

Becker, S.M. (2011). "Risk communication and radiological/nuclear terrorism: a strategic view." *Health Physics* 101(5):551–558.

Benoit, W.L. (1995). *Accounts, Excuses, and Apologies: A Theory of Image Restoration*. Albany, NY: State University of New York Press.

Benoit, W. L. (2005). "Image restoration theory," in *Encyclopedia of Public Relations*, Vol 1, ed. R. L. Heath. Thousand Oaks, CA: Sage.

Benoit, W.L., Pang, A. (2008). "Crisis communication and image repair discourse," in *Public Relations: From Theory to Practice*, eds. T. L. Hansen-Horn and B. D. Neff. New York: Pearson.

Bennett, P., Caiman, K. (eds.) (1999). *Risk Communication and Public Health*. New York: Oxford University Press.

Berland G., Elliott, M., Morales, L., Algazy, J., Kravitz, R. (2001). "Health information on the Internet: Accessibility, quality, and readability in English and Spanish." *Journal of the American Medical Association* 285:2612–2621.

Bier, V M. (2001). "On the state of the art: risk communication to the public." *Reliability Engineering & System Safety* 71(2):139–150.

California Department of Public Health (2015). *Crisis and Emergency Risk Communication Toolkit*. Sacramento, CA: California Depart of Public Health. Accessed at: http://cdphready.org/wp-content/uploads/2015/03/crisis_and_emergency_risk_communication_toolkit_july_2011.pdf

Centers for Disease Control and Prevention (2009). Pre-Event Message Maps. Accessed at: https://www.calhospitalprepare.org/sites/main/files/file-attachments/pandemic_pre_event_maps.pdf

Campbell, Joseph. (2008). *The Hero with A Thousand Faces*, 2nd ed. Novato, CA: New World Library.

Centers for Disease Control and Prevention (CDC) (2019). *Crisis and Emergency Risk Communication*. Atlanta, GA: Centers for Disease Control and Prevention.

Chess, C., Hance, B.J., Sandman, P.M. (1989). *Planning Dialogue with Communities: A Risk Communication Workbook*. New Brunswick: Rutgers University, Cook College, Environmental Media Communication Research Program.

Coombs, W.T. (1995). "Choosing the right words: The development of guidelines for the selection of the 'appropriate' crisis-response strategies." *Management Communication Quarterly* 8(4):447–476.

Coombs, W.T. (2004). "A theoretical frame for post-crisis communication: Situational crisis communication theory," in *Attribution Theory in the Organizational Sciences: Theoretical and Empirical Contributions*, ed. M.J. Martinko. Greenwich, CT: Information Age Publishing.

Coombs, W.T., Holladay, S.J. (2005). "Exploratory study of stakeholder emotions: Affect and crisis," in *Research on Emotion in Organizations, Volume 1: The Effect of Affect in Organizational Settings*, eds. N.M. Ashkanasy, W.J. Zerbe, C. E.J. Hartel. New York: Elsevier.

Coombs, W.T. (2007). "Protecting organization reputations during a crisis: The development and application of situational crisis communication theory." *Corporate Reputation Review* (3):163–176.

Coombs, W.T. (2009a). "Conceptualizing crisis communication," in *Handbook of Crisis and Risk Communication*, eds. R. L. Heath and H. D. O'Hair. New York: Routledge.

Coombs, W.T. (2009b). "Crisis, crisis communication, reputation, and rhetoric," in *Rhetorical and Critical Approaches to Public Relations*, eds. R.L. Heath, E.L. Toth, and D. Waymer. New York: Routledge.

Coombs, W.T., Holladay, S.J., eds. (2012). *Handbook of Crisis Communication*. Hoboken, NJ: Wiley.

Coombs, W.T. (2014). Crisis Management and Communications—Updated. Institute for Public Relations. https://instituteforpr.org/crisis-management-communications/

Coombs, W.T. (2019). *Ongoing Crisis Communication: Planning, Managing, and Reporting*. Thousand Oaks, CA: Sage Publications.

Coombs, W.T., Holladay, S.J. (2005). "Exploratory study of stakeholder emotions: Affect and crisis," in *Research on Emotion in Organizations, Volume 1: The Effect of Affect in Organizational Settings*, eds. N.M. Ashkanasy, W.J. Zerbe, and C.E.J. Hartel. New York: Elsevier.

Covello, V.T. (1983). "The perception of technological risks: A literature review." *Technological Forecasting and Social Change* 23:285—297.

Covello. V.T. (1984). "Uses of social and behavioral research on risk." *Environment International* 4:17–26.

Covello, V.T. (1991). "Risk comparisons and risk communication: Issues and problems in comparing health and environmental risks," in *Communicating Risks to the Public*, eds R.E. Kasperson and P.J.M. Stallen. Dordrecht, Netherlands: Kluwer Academic Publishers.

Covello V.T. (1992). "Risk communication, trust, and credibility." *Health and Environmental Digest* 6(1):1–4.

Covello V.T. (1993). "Risk communication and occupational medicine." *Journal of Occupational Medicine* 35:18–19.

Covello, V.T. (1998). *Risk Perception, Risk Communication, and EMF Exposure: Tools and Techniques for Communicating Risk Information*. Vienna, Austria: World Health Organization/ICNRP International.

Covello, V.T. (2003). "Best practices in public health risk and crisis communication." *Journal of Health Communication* 8:1–5.

Covello, V.T. (2006). "Risk communication and message mapping: A new tool for communicating effectively in public health emergencies and disasters." *Journal of Emergency Management* 4(3):25–40.

Covello, V.T. (2003). "Best practices in public health risk and crisis communication." *Journal of Health Communication* 8:1–5.

Covello, V.T. (2008). "Strategies for overcoming challenges to effective risk communication," in *Handbook of Risk and Crisis Communication*, eds Heath, R. and O'Hair H. New York: Routledge.

Covello, V.T. (2011a). *Guidance on Developing Effective Radiological Risk Communication Messages: Effective Message Mapping and Risk Communication with the Public in Nuclear Plant Emergency Planning Zones*. NUREG/CR-7033. Washington, DC: Nuclear Regulatory Commission.

Covello, V.T. (2011b). *Developing an Emergency Risk Communication (ERC)/Joint Information Center (JIC) Plan for a Radiological Emergency*. NUREG/CR-7032. Washington, DC: Nuclear Regulatory Commission

Covello, V.T. (2011c). "Risk communication, radiation, and radiological emergencies: strategies, tools, and techniques." *Health Physics* 101(5):511–530.

Covello, V. T. (2014). "Risk communication," in *Wiley-Blackwell Encyclopedia of Health, Illness, Behavior, and Society*, eds. W. Cockerham, R. Dingwall, and S. Quah. Oxford, UK: Blackwell.

Covello, V.T (2016). "Risk communication," in *Environmental Health: From Global to Local*, eds. Frumkin, H. 3rd Edition. San Francisco: Jossey-Bass/Wiley.

Covello, V.T, Allen, F. (1988). *Seven Cardinal Rules of Risk Communication*. Washington, DC: US Environmental Protection Agency.

Covello, V., Becker, S., Palenchar, M., Renn, O., Sellke, P. (2010). *Effective Risk Communications for the Counter Improvised Explosive Devices Threat: Communication Guidance for Local Leaders Responding to the Threat Posed by IEDs and Terrorism*. Washington, DC: Department of Homeland Security.

Covello V.T., Hyer, R. (2014). *Top Questions on Ebola: Simple Answers: Risk Communication Guide*. Arlington, VA: Association of State and Territorial Health Officials.

Covello V., Hyer R. (2020). *COVID-19: Simple Answers to Top Questions Risk Communication Guide. Association of State and Territorial Health Officials, 2020*. Arlington, Virginia. Accessed at: https://www.astho.org/COVID-19/Q-and-A/

Covello, V.T., McCallum, D.B., Pavlova, M.T. eds. (1989). *Effective Risk Communication: The Role and Responsibility of Government and Nongovernment Organizations*. New York: Plenum.

Covello, V.T., Minamyer, S., Clayton, K. (2007). *Effective Risk and Crisis Communication during Water Security Emergencies: Summary Report of EPA Sponsored Message Mapping Workshops*. Cincinnati, OH: US Environmental Protection Agency.

Covello, V.T., Mumpower, J. (1985). "Risk assessment and risk management: An historical perspective." *Risk Analysis* 5(2):103–120.

Covello, V.T., Peters, R.G., Wojtecki, J.G., Hyde, R. (2001). "Risk communication, the West Nile virus epidemic, and bioterrorism: Responding to the communication challenges posed by the intentional or unintentional release of a pathogen in an urban setting." *Journal of Urban Health* 78:382–391.

Covello, V.T., Sandman, P.M. (2001). "Risk communication: Evolution and revolution," in *Solutions to an Environment in Peril*, ed. A. Wolbarst. Baltimore: Johns Hopkins University Press.

Covello, V.T., von Winterfeldt, D., Slovic P. (1987). "Communicating scientific information about health and environmental risks: Problems and opportunities from a social and behavioral perspective," in *Uncertainties in Risk Assessment and Risk Management*, eds. V.T. Covello, A. Moghissi, V. Uppulori. New York: Plenum Press.

Crick, M.J. (2021. "The importance of trustworthy sources of scientific information in risk communication with the public." *Journal of Radiation Research* 62(S1):i1–i6.

Cummings, L. (2014). "The 'trust' heuristic: Arguments from authority in public health." *Health Communication* 29(10):1043.

Davies, C.J., Covello, V.T., Allen, F.W., eds. (1987). *Risk Communication: Proceedings of the National Conference*. Washington: The Conservation Foundation.

Dean, D.H. (2004). "Consumer reaction to negative publicity: Effects of corporate reputation, response, and responsibility for a crisis event." *Journal of Business Communication* 41:192–211.

Dohle, S., Wingen, T., Schreiber, M. (2020). "Acceptance and adoption of protective measures during the COVID-19 pandemic: The role of trust in politics and trust in science." *Social Psychological Bulletin* 15(4):1–23.

Dominic Balog-Way, D., McComas, K., Besley, J. (2020). "The evolving field of risk communication." *Risk Analysis*, 10.1111/risa.13615, 40, S1, (2240-2262), (2020).

Durodié, B. (2006). *The Concept of Risk*. London: The Nuffield Trust and the UK Global Health Programme.

Fang, D., Fang, C.L., Tsai, B.K., Lan, L.C., Hsu, W.S. (2012). "Relationships among trust in messages, risk perception, and risk reduction preferences based upon avian influenza in Taiwan." *International Journal of Environmental Research and Public Health* 9(8):2742–2757.

Fazio, L.K., Brashier, N.M., Payne, B.K., Marsh, E.J. (2015). "Knowledge Not Protection Against Illusory Truth." *Journal of Experimental Psychology* 144(5):993–1002.

Fearn-Banks, K. (2016). *Crisis Communications: A Casebook Approach*. New York: Routledge.

Fink, S. (1986). *Crisis Management: Planning for the Inevitable*. New York: American Management Association.

Fischhoff, B. (1989). "Helping the public make health risk decisions," in *Effective Risk Communication: The Role and Responsibility of Government and Nongovernment Organization*, eds. V.T. Covello, D.B. McCallum, and M.T. Pavlova. New York: Plenum Press.

Fischhoff, B. (1995). "Risk perception and communication unplugged: Twenty years of progress." *Risk Analysis* 15(2):137–145.

Fischhoff B. (2005). "Risk perception and communication," in *Handbook of Terrorism and Counter-terrorism*, ed. D. Kamien. New York: McGraw-Hill.

Fischhoff, B., Slovic, P., Lichtenstein, S. (1978). "How safe is safe enough? A psychometric study of attitudes towards technological risks and benefits." *Policy Sciences* 9(2):127–152.

Fischhoff, B., Slovic, P., Lichtenstein, S. (1981). "Lay foibles and expert fables in judgments about risk," in *Progress in Resource Management and Environmental Planning*, eds. T. O"iordan and R.K. Turner. New York: Wiley.

Fischhoff, B., Lichentenstein, S., Slovic, P. (1981). *Acceptable Risk*. New York: Cambridge University Press.

Fischhoff, B., Bostrom, A., Quadrel, M.J. (2002). "Risk perception and communication," in *Oxford Textbook of Public Health: The Methods of Public Health.* 4th edition., eds. R. Detels, J. McEwan, R. Beaglehole, J. Heinz. New York: Oxford University Press.

Fjaeran, L., Aven, T. (2021) "Creating conditions for critical trust – How an uncertainty-based risk perspective relates to dimensions and types of trust." *Safety Science* 133:105008.

Flynn, T. (2009). *Authentic Crisis Leadership and Reputation Management: Maple Leaf Foods and 2008 Listeriosis Crisis.* Hamilton, Ont.: DeGroote School of Business, McMaster University.

Fox-Glassman, K.T., Weber, E.U. (2016). "What makes risk acceptable? Revisiting the 1978 psychological dimensions of perceptions of technological risks." *Journal of Mathematical Psychology* 75:157—169.

Frandsen, F., Johansen, W. (2017). *Organizational Crisis Communication: A Multi-Vocal Approach.* Thousand Oaks, CA: Sage Publications.

Glik, D.C. (2007). "Risk communication for public health emergencies." *Annual Review of Public Health* 28:33–54.

Grunig, J.E. (1997). "A situational theory of publics: Conceptual history, recent challenges and new research," in *Public Relations Research: An International Perspective*, eds. D. Moss, T. MacManus, and D. Vercic. London: International Thomson Business Press.

Gustafson, P.E. (1998). "Gender differences in risk perception: theoretical and methodological perspectives." *Risk Analysis* 18(6):805–811.

Hampel, J. (2006). "Different concepts of risk: a challenge for risk communication." *International Journal Medical Microbiology* 296(40):5–10.

Health Protection Network. (2008). *Communicating with the public about health risks.* Glasgow: Health Protection Scotland.

Heath, R.L., Palenchar, M.J., Proutheau, S., Hocke, T.M. (2007). "Nature, crisis, risk, science, and society: What is our ethical responsibility?" *Environmental Communication* 1(1):34–42.

Henwood, K., Pidgeon, N., Parkhill, K., Simmons, P. (2010). "Researching risk: narrative, biography, subjectivity." *Forum: Qualitative Social Research* 11(1):1–22.

Holmes, B.J. (2008). "Communicating about emerging infectious disease: The importance of research." *Health, Risk & Society* 10:349–60.

Huang, Y.H. (2006). "Crisis situations, communication strategies, and media coverage: A multi-case study revisiting the communicative response model." *Communication Research* 33:180–205.

Hyer, R.N, Covello, V.T. (2007). *Effective Media Communication in Public Health Emergencies: A Field Guide.* Geneva: World Health Organization.

Hyer, R., Covello, V.T. (2017). *Top Questions on Zika: Simple Answers, Risk Communication Guide.* Arlington, VA: Association of State and Territorial Health Officials.

Infanti, J., Sixsmith, J., Barry, M.M., Núñez-Córdoba, J.M., Oroviogoicoechea-Ortega, C., Guillén-Grima, F.A. (2013). *A Literature Review on Effective Risk Communication for the Prevention and Control of Communicable Diseases in Europe.* Stockholm: European Centre for Disease Control and Prevention.

Janssen, A., Landry, S., Warner, J. (2006). "'Why tell me now?' The public and healthcare providers weigh in on pandemic influenza messages." *Journal of Public Health Management Practice* 12:388–394.

Jamieson, K.H., Lammie, K., Warlde, G., et al. (2003). "Questions about hypotheticals and details in reporting on anthrax." *Journal of Health Communication* 8:121–131.

Johnson, B.B., Covello, V. (1987). *The Social and Cultural Construction of Risk: Essays on Risk Selection and Perception.* Dordrecht, Netherlands: D. Reidel Publishing.

Johnson, B.B. (1999). "Ethical issues in risk communication." *Risk Analysis* 19(3):335–348.

Kasperson, R., Renn, O., Slovic, R, Brown, H., Emel, J., Goble, R., Kasperson, J., Ratick, S. (1988). "The social amplification of risk: A conceptual framework." *Risk Analysis* 8(2):177–187.

Kahan, D, Braman, D., Cohen, G., Slovic, P., Gastil, J. (2010), "Who fears the HPV vaccine, who doesn't, and why: An experimental study of the mechanisms of cultural cognition." *Law and Human Behavior* 34 (6):501–551.

Kahan, D., Slovic, P., Braman, D., Cohen, G., Gastil, J. (2009), "Cultural Cognition of the Risks and Benefits of Nanotechnology," *Nature Nanotechnology*, 4 (2): 87—90.

Keeney, R.L., von Winterfeldt, D. (1986). "Improving risk communication." *Risk Analysis* 6 (4):417–424.

Leiss, W. (1995). "Down and dirty: the use and abuse of public trust in risk communication." *Risk Analysis* 15(6):685–692.

Lin, I.H., Petersen, D. (2008). *Risk Communication in Action: The Tools of Message Mapping.* Washington, DC: US Environmental Protection Agency.

Linn, M.R., Tinker, T.L. (1994). *A Primer on Health Risk Communication Principles and Practices.* Washington, DC: Agency for Toxic Substances and Disease Registry.

Lindell, M.K., Perry, R.W. (2012). "The protective action decision model: Theoretical modifications and additional evidence." *Risk Analysis* 32(4):616–632.

Lindenfeld, L., Smith, H.M., Norton, T., Grecu, N.C. (2014). "Risk communication and sustainability science: lessons from the field." *Sustainability Science* 9(2):119–127.

Löfstedt, R.E. (2008) "What environmental and technological risk communication research and health risk research can learn from each other." *Journal of Risk Research* 11(1–2):141–167.

Lundgren, R.E., McMakin, A.H. (2018). *Risk Communication: A Handbook for Communicating Environmental, Safety, and Health Risks.* Hoboken, NJ: Wiley.

Maslow, A. (1970). *Motivation and Personality.* New York: Harper and Row.

McComas, K.A. (2006). "Defining moments in risk communication research: 1996–2005." *Journal of Health Communication* 11(1):75–91.

Mebane, F., Temin, S., Parvanta, C.F. (2003). "Communicating anthrax in 2001: A comparison of CDC information and print media accounts." *Journal of Health Communication* 8:50–82.

Mileti, D.S. (1990). "Warning systems: A social science perspective," in *Preparing for Nuclear Power Plant Accidents*, eds. D. Golding, J. X. Kasperson, R.E. Kasperson. Boulder, CO: Westview Press.

Mileti, D.S., Fitzpatrick, C. (1991). "Communication of public risk: Its theory and its application." *Sociological Practice Review* 2(1):20–28.

Mileti, D.S., Peek, L. (2000). "The social psychology of public response to warnings of a nuclear power plant accident." *Journal of Hazardous Materials* 75:181–194.

Mileti, D.S., Sorensen, J.H. (1988). "Planning and implementing warning systems," in *Mental Health Response to Mass Emergencies*, ed. M. Lystad. New York, NY: Brunner-Mazel.

Miller, G. (1956). "The magical number seven, plus or minus two: some limits on our capacity for processing information." *Psychological Review* 101(2):343–352.

Miller, M., Solomon, G. (2003). "Environmental risk communication for the clinician." *Pediatrics* 112(1):211–217. Accessed at: https://pdfs.semanticscholar.org/33c9/8bd7f526d727e8249632df23c537 db89cd4f.pdf

Morgan, M.G., Fischhoff, B., Bostrom, A., Atman, C. (2001). *Risk Communication: The Mental Models Approach.* New York: Cambridge University Press.

Mousavi, S.Y., Low, R., Sweller, J. (1995). "Reducing cognitive load by mixing auditory and visual presentation auditory and visual presentation modes." *Journal of Educational Psychology* 87(2):319–334.

Mullin, S. (2003), "The anthrax attacks in New York City: The 'Giuliani press conference model' and other communication strategies that helped." *Journal of Health Communication* 8:15–16.

National Research Council (1989). *Improving Risk Communication. Committee on Risk Perception and Risk Communication.* Washington: National Academy Press.

National Research Council. (1996). *Understanding Risk: Informing Decisions in a Democratic Society.* Washington, DC: National Academy Press.

Person, B., Sy, F., Holton, K., Govert, B., Liang, A.(2004). "Fear and stigma: the epidemic within the SARS outbreak." *Emerging Infectious Diseases* 10(2):358–63.

Peters, R.G., Covello, V.T., McCallum, D.B. (1997). "The determinants of trust and credibility in environmental risk communication: An empirical study." *Risk Analysis* 17(1):43–54.

Pidgeon, N., Kasperson, R., Slovic, P., eds. (2003). *The Social Amplification of Risk.* New York: Cambridge University Press.

Plough, A., Sheldon, K. (1987). "The emergence of risk communication studies: social and political context." *Science, Technology, & Human Values* 12(3/4):4–10.

Powell, D., Leiss, W. (1997). *Mad Cows and Mother's Milk: The Perils of Poor Risk Communication.* Montreal: McGill- Queen's University Press.

Power-Hays, A., McGann, P. (2020). "When actions speak louder than words." *New England Journal of Medicine* 383:1902–1903.

Rader, M.H. (1981). "Dealing with information overload." *Personnel Journal* 60(5):373–375.

Rayner, S. (1992). "Cultural theory and risk analysis," in *Social Theories of Risk*, eds. Krimsky, S. and Golding, D. Westport, CT: Praeger.

Renn, O. (2008). *Risk Governance: Coping with Uncertainty in a Complex World.* London: Earthscan.

Renn, O., Levine, D. (1991). "Credibility and trust in risk communication," in *Communicating Risks to the Public*, eds. R. Kasperson and P. Stallen. Dordrecht, Netherlands: Kluwer Academic Publishers.

Reynolds, B.J. (2010). "Principles to enable leaders to navigate the harsh realities of crisis and risk communication." *Journal of Business Continuity and Emergency Planning* 4(3):262–273.

Reynolds, B.J. (2011). "When the facts are just not enough: credibly communicating about risk is riskier when emotions run high and time is short." *Toxicology and Applied Pharmacology* 254(2):206–214.

Reynolds, B., Seeger, M.W. (2005). "Crisis and emergency risk communication as an integrative model." *Journal of Health Communication* 10(1):43–55.

Sandman, P.M. (1987). "Risk communication: Facing public outrage." *EPA Journal* 5:21–22.

Sandman, P.M. (1998). "Hazard versus outrage in the public perception of risk," in *Effective Risk Communication: The Role and Responsibility of Government and Nongovernment Organizations*, eds. V.T. Covello, D.B. McCallum, and M.T. Pavlova. New York: Plenum Press.

Sandman, P.M., Covello, V.T. (2001). "Risk communication: Evolution and revolution," in *Solutions to an Environment in Peril*, ed. Wolbarst, A. Baltimore, MD: Johns Hopkins University Press.

Santos, S., Covello, V.T., McCallum, D. (1996). "Industry response to SARA Title III: pollution prevention, risk reduction, and risk communication." *Risk Analysis* 16(1):57–65.

Savoia, E., Lin, L., Viswanath, K. (2013). "Communications in public health emergency preparedness: A systematic review of the literature." *Biosecurity and Bioterrorism* 11(3):170–184.

Schultz, F., Utz, S., Goritz, A. (2011)." Is the medium the message? Perceptions of and reactions to crisis communication via twitter, blogs, and traditional media." *Public Relations Review* 37(3):430–437.

Seeger, M.W. (2006). "Best practices in crisis communication: An expert panel process." *Journal of Applied Communication Research* 34(3):232–244.

Seeger, M.W., Sellnow, T.L., Ulmer, R.R. (2003). *Communication and Organizational Crisis.* Westport, CT: Praeger.

Sellnow, T.L., Seeger, M.W. (2013). *Theorizing Crisis Communication.* Malden, MA: Wiley-Blackwell.

Sellnow, T.L., Ulmer, R.R., Seeger, M.W., Littlefield, R.S. (2009). *Effective Risk Communication: A Message-Centered Approach.* New York: Springer.

Sheppard, B., Janoske, M., Liu, B. (2012). *Understanding Risk Communication Theory: A Guide for Emergency Managers and Communicators. Report to Human Factors/Behavioral Sciences Division, Science and Technology Directorate*. College Park, MD: US Department of Homeland Security.

Siegrist, M. (2021). "Trust and risk perception: A critical review of the literature." *Risk Analysis* 41(3):480–490.

Slovic, P. (1986). "Informing and educating the public about risk." *Risk Analysis* 6(4):403–415.

Slovic, P., Fischhoff, B., Lichtenstein, S. (1986). "The psychometric study of risk perception," in *Risk Evaluation and Management*, eds V. Covello, J. Menkes, and J. Mumpower. New York: Springer.

Slovic, P. (1987). "Perception of risk." *Science* 236:280–285.

Slovic, P. (1999). "Trust, emotion, sex, politics, and science: surveying the risk-assessment battlefield." *Risk Analysis* 19(4):689–701.

Slovic, P., Finucane, M., Peters, E., MacGregor, D. (2004). "Risk as analysis and risk as feelings: Some thoughts about affect, reason, risk and rationality." *Risk Analysis* 24(2):1–2.

Sohn, Y.J., Lariscy, R.W. (2014). "Understanding reputational crisis: Definition, properties." *Journal of Public Relations* 26 (1):23–43.

Stallen, P.J.M (ed.) (1991). *Communicating Risks to the Public: International Perspectives*. Dordrecht, Netherlands: Kluwer.

Starr, C. (1969). "Social benefits versus technological risks." *Science* 165(3899):1232–1238.

Thomas, C.W., Vanderford, M.L., Crouse Quinn, S. (2008). "Evaluating emergency risk communications: a dialogue with the experts." *Health Promotion Practice* 9(4):5–12.

Tinker, T.L., Silberberg, P.G. (1997). *An Evaluation Primer on Health Risk Communication Programs and Outcomes*. Washington, DC: Department of Health and Human Services.

Tinker, T.L, Covello, V.T., Vanderford, M.L., Rutz, D., Frost, M., Li, R., Aihua, H., Chen, X., Xie, R., Kan, J. (2012). *"Disaster risk communication" Chapter 141, in Textbook in Disaster Medicine*, ed. S. David. Dordrecht, Netherlands: Wolters Kluwer.

Ulmer, R.R., Sellnow, T.L., Seeger, M.W. (2009). "Post-crisis communication and renewal," in *Handbook of Crisis and Risk Communication*, eds. R. L. Heath and H. D. O'Hair. New York: Routledge.

United Nations Children's Fund/UNICEF (2020). *Vaccine Misinformation Management Field Guide*. New York: United Nations International Children's Emergency Fund. (Accessed at: https://publichealthcollaborative.org/wp-content/uploads/2021/01/VACCINEMISINFORMATIONFIELDGUIDE_eng.pdf

US Department of Health and Human Services (2006). *Pandemic Influenza Pre-Event Message Maps*. Washington, DC: US Department of Health and Human Services Accessed at: https://www.hsdl.org/?abstract&did=484495

Varghese, N.E., Sabat, I., Neumann-Böhme, S., Schreyögg, J., Stargardt, T., Torbica, A. (2021). "Risk communication during COVID-19: A descriptive study on familiarity with, adherence to and trust in the WHO preventive measures." *PLOS ONE* 16(4): e0250872

Veil, S., Reynolds, B., Sellnow T.L., Seeger, M.W. (2008). "CERC as a theoretical framework for research and practice." *Health Promotion Practice* 9(4):26–34.

Walaski, P.F. (2011). *Risk and Crisis Communication*. Hoboken, NJ: Wiley.

Wingen, T., Berkessel, J.B., Englich, B. (2020). "No replication, no trust? How low replicability influences trust in psychology." *Social Psychological & Personality Science*, 11(4):454–463.

World Health Organization. (2004). *Communication Guidelines for Disease Outbreaks. WHO Expert Consultation on Outbreak Communications, Singapore, September 21–23, 2004*. Geneva: World Health Organization.

World Health Organization. (2005). *WHO Outbreak Communication Guidelines*. Geneva: World Health Organization.

Endnotes

1 See, e.g., US Department of Health and Human Services (2006). *Pandemic Influenza Pre-Event Message Maps*. Washington, DC: US Department of Health and Human Services Accessed at: https://www.hsdl.org/?abstract&did=484495.

2 Covello, V., and Hyer, R. (2020). *COVID-19: Simple Answers to Top Questions, Risk Communication Guide*. Arlington, Virginia: Association of State and Territorial Health Officials. Accessed at: https://www.hsdl.org/?abstract&did=835774.

3 See, e.g., Maynard, O. M., McClernon, F. J., Oliver, J. A., Munafò, J. (2018). "Using neuroscience to inform tobacco control policy." *Nicotine and Tobacco Research* 57:1–8; Jimenez, J. C. et al. (2018). "Anxiety cells in a hippocampal-hypothalamic circuit." *Neuron* 97(3):670–683; Kahneman, D. (2011). *Thinking, Fast and Slow*. New York: Macmillan; Rock, D. (2008). "SCARF: A brain-based model for collaborating with and influencing others." *NeuroLeadership Journal* 1:1–9.

4 See, e.g., Hobfoll, S., Watson, P., Bell, C., Bryant, R., Brymer, M., Friedman, M. J., Friedman, M., Gersons, B., de Jong, J., Layne, C., Maguen, S., Neria, Y., Norwood, A., Pynoos, R., Reissman, D., Ruzek J., Shalev, A., Solomon, Z., Steinberg, A., Ursano, R. (2007). "Five essential elements of immediate and mid-term mass trauma intervention: Empirical evidence." *Psychiatry* 70(4):283–315.

5 See, e.g., George S. Everly Jr., G. S., Lating, J. M. (2014). *The Johns Hopkins Guide to Psychological First Aid*. Baltimore, MD: The Johns Hopkins University Press.

6 See, e.g., Ruthig, J. C., Trisko, J., Chipperfield, J. G. (2014). "Shifting positivity ratios: Emotions and psychological health in later life. *Aging and Mental Health* 18(5):547–553.

7 See., e.g., Gladwell, M. (2005). *Blink: The Power of Thinking Without Thinking*. New York: Little Brown and Co.

8 The two message map examples are drawn from more than 40 message maps produced by the author of this book and his colleagues for the Centers for Disease Control and Prevention (CDC) for the global pandemic influenza outbreak in 2009–2010. The full set of message maps can be accessed at: http://www.calhospitalprepare.org/sites/main/files/file-attachments/pandemic_pre_event_maps.pdf.

9 Covello. V. and Hyer, R. (2014). *Top Questions on Ebola*. Arlington, VA: Association of State and Territorial Health Official. Accessed at: https://www.astho.org/Infectious-Disease/Top-Questions-On-Ebola-Simple-Answers-Developed-by-ASTHO/

10 Rock, D. (2008). "SCARF: A brain-based model for collaborating with and influencing others." *NeuroLeadership Journal* 1:1–9; Scarlett, H. Scarlett, H. (2019). *Hilary Scarlett (2019), Neuroscience for Organizational Change: An Evidence-based Practical Guide to Managing Change (2nd Edition)*. London and New York: Kogan Page Limited Publishing Co.

11 See, e.g., Kahneman, D. (2011). *Thinking, Fast and Slow*. New York: Macmillan. Kahneman won the Nobel Prize in Economic Sciences partly on the basis of his behavioral and neuroscience research on System 1 and System 2 thinking and decision-making.

12 See, e.g., Morteza, S., and Jazayeri, M. (2019). "Hierarchical Reasoning by Neural Circuits in the Frontal Cortex." *Science* 17:364, Issue 6441:103–111.

13 Snyder, R.A. (2016). *The Social Cognitive Neuroscience of Leading Organizational Change*. New York: Routledge/Taylor and Francis Group; See also Wojtecki, J., and Peters, R. (2007) "Communicating organizational change: Information technology meets the carbon-based employee unit" in *The Pfeiffer Book of Successful Communication Skill-Building Tools*, ed. J. Gordon. Hoboken, NJ: John Wiley & Sons.

14 See. e.g., Covello, V. T., Minamyer, S., and Clayton, K. (2007). *Effective Risk and Crisis Communication during Water Security Emergencies. EPA Policy Report 600-R07-027.* Washington, DC: US Environmental Protection Agency; Edmund, T. R., and Deco, G. (2010). *The Noisy Brain: Stochastic Dynamics as a Principle of Brain Function.* Oxford, UK: Oxford University Press; Covello, V. T. (2003). "Best practices in public health risk and crisis communication." *Journal of Health Communication* 8(Suppl. 1):5–8.

15 Sridhar, N., and Shahin, A. (2014). "Effect of mental noise on sports performance." *International Journal of Physical Education, Sports and Health* 1(2):17.

16 Centers for Disease Control and Prevention (2009). Pre-Event Message Maps. Accessed at: https://www.calhospitalprepare.org/sites/main/files/file-attachments/pandemic_pre_event_maps.pdf

17 http://www.calhospitalprepare.org/sites/main/files/file-attachments/pandemic_pre_event_maps.pdf. See also https://ftp.cdc.gov/pub/avian_influenza1/Session%20D/HHS%20Pandemic%20Influenza%20Message%20Maps-sample.doc

18 See, e.g., Lin, I., and Petersen, D. (2007). *risk Communication in Action: Message Mapping.* EPA Document 625/R-06/012. Washington, DC: Environmental Protection Agency.

19 See, e.g., Covello, V. T., Minamyer, S., and Clayton, K. (2007). *Effective Risk and Crisis Communication during Water Security Emergencies. Summary Report on EPA Sponsored Message Mapping Workshops. EPA Policy Report 600-R07-027.* Washington, DC: US Environmental Protection Agency; US Department of Health and Human Services (2006). *Pandemic Influenza Pre-Event Message Maps.* Washington, DC: US Department of Health and Human Services Accessed at: https://www.hsdl.org/?abstract&did=484495

20 See, e.g., Covello V. T., and Hyer, R. (2014). *Top Questions on Ebola: Simple Answers.* Arlington, VA: Association of State and Territorial Health Officials; Hyer, R., and Covello, V. T. (2017). *Top Questions on Zika: Simple Answers.* Arlington, VA: Association of State and Territorial Health Officials; Covello, V., Becker, S., Palenchar, M., Renn, O., and Sellke, P. (2010). *Effective Risk Communications for the Counter Improvised Explosive Devices Threat: Communication Guidance for Local Leaders Responding to the Threat Posed by IEDs and Terrorism.* Washington, DC: Department of Homeland Security.

21 See, e.g., Agency for Toxic Substances and Disease Registry (ATSDR)/CDC (2021). *Asbestos and Health: Frequently Asked Questions.* Atlanta, GA: ATSDR. Accessed at https://www.atsdr.cdc.gov/asbestos/docs/Asbestos_Factsheet_508.pdf ; Agency for Toxic Substances and Disease Registry (ATSDR)/CDC (2021). *Environmental Health Resources: Self Learning Module: Risk Communication.* Atlanta, GA: ATSDR. Accessed at https://www.atsdr.cdc.gov/sites/brownfields/pdfs/risk_communications-508.pdf

22 Covello, V., and Merkhofer, M. (1993). *Risk Assessment Methods: Approaches for Assessing Health and Environmental Risks.* New York: Plenum Press.

23 See, e.g., Lundgren, R. E., and McMakin, A. H. (2018). "Chapter 7: Determine Purpose and Objective," in *Risk Communication: A Handbook for Communicating Environmental, Safety, and Health Risks.* Hoboken, New Jersey: Wiley.

24 See, e.g., https://www.orau.gov/cdcynergy/erc/content/activeinformation/resources/Covello_message_mapping.pdf; http://rcfp.pbworks.com/f/MessageMapping.pdf

25 Covello, V. T., Minamyer, S., and Clayton, K. (2007). *Effective Risk and Crisis Communication during Water Security Emergencies. EPA Policy Report; EPA 600-R07-027.* Washington, DC: US Environmental Protection Agency.

26 Covello, V. T. (2011). *Guidance on Developing Effective Radiological Risk Communication Messages: Effective Message Mapping and Risk Communication with the Public in Nuclear Plant Emergency Planning Zones. NUREG/CR-7033.* Washington, DC: US Nuclear Regulatory Commission.

27 Adatto, K. (1990). "The incredible shrinking sound bite." *The New Republic.* May 28,1990: 20. Accessed at: https://scholar.harvard.edu/kikuadatto/publications/incredible-shrinking-sound-bite.

28 A TED and TEDx (Technology, Entertainment and Design) talk is a highly popular presentation model whereby a speaker has 18 minutes or less to present their ideas. This short talk model decreases the chance of the mind wandering. Some of the most popular TED and TEDx talks are less than five minutes long.

Appendices

Appendix 9.1   The 93 most frequently asked questions by journalists and the public following a major crisis, emergency, or disaster.

Journalists are trained to ask six basic questions following a major crisis, emergency, or disaster: (1) Who? (2) What? (3) Where? (4) When? (5) Why? and (6) How? Journalists ask many of the same questions asked by the public.

Based on analysis of over 3,000 news conferences following a major crisis, emergency, or disaster, specific questions asked by journalists and the public include the following:

1) What is your name and title?
2) How do you spell and pronounce your name?
3) What are your job responsibilities?
4) Can you tell us what happened?
5) Where were you when the crisis occurred?
6) How do you know what you are telling us?
7) When did it happen?
8) Where did it happen?
9) Who was harmed?
10) How many people were harmed?
11) Are those that were harmed getting help?
12) How are those who were harmed getting help?
13) How much property and other damage have occurred?
14) Is the situation under control?
15) How certain are you that the situation is under control?
16) Is there any immediate danger?
17) What is being done in response to what happened?
18) Who is in charge?
19) Who has responded?
20) What can we expect next?
21) What are you advising people to do?
22) What can people do to protect themselves, families, and friends?
23) How long will it be before the situation returns to normal?
24) What help has been requested or offered from others?
25) What responses have you received for requests for help?
26) Can you be specific about the types of harm that occurred?
27) What are the names, ages, and hometowns of those that were harmed?
28) Can we talk to people who have been harmed?
29) What other damage may have occurred?
30) How certain are you about the damage?
31) How much damage do you expect?
32) What are you doing now?

33) Who else is involved in the response?
34) Why did this happen?
35) What was the cause?
36) Did you have any forewarning that this might happen?
37) Why wasn't this prevented from happening?
38) Could the crisis have been avoided?
39) How could this have been avoided?
40) What else can go wrong?
41) If you are not sure of the cause, what is your best guess?
42) Who caused this to happen?
43) Who is to blame?
44) Do you think those involved handled the situation well enough?
45) What more could those who handled the situation have done?
46) When did your response to this begin?
47) When were you notified that something had happened?
48) Did you and other organizations disclose information promptly?
49) Have you and other organizations been transparent?
50) Who is conducting the investigation?
51) When will the investigation be completed?
52) When will results from the investigation be reported?
53) How will results from the investigation be shared?
54) What are you going to do after the investigation?
55) What have you found out so far?
56) Why was more not done to prevent this from happening?
57) What are your personal feelings about what happened?
58) What are you telling your own family?
59) Are all those involved in the response in agreement?
60) Are people overreacting?
61) Which laws are applicable?
62) Have any laws been broken?
63) How certain are you about whether laws have been broken?
64) Has anyone made mistakes?
65) How certain are you that mistakes have not been made?
66) Have you told us everything you know?
67) What are you not telling us?
68) What effects will this have on the people involved?
69) What precautionary measures were taken?
70) Do you accept responsibility for what happened?
71) Has this ever happened before?
72) Can this happen elsewhere?
73) What is the worst-case scenario?
74) What lessons have you learned?
75) Have lessons learned been implemented?
76) What can be done to prevent a crisis like this from happening again?
77) What steps need to be taken to avoid a similar event?
78) What would you like to say to those who were victims or were harmed?
79) What would you like to say to the families and friends of those who were victims or were harmed?

80) Is there any continuing danger?
81) Are people out of danger?
82) Are people safe?
83) Will there be inconveniences to employees?
84) Will there be inconveniences to the public?
85) Do you need additional help from individuals and organizations?
86) What can people do to help?
87) How much will all this cost?
88) Are you able and willing to pay the costs?
89) Who will pay the costs?
90) When will we find out more?
91) Why should we trust you?
92) How are you holding up under the stress?
93) What does this all mean?

Appendix 9.2   The 400 plus most frequently asked questions following an active shooter incident.

Active shooter incidents in the United States, tragic as they are, have become nearly commonplace. In 2019, the FBI reported there were 28 active shooter incidents—defined by the FBI as one or more persons trying to kill others with a firearm in a populated area. The shootings resulted in 97 deaths and 150 wounded. (See, e.g., https://www.fbi.gov/file-repository/active-shooter-incidents-in-the-us-2019-042820.pdf/view)

One of the deadliest active shooter events in United States history occurred on the evening of October 1, 2017. Stephen Paddock opened fire upon the crowd attending the Route 91 Harvest Music Festival in Las Vegas. Fifty-eight people were killed and over 400 were wounded by gunfire.

One of the deadliest active shooter events at a school in US history was the February 14·, 2018, mass shooting at the Marjory Stoneman Douglas High School in Parkland, Florida. A teenager armed with a semiautomatic rifle killed 17 students and staffers and injured another 17. In the year that followed the Parkland active shooter incident, there were over 30 active shooter incidents at schools (K–12) in the United States in which someone was shot. (See, e.g., https://www.cnn.com/2019/02/14/us/school-shootings-since-parkland-trnd/index.html). This averages out to a school shooting approximately every 12 days.

The media, the public, and other stakeholders will ask six basic questions during and after an active shooter incident: (1) Who? (2) What? (3) Where? (4) When? (5) Why? and (6) How?

Based on a review of active shooter incidents in the United States done for purposes of this book, listed below are over 400 questions asked by stakeholders, including the media, victims, victim families, witnesses, politicians, emergency managers, and the general public. The questions are organized into six categories, beginning with questions related to casualties and victims. Based on the categorization scheme, some questions may appear in more than one category. Questions marked with a "*" are among the most frequently asked.

I. Questions Related to Casualties and Victims
1) Was anyone killed?*
2) Who was killed?*
3) Was anyone injured?*
4) Who was injured?*

5) How many people were killed?*
6) How many people were injured?*
7) What is the status of the injured?*
8) What is being done to protect people?*
9) Where can friends and family get information about casualties and victims?
10) Do you expect additional deaths?
11) Has everyone present at the location been accounted for?
12) Where have the dead been taken?
13) What is the condition of the injured or wounded?
14) How many people are in critical condition?
15) Where have the injured and wounded been taken?
16) What hospitals are being used?
17) Do the hospitals being used have enough personnel to handle a surge in patients?
18) How many security personnel have been injured?
19) How many security personnel have been killed?
20) Do you believe there are injured or wounded persons who have not been found?
21) Did the shooter kill or injure anyone at a site other than where the shooting occurred?
22) How many shots were fired at individuals?
23) How many people escaped from being killed or injured?
24) Were any of those killed children?
25) Were any of those injured or wounded children?
26) Were any of those killed pregnant women?
27) Were any of those injured or wounded pregnant women?
28) Did anybody shoot back at the shooter?
29) Did any of the victims have guns on them?
30) How did people escape?
31) Did any of those who escaped have guns?
32) Where did people at the location hide?
33) How long did people hide?
34) How did people hide?
35) What did people at the location do to protect themselves?
36) How many people were hurt in the process of escaping?
37) Did any of those who escaped or hid know the shooter?
38) Who is helping those who were hurt?
39) How many people were evacuated?
40) Where did those who escaped or evacuated go on their own?
41) Where were those who escaped or evacuated sent to by authorities?
42) Who is helping those who have escaped or evacuated?
43) What medical assistance was provided at the site of the shooting?
44) What type of help or assistance is being provided to those who escaped or evacuated?
45) What assistance is being provided to the families of victims?
46) Were any people trapped in the building? If so, for how long?
47) Did the shooter take any hostages? If so, how many and where are they now?
48) Were any of the casualties or injured known to the shooter?
49) Were any of the casualties or injured a relative of the shooter?
50) Were any of the casualties or injured a friend of the shooter?
51) Were any of the casualties or injured a present or former lover of the shooter?

52) Were any of the casualties or injured a present or former coworker of the shooter?

53) Were any of the casualties or injured a present or former boss or supervisor of the shooter?

54) Were any law enforcement or security personnel among those who were killed, injured, or wounded?

55) Have families been notified of the death of a relative? If so, who has notified them?

56) Have families been notified of the injury of a relative? If so, who has notified them?

57) What is being told to families of victims?

58) What are the names of the dead?

59) What are the names of those injured or wounded?

60) If you cannot share the names of the dead or wounded, when will the names be released?

61) Can you confirm the name of victims being posted on social media sites by families and friends?

62) Are there any missing persons? If so, who?

63) Are you asking for blood donations?

64) Are any memorials taking place or being planned?

65) Is there anything you would like to say to the victims?

66) Can we speak to any of the victims?

67) Can we speak to the families of any of the victims?

68) Where were the dead found?

69) Where were the injured found?

70) Were any of the victims known to the shooter?

71) Were any of the victims distinguishable based on any characteristic, such as age, race, religion, occupation, or ethnicity?

72) Are crisis counselors and psychologists being made available to the injured or wounded?

73) Are crisis counselors and psychologists being made available to the families of victims?

74) Are crisis counselors and psychologists being made available to those who were present at the site of the shooting?

75) Are counselors and psychologists being made available to the law enforcement personnel?

76) Where can people find a qualified crisis counselor or psychologist?

II. Questions Related to the Shooter(s) or Suspect(s)

77) Who was the shooter?*

78) What is the name of the shooter?*

79) Was there more than one shooter?*

80) Were there other persons with the shooter?*

81) Can you describe the shooter, e.g., the shooter's age, gender, race, nationality, weight, height, hair, clothing, and appearance?*

82) Was the shooter a member of any organization?*

83) Was the shooter known by the victims?*

84) Did the shooter have a license to carry a weapon?*

85) Can you describe the personality of the shooter?

86) Was the shooter an introvert?

87) Did the shooter display any unusual behaviors before the shooting?

88) Do you have a photograph of the shooter you can share?

89) Was there more than one shooter?

90) If there was more than one shooter, how many and what do you know about them?

91) Does the shooter have a criminal record, ever been arrested, or charged with a crime or violence?

92) If the shooter has a criminal or an arrest history, what was the charge and when and where did it happen?
93) Did the shooter have any military or weapons training?
94) Where was the shooter living right before the attack?
95) Where has the shooter lived in the past?
96) Where has the shooter been employed?
97) Did the shooter have any problems or disputes with his or her neighbors?
98) Did the shooter have any problems or disputes with his or her coworkers?
99) Did the shooter have any problems or disputes with his or her family?
100) Does the shooter live with anyone?
101) What type of work does the shooter do?
102) Was the shooter currently employed? If so, who is the employer?
103) What type of work did the shooter do in the past?
104) Who employed the shooter in the past?
105) What is the marital status of the shooter?
106) Does the shooter have any children?
107) Does the shooter have a mental health record or a history of mental health problems?
108) Did anyone notice or report mental health problems in the shooter?
109) Is there evidence of a family history of crime or mental health problems?
110) What do you know about the history of the shooter that would provide clues to the motive?
111) Did the shooter or anyone else possess any forewarning of the shooting?
112) Did the shooter previously work at the location where the shooting occurred?
113) Did the shooter know anyone at the site of the shooting?
114) Did the shooter identify anyone by name as the target for the shooting?
115) Did the shooter have disputes with anyone at the location where the shooting occurred?
116) Did the shooter post any threats on the internet?
117) Did anyone, such as a friend, know of the intentions of the shooter before the shooting?
118) Did the shooter speak any words before, during, or after the shooting?
119) What did the shooter say when apprehended?
120) Did the shooter let anyone go free?
121) What time was it when the shooter started shooting?
122) What time was it when the shooter stopped shooting?
123) How long did the shooting continue?
124) Was the shooting continuous or in bursts?
125) How many rounds of shots did the shooter fire?
126) Did the shooter issue any warnings that a shooting was about to occur?
127) Did anyone issue warnings to authorities that a shooting was about to occur?
128) Was the shooter a user of any social media sites?
129) If the shooter used social media, what social media platforms and what type of content did the shooter post?
130) What information did the shooter have about the location (e.g., maps, floor plan)?
131) Did the shooter receive any assistance during the shooting?
132) Did the shooter have to go through a security check to gain access to the location of the shooting?
133) Was the shooter killed or injured?
134) Did the shooter surrender?
135) Did the shooter put up any resistance to arrest?

136) Did the shooter commit suicide? If so, how?
137) Did the shooter leave a note, video, recording, or other communication?
138) Did the shooter try to escape the scene of the shooting?
139) How did the shooter travel to the site of the shooting?
140) What was the shooter's plan for escape?
141) Did the shooter succeed in escaping?
142) If the shooter escaped, did he/she have any assistance?
143) Was the shooter arrested?
144) If the shooter has been arrested, where has the shooter been taken?
145) Has the shooter been questioned?
146) If the shooter has been questioned, what has been said?
147) Has the shooter asked for legal representation?
148) Is the shooter or coconspirators still at large?
149) Was the shooter wounded or injured?
150) If the shooter was wounded or injured, what is the shooter's condition?
151) If the shooter was wounded or injured, where was he/she taken after the shooting?
152) Where is the shooter now?
153) Is the shooter in custody?
154) If the shooter is in custody, where is the shooter being held?
155) If the shooter was arrested, what are the charges?
156) If the shooter is found guilty of the charges, what are the penalties?
157) How long was the shooter in the location?
158) What were the motives of the shooter?
159) Why did the shooting occur?
160) Were there any triggers or sparks that set off the shooter?
161) Did the shooter injure or kill anyone before going to the site of the shooting?
162) Is the shooter or any accomplices still at large?
163) If the shooter or shooters are still at large, is there a search in progress?
164) Do you know the names of relatives of the shooter?
165) Does the shooter belong to any organizations linked to violence? If so, what are the names of the organizations?
166) Does the shooter have any links to terrorist organizations?
167) Did the shooter have any organizational affiliations?
168) How did the shooter get to the location?
169) If the shooter drove to the location, is the vehicle in custody?
170) How did the shooter get weapons into the location?
171) Did the shooter have any relationship in the past with the location where the shooting occurred?
172) Was the shooter known to anyone in the neighborhood?
173) Where was the shooter born?
174) Do you know anything about the shooter, such as the shooter's hobbies?
175) What possessions did the shooter have other than weapons?
176) Did the shooter say anything to his or her victims?
177) Did the victims of the shooter say anything to the shooter?
178) Did the shooter say anything after being captured?
179) Did the shooter resist arrest?
180) Was the shooter injured during the apprehension or arrest?

181) If the shooter was injured during the apprehension or arrest, where has the shooter been taken?
182) Is there any reason to believe the shooter is a terrorist?
183) Is there anything you would like to say to the shooter?
184) Did the shooter offer any terms for negotiation?
185) If the shooter offered terms for negotiation, what were they?

III. Questions Related to the Weapons Used by the Shooter

186) What weapons did the shooter use?*
187) What are the models of the weapons used by the shooter?
188) How many weapons did the shooter use?
189) How many weapons were in possession of the shooter?
190) Where did the shooter purchase or get the weapons?
191) Were the weapons in possession of the shooter obtained legally by the shooter?
192) Where were the weapons found?
193) If the weapons were owned by someone other than the shooter, who is the owner?
194) Did the person who owns the weapons obtain them legally?
195) Does the person who owns the weapons know they were taken or to be used by the shooter?
196) What type of ammunition did the shooter use?
197) Where did the shooter purchase or get the ammunition?
198) How much ammunition did the shooter have?
199) What kind of ammunition did the shooter have?
200) Was the ammunition specialized in any way, such as for maximum damage?
201) Did the shooter use all of the available ammunition?
202) Did the shooter have any bombs or explosive materials?
203) If the shooter had bombs or explosives, what are the types of bombs or explosives?
204) Is the location being searched for bombs or explosive materials?
205) Who is searching for bombs or explosive materials?
206) Was the shooter in contact with anyone else during the incident?
207) Was the shooter wearing any body armor?
208) Did the shooter have any automated weapons?
209) Did the shooter have a permit to carry a weapon?
210) If the shooter had a gun permit, where was it obtained?
211) Does this incident indicate gun policies need to be changed?

IV. Questions Related to Law Enforcement/Security/First Responders

212) What security measures were in place at the location of the shooting?*
213) What are you doing to apprehend the shooter?*
214) Who responded to the shooting?*
215) When were reports of a shooting first received?*
216) What is being done to prevent another shooting incident?*
217) Who first alerted authorities, law enforcement, or security personnel of the shooting?
218) Was there a 911 call or other call for help?
219) Are there any recordings of the call for help?
220) Who was the first person or persons to respond to the call for help?
221) How much time passed between the call for help and the arrival of security, law enforcement personnel, or other persons?

222) When were shots first heard?

223) Who first heard the shots?

224) Who first called for help?

225) What is the overall timeline of events related to the shooting?

226) What law enforcement personnel responded?

227) How many law enforcement personnel responded?

228) What security personnel responded?

229) How many security personnel responded?

230) Who among law enforcement was closest to the scene of the shooting?

231) Can we speak to any of those who responded to the shooting?

232) How far did law enforcement personnel have to travel to get to the scene of the shooting?

233) Did local security personnel call for help?

234) Were local security personnel armed?

235) Who was in charge of the response to the shooting?

236) What was the chain of command?

237) Did all those who responded obey the chain of command?

238) Who is in charge of the investigation?

239) Was the FBI called in for assistance?

240) Have homicide experts been called in for assistance?

241) Have forensic experts been called in for assistance?

242) Did those who responded to the shooting follow standard protocols and procedures for engagement?

243) Have you or other local authorities planned for an active shooter incident?

244) If you had a plan for an active shooter incident, do you think the plan worked?

245) Were law enforcement and security personnel trained in responding to an active shooter incident?

246) Were there any missteps in the response by law enforcement and security?

247) Do you believe those who first responded to the shooting acted appropriately?

248) Are you expecting to hold any disciplinary hearings related to the incident?

249) Will there be an after-action report, and, if so, when will it be available to read?

250) Did those who responded to the shooting have appropriate weapons and personal protective equipment for this type of event?

251) What was the exact time at which law enforcement personnel arrived at the scene of the shooting?

252) What actions did law enforcement and security personnel take upon arrival at the scene of the shooting?

253) What guidance did response authorities offer to those at the location?

254) What weapons did law enforcement and security personnel have?

255) What weapons did law enforcement and security personnel use?

256) What did law enforcement personnel find when they arrived on the scene of the shooting?

257) Where did law enforcement personnel find the shooter?

258) What were the first actions taken by law enforcement and security personnel on the scene?

259) Did law enforcement personnel apprehend the shooter?

260) Did the shooter make any demands or attempt to negotiate with responders or law enforcement?

261) If the shooter made demands or attempted to negotiate with responders to law enforcement, what were they?

262) What was the response by law enforcement to demands or negotiation terms by the shooter?
263) Did law enforcement or security personnel suffer any deaths or injuries?
264) If there were deaths among law enforcement or security personnel, how many were killed?
265) If there were injuries among law enforcement or security personnel, how many were injured?
266) What is the condition of the injured law enforcement or security personnel?
267) Has the home of the shooter been searched?
268) If the home of the shooter was searched, what was found?
269) What locations are being searched?
270) Did anyone at the location of the shooting try to stop the shooter other than law enforcement or security?
271) What assistance was provided to law enforcement at the site of the shooting?
272) Are the actions of law enforcement and security personnel being reviewed?
273) If a review is taking place of actions by law enforcement and security personnel, who is or will be conducting the review?
274) Are there other persons of interest or suspects?
275) Is anyone else being detained?
276) Who are the heroes among first responders?
277) What would law enforcement like people to know?
278) Is there anything you would like to say to law enforcement?
279) Are you searching for any other suspects?
280) Are you searching for any persons of interest?
281) Have the weapons used been recovered by law enforcement?
282) Was the pattern of the shooting similar to any other shooting?
283) Can we speak to the relatives or family of the shooter?
284) Can we speak to the past and present friends of the shooter?
285) Can we speak to past and present coworkers of the shooter?
286) Can we speak to the persons at the current place of employment of the shooter?
287) Can we speak to the past and present boss or supervisor of the shooter?
288) Can we speak to the past and present lovers or spouses of the shooter?
289) Can we speak to those who were injured or wounded?
290) Can we speak to any of the witnesses to the shooting?
291) Can we speak to the families of the victims or injured?
292) Can we speak with the neighbors of the shooter?
293) Can we speak to a member of the response team?
294) What is your experience with shooting incidents?
295) Have you been involved in any other shooting incidents before this incident?
296) Could the outcome of the shooting have been worse?
297) What was the worst-case outcome of the incident?
298) Based on this shooting and others, do you recommend tighter gun control legislation, laws, and enforcement?
299) How can mass shootings be prevented from happening?
300) Should guns be taken away from potentially dangerous people?
301) Should background checks be improved before a person can buy a gun?
302) Should guns be banned for those who are mentally ill? If so, what is your definition of mental illness?
303) Should guns be banned for those with felony criminal records? If so, what types of felonies would you include in the ban?

304) Should the minimum age for buying a gun be raised?

305) Should assault weapons be banned?

306) How can buildings and other locations be protected against an active shooter?

307) Is there a designated place for people to leave flowers and cards for the victims?

308) What would you like to say to those who are speculating about the motive of the shooter?

309) Are you advising people to stay home until further notice?

310) Is it safe for people to leave their homes or businesses?

311) Should people stay away from the site of the shooting?

312) What reactions are you receiving from leaders and others?

313) What would you like to say to the shooter?

314) What would you like to say to those who were killed or injured?

315) What would you like to say to members of the general public?

316) When will members of organizations affected and the community return to normal?

317) When will the community affected return to normal?

318) What, if any, changes are you making in your procedures?

319) What, if any, law enforcement and security actions and behaviors need to change?

320) What does success look like in responding to an active shooting incident?

321) How will you measure success in responding to an active shooting incident?

322) What knowledge, skills, and abilities are needed to ensure appropriate actions and behaviors by law enforcement and security personnel?

323) What training and tools are available for law enforcement and security personnel to improve their responses to an active shooter incident?

324) What new policies, procedures, and/or processes are needed to support the behavior change by law enforcement and security personnel?

325) What assistance is available to emotionally support law enforcement and security personnel?

326) What consequences need to be in place to sustain behavior change by those responding to an active shooter incident?

327) Are you aware of any protests being organized?

328) What are your recommendations for preventing future active shooting incidents?

329) To prevent future active shooter incidents, do you recommend hiring additional police?

330) To prevent future active shooter incidents, do you recommend vulnerable locations hire additional security personnel and police, including armed guards?

331) To prevent future active shooter incidents, do you recommend additional *security technology* at vulnerable locations, such as surveillance cameras, alarms, lockable doors, reinforced steel doors, "hardening" buildings, safe rooms, automatic locks, hand-held metal detectors, walk-through metal detectors, bag screening, and recognition technology (e.g., face, eyes, or fingerprints)?

332) To prevent future active shooter incidents, do you recommend additional *security measures* at vulnerable locations, such as ID badges, comprehensive employee background checks, weapons and weapon training for employees, lockdown drills, bulletproof vests, active shooter drills, limited entry checkpoints, and clear backpacks?

333) What questions are still unanswered?

334) What is your greatest concern about this incident?

335) When will the next update take place?

336) What are you saying about the shooting incident to your own children?

337) What are you recommending parents say about the shooting incident to their children?

338) Do you have any regrets about how the shooting incident was handled?
339) What did you mean when you said this is a "new normal"?

V. Questions Related to the Location Where the Shootings Took Place

340) Where did the shooting take place?*
341) When did the shooting take place?*
342) What was the exact time of the shooting?*
343) Can you describe the location where the shooting took place?*
344) Where exactly in the location did the shooting(s) take place?
345) What were the movements of the shooter at the location?
346) Where exactly did the shooter enter the location where the shooting took place?
347) How did the shooter gain entry to the location?
348) Were there any security personnel present where the shooter entered the location?
349) Did the location where the shooting took place have security cameras?
350) If the location had security cameras, can you share any of the footage?
351) What security technology existed at the location?
352) If there was security technology, was it operational?
353) If there was security personnel or security technology at the entry locations, how did the shooter get through?
354) Did the shooter have any assistance in gaining access to the location of the shooting?
355) Did the shooter return to the same location at the site of the shooting more than once?
356) Did anyone at the location recognize the shooter?
357) What was happening when the shooting took place?
358) What were people doing when the shooting began?
359) What were people doing before the shooting began?
360) What did people do when they first realized a shooting was happening?
361) Were any major events planned for the location where the shooting took place?
362) Has anything like this shooting ever happened at this location?
363) Have any other notable incidents occurred at this location?
364) What is the address of where the shooting took place?
365) What other structures or buildings are close to the address of the shooting?
366) How secure was the building where the shooting took place?
367) How many doors are there to the building?
368) Are the doors to the building monitored?
369) What security or safety systems exist at the location?
370) Does the location of the shooting have metal detectors?
371) How many security personnel are at the location?
372) Were security personnel at the location armed?
373) Were security personnel at the location trained for an active shooter incident?
374) Did anyone help or assist the shooter to gain access to the place of the shooting?
375) Did anyone try to stop the shooter from gaining access to the place of the shooting?
376) Where in the location did the shooter go first?
377) Where exactly in the location did the shootings take place?
378) Did the shooting also take place in more than one location or site?
379) If shootings took place in more than one location, what were the locations?
380) Were alarms sounded?
381) Is the location in lockdown?

382) If the location is in lockdown, when will it be reopened?
383) Who owns the location where the shooting took place?
384) Who operates the location where the shooting took place?
385) What activities take place in the location where the shooting took place?
386) Who uses or is employed at the location where the shooting took place?
387) Were any children at the location where the shooting took place?
388) What security personnel were present at the location where the shooting took place?
389) How many people were present at the location when the shooting began?
390) Were people at the location instructed on what to do in case of a shooting incident?
391) How did persons at the location first determine an active shooter was present?
392) Were there locks on the doors at the location where the shooting took place?
393) Were any of those at the location trained to respond to this type of event?
394) Did the location hold any drills or exercises related to this type of event?
395) Where can people get updates about the status of those who were at the site of the shooting?
396) Have there been any other shootings or violent events at the location of the shooting?
397) Will the site of the shooting have a memorial place for people who want to leave flowers, notes, or other remembrances?
398) If people are being advised to stay away from the location of the shooting, where can they leave or send notes, flowers, or remembrances?

Appendix 9.3   Change management: frequently asked questions.

I. Why Questions (Questions and Concerns That Relate Largely to Seeing the Big Picture)
1) Why are the proposed changes needed?
2) Why are the proposed changes being considered now?
3) Why were the proposed changes not introduced before?
4) Why do you think the proposed changes are better than the status quo?
5) Why have more people not been involved in developing the proposed changes?
6) Why isn't more time being allowed for the proposed changes to be discussed?
7) Why have you selected these changes as opposed to other changes?
8) Why did we not receive prior notice that changes were being considered?
9) Why are the proposed changes targeted on me and my part of the organization?

II. Who Questions (Questions and Concerns That Relate Largely to People and Roles)
1) Who decided that change is needed?
2) Who decided that all the proposed changes will focus on my part of the organization?
3) Who developed the proposed changes?
4) Who approved the proposed changes?
5) Who agrees with you that the proposed changes are needed?
6) Who will be in charge of managing the proposed changes?
7) Who will implement the proposed changes?
8) Who will be in charge of monitoring the impacts of the proposed changes?
9) Who will decide if the proposed changes are working?
10) Who has the power and authority to reverse the proposed changes if they are not working?
11) Who will be affected by the proposed changes?
12) Who will be affected most by the proposed changes?

13) Who will benefit from the proposed changes?
14) Who will benefit most from the proposed changes?
15) Who will be adversely affected by the proposed changes?
16) Who will be most adversely affected by the proposed changes?
17) Who have you consulted on the proposed changes?
18) Who will be asked to comment on the proposed changes?
19) Who will review comments on the proposed changes?
20) Who has to agree before the proposed changes are implemented?
21) Who will pay for the proposed changes?
22) Whose program will be most impacted?
23) Who will be held responsible and accountable for ensuring the proposed changes are implemented?
24) Who will be held responsible and accountable if the proposed changes fail to achieve their goals and objectives?
25) Who do you think has enough time to devote to the proposed changes given all the other things they have to do?
26) Who else knows about the proposed changes?

III. What Questions (Challenges That Relate Largely to People and Roles)

1) What specific changes are being proposed?
2) What problems are the proposed changes meant to address?
3) What are the proposed changes intended to accomplish?
4) What options and alternatives, other than the proposed changes, have been considered?
5) What process did you use to decide which option or alternative would be best?
6) What persons will be affected by the proposed changes?
7) What departments or divisions will be affected by the proposed changes?
8) What processes or procedures will be affected by the proposed changes?
9) What will be the scope of the proposed changes?
10) What are the goals and objectives of the proposed changes?
11) What changes will be needed before the proposed changes can be installed?
12) What are the specifics of the implementation plan?
13) What are the specifics of the transition plan?
14) What is the schedule and timing for the proposed changes?
15) What do you expect will be the response to the proposed changes from within the organization?
16) What do you expect will be the response to the proposed changes from outside the organization?
17) What incentives, rewards, or compensation will you offer for those who are being asked to adopt the proposed changes?
18) What assistance are you offering to those adversely affected by the proposed changes?
19) What are the major uncertainties about the proposed changes?
20) What type of resistance are you expecting to the proposed changes?
21) What barriers do the proposed changes need to overcome?
22) What happens if the proposed changes do not have the intended benefits?
23) What evidence do you have that the proposed changes will be beneficial?
24) What is the most compelling argument for the need for proposed changes?
25) What is the most compelling argument for maintaining the status quo?
26) What change comes first?
27) What change comes next?

28) What things have to change?
29) What people have to change?
30) What roles have to change?
31) In what direction will the proposed changes take us?
32) What permissions are needed before the proposed changes can be put into effect?
33) What behaviors need to change?
34) What knowledge, skills, and abilities are needed to ensure new behaviors?
35) What systems and tools are needed to support the behavior change?
36) What new policies, procedures, and/or processes are needed to support the behavior change?
37) What training is needed to support the behavior change?
38) What organizational assistance is available to emotionally support those who are highly stressed by the change?
39) What reinforcements/consequences need to be in place to sustain behavior change?

IV. Where Questions (Challenges That Relate Largely to Direction and How Things Fit Together)

1) Where in the organization will the proposed changes take place?
2) Where in the organization did the proposed changes originate?
3) Where can people get more information about the proposed changes?
4) Where can people send their comments on the proposed changes?
5) Where will comments on the proposed changes be posted?
6) Where does (person "x"; department "y"; organization "z") fit into the proposed changes?
7) Where are you going to find the resources to implement and sustain the proposed changes?

V. When Questions (Challenges That Relate Largely to Scheduling and Timing)

1) When was the decision made that these changes are needed?
2) When will the proposed changes take place?
3) When will people be notified about the proposed changes?
4) When will we have time to review the proposed changes?
5) When will the proposed changes become finalized?
6) When will the schedule for the proposed changes be announced?

VI. How Questions (Challenges That Relate Largely to How Things Influence One Another and How to Measure Impacts)

1) How involved have you been in developing the proposed changes?
2) How will the proposed changes be implemented?
3) How will the proposed changes be communicated within the organization?
4) How much have others been involved in developing the proposed changes?
5) How ready are we to change?
6) How will the proposed changes be communicated externally?
7) How will the proposed changes be communicated internally?
8) How much time will we have to review the proposed changes?
9) How long will it take for the proposed changes to be implemented?
10) How much time from other work will be needed to implement the proposed changes?
11) How do you expect those implementing the proposed changes to fulfill their other work responsibilities?
12) How will you deal with those who resist or don't want to change?

13) How will you integrate the proposed changes with activities in other parts of the organization?
14) How many people will be affected by the proposed changes?
15) How will people function during the transition period as the proposed changes are being made?
16) How will you monitor and adjust the effectiveness of the proposed changes?
17) How will you sustain any benefits that will be derived from the proposed changes?
18) How much will the proposed changes cost?
19) How will we pay for the proposed changes?
20) How many resources will need to be diverted to achieve the proposed changes?
21) How many resources will need to be diverted to sustain the proposed changes?
22) How will the costs for the proposed changes affect the budgets of other programs or projects?
23) How much will the benefits of the proposed changes outweigh the costs?
24) How will people who have comments on the proposed changes be informed about the response to their comments?
25) How will you address rumors, misperceptions, and misinformation about the proposed changes?
26) How will you prevent the dissemination of rumors and false information about the proposed changes?
27) How will those making decisions about the proposed changes be impacted themselves by the consequences?
28) How prepared are you to address and mitigate any resistance that will develop in response to the proposed changes?
29) How will the proposed changes affect me, my work, my part of the organization?
30) How will you measure the outcomes and benefits of the proposed changes?
31) How will you measure if a change is taking place?
32) How will you measure success or failure?
33) How will people be informed about progress?

Appendix 9.4   The most frequently asked questions at environmental cleanups and hazardous waste sites.

I. Health Risk Concerns

1) Am I at risk from the contamination?
2) What are the risks to my children?
3) What are the risks to my pets?
4) What are the impacts on natural habitat (i.e. fish and other species)?
5) Can my children and pets play in the soil?
6) What health effects can I expect to see if I've been exposed to site contaminants?
7) What are the short-term effects?
8) What are the long-term effects?
9) Could my recent health problems (e.g., headaches, skin rashes) be the result of exposure to contaminants at the site?
10) Have any health problems been reported so far?
11) How many people have become ill as a result of the site?
12) Are you going to test residents for exposure?
13) Can you set up a temporary, local health center or clinic where we can be tested?
14) I'm pregnant (or planning to be). Will the contaminants affect my unborn child?

15) Is it safe to garden in my yard?
16) Is it safe to eat vegetables grown in my garden?
17) Is it safe to drink the water?
18) Will you provide us with bottled water?
19) Is it safe to bathe or shower in the water?
20) Is it safe to water our lawns with potentially contaminated water?
21) Is it safe to mow our lawns if the soil underneath is potentially contaminated?
22) Is it safe to use the river for fishing and other recreational purposes?
23) Is it safe to eat the fish?
24) What's being done right now to protect my and my family's health?
25) Will capping the site protect my health?
26) How serious is the contamination?
27) What happens if my ventilation system shuts down?
28) Can I get sick from breathing the air?

II. Investigation/Data Concerns

1) Where did the contamination come from?
2) How bad is the problem?
3) How much contamination is there?
4) Is the contamination moving and, if so, in what direction?
5) Are there any other contaminants besides the ones we were told about?
6) How can you be sure there are no other contaminants?
7) Will you conduct testing/sampling to make sure the soil in my yard is free of contaminants?
8) How will you decide where to sample and where not to sample?
9) Who determines what levels of contamination are considered "safe"?
10) Why don't you clean up all of the contamination, instead of allowing some to remain?
11) How do you know whether my drinking water is contaminated?
12) How do you know whether my yard has contaminated soil?
13) How do you know that it's safe to breathe the air?
14) How do you know whether it's safe to go fishing?
15) Why hasn't my well been sampled?
16) Why have some people received bottled water and not others?
17) Can I see the results of the testing you've done on my property?
18) Can I see the results of testing you've done on other properties in the neighborhood?
19) Do I have to give you access to sample my property?
20) What if I refuse access to my property?
21) Do I need to be home and take time off work while you're sampling my property?
22) I'm moving into the area; can I see the results of the sampling that's been done?
23) Who will be doing the sampling?
24) How can we be sure the sampling data is accurate?
25) How can we be sure that future sampling won't find things that you didn't find now?
26) Can you guarantee the accuracy of the sampling results?

III. Cleanup Concerns

1) How exactly are you going to clean up the site?
2) Why was this particular cleanup method chosen over other options?
3) How long will the cleanup take?

4) When are you going to start the cleanup?
5) Who is going to perform the cleanup?
6) What process was used (or will be used) to select contractors to perform the cleanup?
7) How will cleanup performance be monitored or evaluated?
8) How much will the cleanup cost?
9) Who will pay for the cleanup?
10) Will my tax dollars have to pay to address this problem that someone else caused?
11) Can taxpayers be reimbursed?
12) How will you know when everything is "clean"?
13) Why are you going to just "cap" everything and leave the contamination there?
14) Why not dig up the contamination?
15) Is dredging safe?
16) Won't dredging just "stir up" things and contaminate the water even more?
17) What if the cleanup doesn't work?
18) Can you guarantee that all of the contamination will be removed?
19) How will my quality of life be affected during the cleanup (i.e. noise, traffic, odors)?
20) After you finish the cleanup, then what? (What happens next?)
21) After the cleanup, will you continue to test to make sure it's still working?
22) What happens if my water (or soil, etc.) is still contaminated after the cleanup?

IV. Communication Concerns

1) Why did it take you so long to tell us about the contamination?
2) How can I trust what you're telling me about the site?
3) How can I trust what you're telling me about my safety?
4) What happens if you find high concentrations of contaminants near my home? How will I know?
5) How will I be informed of what's going on?
6) Will you share the testing data with residents?
7) Will you let us know if something unexpected happens during the cleanup and things get worse?
8) Is there someone local residents can talk to if we have questions or concerns?
9) Where can I get more information about this site?
10) 10.Where can I get more information about similar sites that have already been cleaned up?
11) If a cleanup plan is selected that residents disagree with, is there an appeal process?
12) How will you address public comments?
13) Will you address all of the public comments?
14) How do you decide which comments not to address?
15) If the majority of residents disagree with how EPA [or another agency] is planning to clean up the site, will EPA [or other agency] change its mind?
16) There's another site down the road; can you tell me what's going on there?
17) When you first discovered there might be a problem, why didn't you tell us then?

V. Economic Concerns

1) If soil is excavated from my yard, will I receive financial assistance to replace plants and shrubbery?
2) My property value has decreased because of the site contamination problem. Will I be compensated for this?

3) I'm concerned that cost will be the driving force behind the agency's selected cleanup option; does community opinion really matter?

4) I was told residents might have to relocate during the site cleanup. Who will pay for my moving costs? What about other expenses I may be forced to incur (i.e. costs of transporting my children to school because they won't be able to take the bus, or daily food costs because I won't have access to my stove and refrigerator, etc.)?

5) The site has placed a "negative stigma" on our community that may affect potential investors, developers, or homeowners; what will be done about this?

6) Will this keep our community from developing?

7) Can we get jobs helping with the cleanup?

8) If we can't eat the fish anymore because of health risks, can you give us a food subsidy?

9) Do you have enough money to cover the cleanup costs?

10) If you discover the cleanup is going to cost more than estimated, what happens then?

10

Communicating Numbers, Statistics, and Technical Information about a Risk or Threat

---

CHAPTER OBJECTIVES

This chapter synthesizes a large and rich body of research on communicating numbers, statistics, and technical information in risk, high concern, and crisis situations. It looks at the implications of that research for principles, strategies, and methods.

At the end of this chapter, you will be able to:

- Explain the factors that influence both comprehension of and reaction to numeric and statistical information.
- Predict barriers to comprehension of statistical and technical information.
- Describe common practices that result in poor communication of numbers and statistics.
- Interpret technical factors in risk comparisons and predict how they can be perceived.
- Apply essential guidelines for presenting numbers and statistics about risk.
- Recognize that perceptions are often driven more by emotions and feelings than by technical facts in high-stress situations.
- Recognize that perceptions are often driven more by feelings and emotions than by facts in high-stress situations.

10.1 Case Diary: A Civil Action

Several years ago, I was called by a colleague who asked for my help. My colleague is a professional toxicologist and epidemiologist. He had been hired as a technical advisor to a community advisory committee (CAC). His role was to help the CAC respond to several highly vocal and concerned citizens who believed there was a cancer cluster in their community; they felt they were seeing too many cases of cancer, including childhood cancer, appearing in their area, and they felt that these cancers resulted from improper management of hazardous waste by a local industrial facility. My colleague sought my help on several communication challenges, including responding to angry people and explaining why cancer clusters are often so difficult to determine.

In response to community concerns, a public health agency had investigated the possibility of a cancer cluster. Their report, which had been made available to the community, explained the following:

- Determining if there is a cancer cluster in a community is a complicated and difficult technical task.
- A disease or illness cluster is defined as a *greater-than-expected* number of cases that occurs within a group of people in a geographic area over a period of time. To be a cancer cluster, a

Communicating in Risk, Crisis, and High Stress Situations: Evidence-Based Strategies and Practice, First Edition. Vincent T. Covello.
© 2022 by The Institute of Electrical and Electronics Engineers, Inc. Published 2022 by John Wiley & Sons, Inc.

group of cancer cases must meet several technical criteria. For example, a greater-than-expected number is when the observed number of cancer cases of a particular type of cancer is higher than one would typically observe in a similar setting (i.e., in a group with similar population, age, race, or gender).

- Numerical and statistical confirmation of a cancer cluster does not necessarily mean there is any single, external cause or risk that can be addressed. A confirmed cancer cluster could be the result of chance, miscalculation of the expected number of cancer cases, differences in the definition between observed cases and expected cases, known causes of cancer (e.g., genetics), or unknown causes of cancer.

After careful consideration of the data, the agency report concluded there was insufficient evidence to declare a cancer cluster.

Many in the community were not satisfied with the report's findings. They thought the community's cancer numbers were too large to be a statistical accident.

The issue of cancer clusters is often emotionally charged – as anyone with a friend or relative who has been ill from cancer can attest. Beyond that, the statistics are often hard to comprehend. Cancer is, unfortunately, a common illness. According to the National Cancer Institute, some 39.5% of men and women in the US will be diagnosed with cancer sometime during their lifetimes.[1] A normal cancer rate when it is manifested in certain situations can be perceived by members of a community as a cluster.

The government agency that performed the cluster investigation had a program for addressing community concerns. The program included assistance to communities to do their own analyses. The agency recognized that citizens typically do not have the technical resources to perform their own assessment about a cancer cluster. Experts possess information that non-experts need, especially when investigating a high-stress issue associated with complex technical information. The government agency had provided a technical assistance grant for the local CAC to hire their own experts; the grant was paying for my colleague's assignment as technical advisor. The specific charge given to the CAC by the government agency was to identify issues where there was disagreement with the government investigation, discuss these issues among themselves, and report the results of their discussions to the larger community.

Membership in the CAC was voluntary. The CAC had had no trouble recruiting its twelve members, as this was a high-profile and controversial issue in the community. Members of the CAC were self-selected or had been nominated by community groups.[2] Members of the CAC included politicians, teachers, parents, first responders, retired journalists, engineers, storekeepers, and leaders from faith-based and voluntary organizations. CAC representatives were active participants in such organizations as the Historical Society, the Senior Citizen Center, and the American Legion.

My colleague had a tough assignment, as the method for determining a cancer cluster is complex and technical. He and I agreed that one of our primary communication goals would be to help members of the CAC understand the methods and assumptions used in a cancer cluster investigation. Since many disagreements about cancer clusters arise from misunderstandings about methods and assumptions, and since many members of the CAC were from local organizations trusted in the community, we wanted the CAC to be the hub for sharing technical information about the possible cancer cluster to the community. It was not possible for everyone in the community to become informed and educated about the technical issues, and we needed to rely on the "ladder of trust" (see Chapter 8); the CAC was well represented to serve as trusted spokespersons.

But sharing information and educating the CAC members to a level that allowed them to feel they owned and understood the data was challenging. To accomplish this, we determined we needed to break down the elements of the problem and form subcommittees on different aspects of a cancer cluster investigation. The subcommittees would then share their understandings and findings with the whole CAC. After review and comments by the full CAC, the government agency, and the industrial facility, the report would be shared with the larger community.

My colleague and I understood that two of the most difficult technical concepts to understand about a cancer cluster investigation are (1) the causes of cancer, and (2) exposures to cancer-causing agents. We formed five subcommittees: one on the causes of cancer, one on sources of exposure to cancer-causing agents, one on possible exposure routes to cancer-causing agents, one on possible doses of cancer-causing agents received through different exposure routes, and one on possible health outcomes of exposure to a cancer-causing agent by different individuals and population groups. For background reading, we provided a primer on risk assessment that I had written as well as two other short primers.[3]

We also needed to design a successful learning process. We began each subcommittee meeting with a presentation on the risk assessment methods most relevant to their subcommittee's task. Since there was limited time, we told people what we had left out. We also reserved two-thirds of the time for questions and discussion. (We based this on the proverbial principle that we have two ears and one mouth, so we should listen for twice as much time as we speak.) The discussion time was important for true understanding. My colleague then asked members of each subcommittee to present what they had learned from the session, and how the technical information we had presented applied to the community cancer cluster investigation. We based this aspect of our learning process on the knowledge that having to teach something technical to others is one of the best ways to ensure one's own learning.

My colleague and I also started with the assumption that non-experts can understand complex technical issues if the issues are properly explained and if the person is motivated. One subcommittee member said he would not attend his subcommittee's presentation because he was bad in math and science, so the presentation would be wasted on him. The schoolteacher on his subcommittee, who taught math at a local school, remembered him as a student. She urged him to participate and told him that "Motivation is key. When people are motivated to learn, they learn." Good teacher that she was, she reminded him about a time when he had explained to her the mathematical odds of winning different hands in poker.

The outcome of CAC subcommittee and full committee deliberations was a final report to the community and a press conference. After careful consideration of the evidence, the CAC concluded there was insufficient evidence to declare a cancer cluster.

Ten of the CAC members signed the report. Two members of the CAC wrote a dissenting report. These members had not attended any of the learning sessions, essentially boycotting the process. They had insisted that they should be able to do their own study from scratch (i.e., rather than analyzing the data collected by the government agency). This was not possible both for privacy concerns regarding health records and because they would not be using standard data-collection processes. For these two individuals, the process had not overcome their distrust of any "official" or "scientific" methods.

The majority CAC report concluded with five recommendations for future action and determined they needed to stay active as a CAC to operationalize the recommendations. While there was not sufficient evidence of a cancer cluster, the CAC would not let the industrial facility off the hook. They developed the following thoughtful recommendations, all of which were accepted by the government agency, the industrial facility, the mayor, and the town council.

Oversight. The CAC would continue to receive technical information and actively participate in any future investigations of cancer clusters, environmental health issues, and hazardous waste operations.

Operation and Maintenance. The government agency and industrial facility agreed to a procedure whereby suggested improvements made by the CAC to the operation and maintenance of the facility would be reviewed and implemented as appropriate.

Waste Management. The industrial facility would invest in new equipment to manage and reduce waste and would provide an annual report on progress to the CAC.

Penalties. Fines for any waste management violations would go into a community trust fund, which would be administered by the CAC.

Property Values. The industrial facility agreed to enter into discussions and possible negotiations about protecting the property values of homes near the facility from any new or unanticipated impacts on those values from being close to the industrial facility.

This case diary illustrates many of the challenges involved in communicating complex technical information about a risk or threat. It also illustrates these challenges can be addressed by understanding how people process technical information and how to best share that information. Not all members of the CAC came to the same interpretation of the data, but ten of the twelve members of the CAC concluded they were well enough informed to produce a highly constructive report and outcome.

10.2   Introduction

Numbers, statistics, and technical information about a risk or threat are shared in a wide variety of situations. These circumstances range from common situations such as organizational change and provider–patient interactions to headline issues such as natural and human-caused disasters and disease outbreaks.

Technical information addresses the many questions people have about a risk or threat. These include the following questions: What can go wrong? How likely is it that it will go wrong? Who will be affected if it goes wrong? What will be the consequences if it goes wrong? What can be done to reduce the likelihood and consequences of it going wrong?

There are three basic formats for communicating risk-related numbers, statistics, and technical information: *verbal formats*, *numerical formats*, and *graphical/visual formats*. A key communication challenge with *verbal formats* is that people vary widely in how they understand words. For example, *verbal formats* often include terms such as "likely" or "unlikely." People often interpret these terms differently.

Numerical formats often include percentages, such as "x" percent of people are obese or are infected with a disease such as COVID-19. A key communication challenge with *numerical formats* is that a significant proportion of people are numerically challenged or have difficulty processing the meaning of probabilities, such as one in a million.

Visual formats include infographics, pictograms, tables, charts, drawings, videos, analogies, metaphors, and photographs. One key communication challenge with *visual formats* is that few generalized best practices can be recommended because of lack of consistency in how visual formats have been tested, lack of testing using randomized controlled studies, and lack of theories about why one *visual format* is more effective than another.

A communication challenge with all three formats is determining how much technical information is needed by the target audience for accurate comprehension and informed decision-making.

A second communication challenge affecting all three formats is determining how much technical information is wanted by the target audience. For example, numbers, statistics, and technical information about a risk or threat may provide the target audience with only one of the many important technical characteristics of the risk or threat. These technical characteristics include the probability or likelihood of an adverse occurrence, the magnitude or severity of an adverse occurrence, and the certainty attached to these numbers and statistics.

The models and frameworks discussed in Chapters 7,8 and 9 help explain how people process technical information about risks and threats. For example, the risk perception model posits that perceptions of risks and threats in high concern and high stress situations are often driven more by emotions and feelings than by technical facts. The mental noise model posits that stress will reduce the ability of people to process – hear, understand, and remember – numbers, statistics, and technical information about risks and threats. Shrinkage of cognitive abilities is particularly high in crisis situations. The negative dominance model posits that people under stress will often focus more on negative technical information than on positive technical information: it typically takes three to four pieces of positive information to offset one piece of negative technical information. The trust determination model posits that people will be more attentive to technical information provided by trusted sources of information. Trust, in turn, will be based on perceptions that risk management authorities are caring, empathic, competent, knowledgeable, honest, open, and transparent. The more people trust an individual or organization, the more likely they will attend to and accept technical information provided by that individual or organization.

The sociocultural model posits that social and cultural factors will shape attitudes and decisions about what technical information is important and what can be safely ignored. These same social and cultural factors, especially as they play out through interpersonal, mainstream, and social media networks, will also amplify or lessen attention and importance attached to technical information about a risk and threat. Amplification will also be affected by sociocultural and media framing – the process of selecting and highlighting particular risks and threats through words, images, phrases, worst-case scenarios, and human interest narratives.

10.3 Case Study: Numbers, Statistics, and COVID-19

During the COVID-19 pandemic in 2020, nearly every day brought a flood of new or revised COVID-19 numbers. Americans heard numbers such as the following: "If 95 percent of Americans consistently wore masks, it could save an estimated 56,000 lives by April 1."[4] People were confused by competing claims. For example, some claimed that the rise in numbers of COVID cases and deaths was due to additional testing, while others assured the public that the rise in cases and deaths was real and statistically accurate.

What was clear to most people was the staggering physical and economic stakes at risk. Mental costs of the pandemic were also staggering. Nearly 8 in 10 adults reported that the pandemic was a significant source of stress in their life.[5] For nearly all of 2020, COVID numbers, statistics, and technical information played a role in virtually every policy decision. For example, numbers determined when to apply or lift social distancing restrictions, when to allow or disallow social gatherings, when to allow or cancel sporting events, and when to open or close businesses and schools.

COVID-19 numbers and statistics reported to policy makers, scientists, journalists, and citizens covered a wide range of actions, events, and behavior. These included COVID-19 deaths, infections, cases, patient hospitalizations, emergency department visits, testing, positive tests (viral and antibody tests), negative tests (viral and antibody tests), effectiveness of viral and antibody tests, shortages of personal protective equipment for health-care workers (e.g., ventilators and surgical masks), mask use, adherence to social distancing guidelines, contact tracing, incubation times, and days needed for effective isolation and quarantine. In some cases, the numerical data were organized by racial and ethnic groups. They were provided by country, state, region, and locality.

Each COVID-19 number represented a piece of the puzzle. For those who tracked and understood numbers shared through mainstream, social media, and interpersonal networks, a picture would emerge of the disease, its spread, and its severity. The general public could follow the disease outbreak in almost real time, watching probabilities and outcomes change hour by hour, day by day, through mainstream and social media channels. The problem was that the numbers and statistics being communicated and used to inform decisions about COVID-19 were often imperfect and incomplete. Any number or statistic was likely to yield an under- or over-estimate because of biases.[6] Their interpretation also depended on a basis of technical knowledge about matters such as testing and the course of the disease, and about rules and regulations being imposed.

For example, due to the long course of incubation and infection, COVID-19 *death numbers* were often difficult to interpret because they typically reflected the state of the outbreak several weeks before; *case/infection numbers* were often difficult to interpret because of limited testing and because of testing errors (e.g., false positives and false negatives). *Hospitalization numbers* were often difficult to interpret because they typically reflected only the most severe cases of infection and the patients who were exposed to the virus several weeks before admission to the hospital. *Emergency room numbers* were often difficult to interpret because they reflected patients with symptoms who were exposed up to two weeks earlier. *Positive test numbers* were difficult to interpret because of the limited availability of testing and because people who were tested were mostly symptomatic (versus asymptomatic or mildly symptomatic) or had been exposed to a person with COVID-19. Numbers of all types were difficult to interpret because of the lack of shared standards, protocols, and resources at the federal, state, and local level for collecting and reporting COVID-19 data.

Communications about COVID-19 were at their best when they incorporated principles of risk, high concern, and crisis communication.[7] For example, numbers were best communicated when they were accompanied by information about their strengths and weaknesses, such as small sample sizes, presence or absence of peer review, and time lags between the event and the reporting of the event; when they acknowledged uncertainties and stated clearly what was known and not known; when they acknowledged when the numbers were estimates based on modeling, that the model used was based on multiple assumptions (e.g., assumptions about transmission rates, incubation times, recovery times, and asymptomatic cases), and that these assumptions may result in large overestimates or underestimates;[8] when they acknowledged when they were overall averages and may therefore hide disproportionate impacts on some groups based on age, location, race/ethnicity, socioeconomic status, and other factors; when they provided consistent information from trusted sources; when they were supported and enhanced by visuals tested for the specific context and personal narratives; and when they incorporated feedback mechanisms to engage stakeholders.

Examples of effective COVID-19 technical communication at the international level are the COVID-19 communications of Prime Minister Jacinda Ardern in New Zealand and the "Hands, Face, Space" COVID-19 public health campaign by the National Health Service in Great Britain. The "Hands, Face, Space" campaign distilled complex technical information about COVID-19 into a three-message verbal and visual mantra repeated through multiple channels and formats.[9] As shown in Figure 10.1, the campaign urged the public to wash their hands, cover their faces, and engage in social distancing to control the spread of COVID-19. A wide and diverse set of social media resources expanded the three key messages. In the plethora of confusing numerical and technical information, it was essential to rise above the technical detail and deliver clear, life-saving health messages that could be understood and acted upon by the public.

Examples of effective COVID-19 communication in the United States were the COVID-19 communications by Dr. Anthony Fauci, director of National Institute of Allergy and Infectious Diseases and a leading scientific member of the White House Coronavirus Task Force; by Dr. Richard Besser, president of the Robert Wood Johnson Foundation and previously head of the US Centers for Disease Control and Prevention; by the Association of State and Territorial Health Officials, which produced frequent COVID-19 updates and blogs;[10] and by Maryland governor Larry Hogan and New York governor Andrew Cuomo in their initial statements about what they knew, what they did not know, and what they were doing to find out. Technical information shared by these sources, separately and collectively, often contradicted overly reassuring messages offered by the White House and others.

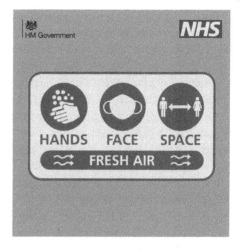

Figure 10.1 Coronavirus (COVID-19) "Hands, Face, Space" Public Health Campaign (Great Britain).

For example, they contradicted the often overconfident messages from the president of the United States.[11]

What these examples of COVID-19 technical communication illustrate is the importance not just of the apparent clarity of a chart or repetition of a statistic, but the approach, purpose, and contextualization of the message that the technical information is intended to convey. What these examples of COVID-19 communication shared in common were six characteristics: (1) learning from history; (2) adapting to a changed social media and stakeholder engagement landscape; (3) recognizing differences between expert and non-expert judgments of risks and threats; (4) setting, guiding, and managing expectations; (5) acknowledging uncertainties; and (6) recognizing the Importance of Special and Vulnerable Populations. Each is briefly described below.

(1) Learning from history. The most effective technical communications about COVID-19 reflected awareness of lessons learned from previous disease outbreaks. For example, communications reflected awareness of lessons learned from the 2003 SARS (severe acute respiratory symptom) outbreak, the 2005 H5N1 (avian influenza) outbreak, and the 2009 H1N1 (pandemic influenza) outbreak. Lessons learned were embedded in various documents, including a nearly 400-page pandemic influenza plan.[12] The communications also reflected awareness of observations about the Spanish flu pandemic published in a paper over 100 years ago in one of the world's leading science journals.[13] The paper argued that three main factors stand in the way of the prevention of disease outbreaks: (1) people do not appreciate the risks they run, (2) it goes against human nature for people to shut themselves up in rigid isolation as a means of protecting others, and (3) people often unconsciously act as a continuing danger to themselves and others.

(2) Adapting to a changed social media and stakeholder engagement landscape. Recognizing that the general public increasingly relies on social media to inform their knowledge about risks and threats, the most effective technical communications about COVID-19 were shared widely through social media channels. Technical information posted on social media channels was enhanced by trust-building messages of caring, empathy, competence, and honesty. Through social media, several goals of successful risk, high concern, and crisis communication were accomplished: facilitating perceptions of control and empowerment through knowledge, encouraging interactive and ongoing stakeholder engagement, broadening the transmission of urgent public health information, and reaching out to diverse audiences.

(3) Recognizing differences between experts and non-experts in judgments of risk. The most effective technical communications about COVID-19 recognized that the way experts process technical information about a risk or threat is often quite different from the way non-experts process the same technical information.[14] These differences are discussed in detail in Chapter 3. For example, emotions play a large role in how the public makes judgments and decisions about a risk or threat. These same emotions also affect, often unconsciously, the judgments and decisions of experts.

(4) Setting, guiding, and managing expectations. The most effective technical communications about COVID-19 focused on setting, guiding, and managing people's expectations. Dr. Glen Nowak, director of the Center for Health and Risk Communication at the University of Georgia and chief of media relations at the Centers for Disease Control and Prevention 2003–2010, noted in an interview about COVID-19 that communications "often get off track by trying to get out as much technical information as possible and 'educating people.' You have to help people understand what's about to unfold. You have to give people an understanding of what government agencies and health care providers will be doing . . . If you set expectations that things will go smoothly and easily, you will probably find that you won't be able to meet that expectation."[15]

(5) Acknowledging uncertainties. The most effective communicators of COVID-19 technical information included explicit information about uncertainties. They noted that guidelines might change quickly as technical knowledge changed. If there were inconsistencies with information provided previously (e.g., because of new research findings), they were transparent about them; they were the first to identify the inconsistencies, and they explained inconsistencies using plain language. For example, the US state health directors continually revised and updated their guidance document with simple answers to over 75 of the most frequently asked questions about COVID-19. The guidance document was almost entirely rewritten every three to four weeks in concert with growing scientific knowledge, the phase of the crisis, and public responses to information.

(6) Recognizing the importance of special and vulnerable populations. The most effective communicators of COVID-19 technical information recognized and emphasized that unique filters affect the way members of many vulnerable, racial, ethnic, and particular socioeconomic groups process technical information.[16] For example, effective COVID-19 communicators in the United States recognized and emphasized that many African Americans and Hispanic Americans viewed technical information about COVID-19 through the lens of past experiences with public health authorities (e.g., the Tuskegee experience for African Americans). In the US, state health directors emphasized the importance of these inequalities in their COVID-19 Q&A risk communication guidance document, noting that data indicated that particular racial and ethnic minorities are at higher risk of COVID-19; that living conditions, work circumstances, underlying health conditions, and access to care contribute to higher rates of COVID-19 disease among particular racial and ethnic minorities; and that risk management authorities need to address more directly racial and minority disparities in COVID-19 disease prevention, protection, and communications.[17]

10.4 Brain Processes That Filter How Technical Information about Risk or Threat Is Received and Understood

Technical information about a risk or threat passes through numerous filters in the brain. Many risk, high concern, and crisis communication efforts proceed without recognizing the functioning of these filters. Not recognizing these filters often means not understanding the audience. These filters affect what an audience hears and what information they take away.

Five of the most important filters are risk perception filters, thought processing filters, mental model filters, emotional filters, and motivational filters. The functioning of these filters depends heavily on context, personality, socio-demographic, and cultural factors.

10.4.1 Risk and Threat Perception Filters

Risk and threat perception filters play a prominent role in how the brain processes numbers, statistics, and technical information. As noted previously in this book, perceptions equal or become reality; that which is perceived as real is real in its consequences. Risk and threat perception factors, sometimes called fear factors, include the 20 factors identified in Chapter 3. Several of the most important are familiarity, perceived trust in risk management authorities, perceived benefit of the activity generating the risk, perceived voluntariness, perceived fairness, perceived personal control, and perceived technical or scientific certainty. These factors, individually and collectively, can lead a person to grossly underestimate or overestimate the probability, frequency, and severity of a risk or threat.

One of the important conclusions coming from over 50 years of risk, high concern, and crisis communication research is the very low correlation between whether something is dangerous and whether it is upsetting. The things that harm people and the things that alarm people are often different. Scientists and technical professionals know something is dangerous, but that in itself does not tell them whether it is upsetting to people or makes them angry.[18]

The more severe, undesirable, or damaging the perceived physical, psychological, or socioeconomic outcomes of a risk or threat, the higher the likelihood a person will search for and process technical information about the risk or threat. Physical adverse outcomes include physical harm and physical effort. Psychological adverse outcomes include anxiety, worry, fear, embarrassment, and peer disapproval. Socioeconomic adverse outcomes include monetary loss and organizational disruption. Information about adverse outcomes can be communicated through a variety of techniques, such as arguments based on authority, reason, or emotions (see Chapter 9).

Because of risk and threat perceptions, information for the public can typically be made more effective by focusing less on the numerical, statistical, and technical characteristics of the risk or threat and more on the psychological and sociocultural characteristics. For example, risk-related technical information about a disease outbreak is more likely to be received and processed by the brain if the communications emphasize: (1) the benefits to the recipient and to others of preventive and protective measures; (2) the alignment of preventive and protective measures with the values and behavior of others; (3) the prospect of social group approval; and (4) the likelihood of encountering misinformation, the various forms that misinformation takes, and how to obtain accurate information.

Technical information about a risk or threat is also more likely to be received and processed by the brain if the information is tailored to the specific concerns of the target audience. For example, polling data focused on the language of vaccine acceptance for COVID-19 found the following:[19]

- Groups least certain about getting vaccinated were rural Americans, young Republicans, and young Black Americans.
- When asked about the biggest concern about taking the COVID-19 vaccine, one-third of all respondents said either long-term side effects or short-term side effects.
- The top three technical statements about side effects that respondents found most reassuring were, "the likelihood of experiencing a severe side effect is less than 0.5 percent," "mild side

effects are normal signs that their body is building protection," and "most side effects should go away in a few days."

- The most convincing technical reason to take the vaccine for all respondents was the vaccine's 95% efficacy.

10.4.2 Thought Processing Filters

Thought processing filters include the "rules of thumb," cognitive biases, and mental shortcuts that the brain uses to simplify decision-making. In the neuroscience, psychology, and behavioral economics literature, they are technically known as *heuristics*. Several of the most important cognitive biases that often come into play are listed in Table 10.1.

Heuristics play a key role in how the brain processes numbers, statistics, and technical information.[20] They make them easier to process. However, heuristics can also result in distortions since the human brain often uses only a small amount of available technical information in making decisions about a risk or threat. One of the most powerful heuristics is the *availability heuristic*. Because of the *availability heuristic*, the brain tends to assign greater probability to events that are visible or of which it is frequently reminded, such as through interpersonal networks, mainstream media, and social media. For instance, holding constant other variables, if a person knows someone who died recently of COVID-19, then that person is more likely than other people to believe that contracting COVID-19 is fatal. The brain tends to assign greater probability to events that are easy to recall or imagine through concrete examples or dramatic images.

Because of the *confirmation heuristic*, the brain often seeks out numbers, statistics, and technical information that supports existing beliefs. The brain tends to ignore or discount numbers, statistics, and technical information that do not conform to existing beliefs.

Because of the *optimism/overconfidence heuristic*, the brain believes that bad things are less likely to befall oneself than others. While the *optimism/overconfidence heuristic* can reduce excessive feelings of stress and anxiety, it can lead the brain to underestimate the probability and magnitude of an event or situation. For example, because of optimism, a person may know the probability of contracting COVID-19 while at the same time underestimating the probability of being personally infected or experiencing consequences; this attitude can in turn lead to more risk-taking behavior.

Table 10.1 Common cognitive biases that affect the understanding of risk-related numbers and statistics.

---

1) *Availability bias* – events are perceived to be more frequent if examples are easily brought to mind, with memorable events seeming more frequent.
2) *Confirmation bias* – once a view has been formed, new evidence is generally made to fit, contrary information is filtered out, ambiguous data are interpreted as confirmation, and consistent information is seen as "proof."
3) *Overconfidence bias* – people often think their predictions or estimates are more likely to be correct than they really are.

10.4.3 Mental Model Filters

A mental model is a representation of how something works. Since the human brain typically cannot keep track of all of the details of how things work, it uses models to simplify complexity into understandable and organizable chunks.

Mental models are based in part on the correct or incorrect technical facts known to a person. Correct and incorrect technical facts, in turn, are based on past experiences, intuition, risk perceptions, and, most generally, whatever a person has heard, read, and viewed. Mental models influence what technical information is attended to. They define how the brain approaches and solves technical problems and serve as the framework into which people fit new technical information.

For communications about a risk or threat to succeed, communicators must first discover what misconceptions exist. Communicators can then attempt to disconnect the erroneous information from the person's mental model and replace it with new technical facts.

One method for correcting an incorrect mental model is to encourage application of *Bayes's theorem,* a thought process whereby the brain takes into account prior relevant probabilities and then updates these probabilities as newer information arrives. Communications should encourage people to ask: How confident am I in this prediction and what probabilistic information will impact my confidence? Each new piece of information will affect the original probability, and that information should be provided. For example, to help individuals understand the risks of a side-effect of a vaccine, they can be encouraged to consider: first, what is the base rate in the population for this risk in general; second, is there sufficient research documenting this risk; third, how much greater than the base rate is this risk after vaccination; and fourth, to what extent do they themselves match the population in which this risk has occurred.

10.4.4   Emotional Filters

Emotional filters play a key role in how the brain processes numbers, statistics, and technical information.[21] For example, the more dreaded and feared an adverse outcome, the more likely the brain will evaluate its risk or threat as high and/or imminent. In many cases, emotions and feeling are more powerful drivers of risk-related behavior than numbers, statistics, and technical information.

Emotions and feelings can result in high levels of anger, frustration, and outrage.[22] The brain tends to overestimate a risk or threat when anger, frustration, and outrage are high, whether or not the actual risk is low. The causal link between emotions and risks creates a challenge for risk managers and communicators. They need to simultaneously manage the risk or threat, manage expectations, and manage emotions. One way of reducing anger, frustration, and outrage is to acknowledge emotional distress, such as fear and anger, and attempt to address the emotional distress by addressing the underlying causes. When people are stressed and feel endangered, emotions often hijack rational thinking; the brain asks reason and logic to take a back seat. Emotions and their causes have deep roots in the brain's fight-freeze-flight response to threats, in risk perceptions, and how the brain processes information in high-stress situations.

10.4.5   Motivational Filters

Motivation is key to the processing of numbers, statistics, and technical information about a risk or threat. Motivation is the readiness and interest of the receiver to process information about a risk or threat.

Motivational filters can arouse, energize, and urge the brain to seek out and process technical information about a risk or threat. Needs that motivate the search for technical information include *physiological needs* (e.g., hunger, thirst, and air); *safety and security needs* (e.g., security, stability, shelter, structure, order, and protection); *support and relationship needs* (e.g., belonging, friendships, children, love, affection, pleasure, and companionship); and *status needs* (e.g., appreciation, achievement, approval, reputation, strength, dignity, wealth, esteem, power, fame, glory, and recognition).[23] Numbers, statistics, and technical information are more likely to be sought out and processed (and produce attitudinal or behavior change) if a trusted relationship has been established between technical information and one or more of these needs.

Abraham Maslow proposed a list of these motivational needs and a hierarchy shaped like a pyramid. He posited that people are motivated to fulfill basic needs (e.g., safety) before moving on to other, more advanced needs. However, later research found limited evidence for Maslow's ranking of needs and that these needs are in a hierarchical order.[24] For example, in particular contexts, needs related to status, support, and relationship may be more important than physiological, safety, or security needs.

10.5 Challenges in Explaining Technical Information About a Risk or Threat

Major challenges exist in presenting and explaining numbers, statistics, and technical information about risks and threats. One challenge is sharing information about an event that has a very low probability of occurring. A classic example is a threat that has a smaller than a one in one million chance of occurring and for which people have little experience. In such circumstances, perceptions are strongly influenced by events that are easy to recall or imagine through concrete examples or dramatic images.

A second challenge is sharing technical information about exponential growth. Many people underestimate how fast a number can increase – a mistake known as the "exponential-growth bias." Exponential-growth bias" is the tendency for individuals to underestimate exponential growth by neglecting the role of compounding.[25] Many people would not realize that if a person invests $1 in an investment club and the club offers to double the investment every three days, in 60 days the person would be a millionaire (exactly $1,048,576), and, after another 30 days, the person would be a billionaire. Misunderstandings about exponential growth underlie many of the mistakes people make in judging a risk or threat. For example, difficulties understanding exponential growth help explain why many people underestimate the speed at which a disease such as COVID-19 can spread – a misunderstanding that can lead to people ignoring public health messages, such as the importance of face masks and social distancing in slowing spread of COVID-19.[26]

A third common challenge arises from the concept of cumulative risk. A particular event might have a low probability of occurring at a particular time or place. However, over time, those probabilities add up and create a significant risk that is easily overlooked.

A fourth challenge arises from the use of qualitative words or expressions to describe numerical probabilities. For example, the term "probably," when used to describe the probability of an uncertain event, can mean different things to different people and can be interpreted in different ways.[27]

A fifth challenge arises from the lack of numerical skills in the general population. Almost one in three Americans lack even the most basic numeracy skills necessary to interpret mathematical information in daily life.[28] Understanding probabilities is even more challenging. For example, an enormous commercial industry has been built entirely on restating and translating the probabilistic information provided in National Weather Service forecasts in ways that a television audience can easily understand. The National Weather Service provides clearly stated probabilities, but the basic assumption behind those probabilities is difficult for many to grasp. For example, for a forecast of 30% probability of precipitation for New York tomorrow, a person may be unsure as to whether that means: (a) it will rain over 30% of the New York area tomorrow; (b) it will rain for 30% of the time tomorrow somewhere in New York; (c) there is a 30% probability it will rain somewhere in New York tomorrow; or (d) at any given location in the New York area, there is a 30% probability that it will rain tomorrow. The definition of a precipitation event used by the National Weather Service is measurable precipitation within the stated time period at any point in the area for which the forecast is valid – i.e. (d) is the correct answer.

Weather forecasts become crisis communications when forecasting severe, potentially life- and property-threatening weather events. Extreme weather events require effective responses by institutions and individuals. However, people frequently underestimate extreme weather events or do not behave in an appropriate manner to protect themselves from the event. For example, the compliance rates with evacuation orders for hurricanes are often significantly less than 100%. Many coastal residents prefer to ride out a hurricane rather than leave.[29] The research literature indicates technical information about a hurricane goes through many filters, each of which can affect compliance with an evacuation order.[30] These include prior hurricane experience, perceived severity of the storm, exposure to severe winds and flooding, household characteristics (e.g., number of persons in the home and their health status), perceived financial costs, concerns about traffic, concerns about protecting property, concerns about the safety of pets and livestock, concerns about family members, lack of transportation, and strong individualist worldviews.

10.6 Framing

Responses to technical information about a risk or threat are highly dependent upon how the information is framed.[31] Framing studies indicate that decisions about a risk or threat are influenced significantly by the manner in which numbers, statistics, and technical information are presented. Variations in the presentation of the same information have a large impact on decisions. Framing studies indicate that the choices and decisions people make about risks and threats depend profoundly on how the information is framed. Some of the most powerful framing effects are observed when a choice or decision is framed in negative or positive terms, such as "lives lost" or "money lost" versus "lives saved" or "money saved."

In a classic study of framing, Tversky and Kahneman explored the framing effect by examining how different phrasing affected choices in hypothetical situations.[32] Participants were asked to choose between two treatments for a deadly disease, each with the same probability of effectiveness. Choices were presented to participants either with positive framing (i.e., how many people would live), or with negative framing (i.e., how many people would die). Participants overwhelmingly chose the treatment with positive framing.

In another study, doctors were presented with a hypothetical choice between two cancer therapies with different framing. Half were told about the relative chances of dying while the rest had the same information presented in terms of survival rates. This change in framing – even though the results were the same – more than doubled the number of doctors choosing the alternative framed in terms of survival.[33] Doctors, for example, are much more likely to prescribe a new medication that saves 30% of its patients than one that loses 70% of them.

Because of framing effects, and because outcomes can be measured against different reference points (as with the bottle half-full or half-empty), no number, statistic, or technical information will ever be the "right number" for every stakeholder. Each way of expressing a number frames it differently, and thus it will have a different impact on the audience.

Framing effects raise important questions about how probabilistic information should be presented to counteract the effect of framing upon decisions. For example, framing affects how

Table 10.2 Different ways to express risk information about possible deaths from exposure to a pollutant.

The expected number of:
1) deaths per million people in the population exposed to the pollutant;
2) deaths per million people living within "x" miles of the source of the pollutant;
3) deaths per unit of concentration of the pollutant;
4) deaths for each source of the pollutant;
5) deaths per ton of pollutant released into the environment.

numbers and statistics concerning deaths are processed by the brain. The various ways of expressing the expected number of deaths, such as the expected number of deaths from an exposure to a toxic substance or virus such as COVID-19, are not equivalent. Consider the issue of how many people will die annually as a result of exposure to a toxic substance in air from a specific industrial accident. As shown in Table 10.2, risk information can be expressed in different ways.

These different ways of presenting information are not psychologically equivalent. Different measures will often strike an audience as being more or less appropriate, frightening, or comprehensible. Expressions of risk should be pretested to understand their effect.

The effect of framing has consistently been shown to be one of the largest factors affecting the processing of numbers, statistics, and technical information. This is a complex factor, as the effects of framing depend greatly on contextual and situational factors, such as knowledge, the level of risk or threat, the type of risk or threat, trust in the source of information, age, experience, the description of options and choices, the cognitive abilities of the sender or receiver, and the topic itself.

10.7 Technical Jargon

To be meaningful, the jargon used by technical professionals to communicate technical information about a risk or threat must be translated into terms that non-technical experts can understand. Technical jargon is typically used by scientists and technical professionals to be precise and accurate when describing a risk or threat. However, technical jargon can also cause confusion when that same information is communicated to the general public. For example, following the nuclear power plant accident in Fukushima, Japan, experts used a variety of ways to describe radiation. A short list of terms included millisieverts, microsieverts, micrograys, beta radiation, gamma radiation, alpha radiation, Bq/m^3, Bq/cm^3, and joules per kilogram. Additional confusion arose from different technical terms for radiation used by scientists and engineering professionals from different parts of the world. For example, as shown in Table 10.3, the US system for describing radiation differs from the system used in other parts of the world.

Translation of technical jargon to plain language is especially challenging for scientific and technical professionals, in part because it is difficult for many experts to remember what they knew and did not know before they became experts. For example, on the first page of a review of the literature on health literacy, the US Department of Health noted:

> Two decades of research indicate that today's health information is presented in a way that isn't usable by most Americans. Nearly 9 out of 10 adults have difficulty using the everyday health information that is routinely available in our health care facilities, retail outlets, media, and communities.[34]

Table 10.3 Technical terms for radiation.

US System For Describing Radiation		International System For Describing Radiation
Rem	=	0.01 Sievert
Rad	=	0.01 Gray
Curie	=	3.7 x 10 10 Becquerel
Roentgen	=	2.5 x 10 -4 Coulombs/kilogram

Aiding this translation task for scientists and technical professionals and communicators are many computer word processors and stand-alone readability software programs that check for readability and comprehension, including tests of vocabulary, polysyllabic words, sentence length, paragraph length, active versus passive construction, and verb and subject placement. These and other characteristics are captured by tests such as the Flesch-Kincaid Readability Test.[35] Based on these readability and comprehension characteristics, communicators of technical information who desire to be better understood by non-experts reduce the number of words per sentence, avoid polysyllable words when possible, use the active voice when possible, avoid the use of acronyms, and use a plain language thesaurus to find substitutes for technical words and phrases.

Readability software programs provide a general idea of how hard a document will be to read. However, they do not ensure comprehension. In the risk, high concern, and crisis communication literature, and holding constant other variables, the goal is to bring the grade level down to two, three, and ideally four grade levels below the average grade level of the audience. As noted by the US Centers for Disease Control and Prevention (CDC), "comprehension levels are often two or more grades below reading or education level. Comprehension drops even more when a person is under stress."[36] When a person is under stress, the average comprehension drops an additional educational grade level. This leads to the AGL-4 (Average Grade Level 4) best practice described in Chapter 3.

Target educational grade levels for technical information about a risk or threat are adjusted based on characteristics of the intended audience. They can vary greatly from person to person, group to group, and location to location. The readability level for risk-related information for the general public is often targeted for the sixth- to eighth-grade educational level (i.e., approximately that of a 12- or 13-year-old).

10.8 Information Clarity

Clarity in communicating technical information about a risk or threat is achieved when the technical information evokes in the receiver the intended meaning of the sender. Clarity reduces or eliminates superfluous technical information but also allows the receiver to easily access more complex information. An early and critical step in communicating technical information about a risk or threat is determining how much technical information the target audience actually needs or wants.

An effective technique for achieving clarity is message mapping, described in Chapter 9. Message maps layer the information in hierarchical tiers ranging from simple to more complex. For example, answers to technical questions about a risk or threat might be written at a lower grade level at the beginning and then increase in complexity, concluding with references to technical reports or documents. This is the format used by the Association of State and Territorial Health Officials for their Q&A risk communication documents about Zika, Ebola, and COVID-19.[37] One advantage of message mapping for achieving clarity is that it gives control to the individual by allowing a person to access the information at whatever level and depth are comfortable for them.

Many organizations have developed guidelines for achieving message clarity based on the research literature on science communication.[38] In addition to readability metrics, the guidelines list multiple ways to enhance the understandability of technical information. Table 10.4, Table 10.5, and Table 10.6 contain short lists of these guidelines. Many of these guidelines also link to dictionaries of technical terms with common, everyday alternatives in plain language. Recognizing that precision is sometimes lost, clarity is achieved by substituting everyday words for technical terms.

Table 10.4 Guidelines for creating clear technical information.

The Communicator:
- avoids the use of technical or bureaucratic jargon, acronyms, and abbreviations;
- uses plain language;
- defines new or unfamiliar terms so the target audience can understand them;
- uses short sentences to define new terms;
- provides a glossary;
- checks frequently for understanding;
- uses new terms only if they are important for the target audience to know and remember;
- avoids new terms that have a different meaning from their common usage.

Table 10.5 Guidelines for delivering clear technical information.

The Communicator:
- assesses the knowledge and understanding of the target audience;
- focuses on what the target audience needs and wants;
- provides no more than three to five key message points;
- tests messages before delivering them;
- uses the "Triple T Model" for presenting complex information: tells the audience briefly what they are going to tell them; tells them more about each point; tells them again briefly what they told them;
- develops technical content that people can interact with, such as through social media platforms;
- uses the active voice for writing and speaking;
- provides complex information in tiers or layers of information that increase gradually in complexity.

10.9 Units of Measurement

Technical information about a risk or threat is often expressed in units of measurement representing different amounts. For example, numerical information about a risk or threat can be expressed quantitatively through numerical expression of concentrations (such as parts per million of a toxic substance); probabilities (such as a one-in-a-million likelihood of an event occurring); and quantities (such as the average number of small and large droplets containing the COVID-19 virus in a sneeze; "x" amount of lead in drinking water; or "x" anthrax spores in a letter sent by a terrorist). Each way communicates different information and serves a different purpose.

Table 10.6 Guidelines for enhancing the clarify of technical information.

The Communicator uses:

- stories to illustrate a point;
- metaphors and analogies;
- visuals (for example, graphics, drawings, maps, charts, flowcharts, paintings, photographs, video, and highlighted text) to enhance attention, recall, and understanding;
- visual cues, such as color, boldface, italics, color, shapes, lines and arrows, font size, spacing, alignment, and headings to enhance attention, recall, and understanding;
- knowledge of local culture to determine if the technical content is consistent with culturally accepted ways of presenting or accessing information;
- knowledge about the target audience to enhance accessibility and understanding (e.g., knowledge about numerical literacy, preferred language or dialect, and preferred style, such as enlarged the type face and font size for target audiences who are elderly or sight-impaired).

One significant challenge related to units of measurement is that the units may paint less than a full picture of the risk or threat. For example, one measure commonly used to express numerical risk is lost life expectancy. However, this measure gives more weight to early deaths and less to deaths in old age. Moreover, in health, safety, and environmental communications, numbers and statistics indicating the presence of a potentially hazardous substance do not necessarily signify that it poses a significant risk. For a potentially hazardous substance or risk agent to pose a threat or risk, there needs to be exposure to the hazardous substance or risk agent. *There is no risk without exposure.* To be complete, technical information about a risk or threat needs to provide numbers and statistics about exposure. Additionally, even if an exposure takes place, the amount, or dose, of the potentially hazardous substance is important. An often-cited statement by risk assessors is: "The dose makes the poison." Exposures to small amounts of a potentially hazardous substance may pose little or no risks. Also, the dose or concentration of a potentially hazardous substance that actually reaches people through inhalation, ingestion, or skin contact is typically far lower than the amounts at the source. The amount often becomes even lower with the passage of time and distance. People therefore need to be provided with understandable numbers and statistics about doses.

A second challenge related to units of measurement is the concept of absolute versus relative risk. Responses to risk messages – especially those communicating increases or decreases in risk – critically depend upon whether the probabilities are presented in absolute terms ("the probability was 2 percent and is now 4 percent") or relative terms (as in "the probability has doubled" or "this group suffers twice the normal risk of . . ."). The latter approach can be seriously misleading. Information about relative risk can result in misperceptions if information about the baseline probabilities is not made clear.

A third challenge related to units of measurement is that transformations of scale can radically change perceptions of risks or threats. For example, in communicating concentrations, the expression "six parts per billion" sounds a great deal larger than 0.006 parts per million, even though they are the same. Scale is also a factor in communicating probabilities. For example, a risk agent expected to result in the death of 1.4 people in every 1,000 can equally be expressed as

- "the risk is 0.0014"
- "the risk is 0.14 percent"
- "in a community of a thousand people, we could expect 1.4 to die as a result of exposure"

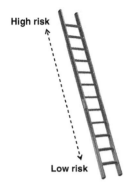

High risk

Low risk

- Heart disease (1 in 6)
- Cancer (1 in 7)
- Motor vehicle crash (1 in 107)
- Motorcycle accident (1 in 899)
- Choking on food (1 in 2,535)
- Cataclysmic storm (1 in 58,669)
- Lightning strike (1 in 138,849)

Figure 10.2   Risk ladder – Comparison of lifetime odds of selected causes of death.
Source: Adapted from National Safety Council (2021). Odds of Dying. https://injuryfacts.nsc.org/all-injuries/
preventable-death-overview/odds-of-dying/

Although these alternatives are equivalent, their meaning to audiences, and hence their effect and influence, may not be identical. The first term may make the risk seem smaller, while the last term may make it seem larger. The second term, which combines both decimal places and a percentage, is a format confusing to many non-experts. Confusion can sometimes be avoided by embedding risk numbers in words that help clarify their meaning. For example, "a risk of 0.047 percent" is comprehensible to only a few people. By comparison, it is much easier to understand that about five people in an auditorium of 100 would be affected.

This embedding process for numbers can also be accomplished through visuals, including graphs, charts, animation, and pictures.[39] An effective visual can increase attention, interest, engagement, understanding, and recall. For example, research indicates that visuals, such as a risk ladder showing where an event falls within a range of risks (e.g., odds of death) from low to high, can be used to emphasize particular risk characteristics, as shown in Figure 10.2.[40]

Comparisons often provide perspective and can serve as useful benchmarks or yardsticks when used appropriately. For example, the concentration comparisons found in Table 10.7 are often used to provide perspective for exposures to toxic chemicals.

Table 10.7   Concentration comparisons for parts per million (ppm), parts per billion (ppb), and parts per trillion (ppt).

Unit	1 part per million	1 part per billion	1 part per trillion
Length	1 inch in 16 miles	1 inch in 16,000 miles	1 inch in 16,000,000 miles
Time	1 minute in 2 years	1 second in 32 years	1 second in 320 centuries
Money	1 cent in $10,000	1 cent in $10,000,000	1 cent in $10,000,000,000
Weight	1 oz. in 31 tons	1 pinch of salt in 10 tons of potato chips	1 pinch of salt in 10,000 tons of potato chips

One commonly used unit for volume is that one part per billion is equal to one drop of ink in "x" number of Olympic-size pools. Such equivalencies are intended to help people understand how "small" an amount of a risk agent really is. However, for some individuals, the use of such equivalences trivializes the problem and prejudges risk acceptability. Concentration units can also be misleading. For example, toxic substances vary widely in their potency – one drop of a highly toxic substance in a community reservoir or pool could kill many people. Additionally, the "one drop in a pool" equivalency may fail because receivers do not consider the effects of dilution (e.g., the concept that "the solution to pollution is dilution"); and because people do not want unknown persons putting one drop of anything in their pool.

10.10 Case Study: Risk Numbers, Risk Statistics, and the *Challenger* Accident

On 28 January 1986, the space shuttle *Challenger* was launched by the National Aeronautics and Space Administration (NASA). Seventy-three seconds after takeoff, the shuttle exploded, killing all seven crew members. The crew consisted of five NASA astronauts, a NASA cargo specialist, and a civilian schoolteacher.

The immediate cause of the accident was the failure, due largely to unprecedented low temperatures, of the O-rings. The O-rings are seals made from rubber meant to seal in flammable fuel and vapors at ignition. They had leaked. The fuel and vapors caught fire. The flames reached the solid rocket booster and then the external tank containing liquid hydrogen and oxygen. The result was a horrendous explosion. The explosion was viewed by millions, including school children around the world who were watching to see a teacher in space.

The O-rings had lost their resiliency because the shuttle was launched on a very cold day for Cape Canaveral, Florida. Temperatures on the day of the launch were in the low 30s and the O-rings themselves were colder, less than 32°F. On the day before the flight, the predicted temperature for the launch day was 26 to 29°F. Concerned that the rings would not seal at such cold temperatures, engineers involved in designing the rocket expressed statistical concerns. Their concerns came from several sources, including statistics describing the history of O-ring damage during previous cool-weather launches of the shuttle, the physics of O-ring resiliency (which declines exponentially with cooling), and experimental data.

The statistical evidence describing the potential for O-ring failure was sent to NASA. NASA engineers and officials noted serious weaknesses in the numbers and statistics and determined that the numbers and statistics were inconclusive. Discussions among officials and engineers lasted well into the night. The decision was made to go ahead with the launch.

Ever since the catastrophe, scholars and commentators have debated what happened and if the tragedy could have been prevented. For example, why were multiple warning signals and statistics ignored about the risk of low temperatures damaging the O-ring seals. Vaughn (1996) proposed a cultural theory of risk explanation. She argued the *Challenger* tragedy was "a mistake embedded in the banality of organizational life and facilitated by an environment of scarcity and competition, elite bargaining, uncertain technology, incrementalism, patterns of information, routinization, organizational and interorganizational structures, and a complex culture."[41]

Among the many explanations cited by the "Report of the Presidential Commission on the Space Shuttle *Challenger* Accident" and others were the following: incomplete statistical analysis by engineers; the violation of safety rules; "group think" – a worldview preference for consensus as opposed to independent thinking; deadlines and a tight project schedule; pressure to conform to bureaucratic norms and decision-making processes (e.g., don't talk back once a decision has been made by a higher-up); pressure to have a successful launch; and organizational values that encouraged risk-taking. Questions posed included the following: Was the decision-making and risk communication process distorted by "group think"? Were the NASA and contract engineers and officials feeling pressure to have a successful launch following the extensive publicity campaign surrounding the launch, including having a schoolteacher on board and encouraging schools to take a break from classes to watch the launch? Were the statistics too complex for non-expert risk managers and decision makers to understand? Were political factors at play that argued against delay of the already delayed launch?[42]

Professor Edward Tufte of Yale University raised another question in his book *Visual Explanations.* Tufte asked: Could the risk numbers and statistics have been communicated more effectively using visual communication tools to convince risk managers and decision makers to delay the launch?[43] The book describes design strategies – the proper arrangement in space and time of images, words, and numbers – for presenting information about motion, process, mechanism, cause, and effect. The book discussed in detail the use of visual evidence in deciding to launch the space shuttle *Challenger.* In the chapter of the book devoted to the *Challenger* tragedy, Tufte reproduces many of the statistics presented to NASA officials. The numbers and statistics are extremely dense and complicated. They included complex statistical explanations of the physics associated with the resiliency of the O-rings.

Tufte took the data the engineers presented to officials and put them into a single visual chart. He argued this visual could have changed history. He reduced the large number of statistical questions down to one question: What is the correlation between temperature and damage to the O-rings? He produced a chart that plotted all the shuttle launches with temperature at launch on one axis and with a "damage index" on the other axis. The chart showed there were several instances of damage occurring at warm temperatures. However, every launch colder than 65° had some damage. Furthermore, the visual showed damage increased as the temperature got colder. The coldest launch before *Challenger* was at 52°. The *Challenger* rocket was launched when the temperature was in the mid to high 20s. The three key messages of the visual were (1) every launch colder than 65° had O-ring damage; (2) damage became worse as it got colder; and (3) expected temperatures on the launch date were significantly colder than other launches.

10.11 Comparisons

Numbers, statistics, and technical information about risks and threats are often communicated through comparisons. The goal of comparisons is to make numbers and statistics more understandable and meaningful by comparing them to other numbers and statistics. Comparisons are also advanced as a means for setting priorities and determining which risks and threats can be ignored, which risks or threats merit concern, and which risks or threats merit protective action.[44]

Table 10.8 shows the lifetime odds of death for selected causes based on calculations available from the National Safety Council as of January 2021. Most of the numbers change only minimally from year to year.

It is often tempting to use the numbers found in Table 10.8 to make the following type of argument:

> The risk of "a" is lower than the risk of "b" (e.g., of being killed in a motor vehicle crash). Since you (the target audience) find "b" acceptable, you are obliged to find "*a*" acceptable.

This argument has a basic flaw in its logic. It could severely damage trust and credibility since some receivers of the comparison will analyze the argument this way:

> I do not have to accept the (small) added risk of "a" just because I accept the (perhaps larger, but voluntary and personally highly beneficial) risk of "b" (e.g., driving my car). In deciding about the acceptability of risks, I consider many factors, only one of them being the size of the risk. I therefore prefer to do my own evaluations of safety or risk and make my own decisions based on these evaluations.

A key principle of risk, high concern, and crisis communication is that risk numbers and statistics are only one of many kinds of information upon which people base their decisions. Comparisons cannot preempt those decisions. Comparisons are unlikely to be successful if the comparison appears to be trying to settle the question of whether something is acceptable.

Table 10.8 Lifetime odds of death for selected causes (United States)

Cause of death	Odds of dying
Heart disease	1 in 6
Cancer	1 in 7
Suicide	1 in 88
Opioid overdose	1 in 92
Motor vehicle crash	1 in 107
Fall	1 in 111
Gun assault	1 in 289
Riding a motorcycle	1 in 899
Drowning	1 in 1,128
Fire or smoke	1 in 1,547
Choking on food	1 in 2,535
Riding a bicycle	1 in 3,825
Accidental gun discharge	1 in 9,077
Sharp objects	1 in 29,483
Cataclysmic storm	1 in 58,669
Hornet, wasp, and bee stings	1 in 59,507
Dog attack	1 in 86,781
Lightning strike	1 in 138,849
Railway passenger/Airline passenger	Too few deaths to calculate odds

Source: Adapted from National Safety Council (2021) *Odds of Dying*. https://injuryfacts.nsc.org/all-injuries/preventable-death-overview/odds-of-dying/

Many variables affect the success of a comparison, including personal experiences, risk percep- tions, perceived benefits, culture, context, and trustworthiness of the source of the comparison. There are also several distinct elements that contribute to the effectiveness or ineffectiveness of a comparison. These include (1) how clear and easy the comparison is to understand, (2) the per- ceived relevance, appropriateness, and helpfulness of the comparison, (3) concerns that the com- parison is being offered to mislead, (4) the level of emotion attached to the issue, (5) acknowledgment that factors other than technical data are relevant to decisions about the issue, (6) perceptions that the comparisons may be meant to minimize or dismiss the issue, and (7) perceptions that the source of the comparison is not trustworthy. If trust is high, emotions are low, the comparison makes intuitive sense, and appropriate caveats are offered, comparisons can be useful in clarifying technical information and putting the information in perspective.

A wide variety of comparative formats are available that can take each of these elements into account. These include the following:

- Comparisons with standards, rules, and regulations
- Comparisons of alternative solutions to the same problem
- Comparisons of the same risk as experienced at different times
- Comparisons with different estimates of the same risk
- Comparisons of the risk of doing something versus not doing something
- Comparisons with the same risk as experienced in other places

All these types of comparisons have some claim to relevance and legitimacy. The most effective comparisons are often comparisons to standards, rules, and regulations. The most ineffective com- parisons, and the most difficult to use, are those that appear to disregard the factors people con- sider important in evaluating information and making decisions about acceptability. Among the most important of these factors are trustworthiness, voluntariness, fairness, benefits, alternatives, control, dread, and familiarity.

Even if the risk managers and communicators have been able to identify and address all relevant emotional and perception factors that affect comparisons, there are still many challenges. For example, one challenge relates to numerical literacy. Numerical literacy is a person's ability to use and understand numbers and statistics. It also includes a person's ability to use numbers and sta- tistics to reason and solve basic and complex problems in real world situations. Numerical literacy is typically needed to understand a comparison. For example, flood emergency policies are typi- cally built around the so-called 100-year floodplain, which is commonly but incorrectly under- stood as an area that would flood once every century. But what it really means is that there is a 1% chance of flooding in any given year.[45]

A second comparison challenge relates to mental models – the brain's representation of how things work. Mental models do two things: they help a person assess how systems work, and they help the person make better decisions. To make use of a comparison, a person needs to have an accurate mental model. Inaccurate mental models function as a lens that can distort understand- ing and the usefulness of comparisons. For example, for an approaching hurricane, what many people think will happen is often very different from what actually happens. Many people have a poor mental model of how high the water can get, how strong the wind can get, how the house will hold up against the pressure of hurricane winds and floodwater, how quickly things can unfold, what is available at evacuation centers, what preparations are needed, and what will be needed if there is no power. Research indicates that many people threatened by major hurricanes have a poor understanding of the threat posed by the storms: they overestimate the likelihood their homes would be subject to hurricane-force wind conditions; they underestimate the potential damage that such winds could cause; they misconstrue the greatest threat as coming from wind rather than water; and they misestimate how long the after-effects of the storm would be. These misperceptions often result in preparations that are not consistent with the nature and scale of a major hurricane.[46]

A third comparison challenge relates to *heuristics* – the mental shortcuts and "rules of thumb" used by the brain to simplify decision-making in problem solving. For example, the brain's use of the *information availability heuristic* – the availability of information about an event that is accessible or easily remembered by the brain – often leads to people overestimating a risk or threat. The brain's use of the *anchoring and adjustment heuristics* – the tendency of the brain to give undue weight to the first piece of information and use it as the reference point, or anchor – also often leads to an overestimation of a risk or threat.

A fourth challenge relates to communicating small numbers, such as a risk or threat with a small probability of occurring in a large population. For example, when a person is told they have a one in one million chance of experiencing harm, the person may focus more on the numerator – the one – than on the denominator – the one million. To help people gain perspective when thinking about small risks, a large number of studies have examined the value of comparisons. For example, in a study of optimal ways to present vaccine-related information, people were asked, "If the risk of a side effect from a vaccine is 0.0001, what is the best way to explain this number?" People expressed a clear preference for a verbal comparison – one person in a town of 10,000 people, followed closely by a visual comparison – one red dot in a field of 10,000 black dots.[47]

A fifth comparison challenge relates to cumulative risks. For example, for environmental risks, the US Environmental Protection Agency has formally defined cumulative risks as "the combination of risks posed by aggregate exposure to multiple agents or stressors in which *aggregate exposure* is exposure by all routes and pathways and from all sources of each given agent or stressor."[48] A particular risk or threat might be associated with low probability of occurring at a particular time or place. However, over time, and from repeated exposures, these probabilities add up and can create a significant risk.

A sixth comparison challenge relates to familiarity. People often have difficulty thinking about risks or threats they have never experienced. They may therefore not be able to relate to the risks or threats included in the comparison. For example, many people are familiar with diseases that can be transmitted by droplets in the air. However, they may be unfamiliar with diseases such as Ebola, which can be transmitted by virtually any bodily fluid, including sweat and tears.

A seventh comparison challenge relates to the numbers and statistics used in comparisons. A critical flaw in many comparative tables is the failure to provide information on the assumptions and uncertainties underlying the calculation of estimates. A related flaw is that many of the numbers and statistics listed in comparative tables are single dimensional, such as the expected number of deaths. The use of a narrow quantitative number can obscure the importance of other significant dimensions, such as injuries, disabilities, concentration, persistence, recurrence, reversibility, and racial, ethnic, or socioeconomic inequities. For example, in radiation risk comparisons, such as for X-rays and other medical procedures, comparisons are often done between radiation doses for X-rays (and other medical procedures) and the equivalent period of exposure to natural background radiation, such as from cosmic radiation. However, background radiation results in whole body exposures while radiation exposure in medical procedures is typically focused on one region of the body.[49] Similarly, radiation doses from medical procedures are sometimes compared to radiation doses from commercial air travel. However, radiation doses from commercial air travel depend on multiple facts, such as flight path, latitude, altitude, and duration.

An eighth comparison challenge relates to comparisons between risks, costs, and benefits, such as trading off the costs of measures to prevent and protect people from a risk against the costs or loss of economic benefits. For example, in the COVID-19 pandemic, many argued that measures promoted to slow the spread of COVID-19 were not commensurate with the costs, such as the costs associated with lockdowns, closing businesses, and closing schools.[50] The dilemma, which is present in most trade-off comparisons, is that finding common measures for comparison is problematic.

10.12   Lessons Learned

Used thoughtfully, well-presented numbers and statistics can help communicators achieve their primary objectives: gaining trust, informing decision-making, and constructively influencing attitudes and behaviors. Numbers and statistics are important. Much of the information available to people about risks and threats is determined mathematically. Numbers and statistics help people stay informed about risks and threats; they provide people with accurate and timely information; they help people decide if information about a risk or threat is relevant to them; they help people sort false information from truthful information. But they can also be confusing and can be misinterpreted.

How to present numbers and statistics effectively is a complicated determination. That determination depends on many factors, including the specific audience and the unique knowledge, perceptions, and worldviews of individuals in that audience. It depends on the specific context or situation, and on the skill of the communicator in presenting and discussing technical information. Communication about a risk or threat is rarely just about numbers and statistics, but also about how those numbers and statistics are received. It is about working with the ways the brain processes numbers and statistics, using research about brain processing of technical information to inform what numbers and statistics to use, and when and how to use them. Good presentation techniques, such as clear charts and graphs, are a pre-requisite to deliver understandable numerical information, but there is a necessary next layer for an audience to constructively and accurately interpret numerical information. That layer is the careful design of the message with the receiver in mind.

10.13   Chapter Resources

Below are additional resources to expand on the content presented in this chapter.

American Psychological Association (2020). *Stress in America, 2020: A National Mental Health Crisis.* Washington, D.C.: American Psychological Association.

Árvai, J., and Rivers, L. III., eds. (2014). *Effective Risk Communication.* Abingdon, UK: Routledge.

Banerjee, I., McNulty, J. P., Catania, D., Maccagni, D., Masterson, L., Portelli, J.L., and Rainford, L. (2019). "An investigation of procedural radiation dose level awareness and personal training experience in communicating ionizing radiation examinations benefits and risks to patients in two European cardiac centers." *Health Physics* 117(7):6–83.

Bennett, P., and Calman, K., eds. (1999). *Risk Communication and Public Health.* New York: Oxford University Press.

Bentkover, J., Covello, V., and Mumpower J. (1986). *Benefit Assessment: The State of the Art.* Dordrecht, the Netherlands: Reidel.

Bier, V. M. (2001). "On the state of the art: risk communication to the public." *Reliability Engineering and System Safety* 71(2):139–150.

Bohnenblust, H., and Slovic, P. (1998). "Integrating technical analysis and public values in risk based decision making." *Reliability Engineering and System Safety* 59:151–159.

Bostrom, A. (2003). "Future risk communication." *Futures* 35:553–573.

Bostrom, A., and Atman, C., Fischhoff, B., Morgan, M.G. (1994). "Evaluating risk communications: Completing and correcting mental models of hazardous processes, part II." *Risk Analysis* 14(5):789–797.

Breakwell, G.M. (2007). *The Psychology of Risk*. Cambridge, UK: Cambridge University Press.

Brosch, T. (2021). "Affect and emotions as drivers of climate change perception and action: a review." *Current Opinion in Behavioral Sciences* 42:15–21.

Caffrey, E. (2021). "Radiation and the skeptical public: Tips and tools for communicating effectively." *Health Physics* 120(6):693–698.

Centers for Disease Control and Prevention (2012). *Emergency, Crisis, and Risk Communication*. Atlanta, GA: Centers for Disease Control and Prevention.

Chess, C., Hance, B.J., and Sandman, P.M. (1986). *Planning Dialogue with Communities: A Risk Communication Workbook*. New Brunswick, NJ: Rutgers University, Cook College, Environmental Media Communication Research Program.

Chess, C., Hance, B.J., and Sandman, P.M. (1998). *Improving Dialogue with Communities*. Trenton (NJ): New Jersey Department of Environmental Protection/Division of Science and Research

Chess, C., Salomone, K.L., and Hance, B.J. (1995). "Improving Risk Communication in Government: Research Priorities." *Risk Analysis* 15 (2):127–135.

Chess, C., Salomone, K.L., Hance, B.J., and Saville, A. (1995). "Results of a national symposium on risk communication: Next steps for government agencies." *Risk Analysis* 15(2):115–120.

Cohrssen, J., and Covello, V. (1999). *Risk Analysis: A Guide to Principles and Methods*. Washington, DC: White House Council on Environmental Quality.

Coombs, W.T. (1998). "An analytic framework for crisis situations: Better responses from a better understanding of the situation." *Journal of Public Relations Research* 10(3):177–192.

Covello, V., and Allen, F. (1988). *Seven Cardinal Rules of Risk Communication*. Washington, D.C.: US Environmental Protection Agency.

Covello, V. (2003). "Best practices in public health risk and crisis communication." *Journal of Health Communication* 8 (Suppl. 1):5–8; discussion, 148–151.

Covello, V. (2014). "Risk communication," in *Environmental Health: From Global to Local*, ed. H. Frumkin. San Francisco: Jossey-Bass/Wiley.

Covello, V. (2006). "Risk communication and message mapping: A new tool for communicating effectively in public health emergencies and disasters." *Journal of Emergency Management* 4(3):25–40.

Covello, V., and Hyer R. (2020). *COVID-19: Simple Answers to Top Questions: Risk Communication Field Guide Questions and Key Messages*. Arlington, VA: Association of State and Territorial Health Officers. Accessed at: https://www.hsdl.org/?view&did=835774

Covello, V., and Hyer, R. (2014). *Top Questions on Ebola: Simple Answers Developed by the Association of State and Territorial Health Officials*. Arlington VA: Association of State and Territorial Health Officials. Accessed at: https://www.astho.org/Infectious-Disease/Top-Questions-On-Ebola-Simple-Answers-Developed-by-ASTHO/

Covello, V., and Merkhofer, M. (1993). *Risk Assessment Methods: Approaches for Assessing Health and Environmental Risks*. New York: Plenum Press.

Covello, V., McCallum, D., and Pavlova, M., eds. (1989) *Effective Risk Communication: The Role and Responsibility of Government and Nongovernment Organizations*. New York: Plenum Press.

Covello, V., Peters, R., Wojtecki, J., and Hyde, R. (2001). "Risk communication, the West Nile virus epidemic, and bio-terrorism: Responding to the communication challenges posed by the intentional or unintentional release of a pathogen in an urban setting." *Journal of Urban Health* 78(2):382–391.

Covello, V., Sandman, P., and Slovic, P. (1988). *Risk Communication, Risk Statistics, and Risk Numbers.* Washington, DC: Chemical Manufacturers Association.

Covello, V., and Sandman, P. (2001). "Risk communication: Evolution and revolution," in *Solutions to an Environment in Peril*, ed. A. Wolbarst. Baltimore, MD: Johns Hopkins University Press.

Covello, V., Slovic, P., and von Winterfeldt, D. (1986). "Risk communication: A review of the literature." *Risk Abstracts* 3(4):171–182.

Covello, V., Minamyer, S., and Clayton, K. (2007). *Effective Risk and Crisis Communication during Water Security Emergencies. EPA Policy Report*; EPA 600-R07-027. Washington, D.C.: US Environmental Protection Agency.

Covello, V., Slovic, P., and von Winterfeld, D. (1987). *Risk Communication: A Review of the Literature.* Washington, DC: National Science Foundation.

Cox, Jr., A.L. (2012). "Confronting deep uncertainties in risk analysis." *Risk Analysis* 32(10):1607–1629.

Crick, M.J. (2021). "The importance of trustworthy sources of scientific information in risk communication with the public." *Journal of Radiation Research* 62 (S1):1–6.

Cvetkovich, G., Vlek, C.A., and Earle, T.C. (1989). "Designing technological hazard information programs: Towards a model of risk-adaptive decision making," in *Social Decision Methodology for Technical Projects*, eds. C.A.J. Vlek and G. Cvetkovich. Dordrecht: Kluwer Academic.

Dietz, T. (2013). "Bringing values and deliberation to science communication." *Proceedings of the National Academy of Sciences* 110:14081–14087.

Druckman, J. (2001). "Evaluating framing effects". *Journal of Economic Psychology.* 22:96–101.

Dunwoody, S. (2014). "Science journalism," in *Handbook of Public Communication of Science and Technology*, eds. M. Bucchi and B. Trench. New York: Routledge.

Dykes, B. (2020). *Effective Data Storytelling: How to Drive Change with Data, Narrative and Visuals.* Hoboken, NJ: Wiley.

Fearn-Banks, K. (2007). *Crisis Communications: A Casebook Approach*, 3rd edition. Mahwah, NJ: Lawrence Erlbaum Associates.

Fischhoff, B. (1995a). "Risk perception and communication unplugged: Twenty years of process." *Risk Analysis* 15(2):137–145.

Fischhoff, B. (1995b). "Strategies for risk communication. Appendix C." in *National Research Council: Improving Risk Communication.* Washington, DC: National Academies Press.

Fischhoff, B. (2012). *Judgment and Decision Making.* New York: Earthscan.

Fischhoff, B. (2013). "The sciences of science communication," in *Proceedings of the National Academy of Sciences* 110 (Suppl. 3): 14033–14039.

Fischhoff, B., Brewer, N.T., and Downs, J.S., eds. (2011). *Communicating Risks and Benefits: An Evidence-based User's Guide.* Washington, DC: Food and Drug Administration.

Fischhoff, B., and Davis, A. L. (2014). "Communicating scientific uncertainty," in *Proceedings of the National Academy of Sciences*, 111 (Suppl. 4): 13664–13671.

Fischhoff, B., and Kadvany, J. (2011). *Risk: A Very Short Introduction.* New York: Oxford University Press.

Germani, F., and Biller-Andorno, N. (2021). "The anti-vaccination infodemic on social media: A behavioral analysis." *PLOS One* 16 (3): e0247642.

Glik D.C. (2007). "Risk communication for public health emergencies." *Annual Review of Public Health* 28(1):33–54.

Hance, B.J., Chess, C., and Sandman, P.M. (1990). *Industry Risk Communication Manual.* Boca Raton, FL: CRC Press/Lewis Publishers.

Halvorsen, P.A. (2010). "What information do patients need to make a medical decision?" *Medical Decision Making* 30 (5 Suppl):11S–13S.

Haight, J.M., ed. (2008). *The Safety Professionals Handbook: Technical Applications.* Des Plaines, IL: The American Society of Safety Engineers.

Heath, R.L., and O'Hair, H.D. (eds.). (2020). *Handbook of Risk and Crisis Communication.* New York: Routledge.

Henstra, D., Minano, A., and Thistlethwaite, J. (2019). "Communicating disaster risk? An evaluation of the availability and quality of flood maps." *Natural Hazards Earth System Science* 19:313–323.

Hess, R., Visschers, V.H.M., Siegrist, M., and Keller, C. (2011). "How do people perceive graphical risk communication? The role of subjective numeracy." *Journal of Risk Research* 14 (1):47–61.

Hutson, M. (2020). "Why modeling the spread of COVID-19 is so damn hard." *IEEE Spectrum.* 22 Sep 2020 Accessed at: : https://spectrum.ieee.org/artificial-intelligence/medical-ai/why-modeling-the-spread-of-covid19-is-so-damn-hard

Hyer, R.N., and Covello, V.T. (2007). *Effective Media Communication During Public Health Emergencies: A World Health Organization Handbook.* Geneva: World Health Organization Publications.

Hyer, R.N., and Covello, V.T. (2017). *Top Questions on Zika: Simple Answers.* Arlington, VA: Association of State and Territorial Health Officials. Accessed at: https://www.astho.org/infectious-disease/top-questions-on-zika-simple-answers-developed-by-astho/

Jardine, C.G., and Driedger, S.M. (2014). "Risk communication for empowerment: An ultimate or elusive goal?" in *Effective Risk Communication*, eds J. Arvai and L. Rivers III. London: Earthscan.

Jong, W. (2020). "Evaluating crisis communication. "A 30-item checklist for assessing performance during COVID-19 and other pandemics." *Journal of Health Communication* 25(12):962–970.

Jong, W., and Dückers, M.L.A. (2019). "The perspective of the affected: What people confronted with disasters expect from government officials and public leaders." *Risk, Hazards & Crisis in Public Policy* 10(1):14–31.

Joslyn, S., and LeClerc, J. (2012). "Uncertainty forecasts improve weather related decisions and attenuate the effects of forecast error." *Journal of Experimental Psychology* 18:126–140.

Joslyn, S., Nadav-Greenberg, L., Taing, M.U., and Nichols, R.M. (2009). "The effects of wording on the understanding and use of uncertainty information in a threshold forecasting decision." *Applied Cognitive Psychology* 23:55–72.

Kahneman, D. (2011). *Thinking, Fast and Slow.* New York: Macmillan Publishers.

Kahneman, D., Slovic, P., and Tversky, A., eds. (1982). *Judgment Under Uncertainty: Heuristics and Biases.* New York: Cambridge University Press.

Kahneman, D., and Tversky, A. (1979). "Prospect theory: An analysis of decision under risk." *Econometrica* 47(2):263–291.

Kasperson, R.E. (2014). "Four questions for risk communication." *Journal of Risk Research* 17(10):1233–1239.

Kasperson, R.E., Renn, O., Slovic, P., Brown, H.S., Emel, J., Goble, R., Kasperson, J. X., and Ratick, S. (1987). "Social amplification of risk: A conceptual framework." *Risk Analysis* 8(2):177–187.

Kasperson, R.E., and Stallen, P.J.M. eds. (1991). *Communicating Risks to the Public: International Perspectives.* Dordrecht: Kluwer.

Kochenderfer, M. (2015). *Decision Making Under Uncertainty: Theory and Application.* Cambridge, MA: MIT Press.

Krimsky S., and Plough A. (1988). *Environmental Hazards: Communicating Risks as a Social Process.* Dover (MA): Auburn House.

Lindell, M.K., and Perry, R.W. (2012). "The protective action decision model: theoretical modifications and additional evidence." *Risk Analysis* 32:616–632.

Lipkus, I.M. (2007). "Numeric, verbal, and visual formats of conveying health risks: suggested best practices and future recommendations." *Medical Decision Making* 27(5):696–713. doi: 10.1177/0272989X07307271.

Liu, T., Zhang, H., and Zhang, H. (2020). "The impact of social media on risk communication of disasters." *International Journal of Environmental Research and Public Health* 883:1–17.

Löftstedt, R. (2003). "Risk communication. pitfalls and promises." *European Review* 11(3):417–435.

Löftstedt, R.E., and Bouder, F. (2014). "New transparency policies: Risk communication's doom?," in *Effective Risk Communication*, eds. J. Àrvai and L. Rivers III. Abingdon, UK: Routledge.

Lum, M.R., and Tinker, T.L. (1994). *A Primer on Health Risk Communication Principles and Practices.* Atlanta, GA: Agency for Toxic Substances and Disease Registry.

Lundgren, R.E., and McMakin A.H. (2018). Risk Communication: *A Handbook for Communicating Environmental, Safety, and Health Risks*. Hoboken, NJ: Wiley.

Malecki, K.M.C., Keating, J.A., and Safdar, N. (2021). "Crisis communication and public perception of COVID-19 risk in the era of social media." *Clinical Infectious Diseases* 72(4):697–702.

Markwart, H., Vitera, J., Lemanski, S., Kietzmann, D., Brasch, M., and Schmidt, S. (2019). "Warning messages to modify safety behavior during crisis situations." *International Journal of Disaster Risk Reduction* 38:101235.

Maslow, A.H., Frager, R., and Fadiman, J. (1987). *Motivation and Personality. 3rd ed*. New York: Harper & Row.

Maxi, R., Tomczyk, S., Schopp, N., and Schmidt, S. (2021). "Warning messages in crisis communication: Risk appraisal and warning compliance in severe weather, violent acts, and the COVID-19 pandemic." *Frontiers in Psychology*, 12: 557178.

McComas, K.A. (2006). "Defining moments in risk communication research: 1996–2005." *Journal of Health Communication* 11(1):75–91.

Mileti, D., Nathe, S., Gori, P., Greene, M., and Lemersal, E. (2004). *Public Hazards Communication and Education: The State of the Art*. Boulder, CO: Natural Hazards Center.

Morgan, M.G., Fischhoff, B., Bostrom, A., and Atman, C.J. (2002). *Risk Communication: A Mental Models Approach*. Cambridge, UK: Cambridge University Press.

National Research Council. (1983). *Risk Assessment in the Federal Government: Managing the Process*. Washington, DC: National Academies Press.

National Research Council (1989). *Improving Risk Communication*. Washington, D.C.: National Academies Press.

National Research Council (1994) *Science and Judgement in Risk Assessment*. Washington, DC: National Academies Press.

National Research Council (1996). *Understanding Risk: Informing Decisions in a Democratic Society*. Washington, DC: National Academies Press.

National Research Council/National Academy of Sciences (2014). *The Science of Science Communication II: Summary of a Colloquium*. Washington, DC: The National Academies Press.

National Research Council/National Academy of Sciences (2017). *Communicating Science Effectively*. Washington, D.C.: The National Academies Press.

National Oceanic and Atmospheric Administration, US Department of Commerce. (2016). *Risk Communication and Behavior. Best Practices and Research*. Washington, DC: National Oceanic and Atmospheric Administration.

Oh, S., Lee, S.Y., and Han, C. (2021). "The Effects of social media use on preventive behaviors during infectious disease outbreaks: The mediating role of self-relevant emotions and public risk perception." *Health Communication* 36(8):972–981.

Olsson, E. (2014). "Crisis communication in public organizations: Dimensions of crisis communication revisited." *Journal of Contingencies and Crisis Management* 22(2):113–125.

Paling, J. (2003). "Strategies to help patients understand risk." *British Medical Journal.* 327(7417):745–8.

Peters, E. (2012). "Beyond comprehension: The role of numeracy in judgments and decisions." *Current Directions in Psychological Science* 21(1):31–35.

Peters, R., McCallum, D., and Covello, V.T. (1997). "The determinants of trust and credibility in environmental risk communication: An empirical study." *Risk Analysis* 17(1):43–54.

Pew Research Center (2021). *Social Media Use in 2021.* Accessed at: https://www.pewresearch.org/internet/2021/04/07/social-media-use-in-2021/

Pidgeon, N., Kasperson, R., Slovic, P. (2003). *The Social Amplification of Risk.* Cambridge, UK: Cambridge University Press.

Plous, S. (1993). *The Psychology of Judgment and Decision Making.* New York: McGraw-Hill.

Prasad, A. (2021). "Anti-science misinformation and conspiracies: COVID–19, Post-truth, and Science & Technology Studies (STS)." *Science, Technology and Society April* 2021: 1–25.

Presidential Commission on Risk Assessment and Risk Management. (1997). Risk Assessment and Risk Management in Regulatory Decision-Making. [Washington, DC.: US Government Printing Office.

Ratzan, S.C., and Moritsugu, K. P. (2014). Ebola crisis-communication chaos we can avoid. *Journal of Health Communication* 19(11):1213–1215.

Renn, O. (2008). *Risk Governance: Coping with Uncertainty in a Complex World.* London, UK: Earthscan.

Reynolds, B. (2014). *Crisis and Emergency Risk Communication.* Atlanta, GA: US Centers for Disease Control and Prevention.

Reynolds B., and Seeger, M.W. (2005). "Crisis and emergency risk communication as an integrative model." *Journal of Health Communication* 10:43–55.

Richter, R., Giroldi, E., Jansen, J., and van der Weijden, T. (2020). "A qualitative exploration of clinicians' strategies to communicate risks to patients in the complex reality of clinical practice." *PLOS ONE* 15(8):0236751

Rodrıguez, H., Dıaz, W., Santos, J., and Aguirre, B. (2007). "Communicating risk and uncertainty: Science, technology, and disasters at the crossroads," in *Handbook of Disaster Research.* New York: Springer.

Sandman, P.M. (1987). "Explaining risk to non-experts: A communication challenge." *Emergency Preparedness Digest* (October-December):25–29.

Sandman, P.M. (1989). "Hazard versus outrage in the public perception of risk," in *Effective Risk Communication: The Role and Responsibility of Government and Non-Government Organizations,* eds. V.T. Covello, D.B. McCallum, M.T. Pavlova. New York: Plenum Press.

Seeger, M.W. (2006). "Best practices in crisis communication: An expert panel process." *Journal of Applied Communication Research* 34(3):232–244.

Sellnow, T.L., Matthew W., and Seeger, M.W. (2020). *Theorizing Crisis Communication.* 2nd edition. Hoboken, New Jersey: Wiley.

Sheppard, B., Janoske, M., and Liu, B. (2012). "Understanding risk communication theory: A guide for emergency managers and communicators." *Report to Human Factors/Behavioral Sciences Division, Science and Technology Directorate, US Department of Homeland Security.* College Park, MD: US Department of Homeland Security.

Silverman, C. (2020). *Verification Handbook for Disinformation and Media Manipulation, 3rd Edition.* Brussels, Belgium and Maastricht, the Netherlands: European Journalism Centre.Accessed at: https://datajournalism.com/read/handbook/verification-3

Slovic, P. (1987). "Perception of risk." *Science* 236 (4799):280–285.

Slovic, P. (1993). "Perceived risk, trust, and democracy." *Risk Analysis* 13(6):675–682.

Slovic, P. (1999). Trust, emotion, and sex. *Risk Analysis* 19(4):689–701.

Slovic, P. (2000). *The Perception of Risk*. London, UK: Earthscan.

Slovic, P. (2010). *The Feeling of Risk: New Perspectives on Risk Perception*. London: EarthScan.

Slovic, P. (2016). "Understanding perceived risk: 1978-2015." *Environment: Science and Policy for Sustainable Development* 58(1):25–29.

Slovic, P., Finucane, L., Peters, E., and MacGregor, D.G. (2004). "Risk as analysis and risk as feelings: Some thoughts about affect, reason, risk, and rationality." *Risk Analysis* 24(2):311–322.

Slovic, P., Fishoff, B., and Lichtenstein, S. (1980). "Facts and fears: Understanding perceived risk," in *Societal Risk Assessment: How Safe Is Safe Enough?*, eds. Schwing JR, Albers WA. New York: Plenum. pp. 181–216.

Steelman, T.A. and McCaffrey, S. (2013). "Best practices in risk and crisis communication: Implications for natural hazards management." *Natural Hazards* 65(1):683–705.

Subramanian, V. and Kattan, M.W. (2020). "Why is modeling coronavirus disease 2019 so difficult?" *Chest Journal* 158(5):1829–1830.

Substance Abuse and Mental Health Services Administration (2020). *Communicating in a Crisis: Risk Communication Guidelines for Public Officials*. Rockville, MD: US Substance Abuse and Mental Health Services Administration.

Swire-Thompson, B. and Lazer D. (2020). "Public health and online misinformation: Challenges and recommendations." *Annual Review of Public Health* 41(1):433–451.

Thaler, R., and Sunstein, C. (2009). *Nudge: Improving Decisions about Health, Wealth, and Happiness*. New York: Penguin Books.

Thompson, K.M., and Bloom, D.L. (2000). "Communication of risk assessment information to risk managers." *Journal of Risk Research* 3:333–352.

Tinker, T.L., and Silberberg, P.G. (1997). *An Evaluation Primer on Health Risk Communication Programs and Outcomes*. Washington, DDL US Department of Health and Human Services.

Tversky, A., and Kahneman, D. (1974). "Judgment under uncertainty: Heuristics and biases." *Science* 185 (4157):1124–1131.

Tversky, A., and Kahneman, D. (1981). "The framing of decisions and the psychology of choice". *Science*. 211 (4481):453–458.

US Environmental Protection Agency. (2016). *Superfund Community Involvement Handbook. EPA 540-K-05-003*. Washington, D.C.: US Environmental Protection Agency. https://semspub.epa.gov/work/HQ/100000070.pdf.

US Environmental Protection Agency (2019). *Superfund Community Involvement Tools and Resources*. Washington, D.C.: US Environmental Protection Agency.https://www.epa.gov/superfund/superfund-community-involvement-tools-and-resources

US Department of Health and Human Services. (2006). *Communicating in a Crisis: Risk Communication Guidelines for Public Officials*. Washington, DC: US Department of Health and Human Services.

US Occupational Safety and Health Administration. (1996). *Hazard Communication*. 29 Code of Federal Regulations, Part 1910.1200.

Walaski, (Ferrante), P. (2011). *Risk and Crisis Communications: Methods and Messages*. Hoboken, NJ: Wiley.

Weinstein, N.D. (1987). *Taking Care: Understanding and Encouraging Self-Protective Behavior*. New York: Cambridge University Press.

Wieder, J. (2019). "Communicating radiation risk: The power of planned, persuasive messaging." *Health Physics* 116(2):207–211.

Wood, M.M., Mileti, D.S., Kano, M., Kelley, M.M., Regan, R., and Bourque, L.B. (2011). "Communicating actionable risk for terrorism and other hazards." *Risk Analysis* 34(4):601–615.

World Health Organization (2020). *Call for Action: Managing the Infodemic--A Global Movement to Promote Access to Health Information and Harm from Health Misinformation Offline Communities.* Accessed at: https://www.who.int/news/item/11-12-2020-call-for-action-managing-the-infodemic

Yoe, C. (2012). *Primer on Quantitative risk analysis: Decision Making Under Uncertainty.* Boca Raton, FL: Taylor and Francis.

Endnotes

1 Accessed at: https://www.cancer.gov/about-cancer/understanding/statistics#:~:text=Approximately%20 39.5%25%20of%20men%20and,will%20die%20of%20the%20disease

2 See, e.g., Vari, A. (1995). "Citizens' advisory committee as a model for public participation: A multiple-criteria evaluation," in *Fairness and Competence in Citizen Participation*, eds. Renn, O., Webler, T., and Wiedemann, P. Dordrecht: Springer Publishing Co.

3 Cohrssen, J. J., and Covello, V.T. (1989). *Risk Analysis: A Guide to Principles and Methods for Analyzing Health and Environmental Risks.* Washington, DC: White House Council on Environmental Quality;US Environmental Protection Agency (1990). *Risk Assessment in Superfund: A Primer.* Washington, D.C:US Environmental Protection Agency;Government of Canada/Health Canada. (2010). *A Primer on Scientific Risk Assessment at Health Canada.* Ottawa: Health Canada. Accessed at: https://www.canada.ca/en/health-canada/services/science-research/reports-publications/about-science-research/primer-scientific-risk-assessment-health-canada-health-canada-2010.html

4 https://www.cnn.com/2020/12/14/health/us-covid-deaths-300k/index.html

5 American Psychological Association (2020). *Stress in America 2020: A National Mental Health Crisis.* Washington, DC: American Psychological Association. Accessed at: https://www.apa.org/ news/press/releases/stress/2020/report-october#:~:text=Despite%20several%20months%20of%20 acclimating,of%20stress%20in%20their%20life%20.

6 See., e.g., National Academies of Sciences, Engineering, and Medicine (2020). *Evaluating Data Types: A Guide for Decision Makers Using Data to Understand the Extent and Spread of COVID-19.* Washington, DC: The National Academies Press.

7 See, e.g., *The COVID Tracking Project*, which collects and reports data on COVID-19 testing and patient outcomes from all 50 states, five territories, and the District of Columbia. Accessed at: https://covidtracking.com

8 See, e.g., Hutson, M. (2020). "Why Modeling the Spread of COVID-19 Is So Damn Hard." *IEEE Spectrum.* Accessed at: https://spectrum.ieee.org/artificial-intelligence/medical-ai/why-modeling-the-spread-of-covid19-is-so-damn-hard; Subramanian,V., and Kattan, M. W. (2020). "Why Is Modeling Coronavirus Disease 2019 So Difficult?" *Chest Journal* 158 (5):1829–1830. Accessed at: https://pubmed.ncbi.nlm.nih.gov/32565270/

9 https://www.gov.uk/government/news/ new-campaign-to-prevent-spread-of-coronavirus-indoors-this-winter

10 See, e.g., Association of State and Territorial Health Officials (2020). "The 2019 Novel Coronavirus: What We Know." Accessed at: https://www.astho.org/StatePublicHealth/What-We-Know-2019-Novel-Coronavirus/01-23-20/?terms=COVID

11 For example, in October 2020, with COVID-19 deaths and cases rising each day, President Donald Trump complained about the news media's intensive coverage of the coronavirus pandemic. At a presidential campaign rally in North Carolina, he attacked two television networks, CNN and MSNBC, stating that the pandemic was "rounding the corner" and insisting that "All you hear is

Covid, Covid, Covid, Covid, Covid, Covid, Covid, Covid, Covid, Covid, Covid." *New York Times.* October 21, 2020. "Trump in North Carolina: 'All you hear is Covid, Covid, Covid, Covid, Covid.'" Accessed at: https://www.nytimes.com/2020/10/21/us/elections/trump-in-north-carolina-all-you-hear-is-covid-covid-covid-covid-covid.html

12 See US Department of Health and Human Services (2005). HHS Pandemic Influenza Plan. Washington, DC: Department of Health and Human Services. Accessed at: https://www.cdc.gov/flu/pdf/professionals/hhspandemicinfluenzaplan.pdf

13 Soper, G. A. (1919). "The lessons of the pandemic." *Science* 49:501–506 (1919); see also: Hatchett, R. J., Mecher, C. E., and Lipsitch, M. (2007). "Public health interventions and epidemic intensity during the 1918 influenza pandemic." *Proceedings of the National Academy of Sciences.* May 2007 104(18):7582–7587; Qualls, N., Levitt, A., Kanade, N., et al. (2017). "Community Mitigation Guidelines to Prevent Pandemic Influenza—United States, 2017." Morbidity and Mortality Weekly Report (MMWR) 6 (RR-1):1–34; Merkel, H. (2004). *When Germs Travel: Six Major Epidemics That Have Invaded America and the Fears They Have Unleashed.* New York: Pantheon Books; Barry, J. (2005). *The Great Influenza: The Epic Story of the Deadliest Plague in History.* New York: Penguin Books.

14 See, e.g., Covello, V., and Sandman, P. (2001). "Risk communication: Evolution and revolution," in *Solutions to an Environment in Peril*, ed. A. Wolbarst. Baltimore, MD: Johns Hopkins University Press; Slovic, P. (1987). "Perception of risk." *Science* 236 (4799):280–285.

15 Interview with Dr. Glen Nowak, in Wetsman, N. (2020). "Effective communication is critical during emergencies like the COVID-19 outbreak: Mixed messaging from the CDC and the White House makes it hard for people to know what to trust." Accessed at: https://www.theverge.com/2020/3/4/21164563/coronavirus-risk-communication-cdc-trump-trust

16 Vaughn, E., and Tinker, T. (2008). "Effective health risk communication about pandemic influenza for vulnerable populations." *American Journal of Public Health* 99:S324–S332.

17 Covello, V., and Hyer, R. (2020). *COVID-19: Simple Answers to Top Questions, Risk Communication Guide.* Arlington, VA: Association of State and Territorial Health Officials. Accessed at: https://www.waukeshacounty.gov/globalassets/health--human-services/public-health/public-health-preparedness/covid/asthocovid19qanda.pdf

18 See, e.g., Covello, V., and Sandman, P. (2001). "Risk communication: Evolution and revolution," in *Solutions to an Environment in Peril*, ed. A. Wolbarst. Baltimore, MD: Johns Hopkins University Press, pp. 164–178.

19 De Beaumont. (2020). "Vaccine Acceptance." Accessed at: https://www.debeaumont.org/covid-vaccine-poll/

20 See, e.g., Kahneman, D., Slovic, P., and Tversky, A., eds. (1982). *Judgment under uncertainty: Heuristics and biases.* New York: Cambridge University Press; Kahneman, D., and Tversky, A. (1979). "Prospect theory: An analysis of decision under risk." *Econometrica* 47(2):263–291.

21 See, e.g., Slovic, P. (1999). Trust, emotion, and sex. *Risk Analysis* 19(4):689–701; Slovic, P. (2010). *The Feeling of Risk: New Perspectives on Risk Perception.* London: EarthScan; Slovic, P., Finucane, L., Peters, E., and MacGregor, D.G. (2004). "Risk as analysis and risk as feelings: Some thoughts about affect, reason, risk, and rationality." *Risk Analysis* 24(2):311–322; Brosch, T. (2021). "Affect and emotions as drivers of climate change perception and action: a review." *Current Opinion in Behavioral Sciences* 42:15–21; Oh, S., Lee, S.Y., and Han, C. (2021). "The effects of social media use on preventive behaviors during infectious disease outbreaks: The mediating role of self-relevant emotions and public risk perception." *Health Communication* 36(8):972–981.

22 See, e.g., Sandman, P.M. (1989). "Hazard versus outrage in the public perception of risk," in *Effective Risk Communication: The Role and Responsibility of Government and Non-Government Organizations,* eds. V.T. Covello, D.B. McCallum, and M.T. Pavlova. New York: Plenum Press; See also Covello, V.T., and Sandman, P.M. (2001). "Risk communication: Evolution and revolution," in *Solutions to an Environment in Peril*, ed. A. Wolbarst. Baltimore, MD: Johns Hopkins University Press.

23 These needs are discussed by Abraham Maslow in his seminal 1943 paper "A Theory of Human Motivation" and in subsequent articles and books (see, e.g., Maslow, A., Frager, R., and Fadiman, J. (1987). *Motivation and Personality. 3rd Edition*. New York: Harper & Row).

24 See., e.g., Wahba, M. A., and Bridwell, L. G. (1976). "Maslow reconsidered: A review of research on the need hierarchy theory." *Organizational Behavior and Human Performance* 15(2):212–240.

25 See, e.g., Levy, M., and Tasoff, J. (2019). "Exponential-growth bias in experimental consumption decisions." *Economica* 87(345):52–80.

26 David Robson/British Broadcasting Company (BBC) (2020) . "Exponential-growth bias (EGB) is the tendency for individuals to underestimate exponential growth." Accessed at: https://www.bbc. com/future/article/20200812-exponential-growth-bias-the-numerical-error-behind-covid-19

27 See, e.g., Karelitz, T.M., and Budescu, D.V. (2004). "You say 'probable' and i say 'likely': Improving interpersonal communication with verbal probability phrases." *Journal of Experimental Psychology* 10(1):25–41.

28 National Center for Education Statistics. (2020). Adult Numeracy in the United States. NCES 2020-025. Washington, DC: US Department of Education. Accessed at https://nces.ed.gov/ datapoints/2020025.asp

29 See, e.g., Handmera, J., and Proudley, B. (2007). "Communicating uncertainty via probabilities: The case of weather forecasts." *Environmental Hazards* 7(2):79–87; Martín, Y., Li, Z., and Cutter, S. L. (2017). "Leveraging Twitter to gauge evacuation compliance: Spatiotemporal analysis of Hurricane Matthew." *PLOS One* 12 (7):1–22.

30 See, e.g., Morss, R. E., Demuth, J. L., Lazo, J. K., Dickinson, K., Lazrus, H., and Morrow, B. H. (2016). "Understanding public hurricane evacuation decisions and responses to forecast and warning messages." *Weather and Forecasting* 31(2):395–417.

31 See, e.g., Tversky, A., and Kahneman, D. (1981). "The framing of decisions and the psychology of choice." *Science.* 211(4481):453–458; Druckman, J. (2001). "Evaluating framing effects." *Journal of Economic Psychology* 22:96–101; Plous, S. (1993). *The Psychology of Judgment and Decision Making*. New York: McGraw-Hill. See also Thaler, R., and Sunstein, C. (2009). *Nudge: Improving Decisions about Health, Wealth, and Happiness*. New York: Penguin Books; Thomas, A.K., and Millar, P.R. (2011). "Reducing the framing effect in older and younger adults by encouraging analytic processing." *The Journals of Gerontology: Psychological Sciences and Social Sciences.* 67B (2):139– 149; Reyna, V.F., and Farley, F. (2006). "Risk and rationality in adolescent decision making: Implications for theory, practice, and public policy." *Psychological Science in the Public Interest* 7 (1):1–44.

32 Tversky, A., and Kahneman, D. (1981) "The framing of decisions and the psychology of choice." *Science* 211(4481):453–458.

33 Marteau, T. M. (1989). "Framing of information: Its influence upon decisions of doctors and patients. *British Journal of Social Psychology* 28(1):89–94; Bornstein, B.H., and Emler, A.C. (2008). "Rationality in medical decision making: A review of the literature on doctors' decision-making biases." *Journal of Evaluation in Clinical Practice* 7(2):97–107.

34 US Department of Health and Human Services, Office of Disease Prevention and Health Promotion. (2010). *National Action Plan to Improve Health Literacy*. Washington, DC: US Department of Health and Human Services. Page 1. Accessed at: https://health.gov/sites/default/ files/2019-09/Health_Literacy_Action_Plan.pdf; see also Institute of Medicine. (2004). *Health Literacy: A Prescription to End Confusion*. Washington, DC: The National Academies Press; Kutner,

M., Greenberg, E., Jin, Y., and Paulsen, C. (2006). *The Health Literacy of America's Adults: Results from the 2003 National Assessment of Adult Literacy* (NCES 2006–483). Washington, DC: National Center for Education Statistics.

35 See, e.g.,Flesch, R. (1948). "A new readability yardstick." *Journal of Applied Psychology* 32(3):221–233. See also Solnyshkina, M., Zamaletdinov, R., Gorodetskaya, L., and Gabitov, A. (2017). "Evaluating text complexity and Flesch-Kincaid grade level." *Journal of Social Studies Education Research* 8(3):238–248.

36 Centers for Disease Control and Prevention. (2010). *Simply Put: A Guide for Creating Easy-to-Understand Materials. Third Edition*. Atlanta, GA: Strategic and Proactive Communication Branch, Centers for Disease Control and Prevention. Page 29. Accessed at: https://www.cdc.gov/healthliteracy/pdf/simply_put.pdf

37 See, e.g., Covello, V., and Hyer, R. (2014). *Top Questions on Ebola: Simple Answers Developed by the Association of State and Territorial Health Officials*. Arlington VA: Association of State and Territorial Health Officials. Accessed at: https://www.astho.org/Infectious-Disease/Top-Questions-On-Ebola-Simple-Answers-Developed-by-ASTHO/; See also Hyer, R and Covello V. (2017). *Top Questions on Zika: Simple Answers*. Arlington, VA: Association of State and Territorial Health Officials. Accessed at: https://www.astho.org/infectious-disease/top-questions-on-zika-simple-answers-developed-by-astho/

38 See, e.g., Centers for Disease Control and Prevention. (2009). *Simply Put: A Guide for Creating Easy-to-Understand Materials. Third Edition*. Atlanta, GA: Strategic and Proactive Communication Branch, Centers for Disease Control and Prevention. Page 29. Accessed at: https://www.cdc.gov/healthliteracy/pdf/simply_put.pdf. See also Centers for Disease Control and Prevention (2020). CDC Clear Communication Index https://www.cdc.gov/healthliteracy/pdf/clear-communication-user-guide.pdf; see also Center for Medicare and Medicaid Services. (2020). Toolkit for Making Written Material Clear and Effective. Accessed at: https://www.cms.gov/Outreach-and-Education/Outreach/WrittenMaterialsToolkit/index?redirect=/WrittenMaterialsToolkit

39 See, e.g., Lipkus, I. M. (2007). "Numeric, verbal, and visual formats of conveying health risks: Suggested best practice and future recommendations." *Medical Decision Making*. September–October:696–713; Cleveland, W., and McGill, R. (1984). "Graphical perception: Theory, experimentation, and application to the development of graphical methods." *Journal of the American Statistical Association*. 77:541–547; Lipkus, I. M., and Hollands, J. G. (1999). "Visual communication of risk." *Journal of the National Cancer Institute Monographs* (25):149–63; Tufte, E. (1998) *Visual Explanations* (Third Edition). Cheshire, CT: Graphics Press. See also Tufte, E. (2006). *Beautiful Evidence*. Cheshire, CT: Graphics Press; Tufte, E. (2001). *The Visual Display of Quantitative Information. 2nd Edition*. Cheshire, CT: Graphics Press. Tufte, E. (1990). *Envisioning Information*. Cheshire, CT: Graphics Press.

40 See, e.g., Waters, E. A., Maki, J., Liu, Y., Ackermann, N., Carter, C. R., Dart, H., Bowen, D. J., Cameron, L. D., and Colditz, G. A. (2021). "Ladder, table, or bulleted list? Identifying formats that effectively communicate personalized risk and risk reduction information for multiple diseases." *Medical Decision Making* 41(1):74–88. See also, Sandman, P. M., Weinstein, N. D., and Miller, P. (1994). "High risk or low: How location on a 'risk ladder' affects perceived risk." *Risk Analysis* 14(1):35–45.

41 Vaughn, D. (1996). *The Challenger Launch Decision*. Chicago: University of Chicago Press. p. xiv.

42 Presidential Commission on the Space Shuttle *Challenger* Accident (1986). *Report of the Presidential Commission on the Space Shuttle* Challenger *Accident*. Washington, DC: Presidential Commission on the Space Shuttle *Challenger* Accident.See also: Vaughn, D. (1996). *The* Challenger *Launch Decision*. Chicago: University of Chicago Press; Dalal, S. R., Fowlkes, E. B., and Hoadley,

B. (1989). "Risk analysis of the space shuttle: Pre-*challenger* prediction of failure." *Journal of the American Statistical Association* 84(408) (December 1989):945–957; Esser, J. K., and Lindoerfer, J. S. (1989). "Groupthink and the space shuttle *challenger* accident: Toward a quantitative case analysis." *Journal of Behavioral Decision Making* 2(3):167–177.

43 Tufte, E. (1998) *Visual Explanations* (Third Edition). Cheshire, CT: Graphics Press. See also Tufte, E. (2006). *Beautiful Evidence.* Graphics Press; Tufte, E. (2001). *The Visual Display of Quantitative Information. 2nd Edition.* Graphics Press. Tufte, E. (1990). *Envisioning Information.* Graphics Press.

44 See, e.g., Cohen, B., and Lee, I. S. "A catalog of risks." *Health Physics*, vol. 36, pp. 707–722, 1979.

45 Federal Emergency Management Agency. (2020). Flood Zones. Accessed at: https://www.fema.gov/glossary/flood-zones

46 Meyer, R. J., Baker, J., Broad, K., Czajkowski, J., and Orlove, B. (2014). "The dynamics of hurricane risk perception: real-time evidence from the 2012 Atlantic hurricane season." *Bulletin of the American Meteorological Society* 95(9):1389–1404; see also "The Faulty 'Mental Models' That Lead to Poor Disaster Preparation." Accessed at https://knowledge.wharton.upenn.edu/article/wind-rain-worse/

47 Kennedy, A., Glasser, J., Covello, V., and Gust, D. (2008). "Development of vaccine risk communication messages using risk comparisons and mathematical modeling." *Journal of Health Communication* 13(8):793–807.

48 US Environmental Protection Agency. National Research Council, Committee on Improving Risk Analysis Approaches Used by the US EPA. Science and Decisions: Advancing Risk Assessment. (2009). Washington (DC): National Academies Press (US), Page 7. Accessed at https://www.ncbi.nlm.nih.gov/books/NBK214617/

49 See, e.g., World Health Organization. (2012). Communicating radiation risks in pediatric imaging: Information to support health care discussions about benefit and risk. Geneva, Switzerland: World Health Organization.

50 Lord, P. (2020). "COVID-19: Is the Cure 'Worse Than the Problem Itself'?" *McGill Journal of Law and Health Online.* Accessed at: https://ssrn.com/abstract=3573397

11

Evaluating Risk, High Concern, and Crisis Communications

CHAPTER OBJECTIVES
This chapter is an overview of the principles and approaches for evaluating risk communications, assessing the effectiveness of the messages, the messengers, and the delivery channels.   At the end of this chapter, you will be able to: • Discuss the necessary role and value of evaluation in all phases of risk, high concern, and crisis communication, from developing materials to in-process progress checking, to outcome assessment. • Discuss challenges in effective evaluation. • Apply helpful tips for evaluating high concern communication programs and messages. • Determine how to find appropriate evaluation tools.

11.1   Case Diary: Finding the Road to Rio

The XXXI Olympic Games – held in August 2016 in Rio de Janeiro – were only six months away when I arrived in Brazil in February of that year. I had been a crisis communication consultant at several previous Olympics, and each had its own story. Now in Brazil, an enemy appeared that threatened the health of adults, the lives of children, and the viability of the Olympic Games. The enemy was a familiar one – mosquitoes. Hundreds of thousands of people worldwide die every year from mosquito-borne diseases. The diseases include malaria, yellow fever, dengue fever, Japanese encephalitis, chikungunya, and the West Nile virus. The newly revealed threat from these tiny, dangerous creatures was the Zika virus they were now carrying.

Brazilian health authorities believed that mosquitoes carrying the Zika virus were the cause of a spike in Brazil of microcephaly, a birth defect marked by an abnormally small head and incomplete development of the brain. On February 1, 2016, the World Health Organization (WHO) had declared an international health emergency. The US Centers for Disease Control and Prevention (CDC) was advising pregnant women or those considering becoming pregnant to avoid travel to places with Zika outbreaks, especially Brazil.

This was the first time in history that a South American country had hosted the Olympic Games. Billions of dollars were at stake. Brazil had spent enormous sums building an infrastructure for Olympic events. At the same time, Brazil was experiencing an economic recession and high unemployment. Distrust of government was increasing because of alleged corruption at the state oil

Communicating in Risk, Crisis, and High Stress Situations: Evidence-Based Strategies and Practice, First Edition.
Vincent T. Covello.
© 2022 by The Institute of Electrical and Electronics Engineers, Inc. Published 2022 by John Wiley & Sons, Inc.

company, Petrobras. Brazil desperately needed the economic boost that a successful Olympic Games would provide.

The challenge that Brazilian risk managers and communicators faced was that in six months Brazil would welcome thousands of athletes and at least a half-million visitors. Yet Brazil was experiencing the largest outbreak of the mosquito-borne Zika virus in the world. There was no vaccine for the virus. It had already infected hundreds of thousands of Brazilians. Experts had evaluated the safety of holding the games in Brazil and determined that it would be safe if proper mosquito control efforts were implemented.

The Brazilian challenge with the Zika threat had to be addressed on three fronts: (1) fighting mosquitoes to the degree necessary to ensure safety, (2) engaging constructively with Brazilian citizens concerned about Zika, including those critical of the government's mosquito-control efforts, and (3) engaging constructively the Olympic athletes and visitors to the games fearful of contracting Zika. My colleagues and I were asked to assist with communications with the Brazilian public, with visitors, and especially with the Olympic Games athletes.

11.1.1 The Mosquito Front

We kept in close touch with the technical experts on their progress in the mosquito-control front so that our messages would be accurate, timely, and relevant to the concerns being voiced.

11.1.2 The Citizen Front

On the citizen front, public meetings on mosquito control organized by the government were not going well. They were repeatedly interrupted by people inside and outside the meeting shouting and holding signs saying, in Portuguese, Spanish, and English: "No more spraying." "No more pesticides." "Stop Poisoning Our Air and Water with Pesticides." Many in the audience were wearing gas masks. Children were wearing their Day of the Dead skeleton costumes. Despite the seriousness of the Zika threat, protesters demanded that the government stop spraying pesticides and not restart.

In multiple locations throughout Brazil, but especially in Rio, the fight to stop spraying for mosquitoes pitted risk management authorities against angry crowds made up of community leaders and residents fearful about pesticide spraying. One of the government officials at the meeting I attended said: "I don't like spraying either. However, the dose makes the poison. You might not like the poisonous pesticide we are using, but you have to take a little if we are to get Zika under control."

Opponents of spraying with pesticides said they were afraid for their community's health, their own health, their children's health, and the health of pregnant women, beneficial insects, and marine life. Several speakers questioned whether Zika was really a problem and really caused birth defects. Several speakers said the actual cause was the use of pesticides by farmers. Another speaker said he believed Zika was part of a genocidal conspiracy by the US and its allies to kill Latinos.

The voices raised at the meeting demonstrated the depth of emotion generated by the mosquito-control program. It also demonstrated a deep mistrust of both science and the government felt by many citizens. Protesters urged the government to stop spraying even as scientific and technical experts told the meeting attendees and the listening public that spraying was safe and few alternatives were available.

11.1.3 The Olympic Athlete and Visitor Front

On 8 February, Reuters reported that the United States Olympic Committee had told US sports federations that athletes and staff concerned for their health over the Zika virus should consider not going to the Rio 2016 Olympic Games.[1] On February 9, newspapers reported that US soccer star goalkeeper Hope Solo had concerns about competing in the Olympics because of the spread of the Zika virus in Brazil.[2] She was quoted as saying: "I would never take the risk of having an unhealthy child. . . Competing in the Olympics should be a safe environment for every athlete, male and female alike. Female athletes should not be forced to make a decision that could sacrifice the health of a child." Solo was one of the first high-profile athletes to say she might not be attending the Olympic Games.

It was against this backdrop that my colleagues and I arrived in Brazil.

11.1.4 Communication Strategy: The Citizen Front

Our team joined with our Brazilian colleagues and conducted in-depth focus groups with officials involved with mosquito control to learn about their experience working with and hearing from community leaders, key opinion leaders, and the public. We also conducted in-depth focus groups to gather information about the perceptions and opinions of citizens, including community leaders and key opinion leaders, to understand the scope of their knowledge and the nature of their concerns. The focus groups were also a venue to evaluate messages and test message maps and different communication formats. Key messages emerged for message maps, such as removing standing water where mosquitoes breed and using only the recommended insect repellents. We engaged with diverse stakeholders, allowing us to evaluate the success of stakeholder engagement practices and communication products published on the Brazilian Ministry of Education website with the catchy name #ZIKAZERO. The site contained print materials and a video gallery. Figure 11.1 is an example of one of the posters.[3] We held multiple open houses and sessions where experts were available. We trained all speakers to use plain language.

After evaluating all aspects of the communications program and working through messages and issues, we gained citizen support for a revised spraying program that would reduce aerial spraying and increase ground spraying; the ground spraying would be done at agreed-upon, pre-announced times. Although aerial spraying could deliver wider coverage, especially in difficult-to-reach areas, it was fear-inducing, triggering reactions that included the anxiety and anger brought about by lack of control and unknown/unknowable threats. With insights gained through evaluation, the government continued its essential mosquito-control program with much greater public acceptance and wider public knowledge.

11.1.5 Communication Strategy: Olympic Athlete and Visitor Front

We developed a comprehensive risk communication strategy and plan with clear objectives.[4] To evaluate the likely success of the plan, we conducted in-depth focus groups with athletes to learn what they already knew and believed about Zika and mosquito control. We identified whom they trusted as sources of information. We maintained a continuous dialogue with opinion leaders – "influencers" – among

Figure 11.1 Zika mosquito poster (Brazil).

the athlete groups. We produced visual products with their assistance. We asked them to share their personal stories and prevention/protection tips, and asked permission to share their stories through television, radio, print, and social media channels such as #ZikaZero. We asked Olympic athletes to meet with and accompany government workers, volunteers, and 200,000 Brazilian soldiers recruited for the Zika mosquito-eradication effort; it was an opportunity for the athletes to be eyewitnesses to the determination, dedication, and commitment invested in bringing the Zika outbreak under control.

Hope Solo competed in the XXXI Olympic Games in Rio. She and her soccer team made it to the quarterfinals. At the end of the Olympics in August, the World Health Organization reported that there were no laboratory-confirmed cases of Zika virus in spectators, athletes, or anyone associated with the Olympics.[5]

This case study illustrates several of the benefits of evaluation during the course of a communication process, including in-process review of current communication activities to inform changes and new directions; collecting attitudinal and behavioral data to inform communication strategies; and the need for baseline data to assess progress toward plan objectives. In this case, we needed baseline information, in-process data, and a continual evaluation loop to be able to get key messages to intended audiences.

11.2   Introduction

This chapter summarizes the principles for evaluating risk, high concern, and crisis communications. In this book, *evaluation* is defined as any effort to determine the effectiveness of a risk, high concern, or crisis communication. Because of the importance of evaluation in decision-making, the theory and practice of evaluation have evolved into a separate discipline with its own journals, definitions, methods, approaches, and applications. The chapter focuses on the role evaluation plays in best practices for risk, high concern, and crisis communication, and considers approaches, benefits, and challenges in evaluating a communication program. The chapter summarizes key elements in the evaluation of messages, messengers, and delivery channels that serve as the means for communication.

The many forms of risk, high concern, and crisis communication evaluation research can be grouped by the purpose of the communication effort; there are four areas of focus: (1) *providing information* – where the primary goals include promoting the transfer, exchange, and discussion of information to support informed decision-making, (2) *changing behavior* – where the primary goals include influencing behavioral intentions and actual behavior in response to a risk or threat, (3) *providing warnings and guidance* – where the primary goals include providing information and directions that can be understood and acted on, and (4) *engaging stakeholders in constructive dialogue and joint problem-solving* – where the primary goals include resolving conflicts and encouraging and empowering stakeholders to inform and influence each other.

For effective evaluation, it is necessary to know the specific objectives of the communications and determine whether these objectives were achieved. Objectives should be developed with evaluation in mind. As shown in Table 11.1, objectives should be Specific, Measurable, Attainable/Achievable, Relevant/Realistic, and Time-bound/Timely (SMART).

Evaluation of a risk, high concern, or crisis communication can be conducted before, during, or after a risk-related event or situation. Evaluation is ideally an integral part of risk, high concern, and crisis communication from the outset. During the planning, pretesting, and pilot phase,

Table 11.1 Risk, high concern, and crisis communication objectives.

Are the communication objectives *SMART*:
(S) Specific: e.g., Are the communication objectives explained in clear, detailed, and well-defined terms?
(M) Measurable: e.g., Are the communication objectives quantifiable?
(A) Attainable/Achievable: e.g., Are the communication objectives achievable and not overly ambitious given available resources?
(R) Relevant/Realistic: e.g., Are the specific communication objectives directly related to the single overriding, overarching communication goal from your communication plan?
(T) Time-bound/Timely: e.g., Have the communication objectives been assigned deadlines for achievement?

evaluation provides data critical to effective program design, including information about the needs of the intended audience and how to meet those needs. Through surveys, questionnaires, focus groups, and other evaluation methods, evaluation (1) identifies relevant target audiences, (2) measures audience opinions and reactions, (3) finds out what people see as important, (4) finds out what issues and events people are aware of, and (5) identifies likely reactions to different messages, messengers, and delivery channels. A key objective in the planning, pretesting, and pilot phase of evaluation is to identify population segments that will likely benefit most from communication and engagement.

One approach for evaluating the impact of a risk communication on perceptions, behaviors, and behavioral intentions in the planning, pretesting, and pilot phase is to observe the effects of communications on *mental models* – representations of what a person believes about a risk or threat.[6] Evaluators have constructed mental models for a wide variety of risks and threats, ranging from health issues, such as HIV/AIDS, to electromagnetic fields, earthquakes, radon, and climate change. Through intensive interviews, researchers construct mental models of what experts and non-experts believe about the same risk or threat, compare these mental models, construct messages that address differences in expert and non-expert mental models, and evaluate the constructed messages to see if they increase the knowledge, understanding, and informed decision-making of target audiences.

Another common approach for evaluating the potential impact of a risk communication in the planning, pretesting, and pilot phase is to use data collected through focus groups and surveys. A detailed example of this approach is provided in a case study at the end of this chapter describing how information collected from focus groups and interviews was used to improve communications about mosquito control. The study, conducted in 2018–2019 by the author and a colleague, employed pretesting and pilot testing to help determine what information was most needed, most wanted, and most useful to stakeholders. Pretesting and pilot testing also provided feedback on how people were likely to process and interpret information about mosquito control. Results from pretesting and pilot testing were then combined with information about program resources to identify risk communication strategies and messages most likely to be effective.

Once the communication program is operational, evaluation studies address questions of accountability and performance. They (1) determine what people received, what people learned, and what changes, if any, took place; (2) determine whether communication materials and activities reached the intended audience; (3) provide feedback to communicators on whether the communications are achieving the intended objectives; (4) identify communication program strengths and weaknesses; (5) indicate ways to communicate more effectively; and (6) determine whether the communication strategies are being implemented appropriately.

Besides evaluating performance, evaluation research in the operational phase addresses logistical questions: How many communication materials were produced? How many communication materials were distributed? How many communication materials were used? How many people accessed a site? How many people downloaded materials? How well was a site ranked in site engine optimization? How many other sites link to the site? How many people resent a posted message to others? How long did it take to produce communication materials? How much did it cost to produce communication materials and hold communication events?

Post-event evaluations help answer critical questions about outcomes. For example, to what degree did communications achieve the desired effect, such as raising awareness about the risk or creating feelings of personal control? To what extent did communications change knowledge, beliefs, or behaviors? To what extent did the target audience understand scientific and technical uncertainties? To what extent did scientific and technical data affect judgments, decisions, and behaviors? Post-event evaluations establish lessons learned to improve future communications work.

11.3 Benefits of Evaluation

Evaluation provides information on whether communication materials and activities are achieving identified, expected, or desired goals. As listed in Table 11.2, several benefits flow from the evaluation. These benefits are especially clear when evaluation is done well and starts early in the planning process.

Without evaluation, it is virtually impossible to determine how to use limited communication resources effectively. Outcome evaluation measures are particularly useful here. Depending on the goals of the communications, evaluation can measure whether the risk, high concern, or crisis communications produced changes in awareness, knowledge, understanding, concerns, attitudes, intentions, behaviors, dialogue, cooperation, and trust. Evaluation researchers ask the following: To what extent did the target audience pay attention to the communications? To what extent did the audience find the communications culturally appropriate, accurate, understandable, clear, attention-getting, credible, easy to recall, and

Table 11.2 Benefits of evaluation.

6) Encourages clear thinking about communication goals and objectives

7) Enhances understanding of which communication strategies and techniques work and which ones do not

8) Helps decision-makers determine whether the costs of the communication activities are cost-effective – worth the money, time, and effort

9) Helps ensure individuals and organizations are using the most appropriate and effective communication methods and tools

10) Allows for on-going monitoring of the effects of communication efforts.

11) Allows for feedback, continuous learning, and opportunities to improve and correct mistakes

12) Helps determine whether messages have been received, understood, and result in positive changes in attitudes and/or behaviors

13) Helps ensure that communication projects and programs are aligned with the mission and objectives of the group or organization

relevant? To what extent did the communications address risk perception and emotional factors, such as perceived fairness.

In pretesting before a risk-related event, specific messages can be incorporated into draft materials, such as fact sheets, brochures, and social media content. Knowledge obtained through pretesting often leads to revised communication objectives. For example, answers to the following questions frequently result in changes regarding the measurability and attainability of communication objectives.

- Which messages gave stakeholders the most important new information about the risk or threat (i.e., to explore how attention-getting and how helpful the messages were)?
- What did stakeholders believe the messages said or were trying to say about the risk or threat (i.e., to explore whether the messages are understandable)?
- What was it about the messages that made them relevant or not relevant to stakeholders (i.e., to explore usefulness)?
- What messages most affected decisions, intentions, and motivations to do something or take actions (i.e., to explore the impact on intentions and behaviors)?
- What can risk management authorities do better to address stakeholder questions and concerns (i.e., whether the messages answered the questions asked by stakeholders)?

Evaluation studies conducted after a risk-related event typically ask stakeholders the following: What communications did you receive? What did you like or dislike about the communications? What actions did you take because of the communications? As an example, people evacuated from near the site of the 2011 Japanese Daiichi/Fukushima nuclear power plant accident were asked about the communications they received soon after the accident. Their answers revealed high levels of dissatisfaction that created high levels of stress and distrust toward technical experts and risk management authorities.[7] Respondents said they had been misled about radiation risks by medical experts and by government assurances that "levels of radiation would not immediately affect the human body." Respondents complained they received an overload of communications about radiation risks, but they did not know whom or what to believe.

Other major types of evaluation studies are those that seek to examine processes as opposed to outcomes.[8] A fundamental assumption of this approach is that effective risk, high concern, and crisis communication is a dynamic, interactive, ongoing, and adaptive process and can be evaluated as such. Based on this approach, specific communication activities can be evaluated based on measured levels of stakeholder involvement, engagement, and participation. Evaluation questions include the following: Did stakeholders participate in communication planning? Were stakeholders consulted about their preferred channels of communication, about their preferred or trusted source of information, and about any specific features of their culture – such as socioeconomic status, age, gender, education, personal networks, and scientific literacy – that needed to be addressed in communications? Did stakeholders feel they had the levels of knowledge and power needed to enable meaningful engagement?

11.4 Evaluation Practices for Risk, High Concern, and Crisis Communication

Evaluation methodologies comprise a wide range of processes grounded in the behavioral science literature. The most commonly used methods for evaluation of communication programs include attitudinal and behavioral surveys, focus groups, interviews, and analyses of quantitative measures

Table 11.3 Best practices: risk and high concern communication.

---

1) Accept and involve stakeholders as partners
2) Carefully plan, coordinate, and evaluate communication efforts
3) Listen to the specific concerns of stakeholders
4) Be transparent and honest
5) Include other trusted sources
6) Meet the needs of the media
7) Speak clearly and compassionately

Source: Covello, V.T., and Allen, F.W. (1988). *Seven Cardinal Rules of Risk Communication.* Washington, DC: United States Environmental Protection Agency. https://archive.epa.gov/publicinvolvement/web/pdf/risk.pdf

of behavior, such as intake of and responses to specific messages. Section 11.10 of this chapter provides a selection of tools introducing evaluation methods.

An emerging field is the application of neuroscience techniques to understand an individual's response to messages. These techniques include direct measures of brain responses, such as functional magnetic resonance imaging and electroencephalogram (EEG) recording, as well as measuring secondary/surrogate responses such as eye-gaze tracking and coding of facial expressions. These techniques are moving from the research laboratory to consumer applications,[9] such as brand marketing, and also to testing of major public health service messages.[10]

Among the most common types of evaluation studies are case studies that describe a specific set of communication activities and compare these activities against established risk, high concern, and crisis communication best practices, such as those listed in Tables 11.3 and 11.4.[11]

These types of studies are a valuable addition to the risk, high concern, and crisis communication literature. They articulate lessons learned and point the way to better preparation and implementation of risk, high concern, and crisis communication plans. Three examples of evaluation studies that compare a case to best practice communication are provided in the next section: *Hurricane Katrina, COVID-19 and Vaccination Hesitancy,* and *Outbreak of COVID-19 in Wuhan, China.*

Table 11.4 Best practices: crisis communications.

---

1) Strategic Planning
    a) Plan pre-event logistics
    b) Coordinate networks
    c) Accept uncertainty
2) Proactive Strategies
    a) Form partnerships
    b) Listen to public concern
    c) Be open and honest
3) Strategic Response
    a) Be accessible to the media
    b) Communicate compassion
    c) Provide self-efficacy

Source: Seeger, M.W. (2006). "Best practices in crisis communication: An expert panel process." *Journal of Applied Communication Research* 34(3):232–244.

11.5 Case Studies of Evaluation Comparison to Best Practice: Hurricane Katrina, COVID-19 and Vaccination Hesitancy, and Outbreak of COVID-19 in Wuhan, China

11.5.1 Hurricane Katrina

Hurricane Katrina was a Category 5 storm that created havoc in New Orleans and the Gulf Coast of the United States in August 2005. Katrina was, at that time, the most lethal and costly hurricane in American history, devastating sizable areas of the Gulf Coast, especially New Orleans. Responses to the hurricane were failures piled on failures in emergency preparation, management, and response, with dire consequences. Poor communication to residents of New Orleans before, during, and after the crisis was a large contributor to the failure.

As summarized by Cole and Fellows (2008) in one of the many substantive evaluations and literature reviews of communications related to Hurricane Katrina, "inadequate clarity, insufficient credibility, and a failure to properly adapt to critical audiences" left residents of New Orleans without essential guidance for their own protection and recovery.[12] At the heart of the failure was the lack of preparation to develop messages that were clear, credible, and meaningful to the multiple, diverse audiences that needed to receive them. Such messages typically arise only from proactive stakeholder engagement and evaluation processes with cultural constituencies. This was absent for Katrina. Logistical planners and risk and crisis communicators did not take into account how difficult evacuation orders are to be heeded if they are not coming from a fully trusted source and do not provide clear, consistent, and feasible instructions along with actionable assistance for physical evacuation. Logistical planners and risk and crisis communicators did not sufficiently consider how difficult evacuation is for poor communities lacking the means of transportation or the resources to relocate; did not consider the mistrust of government officials by local communities, a mistrust based on their experience of a lack of assistance and understanding in many aspects of their lives; and did not consider that for many individuals, especially those of limited means, their homes were their sanctuaries, the place to stay during a crisis, not to abandon. And, as pointed out forcefully by the evaluation researchers, logistical planners and risk and crisis communicators often overlooked the many differences between communities, especially regarding class and race. Reflecting on communications lessons learned from Katrina, one researcher looking at the Katrina communications and response wrote the following:

> In response, many will ask what more can be done? As an African-American researcher, my response is that we need to begin by expanding our view of risk . . . Significantly less attention has been devoted to the unique risks brought on by racial discrimination and poverty. More than 12 percent of the people—men, women, and, most tragically, children—in the United States live below the poverty line. Almost a quarter of the African-American population in the United States—roughly 24 percent—lives below the poverty line. This imbalance in the research agenda needs to be addressed.[13]

The heart-wrenching and profound example of Hurricane Katrina points to the essential need for evaluation to be perceived as a critical component of the risk assessment, management, and communication process. Lives are often at stake. Lessons learned from the Katrina response established more effective policies and practices for all aspects of government hurricane preparation and response, including communications policies and practices – all put to successful use in subsequent storm crises.[14]

11.5.2 COVID-19 and Vaccination Hesitancy

Vaccine hesitancy is the delay in the acceptance of, or refusal of, vaccines, despite the availability of vaccine services. Vaccine hesitancy varies along a continuum, from those who fully accept vaccination to those who delay vaccination to those who completely reject vaccination.

Vaccine hesitancy is an issue for nearly all vaccines. For example, in the United States, approximately one in five children have a parent who is hesitant to have their child vaccinated for childhood diseases such as measles.[15] For seasonal influenza vaccination, in 2020, less than half (48.4%) of adults received the flu vaccine.[16] For the COVID-19 pandemic that began in 2020, a sizable proportion of the population in the US reported they either did not plan to or were unsure about, getting vaccinated with vaccines approved by the US Food and Drug Administration. Vaccine hesitancy also occurred in a sizable proportion of those working in the health-care industry.[17] Much was at stake: long-term control of the COVID-19 pandemic hinged on the uptake of a preventive vaccine.[18] By the middle of 2021, over half a million people had died in the US from COVID-19, and millions had died worldwide.

Evaluation research indicates vaccine hesitancy for COVID-19 and for other diseases is complex and varies greatly across time, place, and populations. For example, "in some countries, the public and media may be very interested in vaccine efficacy. While in other countries the focus may be primarily on safety. It is also likely that concerns beyond science will emerge, such as trust in the government officials who are recommending vaccination, and this will also vary by country."[19] What is universal is that risk perceptions and levels of concern about vaccination are seldom consistent with the actual risks of vaccination.

Vaccine hesitancy is driven primarily by complacency, convenience, trust, and confidence. Complacency, convenience, trust, and confidence, in turn, are driven by the variables listed in Table 11.5: (1) contextual factors, (2) individual and group factors, and (3) vaccine/vaccination-specific factors.[20]

Each of the factors in Table 11.5 needs to be examined in a comprehensive vaccine hesitancy evaluation study. The complexity and specificity of vaccine hesitancy make evaluation studies following best practice essential to developing a successful communication effort for a vaccination program.

Evaluation research specific to COVID-19 indicates that Americans' perceptions about COVID-19 vaccines and their safety differ significantly by political party, race, age, and geography. Survey data indicate that communications addressing COVID-19 vaccine hesitancy are most effective when delivered by a trusted source and when they explain the benefits of getting vaccinated, not just the consequences of not doing it; focus on the need to return to normal and reopening the economy; avoid judgmental language when talking about (or to) people concerned about the COVID-19 vaccine and vaccine safety; explain the vaccine-development process, emphasizing the transparent and rigorous review process by the Food and Drug Administration and its safety board of leading vaccine experts; focus on the extraordinary effectiveness of approved vaccines; explain that the chance of a severe side effect is extremely small; and explain that when mild side effects occur, they are a normal sign the body is building protection to the virus, and most will go away in a few days."[21]

11.5.3 Outbreak of COVID-19 in Wuhan, China

Ever since the first outbreak of COVID-19 in Wuhan, China, in late 2019, researchers have evaluated communications by risk management authorities on COVID-19 against risk, high concern, and crisis communication best practices. For example, using government documents and media reports, researchers evaluated communications in Wuhan at the beginning of the outbreak. They concluded that ineffective risk communication and the lack of transparency heavily impeded

Table 11.5 Determinants of vaccine hesitancy.

1) **Contextual factors**
   a) Trust in leadership
   b) Mainstream and social media environment
   c) Influential leaders
   d) Vaccination program gatekeepers
   e) Anti- and pro-vaccination lobbies
   f) Historical factors
   g) Religion/culture/gender/socio-economic factors
   h) Political factors
   i) Geographical factors
   j) Perceptions of the pharmaceutical industry

2) **Individual and group factors**
   a) Trust
   b) Personal, family, and/or community experience with vaccination
   c) Knowledge and awareness
   d) Trust and personal experience with the health system and health care providers
   e) Perceived risks, costs, and benefits of vaccination
   f) Social norms
   g) Perception of a high risk for severe infection

3) **Vaccine/vaccination-specific issues**
   a) Trust
   b) Self-perception of high risk for severe infection and hospitalization
   c) Real or perceived pain from vaccination
   d) Real or perceived side effects
   e) Beliefs and attitudes about health prevention
   f) Knowledge and awareness
   g) Vaccine program enrollment and delivery system (e.g., routine or mass vaccination program)
   h) Personal experience
   i) Perceived risk/costs/benefits
   j) New vaccine, new formulation, or new recommendation for an existing vaccine
   k) Mode of administration
   l) Reliability and/or source of supply of vaccine and/or vaccination equipment (e.g., syringes)
   m) Vaccination schedule
   n) Strength of the vaccination recommendation
   o) Perception of knowledge and/or attitude of the health care provider

emergency response in Wuhan.[22] Researchers found that risk managers and communicators in Wuhan violated nearly every best practice of risk, high concern, and crisis communication. They failed to be honest and transparent, failed to address risk perception factors, failed to effectively address rumors, and consistently engaged in communication practices that undermined trust.

In the wake of the ineffective response of the Wuhan provincial government in the early stages of the COVID-19 pandemic, evaluation researchers noted that "the Chinese Central Government developed a series of public health emergency strategies including risk communication to release timely information and stem COVID-19 misinformation through press conferences, and enforced preventive behaviors such as mandated use of masks and handwashing in the general population after deciding to lock down Wuhan city."[23] Evaluation data collected across 31 provinces in China indicated that the risk communication practices of the Chinese government's response to COVID-19 reached a high proportion of the general population, resulting in high levels of adherence to preventive and protective behaviors. Despite being the first place to be hit by COVID-19, China was well-positioned to stop the spread of the disease and implement lessons learned from the initial mistakes in Wuhan.[24] China has a centralized epidemic response system that was revamped because of the 2003 SARS outbreak and the high mortality rate associated with that outbreak. Speed was crucial in China's ability to quickly stop transmission of COVID-19. In January 2020, China began enacting a raft of rigorous national operational and risk communication countermeasures. People in Wuhan were told that the city was going into lockdown, which eventually lasted 76 days. People were told public transport in Wuhan was suspended, that 14,000 health checkpoints were being established at public transport hubs throughout China and that outdoor activities would be severely restricted. Within weeks, China had tested nine million people for COVID-19 in Wuhan and had set up an effective national system of contact tracing.

11.6 Barriers and Challenges to Evaluation

Although evaluation is valuable and beneficial, it faces many barriers and challenges. These barriers are typically rooted in different (1) values, (2) goals, (3) resources, and (4) perceived benefits. Each is briefly discussed below.

11.6.1 Differences in Values

Evaluation is, by its nature, a value-laden undertaking. This derives in part from the vested interests of audiences interested in or affected by the conduct and effectiveness of any communication. Stakeholders with vested interests in risk, high concern, and crisis communication include, but are not limited to, government agencies, nongovernmental organizations (NGOs), business organizations, unions, the media, scientists, professional organizations, voluntary organizations, contractors, vendors, and individual citizens. Each stakeholder has varying needs, interests, and perspectives – and different values.

11.6.2 Differences in Goals

A second barrier and challenge affecting evaluation are identifying and agreeing upon goals. Listed in Table 11.6 are several possible goals of a risk, high concern, or crisis communication activity. Without agreement on goals – whether they are for efforts to gain stakeholder engagement, for the operational implementation of a communication plan, or for the intended impact of messaging – it is not possible to evaluate the success of the communication.

11.6.3 Competition for Resources

Resources, often scarce, need to be allocated for evaluation – including money, people, and time. As a result, evaluation of risk, high concern, and crisis communication activities is often neglected in favor of needs and tasks perceived to be more urgent, such as operational tasks and needs. This is especially the case if the evaluation has not been planned and budgeted in advance. Even when resources are available, many risks, high concern, and crisis communication programs exhaust

Table 11.6 Potential goals of a risk, high concern, or crisis communication activity.

---

1) To identify the information needs of stakeholders
2) To engage stakeholders to achieve greater:
   a) awareness
   b) knowledge and understanding
   c) receptivity to recommended actions and policies
   d) judgment in seeking credible information
   e) cooperation and collaboration
   f) willingness to participate in the decision-making process and constructive dialogue
3) To establish communication links with a wide diversity of stakeholders
4) To involve stakeholders in the design and implementation of communications strategies and messaging
5) To meet a specific stakeholder engagement requirement, such as legal mandates for stakeholder engagement and consultation
6) To follow best practices in risk, high concern, and crisis communication, such as acknowledging emotions and speaking clearly and with empathy
7) To achieve specific objectives identified in the major types of risk, high concern, and crisis communication, such as informing people about a risk or threat, encouraging personal protective actions, and providing direction and behavioral guidance
8) To achieve informed consent

---

available resources on process evaluation – such as evaluating the production and distribution of communication materials – rather than outcome evaluation – determining whether the target audience received and internalized the message.

11.6.4 Ability to Learn from Results

There is an understandable reluctance to engage in evaluation of a risk, high concern, or crisis communication given the potential for the evaluation to show that communication efforts were not well received or did not produce the desired outcomes. In such cases, it is often easier simply to proclaim victory than to conduct an evaluation. Even when organizations conduct an evaluation, they too often do so only in a *pro forma* manner, e.g. only to meet an external or internal requirement, and do not invest in trying to learn from the results.

As shown in Table 11.7, evaluations can identify shortcomings and failures that result from several sources.

Table 11.7 Common shortcomings or failures identified by evaluation.

---

1) Poor design or execution
2) Competition received more attention
3) Insufficient time devoted to preparation
4) Insufficient resources devoted to the effort
5) Inadequate communication skills
6) Ingrained attitudes and beliefs
7) Disagreements about goals, values, and metrics
8) Unrealistic expectations

The communications effort may fail because it was poorly designed and executed; competing or conflicting sources of information may have received greater attention and reception; communication activities may not have lasted for long enough for the desired outcomes to be achieved or observed; communication activities may have lacked resources and support from leadership; communicators and communication staff may have lacked the skills needed to implement the communication strategy or activity; existing attitudes and beliefs about a risk or threat may have become so ingrained they are resistant to change; managers and communicators may have disagreed about goals, values, and definitions of success; or the communication effort may have begun with unrealistic expectations.

Shortcomings and failures identified in an evaluation can have adverse impacts on career advancement, allocation of organizational resources, and performance evaluations. To avoid these outcomes, managers and communicators often favor administrative measures of communication success – such as how many people attended a meeting and how many materials were distributed – instead of outcome measures of success – such as changes in attitudes, behaviors, dialogue, and relationships. For social media, success is typically measured through key performance indicators and social media metrics such as *followers, viewings, likes, shares, comments, reposts* (e.g., *retweets*), *calls to action, referrals*, and *mentions* (e.g., mentions on other social media and mainstream media platforms). Virtually every social media platform has its own performance metrics. For Facebook at the time of the publication of this book, they could be found on the *Insights* tab, which provides demographic data about the page's audience and how people are responding to posts. On Twitter, performance metrics could be found on *Twitter Analytics*. To be successful, social media posts also need to be accessible, trusted, relevant, timely, and understandable.[25]

These challenges and barriers to evaluation argue for a strategic approach to evaluation. Several strategies – *Planning for Evaluation, Developing Checklists*, and *Setting Realistic Expectations* – are discussed below.

Strategy: Planning for Evaluation. Despite the many advantages of conducting evaluations before and during an event, a disproportionate number of evaluation studies take place only after a risk-related event has occurred. Planning for evaluation should strategically occur early in the communication effort. If planned early, evaluation becomes more easily integrated into each stage of the communication program. Time and resources for evaluation should be allocated in advance.

Strategy: Developing Communication Checklists. Checklists – typically presented as a series of questions with a yes or no option – help ensure that essential communication tasks get done.[26] They help avoid distractions by forcing communicators and organizations to focus on the necessary tasks. Table 11.8, Table 11.9, Table 11.10, and Table 11.11 provide examples of such checklists for risk, high concern, and crisis communication. An excessive number of "no" responses indicates changes are or may be needed. An excellent example of a checklist for evaluating crisis communication in a pandemic can also be found in Jong (2020).[27]

Strategy: Setting Realistic Expectations. Setting realistic expectations for evaluation helps prevent disappointments and focuses on achievable objectives. For example, communication programs often fail because of unrealistic expectations for three of the most commonly cited risk, high concern, and crisis communication goals: *changed knowledge, changed beliefs*, and *changed behavior*.[28] One challenge in setting a *changed knowledge* goal is that simply bringing up a risk or threat to inform decision-making, even if the risk or threat is extremely small or an extreme worst case, may cause people to react in unintended and, sometimes, dysfunctional ways. Moreover, people are typically exposed to many different communications about a risk or threat, and it is difficult to isolate the change in knowledge caused by any one communication.

Table 11.8 Checklist for evaluating risk, high concern, and crisis messages.

What to Look for in Core Messages	Y/N	Notes
1) Did you present information clearly and in a manner that could be easily understood by your target audience (e.g. 4 years below the average grade level of your audience)?		
2) Were your sentences short (e.g. 10-12 words on average)?		
3) Did you avoid using acronyms, undefined technical jargon, or bureaucratic language that would not be understood by your target audience?		
4) Did you present only a few overriding, top-line messages in response to a question or concern (e.g. no more than 3-5 key messages)?		
5) Did you state your key messages as briefly as possible, such as a total of no more 27-30 words?		
6) Did you bridge often your key messages or to supporting information?		
7) Did you support your key messages with visual aids, such as graphics, infographics, drawings, charts, props, or photographs?		
8) Did you repeat your key messages several times?		
9) Did you provide the most important information first and last? (Note: According to the primacy/recency effects, people in high-stress situations tend to focus on beginnings and ends)		
10) Are your messages consistent with what other trusted sources are saying about the issue?		
11) Do your messages cover only important points, avoiding superfluous detail?		
12) Did you first share messages that communicate listening, caring, empathy, or compassion when responding to an emotionally charged question or concern?		
13) Are your messages responsive to the questions being asked or the concerns being expressed?		
14) In responding to a negative statement or sharing bad news, did you provide at least three to four positive, solution, or constructive statements? (Note: According to negative dominance/loss aversion theory, it typically takes three to four positive messages to offset one negative.)		
15) Did you avoid offering unconditional guarantees about outcomes or unqualified and/or unnecessary absolutes, such "never," "always," "all," and "every"?		
16) Did you focus your messages on what you can do as opposed to what you can't do?		
17) Did you avoid using the word "but" (or the equivalent) when it might negate, invalidate, or erase what you said before?		
18) Did you avoid using the expression "as you know" (or the equivalent) when it might be perceived as presumptuous or insulting?		

One challenge in setting a *changed beliefs* goal is that many beliefs are set deeply in values and are highly resistant to change. A vivid example of the persistence of beliefs and confirmation bias – the tendency of the brain to seek information that supports existing beliefs – could be seen in the COVID-19 culture wars that erupted in the United States in 2020 over the wearing of face masks.[29] Many people disregarded or refused to believe data on face masks and argued that the requirement to wear a face mask infringed on their personal freedoms and right to choose. For many people, facts did not change their beliefs.

One challenge in setting a *changed behavior* goal is that there is often a lack of consensus on what is the best behavioral response to a potential risk or threat. For many risk-related controversies, such as genetically modified foods, climate change, fracking for oil and gas, and nuclear

Table 11.9 Checklist for evaluating question and answer sessions with stakeholders about a risk, high concern, or crisis issue.

What to Look for in Q&A Exchanges	Y/N	Notes
1) When presented with a question for which you did not know the answer, did you start with what you do know and then move on to what you don't know and what you will do to find an answer?		
Alternatively, if you started with what you don't know, did you follow up with 3-4 positive messages?		
2) Did you provide a reasonable explanation for why you could not provide an answer to a question, such as for security or confidentiality reasons?		
3) Did you offer to assist the questioner in getting information related to questions you could not answer?		
4) Did you provide messages about where to find and how to access additional trusted information related to the topic of inquiry?		
5) Did you provide, with a deadline, for the follow-up actions you would take to address the unanswered question?		
6) Did you bridge from questions you could not answer to the information you could talk about?		
7) Did you engage in active listening (e.g. paying attention to body language, such as eye contact and posture; suspending judgment and withholding criticism; paraphrasing and reflecting back)		
8) Did you avoid attacking anyone perceived to be higher in trust than you with your target audience?		
9) Did you identify or cite at least three trusted third parties (e.g. key opinion leaders or trusted experts) for support?		
10) Did you support your message with examples, stories, or analogies?		
11) Did you avoid going beyond the bounds of your knowledge or responsibilities, such as by speaking for others?		
12) Did you acknowledge uncertainty and dilemmas?		
13) Did you avoid offering inappropriate risk comparisons?		
14) Did you avoid using inappropriate humor?		
15) Did you avoid repeating false allegations or strong negative words?		
16) Did you avoid the words "no," "don't," or "can't"?		
17) Did you avoid saying "no comment" to a question?		
18) Did you avoid over-promising or being over-reassuring?		
19) Did you provide information that gives people a sense of control and something concrete to do?		
20) Did your nonverbal communication contribute to your message?		
21) Did you maintain your composure when responding to an aggressive question or emotionally charged comment?		

Table 11.10 Checklist for elements in a comprehensive crisis communication plan.

Does the plan include the following content?	Y/N	Notes

1) Identify communication roles and responsibilities for staff in different crisis scenarios

2) Identify who in the organization is responsible and accountable for leading the crisis communication response

3) Identify who in the organization is responsible and accountable for implementing various communication actions

4) Identify who within and outside the organization needs to be notified and informed about a communication action.

5) Identify who within and outside the organization needs to be consulted about a communication action.

6) Identify who will be the lead communication spokesperson and backup spokesperson for different scenarios

7) Identify procedures for information verification, clearance, and approval

8) Identify procedures for coordinating with important stakeholders and partners (for example, with other organizations, first responders, law enforcement, and elected officials) .

9) Identify procedures to secure the required human, financial, logistical, and physical support and resources (such as people, space, equipment, and food) for communication operations during a short, medium, and prolonged crisis (24 hours a day, 7 days a week if needed)

10) Identify who releases information and procedures, e.g. regarding employees releasing information through mainstream or social media

11) Include regularly verified and updated media contact lists (including after-hours contact information)

12) Include regularly verified and updated partner contact lists (including after-hours contact information)

13) Identify a schedule for exercises and drills for testing the crisis communication plan

14) Identify trusted subject-matter experts willing to collaborate during a crisis, develop contact lists (day and night), and know their perspectives in advance

15) Identify target audiences

16) Identify preferred channels to communicate with the general public, key stakeholders, partners, and targeted audiences

17) Include message maps for core, informational, and challenge questions

18) Include message maps with the anticipated questions and concerns of key stakeholders, (internal and external), and with key messages and supporting information for each priority question or concern.

19) Include procedures for posting and updating information on the organization's website

20) Include holding statements for different anticipated stages of the crisis

21) Include fact sheets, question-and-answer sheets, talking points, maps, charts, graphics, and other supplementary communication materials for mainstream and social media platforms.

22) Include communication task checklists for the first 2, 4, 8, 12, 16, 24, 48 hours, and 72 hours

23) Include procedures for evaluating, revising, and updating the risk, high concern, or crisis communication plan on a regular basis

24) Include a signed endorsement of the communication plan from the organization's director

25) Identify worst case, low probability high consequence scenarios and their implications for the crisis communication plan

Table 11.11 Checklist for communicating effectively in the first 48 hours after a crisis.

Did you take the following steps?	Y/N	Notes
1) Use the crisis plan's notification list.		
2) Make certain that the chain of command has been notified and know who is involved in communications.		
3) Ensure leadership is aware of the crisis, especially if awareness of the crisis comes from mainstream or social media outlets.		
4) Contact and inform organizational partners.		
5) Appoint and activate a crisis spokesperson as designated in the crisis communication plan.		
6) Contact law enforcement authorities if there is potential for a criminal investigation.		
7) Give leadership an assessment of the crisis from a communication perspective		
8) Call in communication personnel per the crisis communication plan.		
9) Prepare a statement for internal and external distribution that addresses priority questions and concerns		
10) Monitor for rumors and coverage by mainstream and social media outlets		

Source: Adapted from Checklist 4–1: First 48 Hours, Commonwealth of Pennsylvania. https://www.health.pa.gov/topics/Documents/Emergency%20Preparedness/Checklist%20-%20First%2048%20Hours%20-%20CDC%20CERC.pdf

power, there is often disagreement about the extent, nature, or even existence of a risk. For example, many people are confused about whether or not changes in their personal behavior, such as recycling, will have an impact on climate change.[30] As another example of confusion about changing behavior, in the early phases of the 2020 COVID-19 pandemic, the public saw many conflicting messages and disagreements among experts about the importance and effectiveness of preventive and protective measures such as face masks and social distancing. The conflicting messages left individuals confused about whether and how to change behavior, and safety measures were not fully implemented at the outset of the epidemic.[31]

11.7 Evaluation Measures

There are three basic kinds of evaluation measures: *process/implementation evaluation measures, outcome/impact evaluation methods,* and *formative evaluation methods.* Each is briefly described below. Numerous guidance documents provide comprehensive details.[32]

11.7.1 Process/Implementation Evaluation Measures

Process/implementation evaluation measures are used to track and measure administrative and procedural progress. Process/implementation evaluators compile numbers related to the implementation of the communication program or project, including numbers on participant demographics, individual participant attendance, and adherence to the communication work plan. The quantitative metrics shown in Table 11.12 can help communicators determine the achievement of administrative objectives, such as what was done, how many people were reached, and which outreach efforts were most cost-effective.

Additional qualitative measures of process success can be gained from stakeholders. For example, evaluators can ask those attending a risk-related open house a variety of questions, such as

Table 11.12   Examples of *process/implementation* evaluation metrics.

What was done?

For example:

- number of staff working on communications
- time expended by staff on communications
- expenditures on communications
- number of communication activities performed

How many people were reached?

For example:

- amount of coverage time through mainstream broadcast media outlets
- amount of coverage space by mainstream print media outlets
- number of viewers and readers engaged through mainstream media outlets
- number of people engaged through social media outlets (e.g. number of hits on the Internet, number of downloads from the Internet)
- number of social media followers
- number of re-postings of content on social media outlets
- size of the audience for each communication outreach activity
- number of communication materials distributed
- number of presentations or communication events
- number of other organizational and/or persons on notification/contact lists
- number of times messages issued by the organization are repeated through media channels

Did people respond?

For example:

- number of in-person, telephone, mail, email, Internet, and social media inquiries
- where did people go to obtain additional information
- number of organizations that agreed to partner
- demographics of those who responded (e.g. gender, education, income, ethnicity, geographic location)

---

whether the information presented was clear and easy to understand, if the open house was held at a convenient time, and if the open house was held at a convenient location.

11.7.2   Outcome/Impact Evaluation Measures

Process/implementation evaluation measures provide important information about communication activities but do not establish that effective communication has occurred. For example, simply counting how many brochures were distributed or how many followers one has on social media sites does not mean that the communications affected knowledge, beliefs, or behaviors. As shown in Table 11.13, outcome/impact evaluation methods track the achievement of these types of communication goals. Outcome/impact evaluators use tools such as phone and mail surveys, focus groups, observational data, face-to-face meetings, and evaluation forms to determine whether the intended audience learned, acted, or made changes because of the communications.

Detailed guidelines on using outcome/impact evaluation methods are contained in the handbooks listed at the end of this chapter.

Table 11.13 Examples of *outcome/impact* evaluation measures.

1) The percentage of the target audience showing changes in their awareness, knowledge, opinions, attitudes, understanding, beliefs, intentions, cooperation, approval, or support
2) The percentage of the target audience showing changes in their behaviors
3) The percentage of the target audience showing changes in their perceptions of:
    a) trust (e.g. listening caring, empathy; knowledge, competence, expertise; honestly, openness, transparency, consistency)
    b) benefits (e.g. personal, community, and societal)
    c) fairness (e.g. personal, community, and societal)
    d) personal control (e.g. things to do)
4) The percentage of persons identified by age, gender, education, income level, ethnicity, or other characteristics interested in the issue
5) The percentage of risk management authorities that changed their risk-related recommendations, guidance, priorities, and/or policies and the direction of those changes
6) Changes in beliefs about:
    • whether risk management authorities and experts are doing a good or poor job
    • whether risk management authorities and experts are effectively addressing fairness and social equity issues
    • which media outlets have the most accurate, clear, timely, and relevant information
    • which media outlets or persons are having the most influence on attitudes and behaviors

Many communication plans call for outcome/impact evaluation at the end of a project, campaign, or effort. However, if a change is to be demonstrated, it is critical to have baseline information. Baseline information can be obtained from various sources, such as interviews, surveys, and focus groups. Repeating the interviews, surveys, or focus groups at the beginning, middle, and end help demonstrate if a change is occurring or has occurred. Analysis of the interview, survey, and focus group results, when combined with information gathered through other sources, helps indicate if observed changes were caused by communication activities.

Many guides exist laying out pitfalls in outcome/impact evaluation.[33] For example, (1) survey questions should be examined and tested for comprehension and bias before the survey is administered, and (2) questions that contain more than one embedded question should be avoided.

Handbooks on evaluation provide standardized protocols for measuring topics such as agreement, value, relevance, quality, and likelihood. One of the most popular tools for measuring responses is the Likert scale. The Likert scale asks survey respondents to select a rating on a scale that ranges from one extreme to another, such as "strongly agree" to "strongly disagree." The Likert scale typically runs from 1 to 5, with strongly agree at one end of the scale and strongly disagree at the other. For consistency, the same Likert scale should be used throughout the survey. Questions using the Likert scale can be supplemented with additional questions that are open-ended. The Likert scale avoids problems associated with binary "yes or no" survey questions. Such questions provide limited insight into what a person is thinking or how they feel about an issue or topic.

11.7.3 Formative Evaluation Measures

Formative evaluation measures examine the feasibility, appropriateness, and acceptability of communication materials and activities. Formative evaluation measures include observations, experiments (e.g., random assignment of individuals to groups to compare the effects of communication materials or activities), interviews (e.g. individual or group; structured, semi-structured, or conversational), surveys (conducted through the mail, email, online, in-person, or via phone), and focus groups, i.e., guided small-group discussion led by a facilitator. The strengths and weaknesses of each of these evaluation tools are discussed in various guidance documents.[34]

Formative evaluation measures can be taken in any phase of evaluation. However, they are most frequently employed when communication materials or activities are being developed or when existing ones are being adapted or modified. Formative evaluation methods tell the communicator which communication materials or activities have the best chance of making a difference in knowledge, beliefs, or behaviors. Risk, high concern, and crisis communication materials are typically tested for characteristics such as trustworthiness, content, clarity, tone, layout, visuals, appeal, consistency, main points, audience relevance, truthfulness, cultural sensitivities, comprehension, comprehensiveness, and emotional reactions.

Focus groups are one of the most frequently used tools for formative evaluation. They help determine which messages are likely to be most salient to the target audience and which communication materials should be developed into a specific risk, high concern, or crisis communication message or activity. Focus group testing helps determine whether the risk, high concern, or crisis messages and formats are perceived by selected members of the target audience as:

- culturally appropriate
- accurate
- understandable
- clear (e.g. use plain English)
- attention-getting
- credible
- memorable (e.g. recalled easily)
- relevant
- likely to achieve the desired effect on others (e.g. changes in knowledge, beliefs, or behaviors)

Focus groups are extremely valuable for evaluation, but it is important to use them with an awareness of their limitations. The usefulness of data from focus groups can be limited by factors such as possible facilitator bias, group dynamics, skills of the facilitator, small sample size, and the lack of control groups. Further limitations of focus groups are difficulties generalizing results and capturing the views of special populations, especially racial and ethnic groups, vulnerable populations, and geographic regions.

In conducting focus groups, the facilitator should provide simple but complete directions. Facilitators will tell people why they are doing the focus groups and should encourage honest responses. The case study in Section 11.9 is provided as a resource to illustrate an effective use of focus groups with different perspectives.

11.8 An Integrated Approach to Evaluation

As noted in a risk and crisis communication guidebook published by the US Food and Drug Administration:

> Communicators sometimes forgo evaluation as an unnecessary and costly step that merely demonstrates what they think they know already. After devoting time, energy, and passion to creating and disseminating a communication, people deeply desire to believe that it works. But faith is no substitute for scientific evidence of effectiveness. Reliance on intuition and anecdotal observation can cause devoting scarce resources to well-intentioned but ineffective communications.[35]

The effectiveness of risk, high concern, or crisis communication is fundamentally a matter of choices – choices about issues such as target audience, communication objectives, and measures for success. For example, in risk, high concern, and crisis communication, success can be measured in impacts on knowledge, beliefs, attitudes, behavior, trust, and reputation. Success can be measured by whether messages are reaching engaged stakeholders, whether messages are understood, whether collaboration has occurred, and whether the desired attitudinal or behavior changes are taking place. The most important choice is to make evaluation an integral part of the communication process at all stages.

11.9   Resource: Case Study of Focus Group Testing of Mosquito-Control Messages, Florida, 2018–2019[36]

Mosquitoes have endangered humans for thousands of years, spreading disease and killing millions of people every year. As carriers of major human diseases, mosquitoes are one of the most dangerous insects in the world. West Nile virus is among the most common viruses spread by mosquitoes in the continental United States, and there are many others, including dengue, chikungunya, malaria, yellow fever, and Zika.

Following the 2015–2017 Zika outbreak in Brazil, the state of Florida intensified its mosquito-control program. The program was successful in controlling mosquitoes but faced considerable public opposition. To improve its mosquito-control program, the Florida Health Department commissioned an evaluation of messages for the public about mosquitoes that carry Zika and ways to control the spread of the disease.

The weather in Florida is tropical to semitropical, a perfect breeding ground for mosquitoes. The health department recognized that it was one of the most trusted sources of information about disease such as Zika and that it played a critical role in protecting the Florida population from mosquitoes carrying the Zika virus.

Focus groups were conducted in Florida in 2018–2019 with members of the public to learn more about people's knowledge and risk perceptions related to mosquitoes and mosquito-control measures, and to test messages designed to address key questions and concerns.

Key questions and concerns addressed in the focus groups derived from an analysis of mainstream media stories, social media sources, and public documents. The questions included the following: What are the signs and symptoms of Zika and when do they appear? What is the connection between Zika, microcephaly, and Guillain–Barré syndrome? Does Zika affect children and adults differently? Are there long-term effects of Zika? Can pets and livestock be infected with the Zika virus? Could someone get Zika more than once in their lifetime?

For each focus group, evaluators recruited a diverse sample across several population segments including gender, age, income, ethnic background, education, and concerns about mosquitoes. The first set of focus groups was conducted in and around the Orlando, Florida area. The second set of focus groups was conducted in and around the Miami, Florida area.

Evaluators also conducted focus groups with public officials involved in mosquito control to learn more about the questions they received from the public about mosquito control and to get feedback on the messages being developed for this project. Follow-up interviews were conducted to clarify the main themes that emerged from the focus groups. Findings from the focus groups and interviews were reviewed and synthesized into a full report.

Moderator's Guide – Public Focus Groups

Introduction

- Introduce self.
  - Welcome and thank you for joining us. This is an important issue, and we really appreciate the fact that you're taking time out of your day to provide us with your opinions.
- Describe the goal of the project.
  - Sponsored by the Florida Department of Health.
  - Design simple, clear messages for the public to increase knowledge of mosquito control; test those messages to make sure they are understandable and address your key questions and concerns; refine those messages based on your feedback.
  - Today we'll be asking you some questions about mosquito control and getting your feedback on messages designed to address your questions and concerns on this topic. Your feedback will help the health department better inform the public about this important issue.
- Address video recording and guidelines.
  - The session will last for about two hours. We are video recording the session to make sure that we don't miss any of your comments. Please be sure to speak up so that we don't miss any of your comments. The recording will be used to help write a report summarizing what was said. It will not be used or shared in any other way.
  - No names or personal identifiers will be used when writing up findings.
  - Please bear in mind that we are interested in your ideas and opinions – both positive and negative. There are no right or wrong answers, and we want to hear from everyone. Please be respectful of others' opinions even when you disagree. Please keep in mind that my role here today is to facilitate the discussion, so I may not be able to answer your questions. At the end of the session, I will hand out a document on mosquito control that will include links to more information.
  - We do have many topics to discuss in a limited amount of time, so at times I may change the subject or move on to keep us on schedule.
  - Any questions for me so far? Is everyone willing to participate in this focus group?
- Respondent introductions.
  - First name (use only your first name).
  - How long have you lived here?
  - To get folks comfortable talking, ask, What are your favorite things to do in Florida during the summer months?

Knowledge, Beliefs, and Risk Perceptions

- Summer in Florida brings mosquitoes. How would you describe your knowledge level about mosquitoes (very high, decent, don't know much about mosquitoes)?
  - How concerned are you about mosquitoes?
  - How concerned are you about the viruses they spread?
    o Zika? Others?
    o How many are aware that you can be a carrier of Zika and not show symptoms? Does knowing this make you more supportive of mosquito control?
- From what you know or have heard, what are the major ways that communities try to control mosquito populations? When you hear "mosquito control," what comes to mind (aerial/truck spraying, pest control spraying around your house)?
- What about things that you can do to protect yourself (using insect repellent, wearing protective clothes, removing standing water, etc.)?

- Generally speaking, are you supportive of, against, or indifferent to mosquito-control methods?
- Are community mosquito-control efforts needed? Important?
  - What, if any, benefits of mosquito control matter to you?
  - How confident are you in the county or state government's ability to control the number of mosquitoes or where they are present?
  - What types of mosquito control are most effective? Why?
- Are there any types of mosquito control that concern you?
  - If so, what type, how concerned are you, and why?
  - Aerial vs. truck spraying? More supportive of one over the other? Why?
  - What if trucks can't spray your backyard where mosquitoes are?
  - What do you know about Naled?
  - What information would make you feel better about Naled?
  - Do you know how long Naled has been in use?
  - What, if any, information would make you more supportive of mosquito-control measures?

Stakeholder Questions and Concerns

- In terms of mosquito control, what are your biggest questions and concerns?
- *Hand out the list – this is a list of questions for which answers have been generated.*
  Is there anything missing from this list?
  Information Source Preferences

- When it comes to news and information about mosquitoes, including how bad they are, whether they are causing harm, or efforts to control them, how interested are you?
- How much attention, if any, do you pay to what's going on with mosquitoes or mosquito control? Explain.
- Whom do you view as a trusted source of information in terms of which people or organizations you would want to hear from if you had questions about mosquito control?

Channel Preferences

- If you wanted more information about mosquito control, how would you go about getting or accessing that information? (e.g. local TV news, websites, social media)? Which channels or sites?
- Have you actively sought out information about mosquitoes or mosquito control? Why or why not? What causes you or would cause you to be interested in such information?

Testing Message Maps
  Review, one by one, the following message maps:

1) Why are mosquito-control efforts important?
2) Are there health effects on humans from aerial spraying of insecticides for mosquito control?
3) What can I do to keep me/my family safe when aerial spraying occurs?
4) Is aerial spraying alone the most effective way to control mosquitoes?

Hand out the printed version of message map #1 to each respondent.
  Next, I would like you to read along with me as I read the message. As I read aloud and you read along with me, please:

- Underline phrases or sentences that you think are important.
- Circle phrases, sentences, or words that you think are unclear or confusing.
- Draw a wavy line through any information that you are skeptical of or don't believe/trust.

After reading the message map aloud, ask the following questions:

- Does this content address an important question/concern for you? Why or why not?
- Does this content affect your beliefs or opinion about mosquito control at all? Why or why not?
- What did you indicate as important and why?
- What did you indicate as unclear or confusing?
  - How might this be presented in an easier way?
- What information were you skeptical of?
- How could this content be improved?
- Is there anything you would want to know that this item doesn't tell you?
  Repeat above steps for additional message map

Testing Flyers

Hand out the printed version of the following fact sheets:

- *"Did You Know: Important Information about Aerial Spraying for Mosquito Control"*
- *"CDC's Response to Zika: Aerial Spraying with Naled"*

Next, I would like you to take a few minutes to review both fact sheets on your own. As you read through each fact sheet, please:

- Underline phrases or sentences that you think are important.
- Circle phrases, sentences, or words that you think are unclear or confusing.
- Draw a wavy line through any information that you are skeptical of or don't believe/trust.

After participants have reviewed both materials, ask the following questions:

- Overall impressions:
  - Which did you prefer overall and why?
  - What information was most important to you?
  - What information did you find unclear or confusing?
    o How might this be presented in an easier way?
  - What information were you skeptical of?
- Visually:
  - Which did you prefer?
    o Layout: easy to read/follow?
    o Pictures/colors
    o Length: too much information or not enough detail?
      o How much information do you like to see included on a fact sheet like this?
      o What if you need more information?
- How would you improve either of these flyers overall?

Wrap-Up

- Thank you very much for participating. Your input has been extremely helpful.
- As you walk out, a staff member will hand you reimbursement for your time tonight. Also, please be sure to take a CDC fact sheet on mosquito control that can help answer questions you may have after today's session and can connect you with where to go for more information.
- Thank you again.

Evaluation Results

Knowledge, Benefits, and Concerns: The Public (Excerpts)

Focus group participants reported a range of mosquito knowledge, with the majority stating they know only a little about mosquitoes. Concern about mosquitoes was fairly high. When asked about their concerns, participants cited Zika and other diseases (including encephalitis and dengue), and the general nuisance of mosquitoes. Participants also expressed concerns about pets and the impact of Zika on tourism to Florida.

When asked about methods of mosquito control, participants mentioned aerial spraying, truck spraying, and personal control measures including removing standing water and wearing light-colored clothing. Overall, there was agreement among participants on the need for mosquito-control measures, with participants citing benefits including the reduction of disease and the general nuisance of mosquitoes.

Concerns about mosquito-control methods were largely focused on spraying. Most participants wanted more information on the chemicals being used. Participants were asked about whether the chemicals were targeted to mosquitoes or if they were having a broader impact on the ecology and on bees.

Focus group participants expressed a clear preference for measures that used naturally occurring substances. Participants were asked about long-term research or information showing that a chemical had been in use for a long time and was found to be safe. Participants enjoyed hearing that aerial spraying was a "last resort" when things are worse than normal (e.g. following a hurricane or disease outbreak). Each of these findings provides potential points of emphasis for future messaging pertaining to spraying.

Knowledge, Benefits, and Concerns: Government Officials (Excerpts)

Government officials confirmed residents see aerial spraying as the "riskiest" method, and some people assume it is happening even when it isn't. They noted people feel a lack of control when aerial spraying is conducted, as they often can't hear it happening and don't know when it is coming. They emphasized they view aerial spraying as a "last resort," a point of emphasis that resonated with many in subsequent public focus groups.

Government officials noted most of the truck spraying is done at night, and that many in the community view this in a positive light. They emphasized: "People can see and hear it," noting that many people don't want to be surprised by spraying. They offer a "call ahead" list for when truck spraying takes places, and many people have opted to receive those calls.

Frequently Asked and Challenging Questions: The Public (Excerpts)

Each group was presented with a list of key questions and concerns that were developed based on background research, local news coverage, and feedback from mosquito-control experts in Florida. Participants were asked to provide feedback on the list, including whether there were questions that should be added. The consensus was that the list accurately reflected their key questions and concerns; however, a few additional suggestions were made. Several participants emphasized the need to know where to go and whom to contact for more information if they had questions. Many also stated that information on when and where spraying is taking place was most important to them.

Frequently Asked and Challenging Questions: Government Officials (Excerpts)

Government officials were asked about the questions they most frequently receive from members of their community, as well as the most challenging questions. The questions they mentioned focused predominantly on spraying. Officials also mentioned that they often hear from concerned beekeepers. This input led to including message maps pertaining to bees in the evaluation.

Government officials also emphasized frequent questions from members of the community about where they can go for accurate information about mosquito control. Feedback from the public and government officials on the initial list of questions, along with the other key evaluation findings, led to a revised list of message map questions developed for this project.

Message Maps and Information Seeking: The Public (Excerpts)

Each of the message maps and each of the fact sheets presented received positive reviews. Feedback from members of the public resulted in refinements of the message maps and fact sheets.

Few members of the public said they had actively looked for mosquito information. The most common reason was to request mosquito-control measures near their home, with participants reporting that they typically went to their local government's website in search of a number to call or called 311. A theme brought up frequently was the need for educating children about mosquitoes in schools and the potential for those children to educate their parents.

When asked what channels they used to get information on mosquitoes or mosquito control, participants provided a range of responses. Radio was mentioned most frequently. Public radio was mentioned as being particularly reliable, neutral, and unlikely to sensationalize information. Many members of the public also cited local television news as a source of information. Several participants said they would go to a government website for information. Several participants also said they would welcome an app that would provide specific, localized information that could be quickly accessed. This idea generated high support from other members of the public. Most of the members of the public indicated they would like to receive email/text alerts about spraying. Several participants mentioned that if their local health department had a Facebook page with information about mosquito control, they would welcome it.

Overall, members of the public were unclear about who is in charge of decisions about mosquito control. Several people mentioned they did not know if their county has mosquito-control responsibilities. This finding from the evaluation study highlighted the need for improved communications about who makes these decisions at the local level and where people can go for more information.

Message Maps and Information Seeking: Government Officials (Excerpts)

Each of the message maps and each of the facts sheets presented received positive reviews from government officials. Feedback from government officials resulted in refinements of the message maps and fact sheets.

Government officials cited CDC as a trusted source of comprehensive information about mosquito control. They particularly liked the CDC infographics and related materials. Government officials also mentioned the World Health Organization and the Florida Department of Health as useful sources of information. Government officials appreciated that they "did not have to reinvent the wheel" and could rely on existing material with relevant content and useful graphics in a variety of languages.

Government officials gave positive reviews to the "Fight the Bite" and the "Drain and Cover" communication campaigns, but cautioned that there are some who don't trust the government at all. Government officials stressed the importance of consistent, unified messaging from departments across the state. Government officials expressed disappointment in the inconsistent use of social media throughout the state.

11.10   Evaluation Tools

The Community Toolbox, 2020 (http://ctb.ku.edu).

This resource offers over 7,000 pages of practical guidance on a wide range of skills essential for risk communication, including tools for evaluation.

Community Health Assessment and Group Evaluation (CHANGE) Tool and Action Guide (www. cdc.gov/healthycommunitiesprogram/tools/change.htm).

This resource, developed by the Centers for Disease Control and Prevention (CDC), focuses on assessment, planning, and evaluation of health communications.

Lance, P., Guilkey, D., Hattori, A., and Angeles, G. (2014). *How do we know if a program made a difference? A guide to statistical methods for program impact evaluation.* Chapel Hill, North Carolina: MEASURE Evaluation.

This resource provides an overview of core statistical and econometric methods for program impact evaluation (and, more generally, causal modeling).

Mobilizing for Action through Planning and Partnerships (MAPP) (www.naccho.org/topics/ infrastructure/mapp/index.cfm).

This resource, developed by the National Association of County and City Health Officials (NACCH), guides practitioners through the planning process, from beginning organizational steps through assessment and action planning, implementation, and evaluation. The website contains a user handbook and a clearinghouse of resources of evaluation.

Royse, D., Thyer, B. A., and Padgett, D. K. (2015). *Program Evaluation: An Introduction to an Evidence-Based Approach 6th Edition.* Boston: Cengage Learning. This resource is a standard reference guide in the field.

11.11 Chapter Resources

Below are additional resources to expand on the content presented in this chapter.

Agency for Toxic Substances and Disease Registry (ATSDR) (2000). *Evaluation Primer on Health Risk Communication Programs* http://www.atsdr.cdc.gov/risk/evalprimer/

Arkin E. (1991). "Evaluation for risk communicators," in *Evaluation and Effective Risk Communication*, eds. A. Fisher, M. Pavlova, and V. Covello. Pub. no. EPA/600/9-90/054. Washington, DC: U.S. Environmental Protection Agency,

Árvai, J., and Rivers, L., III, eds. (2014). *Effective Risk Communication.* London: Routledge.

Balog-Way, D.H.P., and McComas, K.A. (2020). "COVID-19: Reflections on trust, tradeoffs, and preparedness." *Journal of Risk Research* 23:7–8.

Bandura, A. (1997). *Self-Efficacy: The Exercise of Control.* New York: Freeman.

Berk, R., and Rossi, P. (1999) *Thinking about Program Evaluation.* Thousand Oaks, CA: Sage.

Boholm, Å. (2019). "Risk communication as government agency organizational practice." *Risk Analysis* 39:1695–1707.

Breakwell, G.M. (2007). *The Psychology of Risk.* New York and Cambridge, UK: Cambridge University Press.

Brewer, N.T. (2011). "Chapter 2: Goals," in *Communicating Risks and Benefits: An Evidence-Based User's Guide*, eds. Fischhoff, B., Brewer, N.T., and Downs, J.S (pp. 3–10). Silver Spring, MD: US Food and Drug Administration.

Centers for Disease Control and Prevention. (1999). "Framework for program evaluation in public health." *MMWR* 48:No. RR-11.

Centers for Disease Control and Prevention (2020). *Crisis and Emergency Risk Communication* Atlanta, GA: Centers for Disease Control and Prevention.

Centers for Disease Control and Prevention (2020). "Flu Vaccination Coverage, United States, 2019–20 Influenza Season." Accessed online at: https://www.cdc.gov/flu/fluvaxview/ coverage-1920estimates.htm

Chess, C., Hance, B.J., and Sandman, P. M. (1986). *Planning Dialogue with Communities: A Risk Communication Workbook*. New Brunswick, NJ: Rutgers University, Cook College, Environmental Media Communication Research Program.

Chess, C., Hance, B.J., and Sandman, P.M. (1998). *Improving Dialogue with Communities*. Trenton (NJ): New Jersey Department of Environmental Protection/Division of Science and Research

Chess, C., Salomone, K.L., and Hance, B.J. (1995). "Improving risk communication in government: Research priorities." *Risk Analysis* 15(2):127–135.

Chess, C., Salomone, K.L., Hance, B.J., and Saville, A. (1995). "Results of a national symposium on risk communication: Next steps for government agencies." *Risk Analysis* 15(2):115–120.

Chou, W.S., and Budenz, A. (2020). "Considering emotion in COVID-19 vaccine communication: addressing vaccine hesitancy and fostering vaccine confidence." *Health Communication* 35(14):1718–1722

Cialdini, R. B. (2007). *Influence: The Psychology of Persuasion*. New York: HarperCollins

Chen, H. (2005). *Practical Program Evaluation*. Thousand Oaks, CA: Sage.

Chriqui, J.F., O'Connor, J.C., and Chaloupka, F.J. (2011) *What Gets Measured, Gets Changed. Journal of Law and Medical Ethics* 39(1):21–26.

Covello, V.T. (2003). "Best practices in public health risk and crisis communication." *Journal of Health Communication* 8 (Suppl. 1):5–8; discussion, 148–151.

Covello, V.T. (2014). "Risk communication," in *Environmental Health: From Global to Local*, ed. H. Frumkin. San Francisco: Jossey-Bass/Wiley.

Covello, V.T. (2006). "Risk communication and message mapping: A new tool for communicating effectively in public health emergencies and disasters." *Journal of Emergency Management* 4(3):25–40.

Covello, V.T., and Allen, F. (1988). *Seven cardinal rules of risk communication*. Washington, D.C.: U.S. Environmental Protection Agency.

Covello, V.T., McCallum, D.B., and Pavlova, M.T. (1989). "Principles and guidelines for improving risk communication," in *Effective Risk Communication: The Role and Responsibility of Government and Nongovernment Organizations*. New York: Plenum.

Covello, V.T., McCallum, D.B., and Pavlova, M. (1989). *Effective Risk Communication: The Role and Responsibility of Government and Nongovernment Organizations*. New York: Springer.

Covello, V., and Merkhofer, M. (1993). *Risk Assessment Methods: Approaches for Assessing Health and Environmental Risks*. New York: Plenum.

Covello, V., Minamyer, S., and Clayton, K. (2007). *Effective Risk and Crisis Communication during Water Security Emergencies. EPA Policy Report*; EPA 600-R07-027. Washington, D.C.: US Environmental Protection Agency.

Covello, V., Peters, R., Wojtecki, J., and Hyde, R. (2001). " Risk communication, the West Nile virus epidemic, and bio-terrorism: Responding to the communication challenges posed by the intentional or unintentional release of a pathogen in an urban setting." *Journal of Urban Health* 78(2):382–391.

Covello, V., Sandman, P., and Slovic, P. (1988). *Risk Communication, Risk Statistics, and Risk Numbers*. Washington, DC: Chemical Manufacturers Association.

Covello, V., Sandman, P. (2001). "Risk communication: Evolution and revolution," in *Solutions to an Environment in Peril*, ed. A. Wolbarst. Baltimore, MD: Johns Hopkins University Press.

Covello, V., Slovic, P., and von Winterfeldt, D. (1986). "Risk communication: A review of the literature." *Risk Abstracts* 3(4):171–182.

Covello, V., Slovic, P., and von Winterfeld, D. (1987). *Risk Communication: A Review of the Literature*. Washington, DC: National Science Foundation.

Crick, M.J. (2021). "The importance of trustworthy sources of scientific information in risk communication with the public." *Journal of Radiation Research* 62(S1):1–6.

Downs, J.S. (2011). "Chapter 3: Evaluation," in *Communicating Risks and Benefits: An Evidence-Based User's Guide*, eds. Fischhoff, B., Brewer, N.T., and Downs, J.S. Silver Spring, MD: U.S. Food and Drug Administration, pp. 12–18.

Dror, A.A., Eisenbach, N., and Taiber, S. (2020). "Vaccine hesitancy: the next challenge in the fight against COVID-19." *European Journal of Epidemiology* 35:775–779.

England, K.J., Edwards, A.L., Paulson, A.C., Libby, E.B., Harrell, P.T., and Mondejar, K.A. (2021). "Rethink Vape: Development and evaluation of a risk communication campaign to prevent youth E-cigarette use." *Addictive Behaviors* 113:106664.

Fern, E. F. (2001). *Advanced Focus Group Research*. Thousand Oaks, CA: Sage.

Fern, E.F. (2011). "Chapter 3: Evaluation," in *Communicating Risks and Benefits: An Evidence-Based User's Guide*, eds. Fischhoff, B., Brewer, N.T., and Downs, J.S. Silver Spring, MD: US Food and Drug Administration.

Fearn-Banks, K., (2011). *Crisis Communications: A Casebook Approach*. 4th edition. New York: Routledge Communication Series.

Fisher, A., Pavlova, M., and Covello, V. eds. (2001) *Evaluation and Effective Risk communication*. Pub. No. EPA/600/9-90/054. Washington, DC: US Environmental Protection Agency.

Fischhoff, B., Brewer, N.T., and Downs, J.S., eds. (2011). *Communicating Risks and Benefits: An Evidence-Based User's Guide*. Silver Spring, MD: US Food and Drug Administration.

Fitzpatrick, J.L., Sanders, J.R., and Worthen, B.R. (2017) *Program Evaluation: Alternative Approaches and Practical Guidelines*. Boston: Pearson Education.

Gatson, N., and Daniels, P. (1988). *Guidelines: Writing for Adults with Limited Reading Skills*. Washington, DC: U.S. Department of Agriculture, Food and Nutrition Service.

Germani, F., and Biller-Andorno, N. (2021). "The anti-vaccination infodemic on social media: A behavioral analysis." *PLOS One* 16 (3): e0247642.

Gilchrist, V. J. (1992). "Key informant interviews," in *Doing Qualitative Research*, eds. Crabtree, B.F., and Miller, W.L. London: Sage.

Habib, R., White, K., Hardisty, and D.J. Zhao, J. (2021). "Shifting consumer behavior to address climate change." *Current Opinion in Psychology* 42:108–113.

Hance, B.J. (1992). *Communicating with the Public: Ten Questions Environmental Mangers Should Ask*. New Jersey: Center for Environmental Communication, Rutgers University.

Heath, R.L., and O'Hair, H.D. (eds.). (2020). *Handbook of Risk and Crisis Communication*. New York: Routledge.

Henstra, D., Minano, A., and Thistlethwaite, J. (2019). "Communicating disaster risk? An evaluation of the availability and quality of flood maps." *Natural Hazards Earth System Science* 19:313–323.

Hesse-Biber S, and Leavy P. (2006). The Practice of Qualitative Research. Thousand Oaks (CA): Sage.

Jong W. (2020). "Evaluating crisis communication. A 30-item checklist for assessing performance during COVID-19 and other pandemics." *Journal of Health Communication* 25(12):962–970.

Jong, W., and Dückers, M.L.A. (2019). "The perspective of the affected: What people confronted with disasters expect from government officials and public leaders." *Risk, Hazards & Crisis in Public Policy* 10(1):14–31.

Linfield, K.J., and Posavac, E.J. (2018). *Program Evaluation: Methods and Case Studies*. London and New York: Routledge.

Liu, T., Zhang, H., and Zhang, H. (2020). "The impact of social media on risk communication of disasters." *International Journal of Environmental Research and Public Health*, 883:1–17.

Lundgren, R.E., and McMakin, A.H. (2018). *Risk Communication: A Handbook for Communicating Environmental, Safety, and Health Risks*. Hoboken, NJ: Wiley.

Lum, M. (1991). "Benefits to conducting midcourse reviews," in *Evaluation and Effective Risk Communications*, eds. A. Fisher, M. Pavolva, and V. Covello. Pub. no. EPA/600/9-90/054. Washington, DC: U.S. Environmental Protection Agency.

Malecki, K.M.C., Keating, J.A., and Safdar, N. (2021). "Crisis communication and public perception of COVID-19 risk in the era of social Media." *Clinical Infectious Diseases* 72(4):697–702.

Mark, M., and Donaldson, S., eds. (2011*) Social Psychology and Evaluation*. New York: The Guilford Press.

Markwart, H., Vitera, J., Lemanski, S., Kietzmann, D., Brasch, M., and Schmidt, S. (2019). "Warning messages to modify safety behavior during crisis situations." *International Journal of Disaster Risk Reduction* 38:101235.

Maxi, R., Tomczyk, S., Schopp, N., and Schmidt, S. (2021). "Warning messages in crisis communication: risk appraisal and warning compliance in severe weather, violent acts, and the COVID-19 pandemic." *Frontiers in Psychology* 12:557178.

Møllebæk, M., and Kaae, S. (2020). "Why do general practitioners disregard direct to healthcare professional communication? A user-oriented evaluation to improve drug safety communication." *Basic and Clinical Pharmacology and Toxicology* 128 (3):63–471.

Morgan, M.G., Fischoff, B., Bostrom, A., and Atman, C.J. (2001). *Risk Communication: The Mental Models Approach*. New York and Cambridge, UK: Cambridge University Press.

National Cancer Institute (1992). *Making Health Communication Programs Work: A Planner's Guide*. NIH Publication no. 92-1493. Washington, DC: National Cancer Institute.

National Research Council. (1989). *Improving Risk Communication*. Washington, DC: National Academies Press.

Nielsen, K. S., Clayton, S., Stern, P. C., Dietz, T., Capstick, S., and Whitmarsh, L. (2021). "How psychology can help limit climate change." *American Psychologist* 76(1):130–144.

Nowak, G., Karafillakis, E., and Larson H. (2020. "Pandemic influenza vaccines: Communication of benefts, risk, and uncertainties," in *Communicating about Risks and Safe Use of Medicines*, ed. P. Bahri. pp. 162–178. Singapore: Springer Nature.

Pew Research Center (2021). *Writing Survey Questions*. Accessed at https://www.pewresearch.org/our-methods/u-s-surveys/writing-survey-questions/

Platt, M. (2020). *The Leader's Brain*. Philadelphia, PA: The Wharton School Press (University of Pennsylvania).

Prasad, A. (2021). "Anti-science Misinformation and Conspiracies: COVID–19, Post-truth, and Science & Technology Studies (STS)." *Science, Technology and Society*, April 2021.

Ramírez, A.S, Ramondt, S., Van Bogart, K., and Perez-Zuniga, R. (2019) "Public awareness of air pollution and health threats: Challenges and opportunities for communication strategies to improve environmental health literacy. "*Journal of Health Communication* 24(1):75–83.

Regan, M.J., and Desvousges, W.H. (1990). *Communicating Environmental Risks: A guide to Practical Evaluations*. Pub. no. 230-01-91-001. Washington, DC: US Environmental Protection Agency.

Renn, O. (1998). "The role of risk communication and public dialogue for improving risk management." *Risk Decision Policy* 3:5–30.

Renn, O. (2008). *Risk Governance: Coping with Uncertainty in a Complex World*. London: Earthscan.

Renner, B., and Schupp, H.T. (2011). "The perception of health risk," in *The Oxford Handbook of Health Psychology*, ed. H.S. Friedman. New York: Oxford University Press.

Reynolds, B., and Seeger, M.W. (2005). "Crisis and emergency risk communication as an integrative model." *Journal of Health Communication Research* 10(1):43–55.

Rohemann, B. (1992). "The evaluation of risk communication effectiveness." *Acta Psychologica* 81(2):169–192.

Royse, D., Thyer, B.A., and Padgett, D.K. (2015). *Program Evaluation: An Introduction to an Evidence-Based Approach*. 6th Edition. Boston: Cengage Learning.

Rubin, A. (2020). *Program Evaluation*. Cambridge: Cambridge University Press.

Sandman, P. M., and Lanard, J. (2004). *Crisis Communication: Guideline for Action*. Accessed at: http://www.psandman.com/handouts/aiha/contents.pdf

Santibanez, T.A., Nguyen, K.H., and Greby, S.M. (2020). "Parental vaccine hesitancy and childhood influenza vaccination." *Pediatrics* 146(6):1–10.

Seeger, M.W. (2006). "Best practices in crisis communication: An expert panel process." *Journal of Applied Communication Research* 34(3):232–244.

Sellnow, T.L., and M.W. Seeger (2021). *Theorizing Crisis Communication*. 2nd Edition. Hoboken, NJ: Wiley.

Sellnow, T.L., Ulmer, R.R., Seeger, M.W., and Littlefield, R.S. (2009). *Effective Risk Communication: A Message-Centered Approach*. New York: Springer.

Siegrist, M., and Cvetkovich, G. (2000). "Perception of hazards: The role of social trust and knowledge." *Risk Analysis* 20(5):713–719.

Shadish, W.R., Cook, T.D., and Leviton, L.C. (1991). *Foundations of Program Evaluation: Theories of Practice*. Newbury Park, CA: Sage.

Silverman, C. (2020). *Verification Handbook For Disinformation And Media Manipulation*. 3rd Edition. Brussels, Belgium and Maastricht, the Netherlands: European Journalism Centre. Accessed at: https://datajournalism.com/read/handbook/verification-3

Sin, M.S.Y. (2016). "Masking fears: SARS and the politics of public health in China." *Critical Public Health* 26:88–98.

Swire-Thompson, B., and D. Lazer (2020). "Public health and online misinformation: Challenges and recommendations." *Annual Review of Public Health* 41(1):433–451.

Thompson N, Kegler M, and Holtgrave D. (2006). "Program evaluation," in *Research Methods in Health Promotion*, eds. Crosby R.A., DiClemente R.J., and Salazar L.F. San Francisco (CA): Jossey-Bass; pp. 199–225.

Tinker, T., ed. (1994). *Case studies of applied evaluation for health risk communication*. Workshop proceedings. Washington, DC: US Department of Health and Human Services, Public Health Service.

Tremblay, M. C., Hevner, A. R., and Berndt, D. J. (2010). "The use of focus groups in design science research." *Design Research in Information Systems* 22:121–143.

US Centers for Disease Protection and Prevention. (1998). *Practical evaluation of public health programs*. Atlanta, GA: US Department of Health and Human Services.

Weiss, C.H. (1998). *Evaluation: Methods for Studying Programs and Policies*. 2nd edition. Upper Saddle River, NJ: Prentice Hall.

World Health Organization (2020). "Call for Action: Managing the Infodemic." Accessed at: https://www.who.int/news/item/11-12-2020-call-for-action-managing-the-infodemic

Zhang, L., Huijie Li, H., and Chen, K. (2020). "Effective risk communication for public health emergency: Reflection on the COVID-19 (2019-nCoV) outbreak in Wuhan, China." *Healthcare* 8(64):1–13.

Endnotes

1 Reuters, 8 February 2016. U.S. Athletes Should Consider Skipping Rio if Fear Zika—Officials. Accessed at: https://www.reuters.com/article/us-health-zika-usa-olympics-exlusive-idUSKCN0VH0BJ

2 *USA Today*, 9 February 2016. "Hope Solo says right now she wouldn't go to Rio Olympics because of Zika." Accessed at: https://www.usatoday.com/story/sports/olympics/2016/02/09/report-hope-solo-says-right-now-she-wouldnt-go-rio-zika/80060428/

3 https://www.zikacommunicationnetwork.org/resources/zikazero

4 See, e.g., World Health Organization. (2016). *Risk Communication in the Context of Zika Virus. Interim Guidance*. Geneva, Switzerland: World Health Organization. Accessed at: https://apps.who.int/iris/bitstream/handle/10665/204513/WHO_ZIKV_RCCE_16.1_eng.pdf;jsessionid=5DC1D36E9C8C9928444C06CFAE54D39D?sequence=1

5 Doucleff, M. (2016) "Guess how many Zika cases showed up at the Olympics?" National Public Radio, August 26, 2016. Accessed at: https://www.npr.org/sections/goatsandsoda/2016/08/26/491416709/guess-how-many-zika-cases-showed-up-at-the-olympics

6 See, e.g., Morgan, M.G., Fischoff, B., Bostrom, A., and Atman, C.J. (2001). *Risk Communication: The Mental Models Approach*. Cambridge UK and New York: Cambridge University Press; see also Bostrom, A., Anselin, L., and Farris, J. (2008). "Visualizing seismic risk and uncertainty: A review of related research." *Annals of the New York Academy of Sciences* 1128:29–34; Bostrom, A., Atman, C. J., Fischhoff, B., and Morgan, M.G. (1994). "Evaluating risk communications: Completing and correcting mental models of hazardous processes, part II." *Risk Analysis* 14(5):789–798; Bostrom, A., Fischhoff, B., and Morgan, M.G. (1992). "Characterizing mental models of hazardous processes: A methodology and an application to radon." *Journal of Social Issues* 48(4):85–100.

7 Kuroda, Y. (2017). "Current state and problems of radiation risk communication: Based on the results of a 2012 whole village survey." *PLOS Currents Disasters*. 24 Feb 2017. Edition 1. Accessed at: https://www.semanticscholar.org/paper/Current-State-and-Problems-of-Radiation-Risk-Based-Kuroda/1142d425579034eb33d01fea13049a50ba48732e

8 See, e.g., Sellnow, T.L., Ulmer, R.R., and Seeger, M.W., and Littlefield, R.S. (2009). *Effective Risk Communication: A Message-Centered Approach*. New York: Springer.

9 Harrell, E. (2019). "Neuromarketing: What you need to know." *Harvard Business Review,* January 23, 2019. Accessed at https://hbr.org/2019/01/neuromarketing-what-you-need-to-know

10 Harris, J. M., Ciorciari J., and Goutas, J. (2019). *Behavioral Sciences* 2019 9(4) 41. *Accessed at* https://doi.org/10.3390/bs9040042

11 See, e.g., Seeger, M., and Fearn-Banks, K., (2011). *Crisis Communications: A Casebook Approach*. 4th Edition. New York: Routledge Communication Series; Ahmed, P. K., and Rafiq, M. (1998). "Integrated benchmarking: A holistic examination of select Fearn-Banks, K., (2011). *Crisis Communications: A Casebook Approach*. 4th Edition. New York: Routledge Communication Series; Ahmed, P. K., and Rafiq, M. (1998). "Integrated benchmarking: A holistic examination of select techniques for benchmarking analysis." *Benchmarking for Quality Management and Technology* 5(3): 225–242.

12 Cole, T.W., and Fellows, K.L. (2008). "Risk communication failure: A case study of New Orleans and Hurricane Katrina." *Southern Communication Journal* 73(3):211–228.

13 Rivers L. (2006). "A post-Katrina call to action for the risk analysis community." *Risk Analysis* 26(1):1–2. Page 2.

14 See, e.g., Executive Office of the President. (2006). *Federal Response to Hurricane Katrina: Lessons Learned*. Washington, D.C.: Executive Office of the President. Accessed at: https://georgewbush-whitehouse.archives.gov/reports/katrina-lessons-learned/

15 Santibanez, T.A., Nguyen, K.H., and Greby, S.M. (2020). "Parental vaccine hesitancy and child-hood influenza vaccination." *Pediatrics* 146(6):1–10. Accessed at: https://pediatrics.aappublications.org/content/pediatrics/146/6/e2020007609.full.pdf

16 Centers for Disease Control and Prevention. (2020). Flu Vaccination Coverage, United States, 2019–20 Influenza Season. Accessed at: https://www.cdc.gov/flu/fluvaxview/coverage-1920estimates.htm

17 See., e.g. Dror, A.A., Eisenbach, N., and Taiber, S. (2020). "Vaccine hesitancy: The next challenge in the fight against COVID-19." *European Journal of Epidemiology* 35:775–779. Accessed at: https://doi.org/10.1007/s10654-020-00671-y; Wen-Ying Sylvia Chou, W.S., and Budenz, A. (2020). "Considering emotion in COVID-19 vaccine communication: Addressing vaccine hesitancy and fostering vaccine confidence." *Health Communication* 35(14):1718–1722.

18 See, e.g. United Nations Children's Fund (2020). *Vaccine Misinformation Management Field Guide: Guidance for Addressing a Global Infodemic and Fostering Demand for Immunization.* New Yok: UNICEF/United Nations.

19 Nowak, G., Karafillakis, E., and Larson H. (2020). "Pandemic influenza vaccines: Communication of benefits, risk, and uncertainties," in *Communicating about Risks and Safe Use of Medicines*, ed. P. Bahri. Singapore: Springer Nature, pp. 162–178.

20 MacDonald, N.E., and the SAGE Working Group on Vaccine Hesitancy. (2015). "Vaccine hesitancy: Definition, scope and determinants." *Vaccine* 33(34):4161–4164. See also Olson, O., Berry, C., and Kumar, N. (2020). "Addressing parental vaccine hesitancy towards childhood vaccines in the United States: A systematic literature review of communication interventions and strategies." *Vaccines* 8(4):590.

21 See Beaumont Foundation and the American Public Health Association. (2020). *Changing the COVID-19 Conversation.* Accessed at: https://debeaumont.org/changing-the-COVID-19-conversation/; https://www.debeaumont.org/changing-the-COVID-19-conversation/COVID-19-communications-cheat-sheet/

22 Zhang, L., Huijie Li, H., and Chen, K. (2020). "Effective risk communication for public health emergency: reflection on the COVID-19 (2019-nCoV) outbreak in Wuhan, China." *Healthcare* 8(64):1–13.

23 Xiaomin Wang, Leesa Lin, Ziming Xuan, Jiayao Xu, Yuling Wan, and Xudong Zhou. (2020). "Risk communication on behavioral responses during COVID-19 among general population in China: A rapid national study." *Journal of Infection* 81(6):911–922.

24 Burki, T. (2020). "China's successful control of COVID-19." *The Lancet* 20(11):1240–1241. Accessed at: https://www.thelancet.com/journals/laninf/article/PIIS1473-3099(20)30800-8/fulltext#articleInformation

25 See, e.g. World Health Organization. (2021). *Principles for Effective Communication.* Geneva, Switzerland: World Health Organization. Accessed at: https://www.who.int/about/communications/evaluation/principles-evaluation

26 See, e.g. Gawande, A. (2010). *The Checklist Manifesto: How to Get Things Right.* New York: Picador Publishing.

27 Jong W. (2020). "Evaluating crisis communication. A 30-item checklist for assessing performance during COVID-19 and other pandemics." *Journal of Health Communication* 25(12):962–970.

28 See, e.g. Brewer, N.T. (2011). "Chapter 2: Goals," in *Communicating Risks and Benefits: An Evidence-Based User's Guide*, eds. Fischhoff, B., Brewer, N. T., and Downs, J.S. Silver Spring, MD: US Food and Drug Administration, pp. 3–10.

29 See, e.g., Lewis, M. (2021). *The Premonition: A Pandemic Story.* New York: WW Norton.

30 Nielsen, K. S., Clayton, S., Stern, P. C., Dietz, T., Capstick, S., and Whitmarsh, L. (2021). "How psychology can help limit climate change." *American Psychologist* 76(1):130–144; see also Habib, R., White, K., Hardisty, D.J., and Zhao, J. (2021). "Shifting consumer behavior to address climate change." *Current Opinion in Psychology* 42:108–113.

31 See., e.g., Bendavid, E., Oh, C., Bhattacharya, J., and Ioannidis, J.P.A. (2021). "Assessing Mandatory Stay-at-Home and Business Closure Effects on the Spread of COVID-19." *European Journal of Clinical Investigation. Wiley Online Library.* Accessed at: https://onlinelibrary.wiley.com/ doi/10.1111/eci.13484; see also Balog-Way, D. H. P., and McComas, K. A. (2020). "COVID-19: Reflections on trust, tradeoffs, and preparedness." *Journal of Risk Research* 23:7–8.

32 See, e.g., Lance, P., Guilkey, D., Hattori, A., and Angeles, G. (2014). *How Do We Know If a Program Made a Difference? A Guide to Statistical Methods for Program Impact Evaluation.* Chapel Hill, North Carolina: MEASURE Evaluation; Royse, D., Thyer, B. A., and Padgett, D. K. (2015). *Program Evaluation: An Introduction to an Evidence-Based Approach.* 6th Edition. Boston: Cengage Learning; Centers for Disease Control and Prevention (2016). *CDC Evaluation Resources.* Accessed at: https://www.cdc.gov/eval/resources/index.htm

33 See, e.g., Pew Research Center (2021). Writing Survey Questions. Accessed at https://www. pewresearch.org/our-methods/u-s-surveys/writing-survey-questions/

34 See, e.g., Royse, D., Thyer, B.A., and Padgett, D.K. (2015). *Program Evaluation: An Introduction to an Evidence-Based Approach.* 6th Edition. Boston: Cengage Learning.

35 Downs, J. S. (2011). "Chapter 3: Evaluation," in *Communicating Risks and Benefits: An Evidence-Based User's Guide*, eds. Fischhoff, B., Brewer, N.T., and Downs, J.S. Silver Spring, MD: US Food and Drug Administration. pp. 12–18.

36 Adapted from Hipper, T., and Covello, V. (2018). *Mosquito Control Messaging Project—Final Report.* Submitted to: Florida Department of Health and the Center for Public Issues Education in Agriculture & Natural Resources, Institute of Food and Agricultural Sciences, University of Florida, Gainesville, FL.

12

Communicating with Mainstream News Media

CHAPTER OBJECTIVES

This chapter examines the characteristics of the broadcast and published news media landscape and provides essential context, guidelines, and techniques for communicating with journalists and other mainstream media representatives.

   At the end of this chapter, you will be able to:

- Describe the factors affecting the reporting choices that mainstream media journalists make.
- Interpret the way journalists work.
- Apply the important concept of "bridging" to get key points across in a media Interview.

There is no suffering comparable with that which a private person feels when he is for the first time pilloried in print.

—Mark Twain - Life on the Mississippi (1883)

Suppose it were perfectly certain that the life and fortune of every one of us would one day or other depend upon his winning or losing a game of chess. Don't you think that we should all consider it to be a primary duty to learn at least the names and the moves of the pieces; to have a notion of a gambit and a keen eye for all the means of giving and getting out of check?

—Thomas Henry Huxley (1868) (scientist and educator)

12.1   Case Diary: A High Stakes Chess Game with a News Media Outlet

I listened to my voice mail. A colleague and friend, who was the CEO of a large organization, asked me to call her as back as soon as possible. I could hear the stress in her voice. In her message, she said a major national newspaper had published a very negative article about her organization, an article with multiple inaccuracies. If readers believed these allegations, the organization's reputation would be damaged and employee morale would sink. The article had already been reposted in other newspapers, on television news stations around the country, and multiple social media sites.

Communicating in Risk, Crisis, and High Stress Situations: Evidence-Based Strategies and Practice, First Edition.
Vincent T. Covello.
© 2022 by The Institute of Electrical and Electronics Engineers, Inc. Published 2022 by John Wiley & Sons, Inc.

I returned her call. She said the train with this damaging article had already left the station. What could be done?

I asked her to send me the article. The article had been submitted to the national newspaper by a news agency that reported on the type of work performed by her organization. She told me the article's reporter had never bothered to interview her. She suspected this was because she and her organization had taken part in several previous unpleasant experiences with the reporter. This time, the reporter had gone over the top. He was clearly expressing his personal opinions and views in an article classified as a news story.

She wanted to talk with me and weigh possible actions she was considering. She could send a letter to the editor the next day correcting the errors in the article and exposing the reporter's apparent bias. She could demand that the newspaper publish her letter. She could call the newspaper's editor and demand a retraction of the story. Her demand would cite inaccuracies and her rebuttals.

She was considering meeting with her lawyers the next day to discuss legal options. She wondered about suing the reporter, the news agency that had originated the story, and/or the newspaper for defamation. She would also discuss with lawyers if they could get an injunction prohibiting any additional postings of the story by any media outlet.

She felt it might be good to develop a statement for her organization's website and the social media platforms on which she and her organization had a presence. The postings would point out inaccuracies in the newspaper story and counter each inaccuracy with verifiable facts. She could also send that statement to everyone on her own contact list and everyone on the mailing list of the organization.

She was understandably angry and upset. I hoped she would take a deep breath before actually taking action. I started by asking her lots of questions about the case. I also asked for time to conduct research, promising to offer her suggestions in a few hours.

I conducted my research, and I called back. First, I emphasized that regardless of the action she chose, she would need to be transparent and truthful about any operational or communication errors or omissions made by herself or her organization. If there were errors or omissions, she would need to state her values, take responsibility, sincerely apologize, and make a commitment not to repeat the error or omission.

I agreed with her that the reputation of the organization was at substantial risk, especially if the story had mainstream and social media momentum. I had checked out her facts. I found that the published article contained multiple errors, omissions, and inaccuracies, and those needed to be corrected. The errors were much more than misquotes or getting names, times, and dates wrong. They included incorrect facts, misattributed facts, omissions of major facts, and a misleading headline.

I appreciated my colleague's desire to be proactive and timely in stopping the damage. At the same time, it was possible to deepen the damage by taking on the media, especially a large national newspaper. They have an enormous megaphone.[1]

I likened the situation to a game of chess. I offered my suggestions for openings, moves, and endings. If she challenged the newspaper and they were unpersuaded by her corrections, we would have a powerful opponent. We put together a crisis communication team and began sorting through alternative strategies.

First, we agreed to temporarily hold off sending a statement about the story to her email list and social media. Such a strategy had benefits and downsides. One benefit was that it would declare to a wide audience that she had nothing to hide. Also, anyone who was following the story might expect to see her correct the article if it was inaccurate. However, a widely "blasted" statement would likely reach many people who had not seen, or might never see, the original story, unnecessarily raising concerns.

Second, we agreed to monitor the story and see what would happen to it. In today's 24/7 news cycle, many stories burn out in a few days. Regardless, we needed a strategy.

Despite her indignation, we agreed that the odds of getting the newspaper to retract the article were low, despite the inaccuracies. Admitting to a major mistake can damage the reputation, and possibly the career, of a journalist. It could also damage the reputations of the news agency that submitted the article and the newspaper that published the article. A demand for retraction would not be taken lightly. An angry letter to the editor coupled with a demand for a retraction with a threatened lawsuit would likely establish an adversarial situation with the newspaper and the news agency.

After reviewing the pros and cons of alternative strategies, we agreed on a different opening gambit. The original source of the article was a highly respected specialized news agency. After reading their original article, I did research on the news agency. The retired founder of the news agency was a well-respected figure in journalism. He had established the news agency to give a larger voice to issues he felt were underserved by mainstream news media. He was also a vocal critic of competing cable "news" networks, media sensationalism, editorial or personal views embedded in a news story, and the eroding barrier between news and entertainment. He had recently given a speech emphasizing that no tenet of journalism is as important as the obligation to report the facts accurately. He cited articles and reports pointing out that many journalists fall short of this obligation and public opinion polls reflected this failure.[2] He said committing mistakes without correcting them endangers trust and credibility – which are the most precious assets of professional journalism. He cited a report commissioned by the American Society of Newspapers that concluded that even minor errors fed public skepticism about a newspaper's credibility.[3] He cited professional codes of ethics and guidelines that emphasized the importance of truth and accuracy in journalism.

I suggested to my colleague and the crisis communications team that one potential move would be to contact the news agency's founder. After everyone read his speech, all readily agreed. Clearly, the article issued by his news agency to the newspaper ran counter to the principles he so strongly held. I contacted him. I told him I enjoyed reading his speech about trust and the importance of the three "C"s of good journalism: clear, concise, correct. I laid out the case before me, following the three "C"s. He told me he would investigate.

Within two days, the news agency, and then the newspaper, retracted the article. Armed with the retraction, my colleague and her organization could reassure those concerned about the allegations made in the original newspaper article. She and her organization could also ensure adherence to the enshrined journalistic principle of respect for facts and the right of the public to hear the truth.

12.2 Introduction

For most people, mainstream news media still function as an important source of information about risks and threats. They play an important role in setting agendas and in determining outcomes. Mainstream news media coverage is vital for getting risk-related information to interested and affected populations. Journalists who work for mainstream news media organizations play a critical role in keeping stakeholders informed. Mainstream *broadcast news media* includes radio and television. Mainstream *print/digital news media* includes newspapers and magazines.[4]

Mainstream news media outlets can reach many people. In both crisis and noncrisis situations, mainstream media outlets help keep stakeholders informed about risk management actions, guidance, recommendations, and resources. They can encourage appropriate behaviors and counter rumors and misinformation. Except for breaking news, many mainstream outlets have boosted their engagement with audiences by adopting social media practices such as by encouraging comments or featuring hashtags – a word or phrase preceded by a hash sign (#) to identify messages on a specific topic. Comment sections and hashtags allow people to participate in the conversation. They allow readers or viewers to react and interact with other people interested in the same story.

In both crisis and noncrisis situations, effective interactions with mainstream news media can increase the media's – and the public's – confidence in the ability of risk management authorities to deal with risks and threats. Media interactions can raise awareness of actual or potential risks or threats. They can direct readers and viewers to accurate information and establish risk management authorities as the "go to" place for information.

Journalism is a profession that requires time, resources, and practiced skills. Journalists working for credible media outlets agree to adhere to a code of ethics that includes accuracy, impartiality, and integrity. However, journalism is rapidly changing. These changes are due in part to the Internet, social media, and advances in communications technologies.

To boost their engagement with audiences and compete better with social media news sites, many mainstream news media organizations have adopted social media practices. For example, they have become quicker to report news. They often interrupt their regular schedules with breaking news. They monitor social media sites for trending topics. They encourage their audiences to participate using hashtags.

Even with these adaptations, mainstream media organizations are competitively disadvantaged. For example, social media news sites allow users to immediately react and interact with other people interested in the story. Social media news sites are typically not held to the same standards of accuracy and fact checking. Social media news sites can draft an militia of citizen reporters armed with smartphones to report events as they happen. Social media new sites allow users to tailor information to their specific needs. Social media sites allow users to choose whom they want to believe and cite as trustworthy sources of information.

Because of these competitive advantages and changes in technology, news coverage is rapidly shifting from mainstream news media outlets to social media and smaller Internet-based news outlets. Speed in getting information out is increasingly becoming an overarching priority, especially in crises. As Hartman (2009) pointed out:

> Important changes in the way news is gathered and presented are occurring every day, and it is likely that we are seeing only the first wave in this revolution. People who have few or no journalism credentials in the traditional meaning of the term now are gathering and reporting news—sometimes to significant audiences.. . . "Citizen journalists" turn cameras on and within minutes of an event—sometimes live as it's happening—post their pictures or videos for the world to see, often without a pause to check or verify the truthfulness and accuracy of their reporting.[5]

Mainstream journalists now must compete with the timeliness and "on-the-spot" response that "citizen journalists" can provide.

12.3 Characteristics of the Mainstream News Media

Effective risk, high concern, and crisis communication with mainstream news media require intensive preparation and skill. Effectiveness with mainstream news media is rooted in an in-depth understanding of the mainstream news media landscape. Understanding the context and modes in which journalists work can help make the risk, high concern, or crisis communicator more effective in interactions with journalists.

For risk, high concern, and crisis communication purposes, relevant characteristics of mainstream news media are listed in Table 12.1 and described below.

12.3.1 Content

As noted in the *Global Charter of Ethics for Journalists*: "Respect for the facts and for the right of the public to truth is the first duty of the journalist."[6] Journalists worthy of the name deem it their duty to observe faithfully this and the other 15 principles stated in their ethical code. Serious professional misconduct occurs when facts are distorted or when unfounded accusations are made.

Content for most broadcast and print news stories answers six questions journalists learn in journalism school: Who? What? Where? When? Why? How? For example, in a crisis, a journalist is likely to ask: What happened? What caused it to happen? Who was involved? Where did it happen? When did it happen? How did it unfold? What are you doing to ensure it does not happen again? Content of a news story will typically be less accurate, less factual, and more speculative

Table 12.1 Major characteristics of mainstream news media.

1) Content
2) Clarity
3) Avoiding prejudice
4) Topicality
5) Diversity
6) Subject matter expertise
7) Resources
8) Career advancement
9) Watchdogs
10) Amplifiers
11) Skepticism
12) Source dependency
13) Professionalism and independence
14) Covering uncertainty
15) Legal constraints
16) Special populations
17) Competition
18) Confidentiality and protection of sources
19) Deadlines
20) Trust
21) Storytelling
22) Balance and controversy

when journalists have less information to work with. Communicators must prepare for these questions.

To get accurate answers to these questions, many journalists are not embarrassed to ask the same question several times as they dig for an angle they hope will enliven their story. In response, the person being asked the same question should not be embarrassed to stick to their messages.

12.3.2 Clarity

The story must often be simplified to be clear and reach the targeted or broadest potential audience. To enhance clarity, many journalists use an "inverted pyramid" to structure their stories. The base of the pyramid – the most fundamental facts – appears at the top of the story, in the lead paragraph. Less-essential information appears in the following paragraphs. Technical terms, acronyms, and bureaucratic language used by sources of information are explained and clarified. Space constraints (e.g., total number of column inches of newspaper space), time constraints (e.g., total number of seconds or minutes of time devoted to a story), deadline constraints (e.g., total number of hours to file a story), and word constraints can severely limit what journalists can include in their stories.[7] As a result, direct quotes from information sources are usually not printed in full. Instead, they often appear as snippets selected to represent a greater whole and a particular point of view. For example, the President of the United States might give a speech that runs over two hours. The news report may contain only 27 words, nine seconds, or 3 messages, whichever comes first.

12.3.3 Avoiding Prejudice

For many mainstream news media outlets, controversy, bad news, threats, conflict, wrongdoing, disputes, disagreements, and criticism are ideal for a front page or lead story. Most mainstream news media outlets are in the business of selling news. Financial profit, viewership, attention, and ratings are often powerful driving forces. News stories that attract the most attention are those with clear-cut villains, victims, and heroes.[8] However, as noted in the journalist's Global Charter of Ethics, "the journalist's responsibility towards the public takes precedence over any other responsibility, in particular towards their employers and the public authorities."[9] In addition, journalists have a responsibility to ensure that their stories do not contribute to hatred, prejudice, or the spread of discrimination on grounds such as geographical, social, or ethnic origin, race, gender, sexual orientation, language, religion, or disability.[10]

12.3.4 Topicality

Topics that pose the greatest risks to the public do not necessarily receive the greatest amount of coverage from mainstream news media. The column inches of newspaper space and the minutes of television time devoted to risk-related events seldom match the actual risk of the events. For example, a death from an airplane accident is thousands of times more likely to appear on the front page of a newspaper than a death from cancer.[11] Greenberg et al. (1989) found "there was seven times as much coverage of airplane accidents as there was of smoking/tobacco, and 29 times more coverage of airplane accidents as there was of asbestos." Journalists often give greater coverage (as measured by frequency, size, and prominence of reports) to risk-related stories or events that are most likely to capture the attention of their readers or viewers. These include stories and events that are inspirational or about risks perceived to be unfamiliar, unknown, uncertain, unfair, managed by untrustworthy sources, dreaded, potentially catastrophic, memorable, immoral, associated with few benefits, not under a person's personal control, and involuntary. Risks with these negative characteristics typically receive much greater media coverage. Also, mainstream news media often focus on disagreements, controversy, and conflict. Conflict is often easier to cover than the technical details of a complex risk-related issue.[12]

12.3.5 Diversity

Mainstream news media outlets are not monolithic. There is a wide variety in staffing, the nature and size of the markets, practices, and the tasks, e.g., editorials, headline writing, reporting, columns, and opinion pieces. Because of diversity, a news story may have a different slant depending on the outlet. The story's headline is often written by a person other than the journalist, a person who may have a unique perspective beyond the writer. As a result, the headline may misalign with the content of the story.

12.3.6 Subject Matter Expertise

Many journalists are generalists rather than specialists, even in large mainstream news organizations. Journalists are often shuffled among content areas ("beats"). This shuffling provides staffing flexibility and is especially important when there are few staff journalists. In the absence of staff journalists, either freelancers or wire service news agencies are often called upon for assistance. One consequence of the lack of in-house expertise is that a risk-related situation or event may be covered by a journalist who has little or no experience or specialized knowledge and is unable to provide context for the story. For example, during the COVID-19 pandemic, many stories focused on adverse reactions to vaccination. Few stories mentioned background rates of adverse reactions. Similarly, many stories are written when there is a plane crash. Few contain information about the risks of flying.

Reliance on freelancers has been increasing. This is more so now than in the past. Many mainstream news outlets have been cutting staff. Many outlets no longer have the luxury of budget for specialized reporters. A counterforce at work is that many journalists are quick learners. This is especially important if they lack subject-matter expertise. Journalists need to be skilled in quickly gathering, interpreting, and reporting news. One of the most admired skills of the best journalists is their ability to quickly gather information on almost any topic, interpret the information, and report the information in an accurate, engaging, and balanced manner. The ideal journalist knows how to gather information rapidly (thus meeting deadlines), how to nurture sources (thus ensuring a steady flow of information), and how to synthesize quickly.

12.3.7 Resources

Many mainstream news media organizations do not have the resources needed to prepare, in advance, background information (such as graphics and videotape) for a story, especially during a crisis. In addition, news organizations seldom have the resources needed to maintain offices and reporters in distant sites where events may be occurring. It is often difficult getting reporters and equipment to the site of a breaking story.

12.3.8 Career Advancement

Journalists often advance in their careers by moving from smaller news media markets to larger markets. One result is high staff turnover. The assigned journalist may be unfamiliar with key leaders, experts, authorities, and stakeholders. By the time a journalist develops close working relationships with key players, they may find that they are ready to move to a new job.

12.3.9 Watchdogs

Many of those involved in journalism view themselves as "watchdogs" and public advocates, checking for failures, evasions, misdeeds, and abuses.[13] This stems from the fundamental tenet that a free press is essential to a democratic society. From a democratic perspective, a key function of news media is to "aid citizens in becoming informed."[14] In performing as watchdogs, journalists see their role as one that draws attention to issues, setting agendas, raising awareness, and influencing opinion. For example, as the 2020 COVID-19 pandemic evolved globally and the fear of infection increased, journalists elevated onto the public agenda the lack of government preparedness for the pandemic and the stigmatization of people of Asian descent.[15] Journalists did crucial reporting about what governments were doing wrong and provided – often at great personal risk – updates from places around the world devastated by the pandemic.

12.3.10 Amplifiers

By the stories they file, editors and journalists play a critical role in amplifying or dampening perceptions of risk and, through this, creating secondary effects such as support or opposition to a technology or policy.[16] Information about risks typically ripples in circles, not straight lines. Like a stereo receiver, editors and journalists amplify risk-related events or situations through a variety of means, including the amount of media coverage and the emotional content of the story.[17] Editors and journalists will also intensify messages about a risk-related event or situation if access to information is denied by risk management authorities or if answers to their questions are not forthcoming. Editors and journalists will typically probe for underlying political, social, economic, or personal motives if they believe their questions are being evaded. For example, following the accident at the Three Mile Island nuclear power plant in Pennsylvania, staff from the Philadelphia Inquirer went directly to plant workers to get supplemental information. The reporters copied the license plates of vehicles in the parking lot, traced the owners, and contacted them. Over 50 people agreed to be interviewed about any observations, complaints, or compliments they had about the construction, operation, or repair of the plant.[18]

12.3.11 Skepticism

Many mainstream news media journalists are wary of developing close professional or personal relationships with their sources of information. They do not see themselves as the enemy of risk management authorities but neither as their friends. They often feel a professional obligation to maintain a proper distance and adopt a critical and questioning stance regarding the information they collect and the sources they use. During a crisis, this skeptical attitude is often less pronounced, especially during the first 24-72 hours. In this "grace" period, mainstream news media often function as an extension of crisis management. Editors, producers, and journalists give priority to information that helps people protect themselves, their property, and the things they value. Following this grace period, many journalists focus on issues such as what went wrong, who is to blame, and whether the crisis could have been prevented. Also, while many journalists may be wary of developing a too close personal relationship with their sources of information, some journalists will leverage their existing personal relationships in the effort to get more information from a source.

12.3.12 Source Dependency

Mainstream news media journalists highly depend on individuals and organizations for a steady and reliable flow of newsworthy information. This flow of information makes news production more predictable, efficient, and profitable.

High value is placed on sources of information that are accessible and can provide clear, timely, and useful information. Positive relationships with sources of information are built on this flow. When this flow is blocked (i.e., when a news source is unavailable for comment, when the news source does not respond within the journalist's deadline, or when the news source is inarticulate or unable to use plain language), journalists grow frustrated and seek other sources. These other sources of information may be less authoritative, accurate, responsible, or reliable.

Lack of access to articulate and expert spokespersons often leads to imbalanced news stories. Imbalanced stories often lead to heightened controversy. By selecting sources for comment, the media is transferring trust to that source. As a result, sources of information with little expertise but who are readably accessible and articulate may be given substantial amounts of time and space to offer their views and counterviews. Sometimes, they are given more coverage than true experts. When true experts are not accessible or understandable, nonexpert views may be given exclusive coverage. Moreover, in their struggle to present information quickly in a highly competitive environment, many journalists use information in preprint journal reports and researcher press conferences. These sources provide easy access to expertise but also require less verification than the content of standard peer-reviewed scientific journals.

12.3.13 Professionalism and Independence

Most journalists are committed to refraining from activities or engagements that may put their independence in danger, or that could lead to a conflict of interest. There are many conflicts of interest. They include stories about colleagues, the organization where the journalist works, the organization that owns the organization for which the journalist works, romantic relationships, and friends and family. They also include stories for which the journalist has a financial interest or has or will receive a direct or indirect payment or compensation. Payments and compensation range from direct fees, book deals, and paid speaking engagements to perks such as special access privileges, free products, and free copies of books and films. Conflicts of interest also arise when a journalist takes a position on an issue.

Many journalists deal with conflicts of interest by avoiding stories that might compromise their independence or by declaring the conflict of interest at the beginning or end of the story. They also may include in the story an explanation about why they believe the conflict of interest does not compromise their work.

Trust is easily lost when a competing interest is discovered that was not disclosed. According to the NPR (National Public Radio Program) Ethics handbook,

> To secure the public's trust, we must make it clear that our primary allegiance is to the public. Any personal or professional interests that conflict with that allegiance, whether in appearance or in reality, risk compromising our credibility. We are vigilant in disclosing to

both our supervisors and the public any circumstances where our loyalties may be divided—extending to the interests of spouses and other family members—and when necessary, we recuse ourselves from related coverage. Under no circumstances do we skew our reports for personal gain, to help NPR's bottom line, or to please those who fund us. Decisions about what we cover and how we do our work are made by our journalists, not by those who provide NPR with financial support.[19]

12.3.14 Covering Uncertainty

Mainstream news media journalists vary in how they respond to acknowledgments or admissions of uncertainty. Science is a work in progress. Advances in science are generally incremental and open to revision. The nature of science often conflicts with the needs of journalists for clear, bottom-line conclusions. Unless great care is taken, reporting uncertainty can result in the production of misinformation. Journalists vary in how they respond to acknowledgments or admissions of uncertainty. Some view acknowledgment of uncertainty as an indicator of trustworthiness and transparency. Others view acknowledgment of uncertainty more negatively. For example, they may be suspicious and question the cautious and hedging statements often offered by technical experts and risk management authorities (e.g., "On the one hand. . .On the other hand. . ."). As a result, journalists may seek less reputable or less informed sources of information willing to speak out on an issue with greater certainty, greater speculation, and less caution, even though that certainty may be unfounded.

12.3.15 Legal Constraints

Mainstream news media journalists are less constrained than risk management authorities by legal requirements designed to protect privacy and personal information. For example, journalists will often ask crisis managers for the names and addresses of victims even before family members have been notified. They may seek employees outside the chain of command to answer questions about blame and responsibility. These types of questions are a frequent source of tension between risk management authorities and journalists. Risk management authorities often struggle with how best to balance transparency with the right to privacy and confidentiality. If information is not forthcoming from authorities, journalists have often resorted to creative and extraordinary measures to obtain such information.

12.3.16 Special Populations

Mainstream news media journalists are often not positioned to meet the information needs of special populations.[20] Special populations include the elderly, disabled people, homeless people, housebound populations, vulnerable populations, transient populations, travelers, institutionalized populations, illiterate populations, and people experiencing language barriers. One exception is when the special population is the target audience for the news media outlet or if the special population is particularly affected by the issue.

12.3.17 Competition

Competition for stories within and among mainstream news media organizations (and among journalists) is often intense, especially in a crisis and in larger media markets. Many news organizations compete zealously with one another for viewers, listeners, or readers. This competition is centered on getting the story out first or reporting information that the competition does not have. Competition and speed are among the major causes of media sensationalism and inaccuracies.

12.3.18 Confidentiality and Protection of Sources

Most journalists will make efforts to protect confidentiality if their source wants to remain anonymous or if the information provided was based on a mutual understanding that the information was "off the record" or as background. Journalists, editors, and producers have been willing to go to jail to protect the confidentiality of their sources of information. However, given the choice between maintaining a personal relationship or reporting a significant new piece of information about a topic currently being reported by other journalists, some journalists may report information understood or perceived to be confidential. Furthermore, some journalists may reveal detailed personal information about an individual mentioned in their story, including their age, sex, residence, workplace, and behavioral history even when they do not identify the individual by name.[21]

12.3.19 Deadlines

Mainstream news media journalists typically face hard, fixed, and relentless deadlines. As a result, journalists assign a high priority to meeting deadlines. Deadlines assume even greater importance for 24/7 news media outlets. Updates often appear every few minutes. Journalists are given a highly defined and limited time within which to gather, interpret, and report a story. This restricts what can be covered. Sources that make journalists miss their deadlines are generally looked upon with disfavor. A too-slow information source may be bypassed. Sometimes, risk management authorities may be given only minutes or hours to get back quickly to a reporter to meet a deadline. Failure to meet a journalist's deadline often results in an unbalanced story. Risk management authorities and journalists often struggle with how best to balance the demands associated with deadlines with the need for more time to gather accurate facts and information.

12.3.20 Trust

Most journalists recognize that trust in what they report is based on perceptions that they are providing accurate information without the use of manipulative strategies. Trust, like beauty, is in the eyes of the beholder. It is given or withheld by receivers of information during the communication process. It is a social construct made up of factors such as perceived accuracy, fairness, and completeness.[22]

That social construct has changed for mainstream media; trust in traditional, mainstream media has been declining.[23] For example, only four in ten US adults say they have "a great deal" (9%) or "a fair amount" (31%) of trust and confidence in the media to report the news "fully, accurately, and fairly," while six in ten have little or no trust (27% reported "not very much" trust and 33% reported "none at all"). Trust ranged between 68% and 72% in the 1970s. It remained above 50% until 2004 when it dipped to 44%.

This downward slide stems from a variety of factors, including increased competition from other news sources, people selecting news sources that reinforce their existing beliefs, politics supplanting science, and the erosion of trust in institutions. Trust erosion slides further downward as politicians accuse media stories they disagree with as biased and attack the media as purveyors of "fake news."

According to Richard Edelman, head of one of the largest survey organizations in the world, in 2021, there was

> . . .a run on the Trust Bank. Trust in all news sources has hit record lows, with traditional media down to 53 percent from 65 percent two years ago, and social media down to 35 percent from 43 percent. Sixty-one percent of respondents believe media is politicized and not objective.[24]

The 2021 poll also revealed a remarkable 39-point gap in trust in traditional media between those who voted for Joe Biden in the 2020 Presidential election (57% trusted traditional media) and those who voted for Donald Trump (18% trusted media). The only major exception to the general downward trend in trust in institutions was communications from "My Employer." During the 2020 COVID-19 pandemic, people expressed more faith in their own employer than they did in national government, traditional media, and social media.

Based on 2021 polling data, nearly six in ten US adults think "journalists are purposely trying to mislead by saying things they know are false or grossly exaggerated. The same proportion thinks that most news organizations are putting ideology or political position above informing the public about what is happening in the world."[25]

12.3.21 Storytelling

Many journalists are skilled storytellers. Storytelling is a powerful way to engage an audience and share facts and ideas. Great stories have clear heroes, villains, and victims.[26] Stories that win journalism prizes often have relatable characters that help the audience connect to the story through feelings such as pain, joy, sadness, or anger.[27] Readers and viewers may not remember the facts of the story but are likely to remember how the story made them feel.

12.3.22 Balance and Controversy

Journalists are expected, at least as an ideal, to report the news objectively without bias (thus not alienating sources, compromising credibility, or driving away audiences). Except for editorials and opinion pieces, a common approach to achieving objectivity for mainstream media outlets is to cite multiple sources reflecting diverse, even opposing, viewpoints. Unfortunately, the quest for balance may cause the use of sources who are less well informed but easily accessible and willing to speak. It can also result in giving equal weight to alternative, less factual viewpoints along with true and authoritative information, often in the interests of apparent balance but also to achieve interesting controversy. Also, bias is necessarily introduced by the choices made, such as the topics to cover, the sources to quote, and by what is left in and out of a story.

12.4 Guidelines and Best Practices for Interacting with Mainstream News Media

Best practices for working with and communicating to mainstream media stem from understanding the demands, expectations, and realities of journalistic ethics and practice described in the last section. A core need for journalists is ready access to their sources, and risk managers can prepare the way by establishing relationships before an urgent situation. One way to build relationships with key broadcast and print media outlets before a crisis is to try establishing trusting relationships with specific editors and journalists, such as by offering story ideas and background for non-crisis topics, or by offering information related to stories they have done. It is also important to have frequently updated contact lists for key broadcast and print media outlets at the ready, including contact information (e.g., email addresses, mobile phone numbers, and social media accounts) for editors, producers, and key journalists. Achieving good rapport with journalists will not guarantee favorable or accurate coverage, but it is a factor in keeping journalists turning to an organization as a trusted source of information. It is considered wise to treat journalists neither as a friend nor enemy while demonstrating respect for the task they are trying to accomplish. It is essential to be open with and accessible to journalists, respecting their deadlines, and giving frequent updates. Take time to anticipate questions and tailor your answers to the specific needs of the journalist or type of media.[28]

A particularly tough challenge is being honest about uncertainty without appearing to withhold information. There is uncertainty that may be part of the situation being discussed and there is one's personal uncertainty or lack of knowledge. Many risk managers and technical professionals find it hard to say, "I don't know." One of the most effective ways to say "I don't know" is to use the IDK (I Don't Know) tool. When you don't know, can't answer, or are not the best source for information, you can: repeat the question (trying to avoid negative words or allegations); say that you would like to answer the question; say why you can't answer; give a follow-up; bridge to what can be said (convey your prepared messages). Alternately, and often preferably because of negative dominance, you can use the KDK (Know/Don't Know) tool: say what you know; say what you don't know and why you don't know it; say what you are going to do to follow up and find out.

When confronting uncertainty, journalists will try to fill an information void, even if it means they then turn to speculation by a less authoritative source. This is particularly the case when a source of information does not return calls or answers questions with "No comment." To be considered a trusted source by a journalist, individuals and organizations need to be proactive, avoiding the temptation to wait until the full picture is clear. Being first matters, often as much as being clear, concise, and correct.

Journalists are often not experts in the subject areas of a risk-related issue; even if they are, they typically appreciate help in making content clear and understandable. A good way to help journalists better understand a complex risk-related issue is to provide them with clear, concise, and accurate background material, such as fact sheets, maps, graphics, biographical data for key responders, and FAQ (frequently asked questions) documents. In addition to factual background materials, a core best practice technique is to focus messages less on statistics and data than on factors that influence perceived risk, such as trust, benefits, fairness, and control. Journalists also typically appreciate receiving names of other potential sources of information, even if the names provided are selective.

Risk-related issues create many opportunities for media sensationalism. What editors, producers, and journalists see as newsworthy may not be the same as what experts view as newsworthy. Many risk-related issues carry with them characteristics that attract public attention and encourage media interest, such as risks or threats that involve conflict, errors, mistakes, or are associated with negative attributes such as unfamiliarity, unfairness, high scientific uncertainties, or involuntariness. Sensationalism can be tempered by providing timely, accurate, transparent, and credible information; by acknowledging uncertainties; by being willing to admit mistakes; by expressing authentic empathy; and by acknowledging emotions. For example, in his first news conference on pandemic influenza, Dr. Richard Besser, then Acting Director of the Centers for Disease Control and Prevention, said "First I want to recognize that people are concerned about this situation. We hear from the public and from others about their concern. We are concerned as well."[29]

Relationships with the media should be ongoing. They require continual care and maintenance. Following up on broadcast and print media stories with praise, or fair criticism, as warranted can continue to build trust and understanding. Table 12.2 summarizes best practices for interacting with the media.

12.5   The Media Interview

Writing in 1871, Mark Twain closed a letter to a colleague with the following words: "You'll have to excuse my lengthiness—the reason I dread writing letters is because I am so apt to get to slinging wisdom & forget to let up. Thus much precious time is lost."[30] Twain's words illustrate one of the most important principles for a successful news media interview: a set of brief messages developed before the interview which arise from anticipation, preparation, and practice.

Media interviews are typically done face-to-face, by phone, or through a video Internet connection linking the interviewee and the interviewer. To be done well requires intensive preparation. From the perspective of the interviewee, the primary purpose of a media interview is to share key messages, respond effectively to questions, and shape thinking about an issue. It is essential for the interviewee not to lose track of those objectives during the stress – or lure of pleasant engagement – of an interview.

A first step in developing targeted brief messages is to determine what the reporter wants from the interview. This can be achieved by asking the reporter questions – both topical and procedural – before the interview. Several of the most important questions are listed in Table 12.3. Since reporters are typically pressed for time and may find some of these questions inappropriate, judgment should be exercised in selecting three or four questions appropriate to the situation, available time, and the specific reporter.

Table 12.2   General guidelines and best practices for interacting with mainstream news media.

1) Focus on building relationships with key broadcast and print media outlets before a crisis: try establishing trusting relationships with specific editors and journalists, such as by offering story ideas and background for noncrisis topics.

2) Be open with, and accessible to, journalists, respecting their deadlines and giving frequent updates.

3) Create contact lists for key broadcast and print media outlets, including contact information for editors, producers, and key journalists.

4) Treat journalists neither as a friend nor as an enemy.

5) Focus messages less on technical facts than on factors that influence perceived risk, such as trust, benefits, fairness, and control: what you and editors, producers, and journalists see as newsworthy may not be the same.

6) Anticipate questions from journalists: tailor answers to the specific needs of the journalist or type of media.

7) Prepare clear, concise, and accurate background material, such as fact sheets, maps, graphics, biographical data for key responders, and FAQ (frequently asked questions) documents.

8) Follow up on broadcast and print media stories with praise or criticism, as warranted.

9) Be proactive, not waiting until you know more: be first, accurate, truthful, relevant, and credible.

Once answers to selected questions in Table 12.3 are obtained, key messages can be prepared. Key messages should be driven by the *Single Overriding Communications Objective, or SOCO.* The *SOCO* is the key point you want to make in the interview. It reflects what the communicator envisions as the lead in the story and the three or four facts or pieces of information key stakeholders should remember after reading or hearing the interview.

Ideally, messages shared during the media interview should be short enough to be used as a sound bite. The average sound bite in the year 2021 was approximately 27 words, 9 seconds, or 3 to 5 messages.[31] The average size of the media sound bite has shrunk. For example, in 1968, the average political sound bite was 43 seconds. By 1988, the average political sound bite had shrunk to 9 seconds.[32]

Listed in Table 12.4, Table 12.5, and Table 12.6 are tips for before, during, and after a media interview.

Successful media interviews are message driven, not question driven. An essential tool for accomplishing this goal is bridging, i.e. linking the reporter's question to a key message that you want to communicate. Examples of bridging statements are listed in Table 12.7.

Bridging is integral to getting critical information to stakeholders. If the reporter's question is taking the interview off track, it should be answered followed by a bridging statement and a key message. Journalists are aware of bridging. They will generally accept its use unless it becomes excessive. Bridging, while helpful, can be overdone. For example, politicians often overuse bridging. In some cases, they ignore the reporter's question entirely. As a result, they often come across as untrustworthy and evasive.

Besides emphasizing key messages, bridging can be used to respond to difficult questions. One example of a difficult question is the "what if" question–a hypothetical question that asks about a future state or asks for speculation. Another example is the guarantee question – "Can you

Table 12.3 Questions to ask a journalist before a media interview.

I. Background Questions

1) What is the reporter's name, organization, and telephone number?
2) What stories has the reporter previously covered?
3) Who generally views, reads, or hears information shared by the media outlet?

II. Logistical/Procedural Questions

1) Where and when will the story appear?
2) What is the deadline for the story?
3) Where will the interview take place?
4) How long will the interview take?
5) How long will the story be?

III. Topical Questions

1) What the story is about and why you have been selected to be interviewed.
2) What specific topics does the reporter want to cover in the interview?
3) Can the reporter provide the first question and examples of other questions?
4) Has the reporter done background research for the interview? If so, what have they learned?
5) Would the reporter like to receive background material before or after the interview?
6) Who else has been interviewed. If others have been interviewed, can you tell me what was said?
7) Who else do you expect to invite for an interview? If others will be interviewed, can you give me examples of questions you may be asking?

Table 12.4 Tips for before a media interview.

Do:

1) Be quick to respond to the interview request.
2) Ask who will be conducting the interview.
3) Ask which subjects they want to cover.
4) Inquire about the format and duration.
5) Ask who else will be interviewed.
6) Ask yourself: Are you qualified to discuss the topic? Do you have the right information? Is the information requested classified or confidential? Has the accuracy of the requested information been verified? Have you checked to see what others have said about the requested information? Are you permitted to release the requested information?
7) Report contact with the media immediately to management and communications personnel.
8) Anticipate questions based on the issue and a review of other stories done by the reporter.
9) Prepare answers to anticipated questions and the key takeaway messages.
10) Rehearse and practice with an experienced professional.
11) Respect deadlines.

Don't:

1) Tell the news organization which reporter you prefer.
2) Ask for all the questions in advance.
3) Insist the journalist not ask about certain subjects.
4) Demand your remarks not be edited or demand the right to edit your remarks.
5) Insist an adversary not be interviewed.
6) Think keeping a fact secret will prevent the media from finding out.
7) Assume the interview will be easy.
8) Assume you are the right person to be interviewed

---

guarantee that . . ..” One response to these types of questions is to bridge to facts and current knowledge. For example, the communicator might say, “Instead of talking about what if, it might be better to talk about what is and what is known from the past.” One option for responding to the “guarantee” question is: “You've asked me about the future. The best way I know to forecast the future is to emphasize what we know now from the past and present. What the past and present suggest is. . .”

Bridging provides risk, high concern, and crisis communicators a means for taking charge of and controlling media interviews. Bridging also functions as means for controlling the flow of information in a news conference. The primary goals of a news conference – also called press conferences – are to answer questions effectively and share few overarching key messages that are accurate, clear, concise, credible, relevant, and timely. News conferences are typically organized when there is a great deal of media interest, when there are many reporters requesting attention, and when it is important to ensure that the reporters receive consistent messages.

News conferences typically begin with introductory remarks by a moderator, followed by speaker presentations. Speakers should be prepared to offer three to five key points in their presentation. Speakers should also be trained in anticipating questions, message delivery, and bridging techniques. Each presentation should be rehearsed and timed to last no more than 3–5 minutes.

Table 12.5 Tips for during a media interview.

Do:

1) Be helpful.
2) Be open, honest, accurate, and truthful, sharing what you know and don't know.
3) Be first to share bad news.
4) Express caring, listening, and empathy regarding concerns.
5) Acknowledge emotions and fears.
6) Keep messages/talking points short and simple: no more than 27 words, 9 seconds, or 3 messages (i.e. follow the KISS principle: Keep It Short and Simple).
7) Stick to your key message(s).
8) State your conclusions first, and then provide supporting information: follow the "bottom lines up front" model or the Triple T model (tell people what you want them to know, tell more, tell it again).
9) Cite credible sources of information besides yourself.
10) Offer to get information you don't have.
11) Give a reason if you can't discuss a subject.
12) Acknowledge uncertainty but don't over-reassure.
13) Correct mistakes by stating you would like an opportunity to clarify.
14) Use plain English.
15) Be attentive to body language.

Don't:

1) Lie or try to cloud the truth.
2) Improvise or dwell on negatives.
3) Raise issues you don't want to see in the story.
4) Guess: If you don't know the answer, when appropriate, say you will try to find out.
5) Use jargon or assume facts speak for themselves.
6) Speculate or discuss hypothetical or unrealistic situations.
7) Lose your temper or composure.
8) Say, "no comment."
9) Demand an answer not be used; assume everything you say will be quoted.
10) If a crisis, speculate and attempt to answer questions about the following: monetary estimates of damage, insurance coverage, possible causes, blame or responsibility, or anything that might imply fault or negligence.
11) Repeat allegations, incorrect statements, inflammatory statements, or negative quotes.
12) Display poor body language.

Table 12.6 Tips for after a media interview.

Do:

1) Remember you are still on the record, even after the interviewer says the interview is over, if the information shared was provided "off-the-record," or if the information shared was provided only as background.
2) Be helpful and volunteer to get information you promised to provide.
3) Make yourself available for a follow-up interview.
4) Call the reporter to politely point out inaccuracies, if any.
5) Keep a log or diary notes of information shared with the reporter.
6) File the reporter's contact information, including phone number and email.

Don't:

1) Ask the reporter: "How did I do?"
2) Ask to review or edit the story before publication or broadcast.
3) Complain to the reporter's boss before talking first to the reporter.

Table 12.7 Examples of bridging statements for media interviews.

1) And what's most important to know is. . .
2) However, what is more important to look at is. . .
3) However, the real issue here is. . .
4) And what this all means is. . .
5) And what's most important to remember is. . .
6) With this in mind and looking at the bigger picture. . .
7) With this in mind, if we take a look back at . . .
8) If we take a broader perspective. . .
9) If we look at the big picture. . .
10) Let me put all this in perspective by saying. . .
11) What all this information tells me is. . .
12) Before we continue, let me take a step back and repeat that. . .
13) Before we continue, let me emphasize that. . .
14) This is an important point because. . .
15) What this all boils down to is. . .
16) The heart of the matter is. . .
17) What matters most in this situation is. . .
18) And as I said before. . .
19) And if we take a closer look, we would see. . .
20) Let me just add to this that. . .
21) I think it would be more correct to say. . .
22) Let me point out again that. . .
23) Let me emphasize again. . .
24) In this context, it is essential that I note. . .
25) Another thing to remember is. . .
26) Before we leave the subject, let me add that. . .
27) And that reminds me. . .
28) And the one thing that is important to remember is. . .
29) What I've said comes down to this. . .
30) Here's the real issue. . .
31) While [topic of question] is important, it is also important to remember. . .
32) It's true that [topic of question] but it is also true that. . .
33) What is key here is. . .
34) Before we get too far afield, let's not forget. . .
35) Before I answer that. . .
36) It might help if I put that question in perspective. . .
37) Can we think again about what we talked about in the beginning. . .
38) Your question makes it even more important to. . .
39) What really matters to those listening to us is. . .
40) What I feel has been left out of your question is. . .
41) What matters most right now is . . .

Speaker presentations are usually followed by a question-and-answer session with reporters in person or by phone. Failure to allow time for questions may encourage reporters to go elsewhere for information. It may also result in reporters deciding not to join the news conference.

News conferences are especially important in the initial stages of a crisis when facts are still unclear. Risk management authorities need to be visible and available. They need to establish themselves as the primary "go to" place for information. A sample opening statement by a news conference moderator illustrates the framework for a news conference:

> Welcome, ladies and gentlemen to [insert time: this morning's; this afternoon's; tonight's] news conference. My name is [insert name and title].
>
> We will be presenting information at this news conference on [insert topic].
>
> With us today are [insert names and titles]. Biographical information for each person presenting at this news conference can be found in the media packet at the back of the room or given to you when you entered. We will begin the news conference with brief presentation(s) by [name the individual or individuals; indicate the spelling and pronunciation of the person's name; state their title and the name of their organization]. Each presenter will speak for no more than 5 minutes. After all the presentations are completed, I will ask if there are questions. If so, I will direct the question to the appropriate speaker. I ask that each reporter limit their questions to one question and one follow-up question.

Holding an effective news conference can be stressful and demanding. Not every risk-related situation merits a news conference. However, when circumstances suggest that a news conference is appropriate, all aspects of the news conference need to be carefully considered and prepared. Several handbooks provide detailed information about how to conduct an effective news conference.[33]

12.6 Lessons and Trends

The information presented in this chapter on mainstream news media is based on years of evidence-based research. The chapter described the processes of how journalists work and presented strategies, tools, and skills needed to effectively navigate through the mainstream news media landscape and the many challenges likely to be encountered. The research presented in this chapter addresses many questions: why are some media spokespersons and interviewees more effective than others? Why are some people more quoted by mainstream news media than others? Why are some people selected by an organization to present information to journalists over others? Why do some ideas gain traction through mainstream media news channels and others do not?

What is clear from the research is that mainstream media skills are learnable. They can be practiced, repeated, and made intuitive. What is also clear from the research is a great deal is often at stake when a person interacts with a mainstream media journalist. Until a risk-related newsworthy event occurs, or until the discovery of a new risk or threat, the infrastructures and mechanisms that protect people from harm often go unnoticed and attract little media interest. After a newsworthy risk-related event occurs, or after the discovery of an unknown risk or threat, the situation becomes very different. The demand for information by mainstream media news outlets quickly escalates.

This demand for information, and the ways this demand is satisfied, is rapidly changing. Important changes have taken place in the news media landscape.[34] For example, newsrooms across the country made coverage of the 2020–2021 COVID-19 pandemic a priority. However, because of competition from social media, lack of clarity and consistency in communications by government officials, the politicization of issues, and the continually changing nature of

COVID-19 data, journalists faced a daily uphill battle providing clear, concise, and correct information to the public.[35]

In the last 15–20 years, the world has witnessed a revolution in how news about risks or threats is collected, interpreted, shared, and presented. In 2021, more than eight in ten Americans took their news from digital devices.[36] As discussed in greater detail in the next chapter of this book, these changes are due largely to the rise of networked social media platforms and use of digital communication technologies through computers, tablets, and smartphones.

These changes have radically affected mainstream media newsrooms. New competitors create news aggregations on the web, and social media platforms such as Twitter, Facebook, Instagram, Snapchat, YouTube, and Google now collect and distribute information from other news sites. Over half of US adults either took their news "sometimes" or "often" from social media.[37] Facebook is now a regular source of news for approximately a third of Americans. Social media sites gather news from traditional mainstream news media sites such as the New York Times, USA Today, the Wall Street Journal, and the Washington Post. They also often gather news from nontraditional news sites that vary greatly in credibility.

News will also likely be covered by "citizen journalists" with few or no professional journalism credentials but able to reach a large audience through social media. Often mainstream media are now relying on the videos from smartphones collected from individuals on the scene. For example, the first person to report the 2009 "Miracle on the Hudson," where the now famous pilot Captain Chesley "Sully" Sullenberger safely landed his plane on the Hudson River, was a 23-year-old with no journalism experience who was a passenger on a commuter ferry that rushed to rescue passengers. He snapped a photo with his iPhone and posted it together with a message on Twitter: "There's a plane in the Hudson. I'm on the ferry going to pick up the people. Crazy." He shared the message with his 170 followers, but the photo spread quickly to over 10,000 people. He was soon interviewed and quoted on local and national television.[38] Both social and mainstream media rapidly reach a wide audience. Each magnifies the impact of each other.

Despite the shifts in the news media landscape, mainstream news media still command great attention, exercise significant power, and directly influence the information people have about a risk or threat and the course of events. The ability of a communicator to present clear and compelling information to a journalist is still an essential key to building trust, enlisting support, calming nervous people, providing much-needed information, and encouraging cooperative behaviors. Poor mainstream news media communication about a risk or high concern issue can fan emotions and undermine trust and confidence. Excellent mainstream news media communication can make important information about a risk or high concern issue accessible, make the complex simple, and play an important role in engaging constructive responses.

Risk managers and communicators in a democratic society depend on journalists and mainstream media organizations to perform their traditional functions of disseminating accurate information about risks and threats, presenting the information in an understandable and interesting way, helping people interpret the meaning of that information, and fulfilling their calling to help people see the world more clearly. Through effective communication with mainstream news media, risk managers and communicators can engage stakeholders and help them to make informed and better decisions. Journalists and media organizations depend on risk managers and communicators for timely and accurate information. Effective mainstream media communication requires trust and understanding between risk managers, risk communicators, journalists, and media organizations.

12.7 Case Diary: A Ten-Round Exercise

I was escorted into the beautifully furnished office of the company's Chief Executive Officer. He invited me to sit. We had met several times before. I had done consulting work for his organization and he would often attend meetings where I was present.

He said this time was different: he wanted me to conduct a media training for him in the privacy of his office. He was scheduled to meet the following week with an investigative reporter. He had done many media interviews before. This was new for him as it would be his first interview with an investigative reporter. The reporter would be confronting him with allegations. He believed the facts would prove the allegations were untrue. However, he also knew that having the facts on your side is a necessary but not sufficient condition for succeeding in a high-stress media interview.

In preparation for the interview, he had read my World Health Organization handbook on effective media communication in a crisis as well as other media interview guidance documents. He had his communications director and his team put him through a one-day media training course. As part of the training, they did several rounds of mock interviews and had uncovered multiple opportunities for improvement, such as eliminating technical jargon that would likely not be understood and humor that might be interpreted as trivializing concerns or not taking the issue seriously. They also worked with him on his nonverbal communication, including improved eye contact and body language. They worked with him on anticipated questions and key messages that were clear, concise, and correct. They worked with him on bridging and reframing techniques that would allow him to briefly answer or acknowledge the question and then bring the interview back on track. They advised him to show listening, caring, and empathy, and not say: "no comment." They cautioned him about the pluses and minuses of apologizing. They showed him films of other CEOs falling on their swords. I said this was all great. So why did he want another media training from me? He said the training was not tough enough. He felt the trainers, who were his employees, were wearing kid gloves. He needed someone to go at him full strength.

He said he did not need convincing that successful media interviews are hard to do – especially about high concern or high-stress issues. He noted that almost everyone he had spoken to comes out of such interviews with regrets about what they said and did not say; and that the interview could go terribly wrong. He also recognized that many of the recommended strategies and techniques for successful media interviews are not instinctive. He wanted to do enough training and rehearsal to make these strategies and techniques part of his muscle memory. He and his team were aware of the many opportunities, dangers, and pitfalls in media interviews. Because of this, his organization does not allow their employees to talk with journalists unless they are fully prepped. Even then, and with few exceptions, his preference is to leave interactions with reporters to communications professionals.

If I were willing, he said he had time to go ten rounds with me. I said, "let's go for it."

Round One: I asked him to speculate about worst-case scenarios, to hypothesize or guess about what might happen if the worst case would occur. He was happy to do so. Round one: mine. Advice: Stick to the facts; avoid speculating about highly unlikely worst cases or scenarios that lack firm evidence. Return to the safety of your key messages.

Round Two: I asked him if he was sure about the facts that he had provided to me about past events. He said "absolutely." Round two: mine. Advice. Stay away from absolutes. It takes only one exception to disprove an absolute and raises questions about trustworthiness. Avoid words such as never, nothing, every, all, and none.

Round Three: I asked him to comment on the character of those making allegations. He did, offering some negatives. Round Three: mine. Advice: Don't attack the character of an opponent or adversary – it may be necessary to question the facts or science, but not someone's character.

Round Four: I asked him what he thought others would say after hearing the story. He speculated on reactions. Round four: mine. Advice: Speak for others only after they have spoken for themselves.

Round Five: I asked a multipart question at a quick machine-gun pace. He tried to remember and answer all the parts. Round five: mine. Advice. Pick the question that you most want to answer and answer only that one.

Round Six: I interrupted him in the middle of a sentence. I accused him of being evasive and not answering my question. I repeated my question. He became annoyed, combative, and testy. Round six: mine. Advice: Stay calm and composed. Control your emotions and temper no matter what the reporter says. Journalists are not afraid to ask the same question multiple times, especially if they are not satisfied with the answer. You should be polite and repeat your answer and then bridge the interview to another topic.

Round Seven: I stated the key allegation and asked him if it was true. He responded by repeating the allegation. Round seven: mine. Advice: Respond to an allegation with as few words as possible without repeating the sensational or negative elements. Don't get caught offering a sound bite containing negative language (e.g., President Richard Nixon: "I'm not a crook."). Offer positive language or return to your key messages.

Round Eight: I followed up a question with a long pause. He filled the silence. Round eight: mine. Advice: Stay silent until the next question is asked.

Round Nine: I said what I thought he had said in response to a question and asked if I could use my wording. He said "fine." Round nine: mine. Advice: Stay in control. Say it the way you want it exactly to be said.

Round Ten: I suggested he was probably getting tired and that we should take a break. We chatted for a while about his family and the stress created by the situation on his family. I mentioned that he looked fit and asked if he worked out. He relaxed and shared personal details. Round ten: mine. Advice: Assume the reporter is always at work, including during "testing," and chatting before and after the interview, or getting "background" information. Although it may not be a deliberate trap, many spokespersons have been trapped by an "off-air," "background," or "off-the record" comment captured by a tape recorder, notepad, microphone, or camera. You own what you say.

He did the interview with the reporter the following week. He passed with flying colors.

12.8 Chapter Resources

Below are additional resources to expand on the content presented in this chapter.

Austin, L., and Jin, Y., eds. (2018). *Social Media and Crisis Communication*. 1st Edition. New York: Routledge.

Centers for Disease Control and Prevention (2019). *Crisis and Emergency Risk Communication*. Atlanta, GA: Centers for Disease Control and Prevention.

Coleman, C.L (1993). "The influence of mass media and interpersonal communication on societal and personal risk judgments." *Communications Research* 20(4):611–618.

Covello, V.T. (1993). *Pesticides and the Press: A Journalist's Guide to Reporting on Pesticide Issues*. New York: Columbia University (Center for Risk Communication) Accessed at: https://www.amazon.com/Pesticides-press-journalists-reporting-pesticide/dp/B0006F1206

Covello, V.T. (2011). *Developing an Emergency Risk Communication (ERC)/Joint Information Center (JIC) Plan for a Radiological Emergency*. Washington, DC: US Nuclear Regulatory Commission.

Covello, V.T., Sandman, P.M., and Slovic, P. (1988). *Risk Communication, Risk Statistics, And Risk Comparisons.* Washington, DC: Chemical Manufacturers Association, 1988.

Dunwoody, S. (1999). "Scientists, journalists, and the meaning of uncertainty," in *Communicating Uncertainty: Media Coverage of New and Controversial Science.* Mahwah, NJ: Lawrence Erlbaum Associates.

Dooley, K., and Corman, S. (2002). "The dynamics of electronic media coverage," in *Communication and Terrorism: Public and Media Responses to 9/11.* Cresskill, NJ: Hampton Press.

Dunwoody, S., and Scott, B. (1992). "Scientist as mass media sources." *Journalism Quarterly* 59:52–59.

Edelman, R. (2021). *Declaring Information Bankruptcy.* Chicago: Edelman Public Relations Co. Accessed at: https://www.edelman.com/trust/2021-trust-barometer/insights/declaring-information-bankruptcy

Freundenburg, W. (1996). "Media coverage of hazard events: analyzing the assumptions." *Risk Analysis* 16(1):31–42.

Friedman, S., Dunwoody, and Rogers, S., eds. (1999). *Communicating Uncertainty: Media Coverage of New and Controversial Science.* Mahwah, NJ: Lawrence Erlbaum Associates.

Charles, A., and Gavin, S. (2011). *The End of Journalism: News in the Twenty-First Century.* Bern, Switzerland: Peter Lang AG

Eisenman, D.P., Cordasco, K.M., Asch, S., Golden, J.F., and Glik, D. (2007). "Disaster planning and risk communication with vulnerable communities: lessons from Hurricane Katrina." *American Journal of Public Health* 97 Suppl 1:S109–15.

Glik, D.C. (2007). "Risk communication for public health emergencies." *Annual Review of Public Health* 28:33–54.

Greenberg, M.D., Sachsman, P., Sandman, and Salmone K. (1989). "Risk, drama and geography in coverage of environmental risk by network TV." *Journalism Quarterly* 66(2):267–276.

Hart, P.S., Chinn S., and Soroka S. (2020). "Politicization and polarization in COVID-19 news coverage." *Science Communication* 42(5):679–697.

Hartman, N. (2009). *The Media & You.* Washington, DC: National Public Health Information Coalition.

Hove, T., Paek H., Yoon M., and Jwa, B. (2015). "How newspapers represent environmental risk: The case of carcinogenic hazards in South Korea." *Journal of Risk Research* 18(10):1320–1336.

Hughes, E, Kitzinger J., and Murdock G. (2006). "Risk and the media," in *Risk in Social Science*, ed. Taylor-Gooby, P. and Zinn, J. Oxford: Oxford University Press, pp. 250–270.

Hyer, R., and Covello, V. (2005). *Effective Media Communication During Public Health Emergencies: A World Health Organization Handbook.* Geneva: World Health Organization.

Kasperson, J.X., Kasperson R.E., Pidgeon N., and Slovic P. (2003). "The social amplification of risk: Assessing fifteen years of research and theory," in *The Social Amplification of Risk*, eds. N. Pidgeon, R.E. Kasperson, and P. Slovic. Cambridge, UK: Cambridge University Press, pp. 13–46.

Crossref Google Scholar Kitzinger, J. (1999) "Researching risk and the media." *Health, Risk & Society* 1(1):55–69.

Kovach, B., and Rosenstiel, T. (2007). *The Elements of Journalism.* New York: Three Rivers Press/ Random House.

Lanouette, W. (1994). "Reporting on risk: Who decides what's news?" *Risk* 5, Summer:223–232.

Lewis, M. (2021). *The Premonition: A Pandemic Story.* New York: W.W. Norton and Co.

Lundgren, R.E., and McMakin, A.H. (2018). *Risk Communication: A Handbook for Communicating Environmental, Safety, and Health Risks.* Hoboken, New Jersey: IEEE/Wiley

McCarthy, M., Brennan M., Boer M.D., and Ritson, C. (2008). "Media risk communication—what was said by whom and how was it interpreted." *Journal of Risk Research* 11(3):375–394.

Mitchell, A., Jurkowitz M., Oliphant J.B., and Shearer E. (2021). Pew Research Center. *How Americans Navigated the News in 2020: A Tumultuous Year in Review.* Pew Research Center. Accessed at:

https://www.journalism.org/2021/02/22/how-americans-navigated-the-news-in-2020-a-tumultuous-year-in-review/

Mullin, S. (2003). "The anthrax attacks in New York City: The 'Giuliani press conference model' and other communication strategies that helped." *Journal of Health Communication* 8:15–16.

National Safety Council (2000). *Chemicals, the Press & the Public: A Journalists' Guide to Reporting on Chemicals in the Community*, Washington, DC: National Safety Council.

Natividad, I. (2020). *COVID-19 and the Media: The Role of Journalism in a Global Pandemic*. Berkeley, CA: University of California, Berkeley News. Accessed at: https://news.berkeley.edu/2020/05/06/covid-19-and-the-media-the-role-of-journalism-in-a-global-pandemic/

Neeley, L. (2014). "Risk communication and social media," in *Effective Risk Communication*, ed. Arvai, J., Rivers, L. New York: Earthscan.

Ophir, Y. (2018). "Coverage of epidemics in American newspapers through the lens of the crisis and emergency risk communication framework." *Health Security* 16:147–57.

Paek, H., and Hove T. (2017). "Risk perceptions and risk characteristics," in *Oxford Research Encyclopedia of Communication*. Oxford: Oxford University Press.

Peters, H. (1994). "Mass media as an information channel and public arena." *Risk* 5 Summer:241–250.

Peters, H. (1995). "The interaction of journalists and scientific experts: cooperation and conflict between two professional cultures." *Media, Culture, & Society* 17(1):31–48.

Pew Research Center (2020). *News Use Across Social Media Platforms in 2020*. Washington, D.C.: Pew Research Center.

Pidgeon, N., Kasperson R., and Slovic P. (eds.). (2003). *The Social Amplification of Risk*. London: Cambridge University Press.

Rogers, E.M. (2008). "Diffusion of news of the September 11 terrorist attacks," in *Crisis Communications: Lessons from September* 11. Lanham, MD: Rowman & Littlefield.

Ropeik, D. (2010). *How Risky Is It, Really?: Why Our Fears Don't Always Match the Facts*. New York: McGraw-Hill.

Rossmann, C., Meyer L., and Schulz P.J. (2018). "The mediated amplification of a crisis: Communicating the A/H1N1 pandemic in press releases and press coverage in Europe." *Risk Analysis* 38:357–375.

Rowe, G., Frewer, L. and Sjoberg, L. (2000). "Newspaper reporting of hazards in the UK and Sweden." *Public Understanding of Science* 9(1):59–78.

Russell, C. (1999). "Living can be hazardous to your health: how the news media cover cancer risks." *Journal of the National Cancer Institute Monographs* 1999 (25):167–170.

Sandman, P.M. (1987). "Risk communication: Facing public outrage." *EPA Journal* 2:21–22.

Sandman, P.(1989). "Hazard versus outrage in the public perception of risk," in *Effective Risk Communication: Contemporary Issues in Risk Analysis*, eds. Covello V.T., McCallum D., MT P. *Vol* 4. Boston, MA: Springer

Sandman, P. (1994). "Mass media and environmental risk." *Risk* 5:251–257.

Sandman, P.M., Sachsman D.B., Greenberg M.R., and Gochfield M. (1987). *Environmental Risk and the Press: An Exploratory Assessment*. New Brunswick, NJ: Transaction.

Slovic, P. (1987). "Perception of risk." *Science* 236:280–285.

Sellnow, T.L., Ulmer, R.R., Seeger, M.W., and Littlefield, R.S. (2009). *Effective Risk Communication: A Message-Centered Approach*. New York: Springer.

Singer, E., and Endreny, P. (1993). *Reporting on Risk: How the Mass Media Portray Accidents, Diseases, Disasters and Other Hazards*. New York: Russell Sage Foundation.

Strömbäck, J., Tsfati Y., Boomgaarden H., Damstra A., Lindgren E., Vliegenthart R., and Lindholm T. (2020). "News media trust and its impact on media use: toward a framework for future research." *Annals of the International Communication Association* 44(2):139–156.

Turow, J. (2020). *Media Today: Mass Communication in a Converging World (Seventh Edition)*.
New York: Routledge.

Ulmer, R.R., Sellnow T.L., and Seeger M.W. (2019). *Effective Crisis Communication: Moving From Crisis to Opportunity*. 4th Edition. Thousand Oaks, CA: Sage.

Walaski, P. (2011). *Risk and Crisis Communication*. Hoboken, NJ: Wiley.

Williams, E. (2021). *Can the Media Regain Trust?* Chicago: Edelman Public Relations Co. Accessed at:
https://www.edelman.com/trust/2021-trust-barometer/insights/can-media-regain-trust

Willis, W., and Okunade A. (1997). *Reporting on Risks: The Practice and Ethics of Health and Safety Communication*. Westport, CT: Praeger.

Endnotes

1 See, e.g., Maier, S. (2007). "Setting the record straight when the press errs, do corrections follow?" Journalism Practice 1(1):33–43.

2 Porlezza, C., and Russ-Mohl, S. (2012). "Getting the facts straight in a digital era: Journalistic accuracy and trustworthiness," in Rethinking Journalism, eds. Peters, C. and Broersma, M. (pp. 45–59). London: Routledge.

3 Urban, C. (1998). Why Newspaper Credibility Has Been Dropping. Reston, Virginia: American Society News Editors. See also: Maier, S. (2005). "Accuracy matters: A cross-market assessment of newspaper error and credibility." Journalism & Mass Communication Quarterly 82(3):533–551.

4 There is blurry line between mainstream media, journalism, documentary filmmaking, and nonfiction expose books. Documentary filmmaking, for example, is often fact finding combined with storytelling. Many documentary makers consider themselves to be journalists or at least grounded in the same principles. One key difference is that documentary filmmakers often have greater editorial control of their stories.

5 Hartman, N. (2009). *The Media and You*. Washington, DC: National Public Health Coalition. p. 2

6 International Federation of Journalists (IFJ) (2019). The IFJ Global Charter of Ethics for Journalists. Accessed at: https://www.ifj.org/who/rules-and-policy/global-charter-of-ethics-for-journalists.html

7 With the emergence of online news reporting and bloggers, many of these traditional constraints are becoming less important.

8 See, e.g., Campbell, J. (1968). *The Hero with a Thousand Faces*. Princeton, New Jersey: Princeton University Press. See also Vonnegut on the "Shape of Stories" (https://bigthink.com/high-culture/vonnegut-shapes/).

9 International Federation of Journalists (IFJ) (2019). The IFJ Global Charter of Ethics for Journalists. Accessed at: https://www.ifj.org/who/rules-and-policy/global-charter-of-ethics-for-journalists.html. Page 1

10 See, e.g., Yoshioka, T., and Maeda Y. (2020). "COVID-19 stigma induced by local government and media reporting in Japan: It's time to reconsider risk communication lessons from the Fukushima Daiichi Nuclear Disaster." *Journal of Epidemiology* 30(8):372–373.

11 See, e.g., Hughes, E, Kitzinger J., and Murdock G. (2006). "Risk and the media," in Risk in Social Science, eds. Taylor-Gooby, P. and Zinn, J. Oxford: Oxford University Press, pp. 250–271.

12 See, e.g., Ropeik, D. (2010). *How Risky Is It, Really? Why Our Fears Don't Always Match the Facts*. New York: McGraw-Hill.

13 See., e.g., Knobel, B. (2018). *The Watchdog Still Barks: How Accountability Reporting Evolved for the Digital Age (Donald McGannon Communication Research Center's Everett C. Parker Book Series)*. New York: Fordham University Press.

14 See, e.g., Holbert, R.L. (2005) "Back to basics: revisiting, resolving, and expanding some of the fundamental issues of political communication research. *Political Communication*." 22(4):511–514.

15 See, e.g., Noar, S.M., and Austin L. (2020) "(Mis)communicating about COVID-19: Insights from health and crisis communication." *Health Communication* 35:14:1735–1739; See also Yoshioka, T., and Maeda Y. (2020). "COVID-19 stigma induced by local government and media reporting in Japan: It's time to reconsider risk communication lessons from the Fukushima Daiichi Nuclear Disaster." *Journal of Epidemiology* 30(8):372–373.

16 See, e.g., Kasperson, J.X., Kasperson R.E., Pidgeon N., and Slovic P. (2003). "The social amplification of risk: Assessing fifteen years of research and theory," in *The Social Amplification of Risk*, eds. N. Pidgeon, R.E. Kasperson, and P. Slovic. Cambridge, UK: Cambridge University Press. See also Rossmann, C., Meyer L., and Schulz P.J. (2018). "The mediated amplification of a crisis: Communicating the A/H1N1 pandemic in press releases and press coverage in Europe." *Risk Analysis* 38:357–375. Accidents at nuclear power plants such as Three Mile Island, Chernobyl, are classic examples of amplification. Adverse impacts extended well beyond local harm, including decisions by several countries to entirely abandon nuclear power as a source of energy.

17 See, e.g., McCarthy, M., Brennan M., Boer M.D., and Ritson, C. (2008). "Media risk communication—What was said by whom and how was it interpreted." *Journal of Risk Research* 11(3):375–394.

18 Sandman, P.M., and Paden M. (1979). "The Inquirer goes for broke." *Columbia Journalism Review*, July/August 1979, pp. 48–49
(Sidebar article to "At Three Mile Island"). *Columbia Journalism Review*, 18 (July-August): 48-49. Accessed at: https://www.psandman.com/articles/inquirer.htm.

19 NPR (National Public Radio) (2020). *Ethics Handbook*. Accessed at: https://www.npr.org/about-npr/688405012/independence

20 Many government organizations try to overcome this problem by having a sign language translator and/or language translator attend media briefings.

21 See, e.g., Yoshioka, T., and Y. Maeda (2020). "COVID-19 Stigma Induced by Local Government and Media Reporting in Japan: It's Time to Reconsider Risk Communication Lessons From the Fukushima Daiichi Nuclear Disaster." *Journal of Epidemiology* 30(8):372–373.

22 See, e.g., Kohring, M., and Matthes, J. (2007). "Trust in news media." *Communication Research* 34(2):231–252; See also Gaziano, C., and McGrath, K. (1986). "Measuring the concept of credibility." *Journalism Quarterly* 63(3):451–462.

23 See, e.g., Strömbäck, J., Tsfati Y., Boomgaarden H., Damstra A., Lindgren E., Vliegenthart R., and Lindholm T. (2020). "News media trust and its impact on media use: toward a framework for future research." *Annals of the International Communication Association* 44(2):139–156.

24 Edelman, R. (2021). *Declaring Information Bankruptcy*. Chicago: Edelman Public Relations Co. Accessed at: https://www.edelman.com/trust/2021-trust-barometer/insights/declaring-information-bankruptcy; See also *Edelman Trust Barometer* (2021). Accessed at: https://www.edelman.com/trust/2021-trust-barometer. For over 21 years, the public-relations firm Edelman has analyzed public trust in major institutions. They currently survey more than 33,000 people in 28 countries.

25 Williams, E. (2021). "Can the Media Regain Trust? Chicago: Edelman Public Relations Co. Accessed at: https://www.edelman.com/trust/2021-trust-barometer/insights/can-media-regain-trust

26 See, e.g., Campbell, J. (2008). *Hero with a Thousand Faces*. Novato, CA: New World Library.

27 See, e.g., Dykes, B. (2020). *Effective Data Storytelling: How to Drive Change with Data, Narrative and Visuals*. Hoboken, NJ: Wiley

28 See, e.g., Hyer and Covello (2005) for lists of frequently asked questions in crises, including the "77 Most Frequently Asked Questions by Journalists in an Emergency." Accessed at: https://www. who.int/csr/resources/publications/WHO%20MEDIA%20FIELD%20GUIDE.pdf?ua=1, Pp. 7-8

29 CDC H1N1 News Conference, 24 April 2009.

30 Accessed at: https://www.marktwainproject.org/xtf/view?docId=letters/UCCL00617. xml;style=letter;brand=mtp

31 See, e.g., https://www.calhospitalprepare.org/sites/main/files/file-attachments/pandemic_pre_ event_maps.pdf

32 See, e.g., National Public Radio (2011) *The Incredible Shrinking Soundbite*. Accessed at: https:// www.npr.org/2011/01/05/132671410/Congressional-Sound-Bites.

33 See, e.g., Hyer, R., and Covello V. (2005). *Effective Media Communication During Public Health Emergencies: A World Health Organization Handbook*. Geneva, Switzerland: World Health Organization; See also Centers for Disease Control and Prevention (2019). *Crisis and Emergency Risk Communication*. Atlanta, GA: Centers for Disease Control and Prevention.

34 See, e.g., Charles, A., and Gavin, S. (2011). *The End of Journalism: News in the Twenty-First Century*. Bern, Switzerland: Peter Lang AG.

35 For example, at the beginning of the pandemic, journalists reported the advice of government experts that Americans should not wear face masks unless they were sick. On February 29, 2020, the US Surgeon General tweeted: "Seriously people- STOP BUYING MASKS! They are NOT effective in preventing general public from catching #Coronavirus but if healthcare providers can't get them to care for sick patients, it puts them and our communities at risk!" As supplies improved and scientists learned more, this advice changed. On 3 April 2020, the CDC began recommending the use of cloth face masks in public settings where physical distancing was hard to maintain. See also M. Lewis (2021). *The Premonition: A Pandemic Story*. New York: Norton.

36 Shearer, E. (2021). *More than Eight-in-Ten Americans Get News from Digital Devices*. Arlington, VA: Pew Research Center. Accessed at: https://www.pewresearch.org/fact-tank/2021/01/12/ more-than-eight-in-ten-americans-get-news-from-digital-device

37 Shearer, E., and Griecos E. (2020). *Americans Are Wary of the Role Social Media Sites Play in Delivering the News*. Arlington, VA: Pew Research Center. Accessed at: https://www.journalism. org/2019/10/02/americans-are-wary-of-the-role-social-media-sites-play-in-delivering-the-news/; See also Pew Research Center (2020). "News use across social media platforms in 2020." Accessed at: https://www.journalism.org/2021/01/12/news-use-across-social-media-platforms-in-2020/

38 See, e.g. https://www.cnn.com/2014/01/15/tech/hudson-landing-twitpic-krums/index.html

13

Social Media and the Changing Landscape for Risk, High Concern, and Crisis Communication

---

CHAPTER OBJECTIVES

This chapter describes the role of social media in sharing information and considers both the benefits and challenges of that role for risk, high concern, and crisis communication. Through case studies, a review of the evidence-based literature and practical guidance, the chapter enables you to consider how to incorporate social media into the design and application of communication plans.

At the end of this chapter, you will be able to:

- Describe the multiple critical roles that social media play in risk, high concern, and crisis communication.
- Discuss how to take social media into account in developing a communication plan.
- Apply valuable tactics for effectively using social media in risk, high concern, and crisis situations.

13.1   Case Diary: Myth-Busting: Mission Impossible?

It was an unusual request. A colleague I had worked with in the past – his joking tone belying the seriousness of the need – said: "Your mission, should you choose to accept it, is to be part of a global COVID-19 Social Media MythBusters Group." The World Health Organization (WHO) had initiated the idea for groups such as this. I quickly agreed and found myself working with a group of scientists, communication professionals, and savvy young social media specialists and influencers.

COVID-19 was the first pandemic in history in which technology and social media were being employed on a massive scale. What had begun as a trickle of misinformation about COVID-19 in January 2020 had turned into an out-of-control storm by March. It was becoming increasingly difficult for people to determine fact from fiction. Some of this misinformation was extreme in its claims. Conspiracy theorists accused the Bill and Melinda Gates Foundation, which had given millions of dollars to coronavirus vaccine and treatment research, of wanting to use vaccines to implant tracking devices in people.[1] *Plandemic: The Hidden Agenda Behind Covid-19,* a 26-minute video uploaded to multiple social media sites on 4 May 2020, echoed the Gates conspiracy story, accused leading public health scientists of burying research, and advised people that wearing a

Communicating in Risk, Crisis, and High Stress Situations: Evidence-Based Strategies and Practice, First Edition.
Vincent T. Covello.
© 2022 by The Institute of Electrical and Electronics Engineers, Inc. Published 2022 by John Wiley & Sons, Inc.

face mask would activate the COVID-19 virus.[2] Despite being taken down by YouTube, Facebook, and other social media sites, the video went viral[3] and garnered millions of views.[4]

It seemed like an impossible mission to counter the endless blasts of misinformation and the expanding *infodemic,* an epidemic of misinformation. According to the World Health Organization, an *infodemic* "is an overabundance of information, both online and offline. It includes deliberate attempts to disseminate wrong information to undermine the public health response and advance alternative agendas of groups or individuals."[5]

The usual combatants against an *infodemic* were weakened by distrust. For example, in the United States, people were becoming increasingly distrustful of information about COVID-19 provided by federal agencies and the White House, in part because of a perceived politicization of information about public health. Distrust in the WHO's social media postings was also growing, even extending to WHO's excellent myth-busting website. WHO had been criticized in mainstream and social media for publicly praising China for what WHO called a speedy response to the new coronavirus. WHO had thanked the Chinese government for sharing the genetic map of the virus immediately and said the Chinese government's work and commitment to transparency were impressive. Later reports showed China had not been transparent when the disease first appeared. Disagreement with WHO's handling of this matter had translated into distrust of many of their messages.

Keeping up with social media postings about COVID-19 was the core challenge faced by the myth-busting group. Another monumental task for us was keeping up with the massive amount of information appearing in the scientific literature every day and translating it into content for social media. Scientific knowledge and understanding of COVID-19 were changing and growing exponentially as researchers rapidly rose to meet the urgent need to understand the virus and its spread.

Scientists were discovering new information about COVID-19 every day. These new discoveries were critical to share, but the continual change in information created a landscape ripe for the spread of rumors and misinformation. Separating the scientific wheat from the chaff often required hours of reading published journal articles and assessing the thousands of scientific reports that appeared – in the interest of rapid data sharing – before peer review. It required the endless cross-checking of facts. We had to draw on a wide network of trusted experts to clarify information.

Members of our myth-busting group produced what we believed was high-quality information content for the most visited social media platforms in the world. These platforms included Facebook, Twitter, YouTube, WhatsApp, Instagram, WeChat, Tumblr, TikTok, Reddit, Twitter, Pinterest, Snapchat, Tencent, Myspace, and Weibo, and other online forums. A large proportion of the traffic was originating on Facebook, Instagram, Twitter, YouTube, and WhatsApp. For content, we drew heavily on COVID-19 information produced by the World Health Organization, the US Centers for Disease Control and Prevention, the Johns Hopkins University School of Medicine, and the peer-reviewed compendium of COVID-19 knowledge published by the US Association of State and Territorial Health Officers.[6]

Initially, many in our social media group felt as though we were just yelling into a black hole and should give up. We and other groups working on the *infodemic* received a bit of a break and recharge when several search engines and social media platforms, including Google, Facebook, Twitter, TikTok, and YouTube, agreed to filter out glaring misinformation.

We contacted and asked for assistance from celebrities (and received many refusals along with help). We contacted and asked for assistance from health and science *social media influencers* – social media users who have established credibility on health and science issues and who have access to large audiences of social media followers. We spent hours searching through "social media influencer" databases in a quest for the right influencers. Social media influencer databases helped us in several ways. First, they helped us identify influencers who had a good fit with our mission.

Second, social media databases helped us find the most followed of social media influencers and eliminate people who simply had a public account on a social media platform. Third, social media databases helped us monitor the performance of our social influencers and to track metrics for each influencer. These metrics included the number of postings the influencer made and the number received. We also monitored postings received from the followers of the influencer.

We consulted with experts in social media marketing. Several of our group took crash courses in social media tools and technologies such as Google Ads, Google Analytics, Facebook Ads Manager, WordPress, and Adobe Illustrator. Several of us climbed the rigging and learned the basic ropes of social media strategy, social media campaign development, user acquisition, digital advertising, content marketing, retention strategy, and SEO (Search Engine Optimization). Members of our group quickly developed sophisticated SEO skills, including the ability to create keywords and content that attracted searchers and search engines; to create interesting content that effectively answered the searcher's query; to create content that gave users a positive experience; to create content that earned multiple links and citations; and to create content that had a fast-loading speed. Content included infographics, animations, videos, photos, charts, Q&A fact sheets, quizzes, and live interviews. Besides debunking myths, the social media content repeated over and over again our mantras: Wash your hands. Cover your cough or sneeze. Do not touch your face. Keep your distance. Avoid large social gatherings. Hands, Face, Space.

By the summer of 2020, our monitoring and evaluation activities indicated that our efforts, the efforts of the WHO *infodemic* group, and the work of dozens of other organizations, were successful in plugging many of the holes opened by the spread of COVID-19 myths and misinformation. We and the other groups evaluated success using standard social media performance measures, such as site traffic, "likes," audience growth rate, and user engagement, such as followers, comments, and repostings. We took advantage of the analytic packages available on the social networking sites to determine the number of people engaged and how they engaged. For example, we used Facebook Insights to see demographic information and follow interactions over time. We tracked the amount of traffic being driven to the WHO and the CDC websites on COVID-19. We conducted online surveys through tools such as SurveyMonkey to measure user satisfaction and increases in knowledge.

We toasted (with nonalcoholic drinks for the invaluable underage members of our team) our partners and social media influencers. The job was never ending, but the essential processes were in place and the worst of the *infodemic* appeared to be subsiding.

13.2 Introduction

Social media platforms have radically changed the communication landscape. They have transformed how risk, high concern, and crisis information are collected, distributed, viewed, and shared. Social media platforms have changed how risk managers, technical professionals, and communicators strategize. Risk managers and communicators are rapidly adjusting to the reality that social media platforms have become dominant players in the world of communication.

Social media platforms have also changed the interests of many risks, high concern, and crisis communication researchers. For example, at the beginning of 2021, there were over one million entries in Google Scholar for the keywords "social media and crisis communication." Just 10 years ago, there were little to no scholarly articles on this topic.

Social media platforms have radically sped up the ability to communicate with internal and external stakeholders. They have radically sped up the expectation of rapid communication with internal and external stakeholders. They have forced organizations to supplement their traditional methods for communication and add new guidelines, policies, processes, training, and security to address social media. As organizations address issues such as transparency, right to know, information overload, and evaluation, they must increasingly do so in a pervasive social media environment. The use of social media has highlighted more than ever the necessity of stakeholder engagement. Real-time input from stakeholders has become achievable in many risk, high concern, and crisis situations. Stakeholder engagement has always been needed. However, social media allows organizations opportunities to communicate directly to individuals and groups with whom the previous contact was difficult, if not impossible.

Over the past decade, social media platforms have become a primary source of information for large segments of the public on virtually any issue or topic.[7] For example, more than half of all Americans now receive some portion of their daily news through social media; large numbers of people regularly seek news from two or more social media platforms. The Pew Research Center began tracking social media adoption in 2005. At that time, only 5% of American adults used at least one social media platform. By 2011, that share had increased to half of all Americans. By 2021, seven in ten Americans were using some type of social media.

Pew Center research also indicates that the social media user base has grown more representative of the broader population. Young adults were among the earliest social media adopters and continue to use these sites at high levels.[8] However, usage by older adults and lower-income populations has greatly increased. For example, more than three in ten of those 65 and older report using social media, and more than 50% of those living in the lowest-income households now use social media.

YouTube and Facebook have dominated the social media landscape. Eight in ten Americans report using Youtube, and seven in ten Americans report using Facebook. Four in ten adults report using Instagram and approximately three in ten report using Pinterest or LinkedIn. One-quarter of all Americans report using Snapchat, Twitter, or WhatsApp, and one-fifth of all Americans report using TikTok. More than one in ten Americans report using the neighborhood-focused social media platform Nextdoor.

Younger Americans, especially those who are ages 18 to 24, use a wide variety of social media platforms and use them frequently. Nearly four-fifths of 18- to 24-year-olds use Snapchat, and a majority of these users visit the platform multiple times per day. Similarly, nearly three-quarters of young Americans use Instagram, and close to half are Twitter users. As of the first quarter of 2021, Facebook had nearly a billion active users. As of the first quarter of 2021, Twitter had over 350 million active users.

While the use of social media is especially prominent among younger people, use has also increased among senior citizens, ethnic minorities, rural residents, and individuals from low-income households. This rapid expansion of social media usage has revolutionized communications and information seeking. It has made electronic communication possible in parts of the world that did not have the land-based infrastructure, such as telephones and cable lines. The restrictions on in-person gatherings during the COVID-19 pandemic further extended the breadth and extent of social media use in 2020 and beyond.

The enormous growth of social media is evident not only in the United States but globally. For example, in China, over 800 million people have registered Weibo accounts. This includes over 175,000 governmental accounts.[9] WeChat is the biggest social media platform in China; as many as one billion users open their WeChat accounts daily. The average number of contacts on each WeChat account is doubling every three years. Sixty-three million users aged above 55 years old open their WeChat account at least once a month.[10]

These trends in social media usage are especially evident in crises. Social media use often surges during crises. Within minutes of an event, people search for answers about who, what, where, when, why, and how. Almost simultaneously with an incident, people will be asking: What happened? What caused it to happen? What does it mean? Who is to blame? What is being done to ensure it does not happen again? As examples, there were millions of event-related Tweets following the Haiti earthquake (2010), the Japanese Earthquake, Tsunami, and Nuclear Power Plant Disaster (2011), Superstorm Sandy in the U.S. (2012), the Ebola Outbreaks in Africa (2014–2016), the Brazilian Zika Outbreak (2016), Hurricane Maria in Puerto Rico (2017), the COVID-19 Pandemic (2020–2021), and the 2021 storming of the U.S. Capitol. Crises highlight the importance of understanding the use of social media around the world. For example, global issues such as the COVID-19 pandemic and climate change spotlight the need for consistent messaging around the world.

As people become more reliant on social media in risk, high concern, and crisis situations, public and private sector organizations are increasingly using social media platforms to communicate. Expectations to find information on social media have also increased, resulting in heightened pressure on public and private sector organizations to use social media, and use it effectively. For example, a study conducted by the American Red Cross as long ago as 2012 found that over two-thirds of Americans believe first response organizations should actively take part in social media. Three-quarters of Americans said they expect timely responses through social media from crisis communicators when posting requests for help. Many organizations engaged in health, safety, and environmental communication, such as the World Health Organization (WHO) at the global level, and the Centers for Disease Control and Prevention (CDC) at the national level, are increasingly using social media to communicate crisis and risk-related information to the public. Social media strategies are being integrated into risk, high concern, and crisis communication plans.[11]

Another emerging trend related to social media is the expectation that risk and crisis management authorities will respond to calls for help via social media. This was especially evident, for example, during the 2017 floods in Houston following Hurricane Harvey.[12] Risk and crisis management organizations increasingly recognize the importance of establishing and maintaining an active and engaging social media presence. Especially in crises, social media have become a primary channel of communication. Social media platforms allow critical information and messages to reach broader populations faster and with higher impact than ever before. However, organizations are discovering the negative impacts of this, especially when there is little ability to manage the flow of information or manage the expectations of stakeholders, such as the people wanting help. People are also using social media to advocate for what they want. For example, if they are upset with the performance of an organization, they will create a post with a hashtag about their complaint. This is a benefit for stakeholders but a challenge for organizations.

13.3 Benefits of Social Media Outlets for Risk, High Concern, and Crisis Communication

Several specific benefits flow from the use of social media for risk, high concern, and crisis communication (Table 13.1).

13.3.1 Speed

Social media outlets increase the speed at which information about a risk, high concern, or crisis situation or event is distributed.

Table 13.1 Examples of benefits of social media outlets for risk, high concern, and crisis communication.

1) *Speed*
2) *Access to information*
3) *Reach*
4) *Amplification*
5) *Transparency*
6) *Understanding*
7) *Changes in behaviors*
8) *Relationship building*
9) *Timeliness*
10) *Hyperlocal specificity*
11) *Listening and feedback*

13.3.2 Access

Because most social media networks are free or low in cost, social media outlets allow easy access by stakeholders to information. It allows organizations and members of the general public to create content and redistribute risk, high concern, and crisis information.

13.3.3 Reach

Social media outlets create opportunities to reach and connect with large numbers of people. They open doors to the general public as well as targeted audiences.

13.3.4 Amplification

Social media outlets amplify messages. Information posted on a social media site about a risk, high concern, or crisis situation can create ripples like a stone thrown in a pond. Social media users can easily engage large numbers of others in conversations about risks and threats.

13.3.5 Transparency

Transparency means operating in such a way that it is easy for others to view the decision-making process. If used well, social media outlets can enhance openness, collaboration, public participation, and accountability. For example, in politics, transparency is used as a means of holding public officials accountable. When town hall meetings are open to the public through the Internet and social media, the words and actions of government officials related to a risk or high concern issue can be reviewed and discussed by large audiences. Arguments for and against a proposal, final decisions, and the decision-making process itself can be made public and remain publicly archived. Social media platforms that allow live posting are also a way for organizations to demonstrate transparency and avoid mis-interpretation of events done through mainstream media.

13.3.6 Understanding

Social media outlets help organizations and researchers gain insights and deepen their understanding of knowledge, attitudes, and beliefs about susceptibility to harm, the severity of harm, the benefits of preventive and protective actions, and barriers to performing preventive and protective actions. Knowledge, attitudes, and beliefs affect perceptions of probabilities, potential adverse consequences, trust, and personal control, which in turn affect how seriously a person takes a risk or threat and actions they make take. Social media outlets also help organizations and researchers better understand how communication networks influence risk-related judgments and decision-making. Such understanding is critical to producing effective risk, high concern, and crisis communications. Social media engagement also encourages users to create visual content to enhance interest, understanding, and relevance. Visual content includes drawings, graphs, diagrams, symbols, maps, photos, pictures, displays, models, videos, memes, infographics, and storytelling. Visual content can address core challenges of risk, high concern, and crisis communication: limited attention span and limited recall of content; when done well, visual content is grasped quickly.

13.3.7 Changes in Behaviors

Social media platforms can have a major impact on risk-related behaviors. For example, platforms such as Facebook and Twitter have been used successfully to promote healthy behaviors, such as nutrition, lifestyles, and physical activity.[13] Social media platforms have also been used successfully to promote healthy behaviors in vulnerable populations, including low-income populations, those living in rural areas, and minority ethnic groups.

13.3.8 Relationship Building

Social media outlets allow individuals and organizations to link, share ideas, and collaborate with like-minded people.

13.3.9 Timeliness

In disasters and emergencies, social media outlets can reach more people through timely information, alerts, warnings, and risk-related messages.[14] There are times during a disaster or emergency when mainstream media networks cannot function. This leaves social media outlets as one of the few available sources of information to stakeholders. When telephone lines are down, social media outlets provide a connection between families seeking news about their loved ones' status. For example, following the massive infrastructure damage in Puerto Rico caused by Hurricane Maria in 2017 – the worst natural disaster on record to affect the island – only one radio station was operating. Similarly, during the massive wildfire of 2011 that affected the Alberta community of Slave Lake, one of the first critical pieces of infrastructure to burn was the town radio station tower. With the primary mainstream means of community messaging rendered inoperable, the community took to social media. Since social media postings about a disaster or emergency are now virtually instantaneous and simultaneous with the event, and since social media sites are fast becoming the go-to place for information, it is critical for crisis managers and communicators to be where key stakeholders – be they employees, employee families, directors, contractors, clients, shareholders, unions, the general public, the media, first responders, government agencies, nongovernmental organizations, community organizations, partner organizations, or others – gain, share, and discuss information.

13.3.10 Hyperlocal Specificity

Within emergency management, social media outlets allow community members to post geographically identified photos and videos captured on smartphones. This content can then be used to create crisis maps, which display social media content by location. Disaster and emergency managers can use social media outlets such as Facebook Live and Twitter to track events as they are happening. Posts using Geospatial Information Systems (GIS) and free online sources such as Google Earth and Google Maps with aggregate data can plot on a map what and where a crisis or event is occurring. For example, during the highly destructive 2012 Hurricane Sandy, Google maps posted key information spatially, such as the location of open gas stations, evacuation routes, and places where people could charge their electronic devices. In this way, social media can become integral to the broader task of crisis management.

13.3.11 Listening and Feedback

Risk, high concern, and crisis communicators can use social media outlets to both push out information and pull in feedback. Social media platforms provide organizations a low-cost mechanism for two-way communication and listening – a way to identify stakeholder questions and concerns, awareness, and knowledge levels. They provide organizations an additional means for gaining feedback from stakeholders on the extent to which key messages are resonating with or reaching them – and thus, whether revisions to the current messaging strategy may be needed. Feedback allows crisis managers and communicators to listen better, engage in ongoing collaborative communication with stakeholders, and make mid-course corrections.

13.3.12 Taking Advantage of the Benefits of Social Media

Adapting to and taking advantage of the benefits described above is essential for crisis communication strategizing. In response to this change, many organizations, and many mainstream broadcast and print media organizations, now have a strong social media presence. Even mainstream media reporters covering an event do not wait for the evening news. They will post snippets of their report while it is being created ahead of the airing deadline.

In crisis situations, social media sites that focus on information content sharing in many forms – such as Facebook, Instagram, Twitter, TikTok, YouTube, Twitter, Pinterest, Snapchat, Periscope, Flickr, WhatsApp, Nextdoor, many Wikis, Chat Rooms, Message Boards, blogs, and podcasts – can be used beneficially to enhance situational awareness in real time through the exchange of information through text, audio, pictures, and videos. Such platforms allow organizations to quickly launch large-scale risk and crisis communication campaigns. These platforms can also post requests for support and help identify missing individuals, survivors, and victims.

In a crisis, social media platforms focused on social networking, such as Facebook, Twitter, Instagram, TikTok, and Snapchat, can provide a place where people can quickly go to share and exchange information about alerts, warnings, updates, and requests for help. If information on a social media platform is critical or could be misunderstood, users can be referred to a trusted website containing more detailed information. For example, during the COVID-19 pandemic, social media users in the United States looking for more detailed answers to their questions were often directed to websites such as those of the Association of State and Territorial Health Officers (ASTHO), the Centers for Disease Control and Prevention (CDC), the World Health Organization (WHO), Harvard Health, John Hopkins University School of Medicine, and the Mayo Clinic. Social media users were also frequently alerted to how to spot false information and check the accuracy of rumors, such as by using only official government or health care websites; investigating links and quoted sources of information for reliability; and searching multiple trusted sites to see if they are sharing the same or similar information.

Social media platforms create opportunities during a crisis to consider "paid ads." For example, Facebook allows organizational subscribers to target a certain geographic area. If one or two communities are facing a crisis, crisis management authorities can target information specifically at those communities.[15]

Perhaps most importantly, social media networking sites allow people to connect with and support one another in a crisis. During a disaster, people frequently turn to others when deciding how to evacuate from a hurricane, flood, wildfire, earthquake, or tornado.[16] The average person typically checks with four to five sources before reaching important decisions during a crisis.

Social media use is changing and expanding social networks, changing the sources of information people use, and changing how people reach decisions. Social media networking sites also help people fulfill their need for comfort, support, and human connections in a crisis. In high-stress, fearful, and emotionally charged situations, people often seek out, lean upon, and depend on the support of others.

13.4 Challenges of Social Media for Risk, High Concern, and Crisis Communication

Along with its many benefits, the social media revolution also poses unique challenges for risk, high concern, and crisis communicators. These are listed in Table 13.2 and described below.

13.4.1 Rising Expectations

Organizations must keep pace with stakeholders' expectations regarding the use of social media. People increasingly demand immediate information. These rising expectations have resulted in an "expectation gap" between organizations and their audiences. It is forcing many organizations to invest internal resources to create an active social media presence.

13.4.2 Repostings/Redistribution

Social media messages are often reposted and redistributed by users in many forms, such as from Tweets, Google+ and Facebook to YouTube, Instagram, Reddit, TikTok, WhatsApp, and Snapchat. This is a benefit for wide reach but also requires that in the world of digital

Table 13.2 Social media challenges.

---

1) Rising expectations
2) Repostings/redistribution
3) Permanent storage
4) Hacking/security
5) Rise and fall of social media platforms
6) Resources
7) Privacy and confidentiality
8) Cognitive overload
9) Players on the field
10) Misinformation, disinformation, and rumors

---

communication, messages must be crafted with expectations of their reuse. Every social media message can be repurposed and recontextualized multiple times; the shorter and clearer a message is, with embedded sourcing and time stamping, the less it will deteriorate in meaning through reuse. Critical information posted on a social media platform needs to be time stamped. If a social media post is not time stamped, someone may see that post hours later and share it. This makes the post seem "current" despite the information being hours old and possibly revised. One strategy used by social media professionals to address this issue is to take pictures or screenshots of a media release or a blog post and attach it as an image with a time stamp embedded in the post.

13.4.3 Permanent Storage

Information posted on social media platforms may be stored for as long as there is access to the site on which it was first displayed. It is hard to refute or recover from a negative story or misinformation posted on a social media site, and even harder to delete the information. As a result, risk, high concern, and crisis communicators need to exercise great care before posting content, including vetting by others and testing. Even if a post is deleted, it is likely that screen captures still exist.

13.4.4 Hacking/Security

Social media sites are highly vulnerable to hacking and security breaches, including breaches that insert false or doctored information. Hacking is nearly at epidemic levels. For example, disinformation is widely spread through multiple routes, including false or misleading news stories, *trolls* (people who deliberately post lies on social media platforms), *trick search algorithms* (invisible links to target sites), *cyberbullies* that target specific individuals or groups, and social media *bots* (automated fake users). Message content needs to be protected with state-of-the-art security practices, just as protection is employed for confidential information.

13.4.5 Rise and Fall of Social Media Platforms

Social media platforms rise and fall rapidly. Continual effort is needed to track the growing number of social media platforms and near-constant changes in their popularity and numbers of followers. As a result, risk, high concern, and crisis communicators need to continually evaluate and prioritize social media platforms, calculate resources and time available for social media engagement, and prioritize attention and resources on social media sites used by target audiences.

13.4.6 Resources

Monitoring social media content can be a full-time activity. Substantial resources are needed to share content with stakeholders, let alone interact with them. *Organizations need to prioritize and budget resources carefully.*

13.4.7 Privacy and Confidentiality

Information shared through social media sites often violates privacy standards and expectations of confidentiality. Social media privacy and confidentiality mean the ability to control (1) interactions with others, and (2) who gathers and disseminates information about oneself or one's group and under what circumstances.[17] Individuals and organizations use multiple techniques to enhance privacy and confidentiality. In a crisis, it is critical that employees be reminded of the organization's social media policies; that what employees post on their social media platforms becomes part of the public and legal record; that journalists may scan social media sites of employees for information about the crisis; that persons with negative intentions, such as terrorists, may scan the social media postings of employees for insider information; and that prosecutors have used risk-related postings of employees on social media platforms in civil and criminal filings.

13.4.8 Cognitive Overload

Search engines and social media platforms such as YouTube provide users direct access to an unprecedented amount of content about risks and threats. However, the proliferation of information from social media sites, and easy access to such information, can also create a cognitive overload.[18] Cognitive overload occurs when a person's working memory receives more information than it can handle comfortably. As information about risks proliferates through social media, the brain's processing abilities can create bottlenecks, favoring the use of mental shortcuts, information that is the most easily accessed, or information that is consistent with, or conforms to, existing beliefs about susceptibility (i.e. vulnerability to the risk), severity, benefits of protective measures, and barriers to the adoption of protective measures. As a result, information about a risk or threat shared through social media channels is typically more effective when it is presented in tiers, with each tier increasing in content and complexity.

13.4.9 Players on the Field

One of the greatest challenges raised by social media for risk, high concern, and crisis communicators is the much broader range of players that can now share and exchange information about risks, threats, high concern, and emotionally charged issues. This exponential widening of the playing field creates immense opportunities but also immense dangers, including concerns about censorship, monopolies, privacy, biases, disinformation, rumors, and campaigns aimed at defaming, dividing, discrediting, and distracting. Awareness of the complexity of this environment and strategies to monitor the social media environment are important as social media communication plans are developed.

13.4.10 Misinformation, Disinformation, and Rumors

Social media platforms are a breeding ground for *misinformation* – false information spread unintentionally, *disinformation* – false information spread with the intent to do harm, and *rumors* – unverified information that can be true or false. The spread of misinformation, disinformation, and rumors through social media platforms typically becomes greater when there is controversy, confusion, or mistrust. *Fake news* – news that is untrue and disseminated under the guise of news reporting – is a particular problem for both social media or mainstream media outlets.[19] According to the Pew Research Center, approximately two-thirds of all adult Americans say fake news causes them a great deal of confusion.[20] Approximately one-quarter of all adult Americans report they have shared a made-up news story – either knowingly or not. Unfortunately, people often believe and share news that confirms their personal beliefs and ignore or discount news that is contrary to their personal beliefs.

False information can quickly go viral through the Internet and social media outlets. Two of the most egregious examples of false information spread widely initially by Facebook and other social media platforms was the claim that Pope Francis had endorsed the candidacy of Donald Trump in the 2016 US presidential election[21] and the bizarre "pizzagate" story that became viral and led a North Carolina man to bring a gun into a popular Washington, D.C. pizza restaurant under the impression that the restaurant was hiding a child prostitution ring. As a result of a tsunami of disinformation spread through social media platforms, important sites need to be continually monitored for misinformation, disinformation, and rumors. If found, they need to be addressed immediately. Content on the organization's website and social media platforms also needs to be regularly updated to prevent speculation. To efficiently respond to social media misinformation and disinformation, managers and communicators need to decide which social media platforms are most important, which platforms exert the most influence on key stakeholders, and which resources (e.g. staff and time) can be devoted to the problem.

Of the challenges described above, the spread of false information by social media platforms is perhaps the most serious. Table 13.3 contains examples of best practices for detecting fake and misleading news.[22]

There is increasing global concern about the proliferation of false or misleading information about risks and high concern issues. To understand better how quickly false information spreads, researchers examined over 126,000 rumors spread by approximately three million people on Twitter from 2006 to 2017.[23] False information reached more people than the truth; lies spread exponentially faster than truth; false information found many more followers than accurate facts. False information was 70% more likely to be retweeted on Twitter than accurate information. False information occurred in virtually all domains, from terrorism and natural disasters to financial information and science. Public and private sector organizations have mounted efforts to combat the spread of false information. Solutions include removing false information from a social media site; labeling false information in much the same way as food is labeled; elevating the search engine ranking of authoritative sites; redirecting advertising money away from social media platforms that spread false information; identifying fake accounts; developing a rumor management/rumor control site to correct misinformation; and suspending or banning social media accounts of people who post false information – recognizing there may be concerns about the trustworthiness of the policing sites. Effective solutions may have to wait for better evidence-based understanding on issues such as how false information spreads and why people spread false information.

Table 13.3 Detecting fake or misleading social media content.

1) Authenticate the domain site, logo, contact information, and office location
2) Check the accuracy, date, and source of quotes
3) Check the publishing date
4) Check links and references for accuracy
5) Check the authenticity of photos
6) Check for the presence of ads for commercial products
7) Check for headlines that do not reflect facts in the story
8) Investigate the reputation of the author or posting organization
9) Check for criticism from reputable sources

13.5   Case Study: Social Media and the 2007 and 2011 Shooter Incidents at Virginia Polytechnic Institute and State University (Virginia Tech)

In April 2007, Virginia Polytechnic Institute and State University (Virginia Tech), a university campus of nearly 30,000 students, was the site of one of the deadliest shooting rampages in US history.[24] Twenty-three-year-old Seung-Hui Cho killed over 30 people and wounded 17 on the Virginia Tech campus in Blacksburg, Virginia. Six other students were injured when they jumped from windows to escape. As police stormed the building where Cho was positioned, Cho shot himself in the head with a pistol, dying instantly. Cho used semiautomatic pistols and ammunition he purchased online from an out-of-state dealer and a pawnshop. He was able to do so despite the fact that he had previously been diagnosed with a severe anxiety disorder.

University officials were criticized for failing to take actions, including communication actions that might have reduced the number of casualties. While campus police were investigating Cho's initial shooting in a dorm, Cho set out on his plan to shoot individuals in academic buildings by first changing his clothes, then heading to the campus post office to drop a video manifesto on his mass-murder intentions in the mail. He carried chains in his backpack in order to lock campus building doors from the inside, which he did immediately after entering the main academic building.

Campus police and Virginia Tech officials were alerted minutes after the first shooting in the dorm took place. University officials concluded the incident was domestic in nature, that the prime suspect had been identified, that the suspect was no longer a threat to campus, and there was nothing more to do or communicate. A key lesson of crisis communication was ignored: *Share the information you have with stakeholders at the time you have it.*

Campus police did not have the authority to send out an alert themselves. All messages went through university officials. University officials were unfortunately overly reticent about sending out an alert. This was due in part because the university had been criticized months earlier for overreacting to reports of an intruder on campus by calling in a SWAT team, which then mistakenly stormed the student center.

The active shooter attack received widespread media coverage and intense debate about gun violence, gaps in health systems for treating the mentally ill, operational and communication responsibilities of college administrations, and the limits of privacy and confidentially laws. The attack also raised debate about communication coverage by mainstream news organizations during crises. For example, mainstream print and broadcast news organizations had aired portions of the Cho's manifesto. Publication of the manifesto by news organizations was criticized by victims' families, law enforcement officials, and professional mental health organizations.

A state-appointed body assigned to review the attack criticized Virginia Tech administrators. The panel's report noted multiple failures, including failing to immediately send out an emergency campus-wide alert that a gunman was on the loose. University officials first informed students via e-mail at 9:26 am, two hours after the first shooting. At the time of the shooting, Virginia Tech was upgrading the campus alert system to include sirens and speakers. On the day of the shooting, the only alert systems in place were rudimentary phone trees and email.

Unfortunately for Virginia Tech, another shooting incident occurred at the university in December 2011. A Virginia Tech police officer was shot during a routine traffic stop. A second victim of the shooter was found at a university parking lot. Fortunately, Virginia Tech had installed one of the nation's most advanced social media and Internet-based security alert systems following the 2007 active shooter incident. The security system was named "VT Alerts." VT Alerts works as follows: When a crisis or emergency arises that the university determines requires immediate attention by students, employees, and others, a VT Alert is issued. Individuals with a Virginia Tech email address automatically receive an email and text message each time a VT Alert is sent out.

A VT Alert was issued for the 2011 shooting incident, and the entire Virginia Tech campus was put on lockdown. The alert system advised everyone on campus to seek shelter or stay where they were. The alerts also described the gunman, whom the police were searching for. The alerts were also shared via the Virginia Tech homepage, the weather/emergency hotline, Twitter, digital display boards in classrooms and other common meeting locations, outdoor sirens with voice enunciators on the main campus, text messages, voice messages, and email.

Immediately following the university alerts, news of the shooting was reposted on multiple social media platforms, including Twitter and Facebook. The postings were virtually instantaneous with the event. The instantaneous nature of the social media platforms made social media platforms the "go-to" place for information. Alerts sent out by the University were reposted on the Twitter and Facebook accounts of students and others. Thousands of people, including students and community members, followed events through Twitter as they were happening. Mainstream broadcast news organizations posted information and photos on Facebook shortly after the first alert. News organizations also asked university officials, students, and others to share their thoughts about what was happening.

News and updates about the 2017 shooting incident were posted almost by the minute by attaching the pound sign, known as a hashtag, to keywords: #vatech, #virginiatech, #shooting and #blacksburg. Tweets from the university, media outlets, and students on or near campus were posted and reposted as the story unfolded. One early Facebook posting said: "Traffic stop — turned into a shooting. Gunman is still at large."

As the hours passed, information was almost continually being pushed out through social media channels regarding the status of the alert, the description of the suspected shooter, and reports of police activity.[25] Beyond Twitter and Facebook, various individuals and organizations posted updates on Wikipedia and photos on photosharing social media platforms. Social media greatly expanded the reach and access of the University's alert system.

Four hours after the initial alerts through social media and the campus system were posted, the university sent out an alert through traditional and social media channels stating law enforcement agencies had determined there was no longer an active threat on the campus. The shooter, a 22-year-old male who attended nearby Radford University, had taken his own life. Later in the day, the Governor of Virginia, Bob McDonnell commended the multiple law enforcement agencies, Virginia Tech leadership, students, faculty, and staff for effectively implementing the University's social media crisis communication strategy.

13.6   Case Study: Social Media and the 2013 Southern Alberta/ Calgary Flood

In June 2013, the City of Calgary and other large parts of the Province of Alberta experienced massive flooding caused by heavy rainfall combined with the rapid melting of the mountain snowpack. The City responded with a comprehensive "push" social media strategy. The strategy was based on the *APP* Strategic Communication Model: *Anticipate, Prepare, and Practice.* The social media strategy: (1) *Anticipated* the flooding scenario, stakeholder questions, and key stakeholders; (2) *Prepared* social media messages (e.g. Tweets), the messengers (e.g. the municipality, the mayor, the police service, and local influencers), and the social media platforms for communication; and (3) *Practiced,* e.g. through role plays, simulations, and exercises.[26]

Thousands of Canadian Forces troops were deployed to help in flooded areas and support local government agencies, the Royal Canadian Mounted Police, and local law enforcement. The City of Calgary received additional support from the Edmonton Police Service and the Vancouver Fire Department. A massive cleanup effort followed the receding waters.

As is now the norm in many crisis situations, within hours of the flooding, large amounts of information were shared through social media platforms. This included texting, emails, chat rooms, Tweets, Facebook posts, blog posts, forum discussions, and the posting of videos and photographs. Calgary's mayor and the city's emergency officials and crisis response departments extensively used social media throughout the crisis. Through social media platforms such as Twitter and Facebook, Calgary crisis managers provided a steady flow of timely and accurate information.

Calgary Mayor Naheed Nenshi played a key role in actively engaging, informing, and updating the community through social media platforms such as Twitter. He was seen nearly everywhere in the city, out in the community getting his hands dirty, physically helping people in need, and conducting frequent press briefings.

During the floods, the Calgary police actively and continuously engaged the community through Tweets with updates, safety tips, rumor control, and responses to inquiries. The police department's massive use of Twitter led to a temporary suspension of their Twitter account. Twitter had a misuse prevention measure that automatically detects when an account tweets over 100 times in an hour, or 1,000 times in the period of one day. As soon as the police department's Twitter account was locked down, individual police officers used their personal accounts to continue tweeting. Virtually every Tweet from the city contained either "#yycflood" or "#yyc."[27] This made information from Twitter extremely easy to search, track, and monitor.

Surges in searches and requests for information from the City's website crashed the website. However, the City was quick to find alternatives, including sending online visitors to the City's news blog. To maintain consistency of messaging and citizen behavior, the City's Information Technology Department redirected the primary city URL directly to the news blog so visitors would not have to learn a new web address. Once the website was back up and running, the City of Calgary embedded a red notification on its homepage, sending people to the right platforms to receive information and updates.

The City of Calgary took full advantage of the fact that many of the 100,000 who were evacuated carried their mobile phones with them. The city encouraged people to download two free apps to receive push notifications, real-time alerts, and other information. For example, the city posted:

> "Your city anywhere, anytime! 311 Calgary APP is your on-the-go connection to The City. Report and track select city services via your smart phone using location-based technology with the option to include a photo of your concern."
>
> "Stay on top of what's happening in your City with the free City of Calgary News app, linking you to City news, services, careers, and more."

The City of Calgary was well prepared for the use of social media in a crisis. Their crisis communication team had been practicing their social media response to various crisis scenarios for several years. This preparation is crucial for effectively managing a crisis. A sophisticated crisis management system, rooted in a practiced social media strategy, can save countless lives.[28] In response to one of the worst floods in Canadian history affecting hundreds of thousands of individuals, the 2013 Calgary/Southern Alberta flood claimed the lives of five people. Many more lives would have been lost if evacuations and emergency warnings had been less successful.

13.7 Best Practices for Using Social Media in Risk, High Concern, and Crisis Situations

Listed in Table 13.4 and described below are examples of best practices for social media engagement.

13.7.1 Create a Social Media Plan

Your social media plan should identify objectives and the actions necessary to achieve them. Key objectives can include: building trust, informing decision-making, and influencing attitudes and behaviors. Your plan should consider how and where you will track social media content that could positively or negatively impact your organization. Your plan should identify who in your organization is accountable and/or responsible for social media monitoring and activity. Other elements your plan should address include:

- identifying how people can access your social media content, including a continually updated list of the social media platforms used by your partner organizations and your key stakeholders.
- testing your plan through a tabletop exercise.
- identifying restrictions, regulations, and rules for the social media platforms you use.

13.7.2 Staff Appropriately for Social Media Communication

Your organization will need to specify the skills for the person or staff put in charge of social media communications. The person or staff in charge of social media must have a thorough understanding of all relevant social media platforms. Social media skills are complex, specialized, and necessary to master. The staff assigned to social media outreach must be able to handle multiple duties, including choosing engaging content, posting content across different selected platforms (e.g. on Facebook, Twitter, Instagram, LinkedIn, and Snapchat), monitoring comments, moderating comments, keeping pages active, keeping users engaged, and monitoring return on investment by measuring key performance indicators (KPIs), such as site traffic, "likes," audience growth rate, audience engagement, repostings, and share of audience. Performing these duties can be highly time-consuming and you need to allocate sufficient staff time and focus on priorities.

Table 13.4 Best practices for social media engagement.

---

1) Create a social media plan
2) Staff appropriately for social media communication
3) Ensure continuous updating
4) Identify your partners
5) Assess and reassess your selection of platforms
6) Create and maintain as many social media accounts as you and your stakeholders need
7) Be prepared for the special social media requirements and pressures in a crisis
8) Provide guidance for employees and engage them in the process
9) Don't skip evaluation

13.7.3 Ensure Continuous Updating

Social media trends frequently change. Platforms rise and fall in popularity and shift in audience. As a result, your social media strategy and outreach plan will require frequent updating. Be alert to stakeholders' changing information needs and to their changing patterns of and modes of social media use. Periodically assess your own changes in resources or priorities. Your social media plan may also need immediate revision in an emerging situation. A carefully planned and practiced plan provides the base that enables needed flexibility.

13.7.4 Identify Your Partners

Your social media plan should promote the social media platforms of partner organizations to build awareness and gain followers. Building a follower base across a network of partners prior to a crisis or disaster should be a key goal.

13.7.5 Assess and Reassess Your Selection of Platforms

Your social media outreach plan should identify which social media platforms you will use routinely, and which ones will be used in a crisis. You need to know where your key stakeholders go for information and track changes in their social media habits. You need to research and understand the potential benefits, risks, and costs posed by different platforms. In selecting platforms, you should aim for trusted platforms and avoid those that attract considerable false or disreputable content, as this can open up an organization's messages to attack or undermining by irresponsible users. Announce and promote your presence on the social media platforms you have selected.

13.7.6 Create and Maintain as Many Social Media Accounts as You and Your Stakeholders Need

Social media habits are so diverse that it is necessary to have multiple social media accounts for any social media platform, e.g. multiple Twitter and Facebook accounts and various accounts across the other platforms that your stakeholders use. Engage stakeholders across accounts, cross-linking between them and efficiently using hashtags. Some hashtags can be preplanned. Others emerge organically and must be adopted by the organizations involved. It is especially important to use and be attentive to hashtags in crisis communications, allowing people to follow updates and to share information. If it becomes necessary to add new platforms, also ensure that regular users receive guidance on how to navigate a new social media platform.

13.7.7 Be Prepared for the Special Social Media Requirements and Pressures in a Crisis

An essential element of your social media plan will lay out strategies for using social media outlets in a crisis. (See Chapter 6 in this book). Poor crisis management and communication can lead people to overreact or underreact, to take inappropriate actions, and to lose trust in crisis management authorities. Effective use of social media platforms by crisis managers and communicators enhances trust by encouraging two-way conversations in which responding organizations inform – and are informed by – those interested and affected. It encourages stakeholders to participate in a constructive dialogue about what is being done to reduce risks and what stakeholders can do for themselves. Risk, high concern, and crisis communication are primarily a two-way exchange of information, not a one-way transfer, and the social media environment can be carefully used to achieve this.

A key best practice in crisis management and communication aided by social media platforms is to be the first to post and share bad news – or others will likely do it for you. Messages must be developed, vetted, and posted as quickly as possible. Ideally, messages should be drafted in the precrisis stage. These messages can later be tweaked during the crisis event. It is also essential to continuously monitor the content posted by social media users related to the situation. False information needs to be corrected and rumors – which may be true or false – need to be verified.

An important new type of target audience in crisis, as well as noncrisis situations, is social media *influencers*. Influencers are people who have built a reputation in particular social media networks for their knowledge, experience, guidance, or expertise. They make regular posts on their preferred social media outlets and may have large followings of enthusiastic, engaged people. Responsible influencers can help get needed attention in a crisis and encourage appropriate behaviors.

Crises can also increase the volume of activity to an organization's website and result in overload. For example, in addition to searches by members of the general public for information about who, what, where, when, and why, victims may do their own social media searches and outreach. As a result, it will be important to monitor postings, post accurate information, and provide links to trustworthy sites. It also is important to evaluate the maximum amount of traffic your website can handle and be prepared for surges and surprises and for use of alternative sites. Another way to prepare content in advance of a crisis is to develop a website – commonly referred to as a "dark" site – or a dark page on an existing website that is hidden in normal times but can be activated in a crisis.

In a crisis, an organization may be deluged with requests for information from mainstream media journalists and bloggers. It is important to let mainstream media journalists and bloggers know where your organization is posting updates. Developing a blog or social media newsroom is a good way to post updates in a crisis. In large crises, crisis communicators from responding organizations will often gather together in a joint information center and establish a single social media newsroom. Working with on-scene and first responder personnel to produce approved content for the social media newsroom is a good way to improve the accuracy, clarity, and usefulness of messages.

Identifying sources of content in advance can speed getting information out via social media platforms in a crisis. Your organizations can create a social media library modeled after the vetted precrisis messages and social media messages in the Social Media Message Library developed by researchers at the Dornsife School of Public Health at Drexel University.[29] The Social Media Message Library is a web-based resource that contains hundreds of vetted social media messages and templates for public health crises and emergencies, including hurricanes. The library contains message content for the preparedness, response, and recovery phases of different disaster scenarios. Social media messages are often most effective when they contain embedded video and other visual material. Consider developing video and other visual content for posting on social media sites in advance. Visual content can garner more attention, enhance understanding, and increase recall. It is especially powerful when combined with visual content for traditional media. However, large visuals can be difficult to load during a crisis, so it is necessary to ensure the same information is available in text.

13.7.8 Provide Guidance for Employees and Engage Them in the Process

Your social media plan should include guidance for employees on how to use and react to social media. It must ensure employees are aware of and understand the plan. The social media plan should also include examples of mistakes and misuse by employees, such as posting unapproved messages or visual content, since staffs often believe they are being helpful without understanding crisis communication issues and principles.

13.7.9   Don't Skip Evaluation

All good social media plans should be refined based on the evaluation. Benchmarks of success and effectiveness can be evaluated using a combination of free and paid social media monitoring tools (e.g. Facebook Insights, Twitter Analytics, YouTube Analytics, and Google Analytics). Social media evaluation metrics include: (1) the number of social media engagements, i.e. the total number of interactions (reactions, comments, and shares) received by your social media posts as well as by social media sites that mention your posts); (2) the most liked posts; (3) the most commented on posts; and (4) the most shared posts. Evaluation is especially important after a crisis. The impact of social media activities should be assessed and lessons learned incorporated in ongoing planning. The interactive, dynamic environment of social media demands continuous learning and adaptation.

13.8   Case Diary: Social Media and the Negative Power of "Junk" Information about Risks and Threats

I started my presentation with a few slides of a Google search result. I was speaking to managers of a conscientiously run hazardous waste site who felt unfairly attacked. They did not understand why they were being criticized by so many members of the public about the risks of birth defects and childhood cancers stemming from their site. Responsible scientific studies had been done and reports published indicating that the site was not causing health risks. I showed site managers the results of the Google search I had done entering only the name of the site and the words "water pollution." My goal was to replicate what an average citizen might do if they were thinking about moving to the area near the hazardous waste site and had heard concerns about water pollution.

All entries on the first three screens of my Google search were from local groups. The entries described (inaccurately) improper disposal of toxic chemicals at the site and the scores of people who believed they and their children had developed cancers and other sicknesses because of the water they had drunk. The first entry on Google was a Wikipedia piece. It characterized the site as one of the worst water contamination sites in that region. The only citations in the Wikipedia piece were local media stories and activist websites. The first mention of the site's own authoritative website appeared on page four of the Google results.

I asked the site managers if they had seen what I had seen on the first three pages of my Google search. Their answer was, "Why would we want to look at that junk? It's nothing but junk, repeated over and over again." "It may be *junk*," I responded, "but it is *junk* that a lot of people see, and consequently believe. In a risk communication landscape dominated by social media, if you're not on the playing field, you're not a player."

After a long discussion and a review of data on *Google Analytics*,[30] we agreed to initiate a major social media effort. We began by checking the health claims posted on the most popular social media platforms against assessments by internal and external scientific and technical experts. After confirmation of the facts and affirmations from trusted third parties, we developed an appropriate social media risk communication plan and implementation. We designed content for social media platforms that was clear, visually interesting, and could compete with false information. We began to correct the errors in Wikipedia. We enlisted the help of local organizations, such as local health care organizations and local libraries, to help people find trusted sites and assist us in our efforts to monitor for misconceptions, false information, and rumors. It was too late to change the minds of many of those sharing false information. However, based on evaluation studies, most residents of the area said the information they received from our social media postings corrected many misconceptions and gave them a better foundation for informed decision-making.

13.9 Lessons Learned and Trends

The events described in this case diary took place in 2012. Today, few organizations and individuals would so quickly dismiss the high impact potential of false ("junk") information about a risk or threat communicated via social media. Yet too many organizations are still unprepared to take advantage of the positive power of social media for building trust, informing decision making, and influencing attitudes and behaviors, or are unprepared to counter the power of social media to do damage.

Posting, updating, and monitoring social media content is itself challenging. Addressing misconceptions, falsehoods, or incomplete information adds another layer of effort, one that can feel like an endless game of *whac-a-mol*. But when the well-being of people and reputations of individuals, organizations, and institutions are at stake, a good social media plan and ongoing social media program can have an immense – and critical – impact.

Social media platforms have become an integral part of people's lives. Despite the many challenges and dangers, especially the spread of false information and threats to privacy, it has revolutionized the study and practice of risk, high concern, and crisis communications. Through a changing and expanding set of channels, social media outlets allow anyone with Internet access to create and share content. Social media platforms connect people and organizations to each other in ways previously impossible. Used with an understanding of the territory and skill in exploiting their possibilities and reach, social media platforms provide new ways for risk managers, technical professionals, and communicators to share information, observe trends, gain insights, solve complex problems, engage in constructive dialogue, and develop positive relationships.

13.10 Chapter Resources

Below are additional resources to expand on the content presented in this chapter.

American Red Cross (2012). *Social Media in Disasters and Emergencies.* Washington, DC: Red Cross.

Apuke, O.D., and Omar B. (2021). "Fake news and COVID-19: Modelling the predictors of fake news sharing among social media users." *Telematics and Informatics* 56, January 2021, 101475. Accessed at: https://doi.org/10.1016/j.tele.2020.101475

Bawden, D., and Robinson, L. (2009). "The dark side of information: Overload, anxiety and other paradoxes and pathologies." *Journal of Information Science* 35:180–191.

Centers for Disease Control and Prevention (2019). *CDC Social Media Tools, Guidelines, & Best Practices.* Atlanta, GA: Centers for Disease Control and Prevention. Accessed at: https://www.cdc.gov/socialmedia/tools/guidelines/index.html

Centers for Disease Control and Prevention (2011). *The Health Communicator's Social Media Toolkit.* Atlanta, GA: Centers for Disease Control and Prevention. Accessed at https://www.cdc.gov/healthcommunication/ToolsTemplates/SocialMediaToolkit_BM.pdf

Conrado, S. P., Neville, K., Woodworth, S., and O'Riordan, S. (2016). "Managing social media uncertainty to support the decision-making process during emergencies." *Journal of Decision Systems 25 supplement* 1 (2016):171–181.

Coombs, W.T. (1998). "An analytic framework for crisis situations: Better responses from a better understanding of the situation." *Journal of Public Relations Research* 10(3):177–192.

Coombs, W. (2019). *Ongoing Crisis Communications: Planning, Managing, and Responding.* Thousand Oaks, CA: Sage Publications, Inc.

Coombs, W.T. (2007). "Protecting organization reputations during a crisis: The development and application of situational crisis communication theory." *Corporate Reputation Review* 10(3):163–176.

Covello V. (1983). "The perception of technological risks." *Technology Forecasting and Social Change: An International Journal* 23:285–297.

Covello V. (1989). "Issues and problems in using risk comparisons for communicating right-to-know information on chemical risks." *Environmental Science and Technology* 23(12):1444–1449.

Covello V. (1992). "Risk communication, trust, and credibility." *Health and Environmental Digest* 6(1):1–4.

Covello V. (1993). "Risk communication and occupational medicine." *Journal of Occupational Medicine* 35:18–19.

Covello, V. (2003). "Best practices in public health risk and crisis communication." *Journal of Health Communication* 8 (Suppl. 1):5–8; 148–151.

Covello, V. (2006). "Risk communication and message mapping: A new tool for communicating effectively in public health emergencies and disasters." *Journal of Emergency Management* 4(3):25–40.

Covello, V. (2014). "Risk communication," in *Environmental Health: From Global to Local*, ed. H. Frumkin. San Francisco: Jossey-Bass/Wiley.

Covello, V., and Allen, F. (1988). *Seven Cardinal Rules of Risk Communication.* Washington, D.C.: U.S. Environmental Protection Agency.

Covello, V., and Hyer R. (2020). *COVID-19: Simple Answers to Top Questions: Risk Communication Field Guide Questions and Key Messages.* Arlington, VA: Association of State and Territorial Health Officers. Accessed at: https://www.hsdl.org/?view&did=835774

Covello V., McCallum D., and Pavlova M. (1989). *Effective Risk Communication: The Role and Responsibility of Government and Nongovernment Organizations.* New York: Plenum Press.

Covello, V., and Merkhofer, M. (1993). *Risk Assessment Methods: Approaches for Assessing Health and Environmental Risks.* New York: Plenum Press.

Covello, V., Minamyer, S., and Clayton, K. (2007). *Effective Risk and Crisis Communication during Water Security Emergencies. EPA Policy Report*; EPA 600-R07-027. Washington, D.C.: U.S. Environmental Protection Agency.

Covello, V., Peters, R., Wojtecki, J., and Hyde, R. (2001). "Risk communication, the West Nile virus epidemic, and bio-terrorism: Responding to the communication challenges posed by the intentional or unintentional release of a pathogen in an urban setting." *Journal of Urban Health* 78(2):382–391.

Covello, V., Sandman, P., and Slovic, P. (1988). *Risk Communication, Risk Statistics, and Risk Numbers.* Washington, DC: Chemical Manufacturers Association.

Covello, V., Sandman, P. (2001). "Risk communication: Evolution and revolution," in *Solutions to an environment in peril*, ed. A. Wolbarst. Baltimore, MD: Johns Hopkins University Press.

Covello, V., Slovic, P., and von Winterfeldt, D. (1986). "Risk communication: A review of the literature." *Risk Abstracts* 3(4):171–182.

Crick, M. J. (2021). "The importance of trustworthy sources of scientific information in risk communication with the public." *Journal of Radiation Research* 62(S1):1–6.

Fung, I.C.H., Tse, Z., Cheung, C.H., Miu, A.S., and Fu, K.W. (2014). "Ebola and the social media." *The Lancet* 384:9961 (Dec. 2014).

Gao, X., and Lee, J. (2017). "E-government services and social media adoption." *Government Information Quarterly* 34, 4 (Dec. 2017):627–634.

Gao, H., Barbier, G., and Goolsby, R. (2011). "Harnessing the crowdsourcing power of social media for disaster relief." *IEEE Intelligent System* 26(3):10–14.

Germani, F., and Biller-Andorno, N. (2021). "The anti-vaccination *infodemic* on social media: A behavioral analysis." PLOS One. 16 (3): 1–14.

Graham, M., Avery, E. (2013). "Government public relations and social media: An analysis of the perceptions and trends of social media use at the local government level." *Public Relations Journal* 7 4 (2013):1–21.

Graham, M.W., Avery, E.J, and Park, S. (2015). "The role of social media in local government crisis communications." *Public Relations Review* 41(3):386–394.

Greenwood, S. (2016). *Social Media Update 2016*. Washington, D.C.: Pew Research Center: Internet, Science & Technology.

Guskin, E., and Hitlin, P. (2012). *Hurricane Sandy and Twitter*. Washington, D.C.: Pew Research Center's Journalism Project.

Hagen, L., Keller, T., Neely, S., DePaula, N., and Cooperman, C. (2017). "Crisis communications in the age of social media: A network analysis of Zika-related tweets." *Social Science Computer Review* 36(5): 523–541.

Hagen, L., Scharf, R., Neely, S., and Keller, T. (2018). "Government social media communications during Zika health crisis." *Proceedings of the 19th Annual International Conference on Digital Government Research: Governance in the Data Age*. New York: Association for Computing Machinery.

Hills, T.T. (2019). "The dark side of information proliferation." *Perspectives on Psychological Science* 14(3):323–330.

Hornmoen, H., and Backholm, K., eds.(2018). *Social Media Use in Crisis and Risk Communication: Emergencies, Concerns and Awareness*. Bingley, UK: Emerald Publishing

Hughes, A.L., and Palen, L. (2012). "The evolving role of the public information officer: An examination of social media in emergency management." *Journal of Homeland Security and Emergency Management* 9 (1): 1–14.

Jin, Y., Fisher, B., and Austin, L. (2014). "Examining the role of social media in effective crisis management: The effects of crisis origin, information form, and source on publics' crisis responses." *Communication Research* 41(1):74–94.

Latonero, M., and Shklovski, I. (2011). "Emergency management, twitter, and social media evangelism." *International Journal of Information Systems for Crisis Response and Management* 3, 4(2011):1–16.

Liu, T., Zhang, H., and Zhang, H. (2020). "The impact of social media on risk communication of disasters." *International Journal of Environmental Research and Public Health* 883:1–17.

Malecki, K.M.C., Keating, J.A., and Safdar, N. (2021). "Crisis communication and public perception of COVID-19 risk in the era of social media." *Clinical Infectious Diseases* 72(4):697–702.

McNutt, K. (2014). "Public engagement in the Web 2.0 era: Social collaborative technologies in a public sector context." *Canadian Public Administration/Administration Publique du Canada 57*, 1(Mar. 2014):49–70.

Mendoza-Herrera, K., Valero-Morales, I., Ocampo-Granados, M.E., Reyes-Morales, H., Arce-Amaré, F., and Barquera, S. (2020). "An overview of social media use in the field of public health nutrition: Benefits, scope, limitations, and a Latin American experience." *Preventing Chronic Disease* 17(E76):1–6.

Merchant, R.M., Elmer S., and Lurie N. (2011). "Integrating social media into emergency preparedness efforts." *New England Journal of Medicine* 365, 4 (Jul. 2011):289–291.

Mergel, I. (2017). "Social media communication modes in government," in *Routledge Handbook on Information Technology in Government*, eds. Y.C. Chen and M.J. Ahn, eds. Milton Park, UK: Routledge. pp. 168–179.

Moorhead, S.A., Hazlett, D.E., Harrison, L., Carroll, J.K., Irwin, A., and Hoving, C. (2013). "A new dimension of health care: Systematic review of the uses, benefits, and limitations of social media for health communication." *Journal of Medical Internet Research* 15(4):85–92.

Murthy, D., and Gross, A.J. (2017). "Social media processes in disasters: Implications of emergent technology use." *Social Science Research* 63:356–370.

National Research Council. (2013). *Public Response to Alerts and Warnings Using Social Media: Report of a Workshop on Current Knowledge and Research Gaps.* Washington, DC: National Academies Press.

Neeley, L. (2014). "Risk communication and social media," in *Effective Risk Communication*, eds. Arvai, J., and Rivers, L. New York: Earthscan.

Overbey, K.N., Jaykus, L., and Chapman, B.J. (2017). "A systematic review of the use of social media for food safety risk communication." *Journal of Food Protection* 80(9):1537.

Palen L., and Hughes A.L. (2018). "Social media in disaster communication," in *Handbook of Disaster Research*, eds. Rodríguez H., Donner W., and Trainor J. New York: Springer, pp. 497–518.

Palen, L., Vieweg, S., Liu, S. B., and Hughes, A. L. (2009). "Crisis in a networked world." *Social Science Computing Review* 27(4):467–480.

Perrin, A. (2015). *Social Media Usage: 2005-2015.* Washington, D.C.: Pew Research Center: Internet, Science & Technology.

Pew Research Center (2021). *Social Media Use in 2021.* Accessed at: https://www.pewresearch.org/internet/2021/04/07/social-media-use-in-2021/

Prasad, A. (2021). "Anti-science misinformation and conspiracies: COVID–19, Post-truth, and Science & Technology Studies (STS)." *Science, Technology and Society* 2021 (2021): 1–25.

Sellnow, T.L., and Seeger M.W. (2021). *Theorizing Crisis Communication.* 2nd Edition. Hoboken, New Jersey: Wiley.

Silverman, C. (2020). *Verification Handbook for Disinformation And Media Manipulation.* 3rd Edition. Brussels, Belgium and Maastricht, the Netherlands: European Journalism Centre. Accessed at: https://datajournalism.com/read/handbook/verification-3

Smith, A., and Anderson. M. (2018). *Social Media Use in 2018.* Pew Research Center. Accessed at: https://www.pewinternet.org/2018/03/01/social-media-use-in-2018.

Stelter, B., and Preston, J. (2012). "In crisis, public officials embrace social Media." *The New York Times* November 12, 2012. Accessed at: https://www.nytimes.com/2012/11/02/technology/in-crisis-public-officials-embrace-social-media.html

Stewart, M.C., and Wilson, B.G. (2016). "The dynamic role of social media during Hurricane." *Computers in Human Behavior* 54(2016):639–646.

Swire-Thompson, B., and D. Lazer (2020). "Public health and online misinformation: Challenges and recommendations." *Annual Review of Public Health* 41(1):433–451.

Tinker, T., Fouse, D. (ed.) (2009). *Special report – Expert Round Table on Social Media and Risk Communication during Times of Crisis: Strategic challenges and opportunities.* Washington, DC. American Public Health Association. Accessed at: file:///Users/Vince/Downloads/Crisis%20Comm%20Social%20Media%20Report.pdf

Veil, S.R., Buehner, T., and Palenchar, M.J. (2011), "A work-in-process literature review: Incorporating social media in risk and crisis communication." *Journal of Contingencies and Crisis Management* 19(2):110–122.

Wendling, C., Radisch, J., Jacobzone, S. (2013). *The Use of Social Media in Risk and Crisis Communication: Working Paper on Public Governance No. 24*. Paris: OECD Publishing.

Wirz, C.D., Xenos, M.A., Brossard, D., Scheufele, D., Chung, J.H., and Massarani, L. (2018). "Rethinking social amplification of risk: Social media and Zika in three languages." *Risk Analysis* 38:2599–2624.

U.S. Department of Homeland Security (2013). *Innovative Uses of Social Media in Emergency Management*. Washington, D.C.: U.S. Department of Homeland Security.

World Health Organization (2020). "Call for Action: Managing the Infodemic--A Global Movement to Promote Access to Health Information and Harm from Health Misinformation Offline Communities." Accessed at: https://www.who.int/news/item/11-12-2020-call-for-action-managing-the-infodemic

Yun, G.W., Morin, D., Park, S., Youngnyo-Joa, C., Labbe, B., Lim, J., Lee, S., Hyun, D. (2016). "Social media and flu: Media Twitter accounts as agenda setters." *International Journal of Medical Informatics* 91(Jul. 2016):67–73.

U.S. Environmental Protection Agency (2019). *Community Involvement Tools: Social Media*. Accessed at: https://semspub.epa.gov/work/HQ/100001966.pdf

Zavattaro, S.M., and Bryer, T.A. (2016). *Social Media for Government: Theory and Practice*. London: Routledge.

Endnotes

1 CNBC (2020)."Bill Gates denies conspiracy theories that say he wants to use coronavirus vaccines to implant tracking devices." 22 July 2020. Accessed at: https://www.cnbc.com/2020/07/22/bill-gates-denies-conspiracy-theories-that-say-he-wants-to-use-coronavirus-vaccines-to-implant-tracking-devices.html

2 See, e.g. McGreal, C. (2020). "A disgraced scientist and a viral video: how a Covid conspiracy theory started." *The Guardian* 14 May 2020. Accessed at: https://www.theguardian.com/world/2020/may/14/coronavirus-viral-video-plandemic-judy-mikovits-conspiracy-theories?utm_term=RWRpdG9yaWFsX1VTTW9ybmluZ0JyaWVmaW5nLTIwMDUxNA%3D%3D&utm_source=esp&utm_medium=Email&utm_campaign=USMorningBriefing&CMP=usbriefing_email

3 See, e.g. Berger, J., and Milkman, K.I. (2021). "What makes online comment viral?" *Journal of Marketing Research* 49(2):192–205.

4 See, e.g., Prasad, A. (2021). "Anti-science misinformation and conspiracies: COVID–19, Post-truth, and Science & Technology Studies (STS)." *Science, Technology and Society*. April 2021; Germani, F., Biller-Andorno, N. (2021). The anti-vaccination infodemic on social media: A behavioral analysis. *PLOS One*. March 3, 2021.

5 World Health Organization (2020. "Managing the COVID-19 *Infodemic*: Promoting healthy behaviours and mitigating the harm from misinformation and disinformation. Joint statement by WHO, UN, UNICEF, UNDP, UNESCO, UNAIDS, ITU, UN Global Pulse, and IFRC." Accessed at: https://www.who.int/news/item/23-09-2020-managing-the-covid-19-infodemic-promoting-healthy-behaviours-and-mitigating-the-harm-from-misinformation-and-disinformation

6 Covello, V.T., and Hyer R.N. (2020). *COVID-19: Simple Answers to Top Questions: Risk Communication Field Guide Questions and Key Messages*. Arlington, VA: Association of State and Territorial Health Officers. Accessed at: https://www.hsdl.org/?view&did=835774. Accessed also at: https://www.waukeshacounty.gov/globalassets/health--human-services/public-health/

public-health-preparedness/covid/asthocovid19qanda.pdf. See also Silverman, C. (2020). *Verification Handbook: For Disinformation and Media Manipulation*. 3rd Edition. Belgium and Maastricht, the Netherlands: European Journalism Centre. Accessed at: https://datajournalism. com/read/handbook/verification-3.

7 Pew Research Center (2021). *Social Media Use in 2021*. Accessed at: https://www.pewresearch.org/ internet/2021/04/07/social-media-use-in-2021/

8 Pew Research Center (2021). *Social Media Use in 2021*. Accessed at: https://www.pewresearch.org/ internet/2021/04/07/social-media-use-in-2021/

9 See, e.g. Swire-Thompson, B., and Lazer D. (2020). "Public Health and Online Misinformation: Challenges and Recommendations." *Annual Review of Public Health* 41(1):433–451; See also Liu, T., Zhang, H., and Zhang, H. (2020). "The impact of social media on risk communication of disasters." *International Journal of Environmental Research and Public Health* 883:1–17.

10 Accessed at: https://www.zdnet.com/article/ daily-active-user-of-messaging-app-wechat-exceeds-1-billion/

11 See, e.g., U.S. Environmental Protection Agency. *Community Involvement Tools: Social Media* (2019). Accessed at: https://semspub.epa.gov/work/HQ/100001966.pdf; See also Centers for Disease Control and Prevention (2019). *CDC Social Media Tools, Guidelines & Best Practices*. Atlanta, GA: Centers for Disease Control and Prevention.

12 See: https://twitter.com/hashtag/sosharvey?lang=en; See also: https://www.nbcnews.com/ storyline/hurricane-harvey/ soshouston-how-apps-social-media-assist-harvey-rescue-efforts-n797841

13 For a review of this literature, see Mendoza-Herrera, K., Valero-Morales, I., Ocampo-Granados, M.E., Reyes-Morales, H., Arce-Amaré, F., and Barquera, S. (2020). "An overview of social media use in the field of public health nutrition: Benefits, scope, limitations, and a Latin American experience." *Preventing Chronic Disease* 17(E76):1–6.

14 U.S. Department of Homeland Security (2013). *Innovative Uses of Social Media in Emergency Management*. Washington, D.C.: U.S. Department of Homeland Security.

15 In crisis situations, social media platforms can also be used to mobilize and coordinate recruitment of volunteers. However, recruitment of volunteers comes with its own risks. These include liability risks, the diversion of scarce resources to an auxiliary activity, and logistical problems if there is a larger than expected response to the call for volunteers.

16 See, e.g. Ripley, A. (2008). *The Unthinkable: Who Survives When Disaster Strikes - and Why*. New York: Three Rivers Press/Crown Publishing.

17 See, e.g. Rothstein, M. (2015). "Privacy and confidentiality," in *Routledge Handbook of Medical Law and Ethics*, eds. Y. Joly and B.M Knoppers. London: Routledge.

18 Hills, T.T. (2019). "The dark side of information proliferation." *Perspectives on Psychological Science* 14(3):323–330; See also Bawden, D., and Robinson, L. (2009). "The dark side of information: Overload, anxiety and other paradoxes and pathologies." *Journal of Information Science*, 35:180–191; Brunken, R., Plass, J., and Leutner, D. (2003). "Direct measurement of cognitive load in multimedia learning." *Educational Psychologist* 38(1):53–56.

19 See, e.g., Tandoc, E.C., Lim, Z.W., and Ling, R. (2018) "Defining 'Fake News." *Digital Journalism* 6(2):137–153.

20 Barthel, M., Mitchell, A., and Holdcomb, B. (2016). "Fake news is sowing confusion." Pew Research Center. Accessed at: https://www.journalism.org/2016/12/15/ many-americans-believe-fake-news-is-sowing-confusion/

21 See, e.g., Wingfield, Issac, M., and Benner, K. (2016). "Google and Facebook take aim at fake news sites." *New York Times*. November 14, 216.

22 See, e.g., Monther Aldwairi, M., and Ali Alwahedi, A. (2018). "Detecting fake news in social media networks." *Procedia Computer Science* 141:215–222.

23 Vosoughi, S., Roy, D., Aral, S. (2018). "The spread of true and false news online." *Science*, 359 (6380): 1146–1151

24 Virginia Tech Review Panel (2007). Mass Shootings at Virginia Tech, April 16, 2007: Report of the Review Panel (the Massengill Report). Richmond, VA: The Commonwealth of Virginia. See also: Agger, B., Luke, T.W., eds (2008). There is a gunman on campus: tragedy and terror at Virginia Tech. Lanham, MD:. Rowman & Littlefield; Worth, R. (2008). Massacre at Virginia Tech: disaster and survival. Berkeley Heights, NJ: Enslow Publishers; Fretz, R. (2007). Lessons learned at Virginia Tech shooting. Accessed at: https://www.policeone.com/school-safety/articles/1473536-lessons-learned-at-virginia-tech-shooting/;

25 There is an increasing trend for people viewing events involving the police to post in real time what they are seeing on social media platforms. Many police agencies are now routinely alerting social media users that these types of posts put the police at risk by potentially alerting suspects of police locations and what the police are doing.

26 See, e.g., Branham, M. (201). "Keys to Risk Communication: Anticipate, Prepare and Practice." E-newsletter Issue #47, May 27, 2010, Knowledge Center, Council of State Governments. Accessed at: https://knowledgecenter.csg.org/kc/content/keys-risk-communication-anticipate-prepare-and-practice

27 "yyc" in the hashtag stands for City of Calgary.

28 See, e.g., Dangerfield, K. (2019). "6 of the worst floods in Canadian history. Global News. April 29, 2019. Accessed at: https://globalnews.ca/news/5216176/worst-floods-canadian-history/

29 https://drexel.edu/dornsife/research/centers-programs-projects/center-for-public-health-readiness-communication/social-media-library/

30 *Google Analytics* is a web analytics service offered by Google that tracks and reports website traffic. Google launched the service in November 2005.

Index

Communicating in Risk, Crisis, and High Stress Situations: Evidence-Based Strategies and Practice, First Edition.
Vincent T. Covello.
© 2022 by The Institute of Electrical and Electronics Engineers, Inc. Published 2022 by John Wiley & Sons, Inc.